THE QUANTUM THEORY OF RADIATION

BY

W. HEITLER
PROFESSOR OF THEORETICAL PHYSICS
IN THE UNIVERSITY OF ZÜRICH

THIRD EDITION

DOVER PUBLICATIONS, INC.
NEW YORK

Published in Canada by General Publishing Company, Ltd., 30 Lesmill Road, Don Mills, Toronto, Ontario.

Published in the United Kingdom by Constable and Company, Ltd., 10 Orange Street, London WC2H 7EG.

This Dover edition, first published in 1984, is an unabridged and unaltered republication of the third edition of the work, as published by Oxford University Press, London, in 1954 (first edition, 1936). The present edition is published by special arrangement with Oxford University Press, 200 Madison Avenue, New York, N.Y. 10016.

Manufactured in the United States of America

Dover Publications, Inc., 180 Varick Street, New York, N.Y. 10014

Library of Congress Cataloging in Publication Data

Heitler, Walter, 1904–
The quantum theory of radiation.

Reprint. Originally published: 3rd ed. Oxford : Clarendon Press, 1954. (The International series of monographs on physics)
Includes bibliographical references and index.
1. Quantum electrodynamics. I. Title.
QC680.H44 1984 537.6 83-5201
ISBN 0-486-64558-4

PREFACE TO THE THIRD EDITION

THE reader who was acquainted with the second edition of this book will notice in the first place the considerable increase in size. By so much, and more, the *Quantum Theory of Radiation* has progressed in the meantime. In order not to let the volume grow unduly, I have limited myself to the electrodynamics of electrons and positrons. The electrodynamics of mesons and nucleons, as well as all nuclear physics, including hyperfine structure, have been excluded.

It was inevitable that the book became, in some parts, more 'difficult'. However, in the sections dealing with the more elementary parts of the theory (on which the more advanced theory is built up throughout) I have endeavoured to retain the style of the first edition. These sections form a sufficient theoretical basis for the understanding of most of the applications in Chapters V and VII. For this purpose also a special section 'Elementary Perturbation Theory' has been added. The reader who is mainly interested in the applications of the theory will find a chart at the end of the introduction that will guide him through the book without being involved in theoretical intricacies.

Certain theoretical details have been referred to appendixes in order not to interrupt too much the general line of reasoning. In the development of the mathematical methods elegance has taken second place when precision and clearness made this seem advisable.

The experiments quoted have been selected more or less arbitrarily with a view to checking the theory at crucial points. There is no claim of completeness in this respect.

My debt to numerous helpers is great: to the Rev. Professor J. McConnell for critically reading most of the new theoretical sections; to Professor H. Wäffler for compiling the experimental literature and critical advice concerning this; to Dr. K. Bleuler for help and several simplifications in §§ 10 and 16; to Dr. E. Arnous and Mr. S. Zienau for help and advice in §§ 15, 16, and 34; to Mr. S. Zienau also for help in proof-reading; to my wife for various stylistic improvements; finally, to a large number of colleagues who have drawn my attention to minor inconsistencies in the second edition or have suggested improvements. I apologize for not mentioning them individually.

W. H.

ZÜRICH
August, 1953

CONTENTS

III. THE ELECTRON FIELD AND ITS INTERACTION WITH RADIATION

IV. METHODS OF SOLUTION

V. RADIATION PROCESSES IN FIRST APPROXIMATION

VI. RADIATIVE CORRECTIONS, AMBIGUOUS FEATURES

INTRODUCTION

It is usually the fate of a good physical theory that, after its initial success, difficulties or limitations of its applicability become apparent. Eventually it is superseded by a better theory in which some of the difficulties are removed or which has a wider field of application, as the case may be. The history of the quantum theory of radiation, or quantum electrodynamics, is remarkable in showing exactly the opposite trend. As time went on the theory became, as it were, more and more correct. When quantum electrodynamics was established by Dirac, Heisenberg, and Pauli, soon after the completion of non-relativistic quantum mechanics, grave difficulties appeared almost immediately and the idea could hardly be entertained that this was a correct physical theory. There was not only the old problem of the diverging self-energy of a charged point particle, now aggravated by an additional transverse self-energy, but it appeared that the answer to almost every physical question was a 'diverging integral' as soon as one wanted an exact result, and not merely a first approximation. However, as a first step in the development, it turned out that the theory yielded good results, in excellent agreement with the experiments, if only the calculation was limited to the first approximation and the interaction between electron and radiation was treated as weak. This first approximation was shown to correspond closely to classical theory. At the same time the discovery of the positive electron placed relativistic quantum mechanics, which can hardly be separated from radiation theory, on a sound basis.

Moreover, the profound analysis of the problem of measurements of field strengths by Bohr and Rosenfeld made it clear that at least the quantum electrodynamics of the vacuum must be correct. It forms, together with particle mechanics, a consistent whole from which neither of the two theories can be omitted.

For a time it seemed that, even if used in first approximation, the theory failed when the energy of the particles and photons concerned was very high. The discovery of the cascade showers, however, showed that this was not the case and that there was essentially no limitation to the validity of the theory, due to high energies.

As a second step, a further useful part could be extracted from the theory, namely a consistent treatment of the line breadth, and later of all other damping phenomena. These go beyond the 'first approximation', but are also closely related to classical theory and correspond to effects due to the damping force of Lorentz.

The next, and certainly last conceivable, step in this development, short of a final solution, has now been taken. A rational treatment of the higher approximations has been made possible when it was realized that all infinities which occur in the theory originate from a few diverging quantities, which are unobservable, namely, a contribution to the mass and to the charge of a particle due to its interaction with the quantum field. They must be embodied in the observed finite values of these quantities. If we pretend not to notice that the contributions in question turn out to be infinite and therefore cannot be handled in a way free from mathematical objections, there is no obstacle in the way of giving an unambiguous answer to every legitimate physical question. The outstanding success of this latest phase of the theory is the quantitative explanation of the radiative displacement of atomic energy levels and the magnetic moment of the electron additional to the usual magneton. These two effects belong to the most accurate confirmations of the theory extant.

It appears therefore that the theory as it stands now must be strangely near to the final solution, yet it cannot possibly be correct, as the mathematical procedure used to extract these results is plainly unacceptable. One can visualize two quite different courses of further development: (i) It may yet turn out that the present theory has exact solutions and that the present difficulties are merely the result of an impermissible mathematical expansion. Or (ii) it may be that the theory has no solutions in the mathematical sense at all and that profound alterations in the physical concepts are required. These, one may hope, will lead to a better understanding of the elementary charge e, which is connected with the other universal constants $\hbar$ and c by the fine structure constant, and also of the masses of the so-called fundamental particles. Even if the first alternative were true one must hope for a further development on the lines of (ii). However that may be, there can be no doubt at all that the present theory, including the somewhat doubtful mathematical procedures which must as yet be employed, will be vindicated and finally emerge as an excellent approximation, with an accuracy that leaves little to be desired.

Logical relationship between sections

The chart shows the way in which each section depends logically on the preceding sections, as far as the theoretical development is concerned.

For example, to understand § 12, the reader should be acquainted with Chapter I, §§ 7, 8, 10 as well as with § 11. V (—20) means Chapter V excepting § 20. 13. 1, 2 means § 13 subsections 1, 2. The left-hand line is the 'elementary line' leading to the chief applications in Chapters V and VII. The right-hand line, leading to VI, is that of the more advanced theory. The central line is specially concerned with line breadth phenomena. The chart is meant to give some guidance and should not be taken too literally.

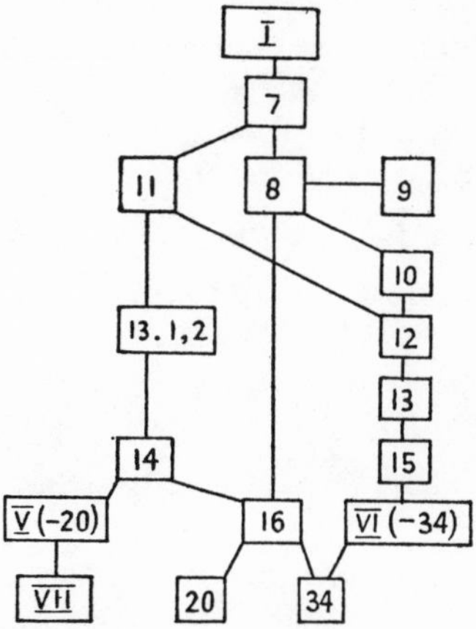

I

CLASSICAL THEORY OF RADIATION

1. The general Maxwell–Lorentz theory

1. *Field equations.* The classical theory of radiation is based on Maxwell's theory of the electromagnetic field. The two fundamental quantities describing the electromagnetic field are the electrical and magnetic field strengths **E** and **H**, which are both functions of space and time. To describe the electric state of matter one needs, besides the field, the charge density ρ and current density **i** which are also functions of space and time. If the velocity of the charge at a given point and a given time is **v**, the current density is

$$\mathbf{i} = \rho\mathbf{v}. \tag{1}$$

For a given distribution of charge and current the field is determined by the *Maxwell–Lorentz equations*

$$\operatorname{curl}\mathbf{E}+\frac{1}{c}\dot{\mathbf{H}} = 0, \tag{2 a}$$

$$\operatorname{div}\mathbf{H} = 0, \tag{2 b}$$

$$\operatorname{curl}\mathbf{H}-\frac{1}{c}\dot{\mathbf{E}} = \frac{4\pi}{c}\rho\mathbf{v}, \tag{2 c}$$

$$\operatorname{div}\mathbf{E} = 4\pi\rho. \tag{2 d}$$

(The dot denotes differentiation with respect to the time t.)

From these equations it can easily be deduced that charge and current satisfy the equation of continuity (conservation of charge)

$$\operatorname{div}(\rho\mathbf{v})+\dot{\rho} = 0. \tag{3}$$

On the other hand, the motion of the charges in a given field is determined by the *Lorentz equation*

$$\mathbf{k} = \rho\left(\mathbf{E}+\frac{1}{c}[\mathbf{vH}]\right), \tag{4}$$

where **k** represents the *density of force* acting on the charge density ρ. This electromagnetic force is in equilibrium with the force of inertia, which is given by the mass distribution of the charges.

For a point charge e (elementary particle) we have to carry out in the equations (2) and (4) the transition to the case where ρ is concentrated in an infinitely small volume. The Lorentz equation (4) can then be integrated over this volume, giving the whole force acting on the particle

$$\mathbf{K} = e\left(\mathbf{E}+\frac{1}{c}[\mathbf{vH}]\right). \tag{5}$$

K has to be equated to the force of inertia, which in non-relativistic dynamics is:

$$\mathbf{K} = \frac{d}{dt}(m\mathbf{v}), \tag{6}$$

m being the inert mass of the particle.

The field which has to be inserted in the Lorentz equation (4) or (5) is the external field produced by other charges (condensers, magnets, etc.) as well as the field produced by the point charge itself. This self-produced field will also react on the motion of the particle. The reaction of the field is in general small, so that we can, to a first approximation, insert in (5) the external field only. The theory of the field reaction will be dealt with in detail in this book. It is also connected with, as yet, unsolved problems, such as the problem of the inert mass of the particle.

2. *Potentials.* The field equations (2) can be reduced to simpler equations between a vector and a scalar function only instead of two vectors. From equation (2 b) it follows that **H** can always be represented as a curl of another vector **A**, which we call the *vector potential*:

$$\mathbf{H} = \operatorname{curl}\mathbf{A}. \tag{7 a}$$

Then (2 a) becomes
$$\operatorname{curl}\left(\mathbf{E}+\frac{1}{c}\dot{\mathbf{A}}\right) = 0$$

or
$$\mathbf{E}+\frac{1}{c}\dot{\mathbf{A}} = -\operatorname{grad}\phi, \tag{7 b}$$

where ϕ represents a scalar function which we call the *scalar potential.* The other two equations (2 c, d) result in two differential equations for these potentials. Making use of the general vector relation

$$\operatorname{curl}\operatorname{curl} = \operatorname{grad}\operatorname{div}-\nabla^2$$

they can be written in the form:

$$\frac{1}{c^2}\ddot{\mathbf{A}}-\nabla^2\mathbf{A}+\operatorname{grad}\left(\operatorname{div}\mathbf{A}+\frac{1}{c}\dot{\phi}\right) = \frac{4\pi}{c}\rho\mathbf{v}, \tag{8 a}$$

$$-\nabla^2\phi-\frac{1}{c}\operatorname{div}\dot{\mathbf{A}} = 4\pi\rho. \tag{8 b}$$

The vector **A** is not determined completely by the magnetic field **H**. Since, for any scalar function χ, $\operatorname{curl}\operatorname{grad}\chi = 0$, we can add to **A** the gradient of an arbitrary function χ. According to (7 b), however, we have to replace ϕ by $\phi+(1/c)\dot{\chi}$ if we replace $\mathbf{A}$ by $\mathbf{A}-\text{grad}\,\chi$, in order that $\mathbf{E}$ should not be changed. This freedom in the choice of the potentials can be used to simplify the field equations (8). If $\mathbf{A}_0$ and ϕ_0 represent certain possible values of $\mathbf{A}$ and ϕ, we determine χ from the equation

$$\nabla^2\chi-\frac{1}{c^2}\ddot{\chi} = \text{div}\,\mathbf{A}_0+\frac{1}{c}\dot{\phi}_0. \tag{9}$$

If we now put

$$\mathbf{A} = \mathbf{A}_0-\text{grad}\,\chi,$$

$$\phi = \phi_0+\frac{1}{c}\dot{\chi},$$

we obtain

$$\text{div}\,\mathbf{A}+\frac{1}{c}\dot{\phi} = 0. \tag{10}$$

(10) represents a relation between the potentials and is called the *Lorentz relation*. The field equations (8) then become simply

$$-\Box\mathbf{A} \equiv \frac{1}{c^2}\ddot{\mathbf{A}}-\nabla^2\mathbf{A} = \frac{4\pi}{c}\,\rho\mathbf{v}, \tag{11 a}$$

$$-\Box\phi \equiv \frac{1}{c^2}\ddot{\phi}-\nabla^2\phi = 4\pi\rho. \tag{11 b}$$

$\mathbf{A}$ and ϕ satisfy, therefore, the *inhomogeneous wave equation*. They are coupled by the Lorentz condition (10) only.

Still $\mathbf{A}$ and ϕ are not yet determined completely by the field strengths $\mathbf{E}$ and $\mathbf{H}$. χ is limited in so far as equation (9) has to be satisfied. We are still free to choose an arbitrary χ which satisfies the homogeneous wave equation

$$\nabla^2\chi-\frac{1}{c^2}\ddot{\chi} = 0. \tag{12}$$

Replacing $\mathbf{A}$ by $\mathbf{A}-\text{grad}\,\chi$ and ϕ by $\phi+\dot{\chi}/c$ the field strengths and the Lorentz condition (10) remain unchanged.

The different possible choices one can make for $\mathbf{A}$ and ϕ, leaving $\mathbf{E}$ and $\mathbf{H}$ unchanged, are called gauges, and the invariance of $\mathbf{E}$ and $\mathbf{H}$ under these transformations is called *gauge invariance*. In particular the class of gauges satisfying (10) will be called the *Lorentz gauge*. Another important gauge which is particularly convenient in quantum theory and which will be called the *Coulomb gauge* is determined by

$$\text{div}\,\mathbf{A} = 0 \tag{13 a}$$

with the field equations (from (8)):

$$-\Box \mathbf{A} + \frac{1}{c}\operatorname{grad}\dot{\phi} = \frac{4\pi}{c}\rho\mathbf{v}, \tag{13 b}$$

$$\nabla^2\phi = -4\pi\rho. \tag{13 c}$$

(13 c) is the Poisson equation. The scalar potential is then determined from the charges as if the latter were at rest. (Hence the name Coulomb gauge.) This gauge will be treated in more detail in § 6 and the following chapters.

3. *Retarded potentials.* The general solutions of the wave equations (11) can easily be written down. As is well known a special solution of the Poisson equation $\nabla^2\phi = 4\pi\rho$ is represented by the Newtonian potential

$$\phi(P) = \int \frac{\rho(P')\,d\tau'}{r_{PP'}},$$

where the integration has to be carried out all over the space. $r_{PP'}$ is the distance between a point P' of the charge distribution and the point P at which we are evaluating the potential ϕ. From this solution one can easily find also a special solution of the time-dependent Poisson equation (11 b):

$$\phi(P,t) = \int \frac{\rho(P',t-r_{PP'}/c)}{r_{PP'}}\,d\tau'. \tag{14 a}$$

This expression has the following meaning: If we wish to know the potential at a point P at a time t, we have to take for each point P' the charge density which was present there at the previous time $t-r_{PP'}/c$. Therefore for each point of the integration space we have to take the density at another time. $t-r_{PP'}/c$ represents the time which the light needs to come from P' to the point P at which the potential is considered. Therefore (14 a) takes account of the finite velocity c of propagation of an electromagnetic field.

In the same way we obtain a special solution of (11 a):

$$\mathbf{A}(P,t) = \frac{1}{c}\int \frac{(\rho\mathbf{v})(P',t-r_{PP'}/c)}{r_{PP'}}\,d\tau'. \tag{14 b}$$

The potentials (14) are called the *retarded potentials*. The Lorentz condition (10) is satisfied by the solution (14) because of the conservation of charge (3), as is easily verified.

(14) represents a special solution of the field equations: namely, the field which arises from the considered charges only. To obtain the

general solution we have to add the general solution of the homogeneous wave equations

$$\left.\begin{aligned} \nabla^2\mathbf{A}-\frac{1}{c^2}\ddot{\mathbf{A}} &= 0 \\ \nabla^2\phi-\frac{1}{c^2}\ddot{\phi} &= 0 \\ \operatorname{div}\mathbf{A}+\frac{1}{c}\dot{\phi} &= 0 \end{aligned}\right\}, \tag{15}$$

representing the field in free space.

For this part of the field, which satisfies the homogeneous wave equations (15), one can, according to (12), choose χ within the Lorentz gauge so that the scalar potential ϕ vanishes. The field which is independent of the charges is therefore simply given by

$$\left.\begin{aligned} \nabla^2\mathbf{A}-\frac{1}{c^2}\ddot{\mathbf{A}} = 0, \qquad \operatorname{div}\mathbf{A} = 0 \\ \mathbf{E} = -\frac{1}{c}\dot{\mathbf{A}}, \qquad \mathbf{H} = \operatorname{curl}\mathbf{A} \end{aligned}\right\}. \tag{15'}$$

The general solution of (15′) is formed by superposing *transverse waves* (see § 6).

The retarded potentials (14) will be evaluated in § 3 for the case of an arbitrary motion of a point charge and then be applied to problems of the emission of light, etc.

4. *Energy and momentum balance.* It is assumed in Maxwell's theory that a volume in which the field is different from zero contains a certain amount of energy and momentum. The energy in a given volume is given by:

$$U = \frac{1}{8\pi}\int (E^2+H^2)\,d\tau = \int u\,d\tau. \tag{16}$$

The density of momentum is assumed to be $1/c^2$ times the energy passing through unit area per unit time. The latter is given by the Poynting vector $\mathbf{S}$.

It is convenient, in all discussions of high energy radiation, to introduce the *quantity* $c\times$ *momentum*, having the dimensions of energy. Therefore, throughout this book, we shall *refer to the latter quantity simply as 'momentum'*, for particles and for radiation.

According to this definition the momentum of the field contained in a certain volume is given by

$$\mathbf{G} = \int \mathbf{g}\,d\tau = \frac{1}{c}\int \mathbf{S}\,d\tau = \frac{1}{4\pi}\int [\mathbf{EH}]\,d\tau. \tag{16'}$$

The assumptions (16) and (16′) are suggested by considerations of the energy and momentum balance of a charge distribution in the field. The identity of momentum and energy flow follows from these considerations. In fact, one can easily prove—with the assumptions (16) and (16′)—the conservation of energy and momentum.

By (4) and (5), the whole force acting on the charges contained in a certain volume is given by

$$\mathbf{K} = \int \mathbf{k}\,d\tau = \int \rho\Big(\mathbf{E}+\frac{1}{c}[\mathbf{vH}]\Big)\,d\tau.$$

Since $\mathbf{K}$ is identical with the force of inertia, it also represents the change of the mechanical momentum $\mathbf{u}$ of the charges per unit time

$$\mathbf{K} = \frac{1}{c}\frac{d\mathbf{u}}{dt}. \tag{17}$$

Furthermore, the change of the kinetic energy T of the charges is given by

$$\frac{dT}{dt} = \int (\mathbf{kv})\,d\tau, \tag{18}$$

where $\mathbf{k}$ represents the density of force (4). Inserting ρ and $\rho\mathbf{v}$ from the Maxwell equations (2 c, d) we obtain

$$\frac{1}{c}\frac{d\mathbf{u}}{dt} = \frac{1}{4\pi}\int\Big(\mathbf{E}\operatorname{div}\mathbf{E}-[\mathbf{H}\operatorname{curl}\mathbf{H}]-\frac{1}{c}[\dot{\mathbf{E}}\mathbf{H}]\Big)\,d\tau, \tag{19 a}$$

$$\frac{dT}{dt} = \frac{c}{4\pi}\int \mathbf{E}\Big(\operatorname{curl}\mathbf{H}-\frac{1}{c}\dot{\mathbf{E}}\Big)\,d\tau. \tag{19 b}$$

Making use of (2 a) the last terms of (19 a, b) can also be written

$$-\frac{1}{c}[\dot{\mathbf{E}}\mathbf{H}] = -[\mathbf{E}\operatorname{curl}\mathbf{E}]-\frac{1}{c}\frac{d}{dt}[\mathbf{EH}],$$

$$-(\mathbf{E}\dot{\mathbf{E}}) = -\frac{1}{2}\frac{d}{dt}(E^2+H^2)-c(\mathbf{H}\operatorname{curl}\mathbf{E}).$$

According to the definitions (16), (16′) the equations (19) become

$$\frac{d\mathbf{u}}{dt} = -\frac{d\mathbf{G}}{dt}+\frac{c}{4\pi}\int(\mathbf{E}\operatorname{div}\mathbf{E}-[\mathbf{H}\operatorname{curl}\mathbf{H}]-[\mathbf{E}\operatorname{curl}\mathbf{E}])\,d\tau, \tag{20 a}$$

$$\frac{dT}{dt} = -\frac{dU}{dt}+\frac{c}{4\pi}\int(\mathbf{E}\operatorname{curl}\mathbf{H}-\mathbf{H}\operatorname{curl}\mathbf{E})\,d\tau. \tag{20 b}$$

The integrals on the right-hand side can be transformed into surface integrals. In (20 b) we simply have, according to Gauss's formula,

$$\frac{c}{4\pi}\int(\mathbf{E}\operatorname{curl}\mathbf{H}-\mathbf{H}\operatorname{curl}\mathbf{E})\,d\tau = -\int \operatorname{div}\mathbf{S}\,d\tau = -\oint S_n\,d\sigma,$$

where n denotes the component normal to the surface of the volume considered. For equation (20 a) we define a tensor which is composed of quadratic terms in the field strengths, the *Maxwell tensor*:

$$\begin{aligned} 4\pi T_{xx} &= \tfrac{1}{2}(E_x^2 - E_y^2 - E_z^2 + H_x^2 - H_y^2 - H_z^2), \\ 4\pi T_{xy} &= 4\pi T_{yx} = E_x E_y + H_x H_y, \\ &\ldots\ldots\ldots \end{aligned} \tag{21}$$

The x-component of the divergence of this tensor is just

$$\begin{aligned} \mathrm{Div}_x T &\equiv \frac{\partial T_{xx}}{\partial x} + \frac{\partial T_{xy}}{\partial y} + \frac{\partial T_{xz}}{\partial z} \\ &= \frac{1}{4\pi}(E_x \operatorname{div}\mathbf{E} + H_x \operatorname{div}\mathbf{H} - [\mathbf{H}\operatorname{curl}\mathbf{H}]_x - [\mathbf{E}\operatorname{curl}\mathbf{E}]_x), \end{aligned}$$

where the second term vanishes, by equation (2 b).

The integral on the right-hand side in (20 a) can be written for the x-component (Gauss's formula also holds for a tensor divergence)

$$\int \mathrm{Div}_x T\, d\tau = \oint T_{xn}\, d\sigma.$$

Thus the equations (20) become

$$\frac{d(u_x + G_x)}{dt} = c \oint T_{xn}\, d\sigma, \tag{22 a}$$

$$\frac{d(T+U)}{dt} = -\oint S_n\, d\sigma. \tag{22 b}$$

In (22) the left-hand side represents the change of energy and momentum of the charged matter and of the field enclosed in a certain volume. On the right-hand side we have an integral of a normal component over the surface of the volume considered. S_n and $-T_{xn}$ have therefore to be interpreted in the following way: S_n represents the energy of the field passing (in the outward direction) through the surface per unit area and unit time. $-T_{xn}\, c$ represents the x-component of the momentum of the field passing through the surface per unit area and per unit time. The equations (22) give therefore a complete account of the conservation of energy and momentum in the field.

2. Lorentz invariance, momentum, and energy of the field

1. *Lorentz transformations.* The equations of classical electrodynamics are invariant with respect to Lorentz transformations.

A Lorentz transformation is an orthogonal transformation between the 4 space and time coordinates

$$x_1, x_2, x_3, x_4 = ict.$$

We denote indices running from 1 to 4 by Greek letters while Latin indices run from 1 to 3. A Lorentz transformation is then

$$x'_\mu = \sum_\nu a_{\mu\nu} x_\nu \quad \text{or} \quad x_\nu = \sum_\mu a_{\mu\nu} x'_\mu. \tag{1}$$

The $a_{\mu\nu}$ form a 4-dimensional orthogonal matrix (notwithstanding the fact that certain elements are complex)

$$\sum_\mu a_{\mu\nu} a_{\mu\lambda} = \sum_\mu a_{\nu\mu} a_{\lambda\mu} = \delta_{\nu\lambda} = \begin{cases} 1 & \text{if } \nu = \lambda \\ 0 & \text{if } \nu \neq \lambda \end{cases} \tag{2}$$

and if no reflections are included in the transformation the determinant is

$$|a_{\mu\nu}| = 1. \tag{2'}$$

(1) represents in general a rotation in space and a uniform translation. A special transformation between x_1 and x_4 only, which signifies a translation along the x-axis with the velocity $v = \beta c$, is given by

$$x'_1 = \frac{x_1 + i\beta x_4}{\sqrt{(1-\beta^2)}} = \frac{x_1 - vt}{\sqrt{(1-\beta^2)}}, \qquad x'_4 = ict' = \frac{x_4 - i\beta x_1}{\sqrt{(1-\beta^2)}} = \frac{i}{c}\frac{c^2t - vx_1}{\sqrt{(1-\beta^2)}}. \tag{3}$$

For $v \ll c$ we obtain the Galileo transformation $x'_1 = x_1 - vt$ and $t' = t$.

A 4-vector is a set of four quantities A_μ ($\mu = 1,..., 4$) which are transformed in the same way as the x_μ:

$$A'_\mu = \sum_\nu a_{\mu\nu} A_\nu. \tag{4}$$

Similarly, a tensor $A_{\mu\nu}$ is defined by the transformation

$$A'_{\mu\nu} = \sum_{\rho,\sigma} a_{\mu\rho} a_{\nu\sigma} A_{\rho\sigma}. \tag{5}$$

From (4) and from the orthogonality (2) it follows that the scalar product of two vectors is invariant, the scalar product of a vector and a tensor represents a vector, and so on. Thus

$$\sum A_\mu B_\mu = \text{inv.}, \qquad \sum_\nu A_{\mu\nu} B_\nu = C_\mu. \tag{6}$$

Therefore the length of a vector also is invariant:

$$\sum_\mu A_\mu^2 = \text{inv.} \tag{6'}$$

If ϕ is an invariant function of $x_1,..., x_4$, we see from (1) that

$$\text{if} \quad \frac{\partial \phi}{\partial x_\mu} = B_\mu, \quad \text{then} \quad B'_\mu = \sum_\nu \frac{\partial \phi}{\partial x_\nu} \frac{\partial x_\nu}{\partial x'_\mu} = \sum_k a_{\mu\nu} B_\nu. \tag{7}$$

Thus B_μ represents a 4-vector. The derivatives of a 4-vector form a tensor:

$$\frac{\partial A_\mu}{\partial x_\nu} = B_{\mu\nu}, \tag{8}$$

therefore the symbol $\partial/\partial x_\mu$ can be considered as the μ-component of a 4-vector. The symbol

$$\nabla^2-\frac{1}{c^2}\frac{\partial^2}{\partial t^2}=\sum_\mu \frac{\partial^2}{\partial x_\mu^2} \tag{9}$$

is invariant. From (6) it follows that if we put $A_\mu = \partial/\partial x_\mu$, then the 4-dimensional divergence of a vector is an invariant, that of a tensor is a vector:

$$\sum_\mu \frac{\partial A_\mu}{\partial x_\mu} = \text{inv.}, \qquad \sum_\nu \frac{\partial A_{\mu\nu}}{\partial x_\nu} = C_\mu. \tag{10}$$

The 4-dimensional volume element

$$dx_1\,dx_2\,dx_3\,dx_4 = \text{inv.} \tag{11}$$

is, according to (2), also invariant, whereas with the 3-dimensional volume element this is not the case.

According to (6′), the length of the infinitesimal vector dx_μ which may represent the displacement of a particle is an invariant

$$\sum_\mu dx_\mu^2 = -ds^2.$$

If we divide this equation by dt^2 we obtain

$$\left(\frac{ds}{dt}\right)^2 = -\sum_{i=1}^{3}\left(\frac{dx_i}{dt}\right)^2+c^2 = c^2-v^2 = c^2(1-\beta^2).$$

Therefore we can define an invariant time element ds

$$ds = c\,dt\,\sqrt{(1-\beta^2)}, \tag{12}$$

which is called the 'proper time'. The derivative of a vector with respect to s is again a vector, etc.

2. *Invariance of the Maxwell equations.* The Maxwell–Lorentz equations are shown to be invariant with respect to Lorentz transformations if we succeed in writing them in covariant form, i.e. as relations between 4-vectors and tensors. A clue for the 4-dimensional interpretation of these equations can be obtained from the following consideration:

From experience it follows that the *electric charge* is an *invariant* quantity. A charge element enclosed in a given volume $dx_1\,dx_2\,dx_3$ is given by

$$de = \rho\,dx_1\,dx_2\,dx_3 = \text{inv.} \tag{13}$$

Since the 4-dimensional volume element (11) is an invariant, the charge density ρ in (13) must have the transformation properties of the fourth component of a 4-vector. We put, therefore,

$$ic\rho = i_4.$$

Furthermore, the x-component of the current density is given by

$$i_x = \rho v_x = \rho \frac{dx_1}{dt} = i_4 \frac{dx_1}{dx_4}.$$

Since i_4 is transformed in the same way as dx_4, i_x is transformed as dx_1 and represents, therefore, the first component of a 4-vector. Therefore charge and current density form together a 4-vector:

$$\rho v_x = i_1, \qquad ic\rho = i_4. \tag{14}$$

We consider now the equations § 1 (11) for the potentials. Both equations have on the left-hand side the operator $\frac{1}{c^2}\frac{\partial^2}{\partial t^2}-\nabla^2$, which according to (9) is invariant. The right-hand sides of the two equations (11) represent just the first three and the fourth components respectively of the charge-current vector (14). Therefore, the scalar and vector potential together must also represent a 4-vector

$$A_x = A_1, \qquad A_4 = i\phi. \tag{15}$$

The condition $\operatorname{div}\mathbf{A}+\dot{\phi}/c = 0$ for Lorentz gauge can be written as a 4-dimensional divergence:

$$\sum_\mu \frac{\partial A_\mu}{\partial x_\mu} = 0. \tag{16}$$

The field equations for the potentials § 1 eq. (11) in Lorentz gauge take the form

$$\Box A_\mu = -\frac{4\pi}{c} i_\mu. \tag{16'}$$

In contrast to the Lorentz gauge, the condition for Coulomb gauge ($\operatorname{div}\mathbf{A} = 0$) is not covariant, nor is this the case for the field equations § 1 eq. (13 b, c). The Coulomb gauge can be retained after a Lorentz transformation only by regauging the potentials again.

The field strengths are obtained from the potentials by differentiation. If we use the 4-dimensional notation (15) we have from § 1 eq. (7 a) and (7 b)

$$\left.\begin{aligned} iE_x &= -\frac{i}{c}\dot{A}_x - i\frac{\partial\phi}{\partial x} = \frac{\partial A_1}{\partial x_4}-\frac{\partial A_4}{\partial x_1} \\ H_x &= \operatorname{curl}_x\mathbf{A} = \frac{\partial A_3}{\partial x_2}-\frac{\partial A_2}{\partial x_3} \end{aligned}\right\}. \tag{17}$$

According to (8) $\partial A_\nu/\partial x_\mu$, and therefore also the difference

$$\frac{\partial A_\nu}{\partial x_\mu}-\frac{\partial A_\mu}{\partial x_\nu} = f_{\mu\nu} = -f_{\nu\mu},$$

is a tensor. It is *antisymmetrical*. Thus the field strengths **E** and **H** together form an antisymmetrical 4-tensor

$$H_x = f_{23}, \qquad H_y = f_{31}, \qquad H_z = f_{12}; \qquad iE_x = f_{41}, \qquad \dots. \tag{18}$$

The general transformation formulae (3) give us immediately the rules according to which the field strengths have to be transformed in a moving system of coordinates. Considering the case of a uniform translation along the x-axis, we obtain, according to (3), (5), and (18):

$$\left.\begin{array}{ll} E'_x = E_x & H'_x = H_x \\ E'_y = (E_y - \beta H_z)\gamma & H'_y = (H_y + \beta E_z)\gamma \\ E'_z = (E_z + \beta H_y)\gamma & H'_z = (H_z - \beta E_y)\gamma \\ \multicolumn{2}{c}{\gamma = 1/\sqrt{(1-\beta^2)}} \end{array}\right\}. \tag{19}$$

In the formulae (19) it has to be understood that **E**, **H** are functions of the four coordinates x_μ, whereas **E**$'$, **H**$'$ are considered as functions of the transformed coordinates x'_μ. The values of x'_μ which have to be inserted as arguments on the left-hand side of (19) are those which are obtained from x_μ on the right-hand side by the transformation (3).

We can now easily write down the Maxwell–Lorentz equations in vector and tensor form. We consider first the two inhomogeneous equations § 1 (2 c) and (2 d):

$$\operatorname{div}\mathbf{E} = 4\pi\rho, \qquad \operatorname{curl}\mathbf{H} - \frac{1}{c}\dot{\mathbf{E}} = \frac{4\pi}{c}\rho\mathbf{v}.$$

Using the 4-dimensional notation (14) and (18) these two equations may be written in the *unified* form

$$\sum_\nu \frac{\partial f_{\mu\nu}}{\partial x_\nu} = \frac{4\pi}{c} i_\mu \quad (\mu = 1,\dots, 4). \tag{20 a}$$

The second homogeneous pair of Maxwell's equations § 1 (2 a, b) is

$$\operatorname{div}\mathbf{H} = 0 \qquad \operatorname{curl}\mathbf{E} + \frac{1}{c}\dot{\mathbf{H}} = 0.$$

In the 4-dimensional notation (18) they become respectively

$$\left.\begin{array}{l} \dfrac{\partial f_{23}}{\partial x_1} + \dfrac{\partial f_{31}}{\partial x_2} + \dfrac{\partial f_{12}}{\partial x_3} = 0 \\[2ex] \dfrac{\partial f_{43}}{\partial x_2} + \dfrac{\partial f_{24}}{\partial x_3} + \dfrac{\partial f_{32}}{\partial x_4} = 0 \end{array}\right\}, \tag{21}$$

etc. $\partial f_{\lambda\mu}/\partial x_\nu = t_{\lambda\mu\nu}$ forms a tensor of the third rank. Applying the transformation formulae (1) one can easily see that the same is also true for the sum $t_{\lambda\mu\nu}+t_{\mu\nu\lambda}+t_{\nu\lambda\mu}$. Therefore the four equations (21) can be written as a single tensor equation

$$t_{\lambda\mu\nu}+t_{\mu\nu\lambda}+t_{\nu\lambda\mu}=0, \quad t_{\lambda\mu\nu}=-t_{\mu\lambda\nu}=\frac{\partial f_{\lambda\mu}}{\partial x_\nu}. \tag{20 b}$$

Thus we have proved the Lorentz covariance of the Maxwell–Lorentz field equations.

3. *The Lorentz force. Momentum and energy of a particle.* To show the invariance of the whole system of equations of classical electrodynamics, we have finally to write the formula for the Lorentz force in 4-dimensional notation. The formula in question is (§ 1 (4))

$$c\mathbf{k}=c\rho\mathbf{E}+\rho[\mathbf{vH}], \tag{22}$$

where $\mathbf{k}$ is the force density. The right-hand side is simply the scalar product of a vector and a tensor, namely $\sum_\mu f_{i\mu}\, i_\mu$ $(i=1,2,3)$. Therefore the force density must represent the spatial part of a 4-vector, and we can write (22):

$$k_i=\frac{1}{c}\sum_\mu f_{i\mu}\, i_\mu. \tag{23}$$

These are the first three components of a 4-vector. The fourth component simply defines k_4: we obtain

$$k_4=\sum_\mu f_{4\mu}\, i_\mu=\frac{i}{c}\rho(\mathbf{Ev}). \tag{24}$$

k_4 therefore represents the work which the field performs on the charge per unit volume and per unit time.

If we insert for the charge-current vector i_μ the value (20 a) taken from the Maxwell equation, the equations (23), (24) become

$$k_\mu=\frac{1}{4\pi}\sum_{\nu,\lambda} f_{\mu\nu}\frac{\partial f_{\nu\lambda}}{\partial x_\lambda}. \tag{25}$$

The right-hand side can be written also as the 4-dimensional divergence of a tensor. For this purpose we define the symmetrical tensor

$$T_{\mu\nu}=\frac{1}{4\pi}\Big[\sum_\lambda f_{\mu\lambda}f_{\lambda\nu}+\tfrac{1}{4}\delta_{\mu\nu}\sum_{\lambda,\rho} f_{\lambda\rho}^2\Big], \tag{26}$$

where
$$\delta_{\mu\nu}=\begin{cases}0 & (\mu\neq\nu)\\ 1 & (\mu=\nu)\end{cases}$$

($\delta_{\mu\nu}$ is a tensor because of the orthogonality of the Lorentz transformation.) The right-hand side of (25) then simply becomes (according to (20 b)):

$$k_\mu=\sum_\nu \frac{\partial T_{\mu\nu}}{\partial x_\nu}. \tag{27}$$

The physical significance of this tensor becomes clear if we write it in the 3-dimensional notation. Substituting the $f_{\mu\nu}$ from (18) we obtain the following scheme:

$$T_{\mu\nu} = \begin{pmatrix} T_{xx} & T_{xy} & T_{xz} & -i/cS_x \\ T_{yx} & T_{yy} & T_{yz} & -i/cS_y \\ T_{zx} & T_{zy} & T_{zz} & -i/cS_z \\ -i/cS_x & -i/cS_y & -i/cS_z & u \end{pmatrix}. \tag{28}$$

The space-time part of this tensor represents the energy flow **S**, the pure time part the energy density u. The pure space part is identical with the Maxwell tensor, introduced in § 1 (21), and represents the flow of momentum. We call (28) the energy-momentum tensor. Its spur vanishes

$$\sum T_{\mu\mu} = 0. \tag{28'}$$

Equation (27) is identical with the equations § 1 (22) which express the energy and momentum balance. To show this, we have only to write down the space and time derivatives separately, for instance

$$k_x = \frac{\partial T_{xx}}{\partial x}+\frac{\partial T_{xy}}{\partial y}+\frac{\partial T_{xz}}{\partial z}+\frac{1}{ic}\frac{\partial T_{x4}}{\partial t} = \mathrm{Div}_x\, T-\frac{1}{c}\frac{\partial g_x}{\partial t}.$$

Integrating over all space we obtain

$$\int k_x\,d\tau = -\frac{1}{c}\frac{\partial G_x}{\partial t}+\oint T_{xn}\,d\sigma = K_x, \tag{29 a}$$

which is identical with § 1 (22 a). In the same way the fourth component k_4 of (27) gives, according to (24),

$$\frac{c}{i}\int k_4\,d\tau = \int \rho(\mathbf{E}\mathbf{v})\,d\tau = -\oint S_n\,d\sigma-\frac{\partial U}{\partial t}, \tag{29 b}$$

which is identical with equation § 1 (22 b) for the energy balance.

If we integrate (27) over a 4-dimensional volume—which is an invariant—we obtain another very important 4-vector, namely, the vector of the *momentum and energy* of the total charge u_μ. If the charge is concentrated in a small volume and if the space integration is extended over the whole of this volume, we may think of the integrated charge as a 'particle' and shall simply speak of the energy and momentum of the particle. The left-hand side of (27) becomes

$$\left.\begin{aligned} c\int k_x\,d\tau dt &= c\int K_x\,dt = u_x = u_1 \\ c\int k_4\,d\tau dt &= i\int \rho(\mathbf{E}\mathbf{v})\,d\tau dt = iT \end{aligned}\right\}, \tag{30}$$

i.e. the *momentum* **u** *and kinetic energy* T (times i) of the *particle*. They form therefore together a 4-vector. The transformation formulae for the kinetic energy and momentum of a particle are then:

$$\left.\begin{aligned} u'_x &= (u_x-\beta T)\gamma, & u'_y &= u_y \\ T' &= (T-\beta u_x)\gamma, & \gamma &= 1/\sqrt{(1-\beta^2)} \end{aligned}\right\}. \tag{31}$$

The length of the vector u_μ is invariant:

$$-\sum_\mu u_\mu^2 = T^2-(u_x^2+u_y^2+u_z^2) = \mu^2. \tag{32}$$

This invariant μ represents obviously the *energy of the particle at rest.* ($u_x = u_y = u_z = 0$.)

From (32) we therefore obtain the momentum and energy as functions of the velocity, if we assume the particle is at rest in one system of coordinates,

$$u_x = \frac{\mu\beta}{\sqrt{(1-\beta^2)}}, \qquad T = \frac{\mu}{\sqrt{(1-\beta^2)}}, \qquad \beta = v/c. \tag{33}$$

These are the famous *Einstein formulae* for the energy and momentum of a moving particle. Since for $v \ll c$ the momentum u_x must be identical with the classical value $mv_x c$, the rest energy μ takes the form

$$\mu = mc^2. \tag{34}$$

Finally we consider the relativistic *equation of motion of a charged particle in a given external field.* The total force acting on the particle was given by

$$\mathbf{K} = e\left(\mathbf{E}+\frac{1}{c}[\mathbf{vH}]\right).$$

In the non-relativistic theory **K** has to be equated to the time derivative of the momentum. As we shall see immediately, one has in relativistic dynamics to insert for **K** the time derivative of our 4-vector u_μ, so that the equation of motion becomes

$$K_x = \frac{1}{c}\frac{du_x}{dt} = \frac{d}{dt}\frac{mv_x}{\sqrt{(1-\beta^2)}} = e\left(E_x+\frac{1}{c}[\mathbf{vH}]_x\right). \tag{35}$$

In fact (35) can be written as a 4-dimensional vector equation. Dividing by $\sqrt{(1-\beta^2)}$ the right-hand side becomes, according to (18) and (33),

$$\frac{e}{\mu}\sum_\nu f_{1\nu}u_\nu.$$

The left-hand side represents the derivative with respect to the proper time s (12) which is invariant. Thus (35) can be written as

$$\frac{du_\mu}{ds} = \frac{e}{\mu}\sum_\nu f_{\mu\nu}u_\nu. \tag{36}$$

(36) represents the relativistic equation of motion of a particle in an external field.

The fourth component of the 4-vector u_μ represents the *kinetic* energy T. For a particle moving in an external field one can also consider the *total energy* E, i.e. kinetic energy+potential energy. The potential energy is simply equal to the scalar potential ϕ multiplied by e.

Since ϕ is the fourth component of a 4-vector the total energy $T+e\phi$ of the particle can also be related to a 4-vector p_μ

$$\left.\begin{aligned} p_\mu &= u_\mu+eA_\mu, \\ p_1 = u_x+eA_x, \quad \ldots, \quad p_4 &= iE = i(T+e\phi) \end{aligned}\right\}. \tag{37}$$

Thus, in an external field with a vector potential $\mathbf{A}$, the space components of our 4-vector u_μ do not represent the total momentum of the particle but a quantity which we may call the *kinetic momentum*. This quantity bears the same relation to the ordinary momentum as the 'kinetic energy' does to the total energy. In the literature $u_x = p_x - eA_x$ (divided by mc) is usually called *4-velocity*.

The space components of p_i representing the total momentum differ from the 'kinetic momentum' u_i by the vector potential, just as E differs from T by the potential energy $e\phi$. If $\mathbf{A}$ is zero, the kinetic momentum $\mathbf{u}$ is identical with the total momentum $\mathbf{p}$.

According to (32) the momentum energy vector p_μ (37) satisfies the important equation

$$\sum_\mu u_\mu^2 = \sum_\mu (p_\mu - eA_\mu)^2 = -\mu^2. \tag{38}$$

4. *Non-electromagnetic nature of the inertial mass.* The right-hand side of equation (27) integrated over a 4-dimensional volume represents also a 4-vector, viz.

$$\int \sum_\nu \frac{\partial T_{\mu\nu}}{\partial x_\nu}\, dx_1 \ldots dx_4. \tag{39}$$

At first sight one would perhaps be inclined to call this 4-vector the energy-momentum vector of the field. But to do so would in general have no significance. The expression (39) consists of two parts. Separating the time and space derivatives we obtain from (28) for the x- and t-components the time integrals of (29)

$$c\int \sum_\nu \frac{\partial T_{x\nu}}{\partial x_\nu}\, dx_1\, dx_2\, dx_3\, dt = -G_x + c\int dt \oint T_{xn}\, d\sigma, \tag{40 a}$$

$$ic\int \sum_\nu \frac{\partial T_{4\nu}}{\partial x_\nu}\, dx_1\, dx_2\, dx_3\, dt = +U + \int dt \oint S_n\, d\sigma. \tag{40 b}$$

The first terms, U and $\mathbf{G}$, represent energy and momentum of the field, whereas the second terms represent the time integrals of the flow of energy and momentum through the surface. The latter integrals depend, for a given instant of time t, not only upon the field at this time, but also upon the field at all previous times. But a quantity which depends upon the past history of the system cannot, of course, be called the energy or momentum of the field. On the other hand, the quantities U and G_x, which for good reasons we have interpreted as energy and momentum of the field, do *not* form a 4-vector, but are parts of a 4-tensor. They behave quite differently from the corresponding quantities for a particle. Thus, *in general the field has not the properties of a particle with regard to the Lorentz invariance of energy and momentum.*

This is the reason for the lack of success of some older theories which tried to explain the inertial mass of an electron by its pure electromagnetic properties. The idea of these theories, which are chiefly due to Abraham, was this:† the electron is assumed to have no *mechanical* inertial mass at all. But it produces a field which has a certain amount of energy and momentum, and therefore also a certain inertia. The inertia, which we observe as mass if we accelerate an electron, should be entirely due to its field. From the relativistic point of view it becomes clear that this idea cannot be correct. It has been shown by many experiments that $\mathbf{u}$ and T for an electron actually behave like a 4-vector (velocity dependence of the inertial mass of the electron). But the self-produced field of an electron has quite a different transformation character and depends upon the velocity in a different way. In classical theory we must therefore not only attribute to the electron a mechanical inert mass but we have also to see to it that the wrong relativistic properties of its field are compensated (internal mechanical stress, etc.), so that the electron as a whole behaves like a particle. In quantum theory the same problem will face us but we shall see (§ 29 and appendix 7) that it takes a somewhat different form, and that the field of the electron has, in a certain sense, indeed the required particle properties (although the solution of the problem is not quite satisfactory yet).

5. *'Particle properties' of light waves.* There are, however, cases where the total energy and momentum of a field themselves form a 4-vector. This is true, for instance, for a light wave of any shape and finite extension. We shall prove in general: *If the field differs from zero only within a certain given volume V, and if no charges are present inside this volume, then the*

† M. Abraham, *Theorie der Elektrizität*, II, 5th edition, Leipzig 1923, and 6th edition by R. Becker, Leipzig 1933.

total energy and momentum of the field form a 4-vector.† Since the density of force then also vanishes, we have according to (27)

$$\sum_\nu \frac{\partial T_{\mu\nu}}{\partial x_\nu} = 0, \qquad T_{\mu\nu} = 0 \text{ on the boundary.} \tag{41}$$

To prove the above statement we first consider an arbitrary vector A_μ which also satisfies these conditions (41), namely,

$$\sum_\nu \frac{\partial A_\nu}{\partial x_\nu} = 0$$

and $A_\nu = 0$ on the boundary. From Gauss's formula applied to the 4-dimensional divergence of a 4-vector, it follows that the surface integral of the normal component of A_ν over the surface of a 4-dimensional volume vanishes. For this 4-dimensional volume we choose a cylinder which is parallel to the x_4-axis. The base is the 3-dimensional volume considered. The cylinder is bounded by a section $x_4 =$ const. and by another section $x'_4 =$ const., where x'_4 denotes the time referred to a *moving* system of coordinates. (See Fig. 1; the x_2 and x_3 coordinates are omitted.)

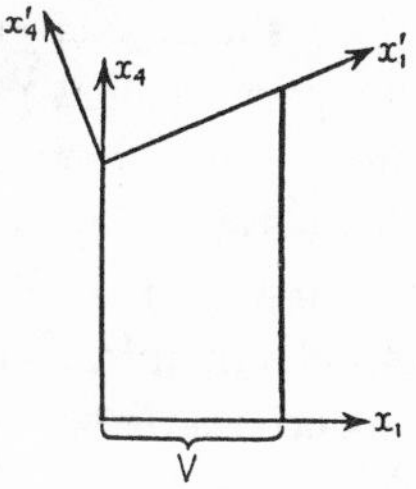

Fig. 1. A 4-dimensional integration.

The normal components of A_ν are $-A_4$ and $+A'_4$ respectively on these two sections and zero on the walls of the cylinder. Gauss's formula gives

$$-\int A_4\,d\tau + \int A'_4\,d\tau' = 0. \tag{42}$$

Thus the integral of A_4 over the volume V is according to (42) an invariant. Putting now

$$A_\mu = \sum_\nu T_{\mu\nu} b_\nu,$$

where b_ν is an arbitrary constant vector, we obtain from the invariance of $\int A_4\,d\tau$

$$\int T_{4\mu}\,d\tau = \mu\text{-component of 4-vector.} \tag{43}$$

The conditions (41) for which (43) holds are satisfied for any light wave with a finite extension which may have been emitted from a light source some time ago. In this case the *momentum* **G** *and energy* U of the field form a *4-vector* G_μ which behaves as regards its transformation properties like the *energy and momentum* of a particle. In particular, the transformation formulae for a moving system of coordinates are

$$G'_x = (G_x - \beta U)\gamma, \qquad G'_y = G_y, \qquad U' = (U - \beta G_x)\gamma$$
$$1/\gamma = \sqrt{(1-\beta^2)}, \qquad \beta = v/c. \tag{44}$$

† Cf. for instance, Abraham–Becker, ibid., p. 308

In the case of a plane wave, $E_y = a \sin \nu(x/c - t) = H_z$, the length of the vector G_μ is zero. Since the momentum and energy contained in a volume V are given by

$$G_x = \frac{V}{8\pi} a^2, \qquad U = \frac{V}{8\pi} a^2,$$

we have
$$\sum_\mu G_\mu^2 = G_x^2 - U^2 = 0. \tag{45}$$

This can also be expressed as follows: the rest energy of a plane wave is zero. A plane wave cannot be transformed to rest.

Equation (45) does not, however, hold in the general case. A spherical light wave emitted from a point source (the source may be removed after the emission) has a momentum zero, but a finite energy.

We shall see later in § 6 that every field can be composed of two parts: one part which is given by superposing light waves and for which therefore energy and momentum form a 4-vector, and another part containing the static field for which this is not the case. In the quantum theory only the first part is subjected to a quantization, giving rise to the existence of light quanta which behave also in some other ways like particles; the second, static part of the field remains, on the other hand, unquantized.

3. Field of a point charge and emission of light

The general Maxwell–Lorentz theory of § 1 gives immediately the field which is produced by any given charge distribution. For the applications to atomic physics the most important case is where the charge producing the field is a point charge. We shall therefore first work out the field produced by a *point charge* moving in an arbitrary way (we need this for several purposes). We shall then apply the formulae to a simple model of a light source.

1. *The Wiechert potentials.* The field of a charge distribution ρ is in general given by the equations § 1 (14) for the retarded potentials (in Lorentz gauge):

$$\phi(P) = \int \frac{\rho(P', t')}{r_{PP'}} \, d\tau' \qquad (t' = t - r_{PP'}/c), \tag{1a}$$

$$\mathbf{A}(P) = \frac{1}{c} \int \frac{\rho \mathbf{v}(P', t')}{r_{PP'}} \, d\tau', \tag{1b}$$

where t' represents the retarded time of the point P'. If the charges are in motion, we must be careful in carrying out the transition to a point charge. We may not, for instance, write simply

$$\left. \frac{e}{r} \right|_{t-r/c}$$

for the integral (1 a). Since we have to insert in (1) for each point P' another retarded time, the integral $\int \rho(P', t')\, d\tau'$ would not represent the total charge. Before carrying out the transition to a point charge we must transform the integrals (1) into integrals over the charge elements de. This can be done in the following way: we assume that the charge elements are rigidly connected with each other and have at a given time t' the same velocity $\mathbf{v}(t')$. We then consider a spherical shell of thickness dr at a distance r from P. A volume element of this shell will be $d\tau = d\sigma dr$. The contribution to the integral (1 a) is seen to be the charge density ρ (divided by r) which a spherical light wave meets if it contracts with velocity c and arrives at P at the time t. This wave will pass the outer surface of the shell at a time t'. During the time $dt = dr/c$ which this light wave would take to pass the shell dr, a certain amount of charge will stream through the inner surface of the shell. This amount of charge (per unit area) is given by

$$-\rho \frac{(\mathbf{vr})}{r} dt = -\rho \frac{(\mathbf{vr})}{r} \frac{dr}{c}$$

if the direction of $\mathbf{r}$ is chosen to be that of the line PP'. The charge element de collected by the light wave during the time dt is therefore

$$de = \rho\left(1+\frac{(\mathbf{vr})}{rc}\right) d\tau.$$

The integrand of (1) becomes then

$$\rho\, d\tau = \frac{de}{1+(\mathbf{vr})/rc}. \tag{2}$$

Inserting this expression in the integrals (1) we can now carry out the transition to a point charge and obtain

$$\phi(P, t) = \left.\frac{e}{r+(\mathbf{vr})/c}\right|_{t-r/c}, \tag{3 a}$$

$$\mathbf{A}(P, t) = \left.\frac{1}{c}\frac{e\mathbf{v}}{r+(\mathbf{vr})/c}\right|_{t-r/c}. \tag{3 b}$$

All quantities in (3) have to be taken at the retarded time $t-r/c$. Also r is of course the distance at this retarded time t'. The potentials (3), therefore, contain the time t implicitly in a rather complicated way. It is easy to verify that (3) satisfies the Lorentz condition § 1 eq. (10).

The expressions (3) were first obtained by Lienard and Wiechert.

2. *Field strengths of an arbitrarily moving point charge.* The field strengths **E**, **H** can be deduced from (3) by differentiation:

$$\mathbf{E} = -\operatorname{grad}\phi - \frac{1}{c}\dot{\mathbf{A}},$$

$$\mathbf{H} = \operatorname{curl}\mathbf{A}.$$

The derivatives have to be taken with respect to the time t and coordinates at P. But since the motion of the particle at the time t' is given by $\mathbf{r}(t')$ and $\mathbf{v}(t') = \partial\mathbf{r}/\partial t'$ the quantities $\mathbf{r}$, $\mathbf{v}$ occurring in (3) are given as functions of the retarded time t'. We must, therefore, first express the derivatives with respect to t in terms of the derivatives with respect to t'. The retarded time t' is defined by the distance r at t'

$$r(t') = c(t-t'). \tag{4}$$

Thus, differentiating with respect to t,

$$\frac{\partial r}{\partial t} = \frac{\partial r}{\partial t'}\frac{\partial t'}{\partial t} = \frac{(\mathbf{rv})}{r}\frac{\partial t'}{\partial t} = c\left(1-\frac{\partial t'}{\partial t}\right),$$

or

$$\frac{\partial t'}{\partial t} = \frac{1}{1+(\mathbf{vr})/rc} = \frac{r}{s}, \tag{5}$$

where we have used the abbreviation

$$s = r + \frac{(\mathbf{vr})}{c}. \tag{6}$$

Also because of equation (4) t' is a function of the coordinates of P. Therefore we have ($\mathbf{r}$ having the direction from P to P'):

$$\operatorname{grad} t' = -\frac{1}{c}\operatorname{grad} r(t') = -\frac{1}{c}\left(\frac{\partial r}{\partial t'}\operatorname{grad} t' - \frac{\mathbf{r}}{r}\right)$$

or, according to (6),

$$\operatorname{grad} t' = \frac{\mathbf{r}}{cs}. \tag{7}$$

For the derivatives of s we obtain, according to (5), (7),

$$\left.\begin{aligned} \frac{\partial s}{\partial t} &= \frac{\partial s}{\partial t'}\frac{\partial t'}{\partial t} = \frac{r}{s}\left(\frac{(\mathbf{rv})}{r}+\frac{v^2}{c}+\frac{(\mathbf{r}\dot{\mathbf{v}})}{c}\right) \\ \operatorname{grad} s &= -\frac{\mathbf{r}}{r}-\frac{\mathbf{v}}{c}+\frac{\mathbf{r}}{cs}\left(\frac{(\mathbf{rv})}{r}+\frac{v^2}{c}+\frac{(\mathbf{r}\dot{\mathbf{v}})}{c}\right) \end{aligned}\right\}. \tag{8}$$

In these formulae $\dot{\mathbf{v}}$ denotes the derivative of $\mathbf{v}$ with respect to t'. Also $\mathbf{v}$ depends only upon t'. Thus

$$\frac{\partial \mathbf{v}}{\partial t} = \dot{\mathbf{v}}\frac{r}{s}, \qquad \operatorname{curl}\mathbf{v} = -[\dot{\mathbf{v}}\operatorname{grad} t'] = \frac{[\mathbf{r}\dot{\mathbf{v}}]}{cs}. \tag{9}$$

For the field strengths we have, according to (3),

$$\frac{\mathbf{E}}{e} = -\text{grad}\frac{1}{s}-\frac{1}{c^2}\frac{\partial}{\partial t}\frac{\mathbf{v}}{s} = \frac{1}{s^2}\text{grad}\,s-\frac{1}{c^2 s}\frac{\partial \mathbf{v}}{\partial t}+\frac{\mathbf{v}}{c^2 s^2}\frac{\partial s}{\partial t},$$

$$\frac{\mathbf{H}}{e} = \frac{1}{c}\,\text{curl}\frac{\mathbf{v}}{s} = \frac{1}{sc}\,\text{curl}\,\mathbf{v}+\frac{1}{cs^2}[\mathbf{v}\,\text{grad}\,s].$$

Inserting our formulae (8), (9), and (6), we obtain finally

$$\frac{\mathbf{E}}{e} = -\frac{1-\beta^2}{s^3}\Big(\mathbf{r}+\frac{\mathbf{v}}{c}r\Big)+\frac{1}{s^3c^2}\Big[\mathbf{r}\Big[\mathbf{r}+\frac{\mathbf{v}}{c}r,\dot{\mathbf{v}}\Big]\Big], \tag{10a}$$

$$\frac{\mathbf{H}}{e} = \frac{[\mathbf{E}\mathbf{r}]}{r}, \qquad \beta = \frac{v}{c}. \tag{10b}$$

The magnetic field strength is always perpendicular to $\mathbf{E}$ and to $\mathbf{r}$, whereas the electric field strength has, besides a component which is perpendicular to $\mathbf{r}$, a component also in the direction of $\mathbf{r}$.

In these expressions (10), all quantities are of course understood to refer to the retarded time $t' = t-r/c$, where r itself is the distance at the retarded time.

The formulae (10) are general, in as much as they are valid for any given motion of the particle with any velocity, but they hold only as long as the particle can be considered as a point charge. They do not hold at distances which are comparable with the electronic radius.

The field (10a) is composed of two parts which behave quite differently. The first part depends only upon the velocity and decreases with r^{-2} for long distances from the particle. It represents the static (Coulomb) part of the field and reduces simply to $-\mathbf{r}/r^3$ for $\mathbf{v} = 0$. For $\mathbf{v} \neq 0$ this part can readily be deduced from the Coulomb field by means of a Lorentz transformation (application in appendix 6). The second part is proportional to the *acceleration* $\dot{\mathbf{v}}$, and decreases only as r^{-1} for long distances from the particle. This part of the field is *transverse*, the field strengths being perpendicular to the radius vector $\mathbf{r}$. We shall see that it gives rise to emission of light. We call the region where the second part preponderates the '*wave zone*'. The distinction between these two parts of the field has, however, only a meaning if the region which the electron passes during the time r/c is small compared with r itself (quasi-periodic motion, or if $v/c \ll 1$).

3. *The Hertzian vector of a system of charges. Dipole and quadripole moment.* As a first application we shall consider the emission of light by a system of point charges. We assume that a certain number of particles with charges e_k are all situated in the neighbourhood of a

centre Q which is fixed. The distance of Q from the point P at which we consider the field will be denoted by $\mathbf{R}$ (direction $P \rightarrow Q$) and the distance $P \rightarrow e_k$ by $\mathbf{r}_k$. Each charge can then be described by a vector representing the displacement of e_k from the centre Q

$$\mathbf{x}_k = \mathbf{r}_k - \mathbf{R}. \tag{11}$$

We assume that all displacements are small compared with R, and furthermore that the velocities of all particles are small compared with the velocity of light

$$|x_k| \ll R, \qquad v_k \ll c. \tag{12}$$

The field is of course the result of the superposition of the fields arising from all single charges.

In the expressions (10) one has, however, to insert for each particle another retarded time $t'_k = t - r_k/c$. But since all displacements are small, it will be convenient to introduce a new time T, the retarded time of the centre Q, viz.

$$T = t - \frac{R}{c}, \tag{13}$$

and to express the field strengths as functions of this time T.

If $\mathbf{x}_k$ is small the difference between the common time T and the times t'_k of the individual particles will also be small.

We first express $\mathbf{v}$ and $\dot{\mathbf{v}}$ as derivatives with respect to T instead of t', using the connexion

$$c(t'_k - T) = R - r_k(t'_k). \tag{14}$$

Then we express the electric field strength $\mathbf{E}$ implicitly as a function of T rather than of the t'_k. This can be done without using yet the approximations (12). We shall only be interested in the wave zone part of $\mathbf{E}$; all contributions behaving like r^{-2} or smaller do not contribute to the emission of light.

(14) gives (for each k), by differentiation with respect to t':

$$\frac{\partial T}{\partial t'} = 1 + \frac{(\mathbf{rv})}{rc} = \frac{s}{r} \tag{15}$$

according to (6). Hence the velocity is

$$\mathbf{v} \equiv \frac{\partial \mathbf{r}}{\partial t'} = \frac{\partial \mathbf{r}}{\partial T}\frac{\partial T}{\partial t'} = \mathbf{r}'\frac{s}{r} = \mathbf{x}'\frac{s}{r}, \tag{16}$$

denoting derivatives with respect to T by a dash. Similarly, omitting terms r^{-1}, which would not contribute to the wave zone, $\dot{\mathbf{v}}$ is given by

$$\dot{\mathbf{v}} = \mathbf{x}''\frac{s^2}{r^2} + \mathbf{x}'\frac{\partial}{\partial t'}\frac{s}{r} = \mathbf{x}''\frac{s^2}{r^2} + \mathbf{x}'\frac{(\mathbf{r}\dot{\mathbf{v}})}{rc}. \tag{16'}$$

Multiplying (16) by **r** and comparing with (6) we get

$$\frac{r}{s} = 1 - \frac{(\mathbf{r}\mathbf{x}')}{rc}, \qquad \mathbf{v} = \frac{\mathbf{x}'}{1-(\mathbf{r}\mathbf{x}')/rc}. \tag{17}$$

Similarly, multiplying (16′) by **r** and eliminating $(\mathbf{r}\dot{\mathbf{v}})$ we obtain

$$\dot{\mathbf{v}} = \frac{s^3}{r^3}\left[\mathbf{x}''\left(1-\frac{(\mathbf{r}\mathbf{x}')}{rc}\right)+\frac{\mathbf{x}'(\mathbf{r}\mathbf{x}'')}{rc}\right]. \tag{18}$$

Inserting now (17) and (18) into the wave zone part (second term) of (10a), the contribution from one particle k to **E** reduces to

$$\mathbf{E}_k = \frac{e_k}{r_k^3 c^2}[\mathbf{r}_k[\mathbf{r}_k \mathbf{x}_k'']]. \tag{19}$$

If the displacements $\mathbf{x}_k$ are small compared with R we can replace $\mathbf{r}_k$ in (19) by **R**, the difference does not contribute to the wave zone.

We then define a vector, called the *Hertzian vector*, representing the algebraic sum of all displacements

$$\mathbf{Z}(T) = \sum_k e_k \mathbf{x}_k(t_k'). \tag{20}$$

It is to be understood that in (20) the $\mathbf{x}_k$, which are primarily functions of t_k', are to be expressed as functions of T with the help of (14), which is a rather implicit connexion. In this sense differentiations like $\mathbf{Z}''$ are to be carried out. We obtain for the field strengths in the wave zone

$$\begin{aligned} \mathbf{E} &= \frac{1}{R^3c^2}[\mathbf{R}[\mathbf{R}\mathbf{Z}'']]_{T-R/c}, \\ \mathbf{H} &= \frac{1}{R^2c^2}[\mathbf{R}\mathbf{Z}'']_{T-R/c}. \end{aligned} \tag{21}$$

We have now to express $\mathbf{x}_k(t_k')$ as a function of T explicitly. Since t_k' and T differ only slightly, we can develop, using (14), (15), (17),

$$\mathbf{x}_k(t_k') = \mathbf{x}_k(T)+\mathbf{x}_k'\frac{(\mathbf{x}_k\mathbf{R})}{Rc}+\dots. \tag{22}$$

If the conditions (12) are fulfilled the series converges. Thus

$$\mathbf{Z} = \mathbf{Z}_1+\mathbf{Z}_2+\dots,$$

where $$\mathbf{Z}_1 = \sum_k e_k \mathbf{x}_k(T), \qquad \mathbf{Z}_2 = \sum_k e_k \mathbf{x}_k'(T)\frac{(\mathbf{R}\mathbf{x}_k(T))}{Rc}. \tag{22′}$$

The first term Z_1 represents the dipole moment of the charge system. The second term is due to the effect of retardation, since $(\mathbf{x}_k\mathbf{R})/Rc$ represents the difference between the retarded times for the kth particle and for the centre. It includes the *electric quadripole* and *magnetic dipole moment* of the charge system.† It plays, in general, an important role only if the arrangement of the charge is so symmetrical that the dipole moment vanishes. The decomposition of $\mathbf{Z}_2$ into the electric quadripole and magnetic dipole parts is as follows:

We define the tensor of the quadripole moment of the charge system

$$q_{ij} = \sum e_k x_i^k x_j^k \quad (i, j = 1, 2, 3) \tag{23 a}$$

(writing the index k denoting the particle on top) and the magnetic dipole moment

$$\mathbf{m} = \frac{1}{2c}\sum_k [\mathbf{x}^k \mathbf{i}^k] = \frac{1}{2c}\sum_k e_k[\mathbf{x}^k \mathbf{x}'^k] \quad (\mathbf{i}^k = e\mathbf{x}'^k). \tag{23 b}$$

Then

$$\mathbf{Z}_2 = \mathbf{Z}_q + \mathbf{Z}_{\text{md}}, \tag{24 a}$$

$$Z_{qi} = \frac{1}{2Rc}\sum_j q'_{ij} R_j, \tag{24 b}$$

$$\mathbf{Z}_{\text{md}} = \frac{1}{R}[\mathbf{mR}]. \tag{24 c}$$

In all these formulae $\mathbf{x}$ and $\mathbf{x}'$ are to be taken at the time T.

4. *Emission of light.* The field given by the equations (21) decreases with the first power of R. Any quadratic expression in the field strengths, for instance the Poynting vector of the energy flow, will decrease as R^{-2} and the integral over a spherical surface will therefore be finite and independent of R. Thus the field given by equations (21) gives rise to a finite energy flow through any distant sphere, i.e. to an *emission of radiation*. The radiation is *transverse*, both field strengths being perpendicular to $\mathbf{R}$ and to each other. The energy flow per unit time through the area $R^2\,d\Omega$ becomes

where $$\mathbf{S}R^2\,d\Omega = \frac{c}{4\pi}[\mathbf{EH}]R^2\,d\Omega = \frac{d\Omega}{4\pi c^3}\frac{\mathbf{R}}{R}Z''^2\sin^2\theta, \tag{25}$$

where θ represents the angle between the direction of the Hertzian vector $\mathbf{Z}''$ and the direction of observation. The energy flow (25) is normal to a spherical surface and proportional to the square of the second time derivative of the Hertzian vector. It has a maximum at right angles to $\mathbf{Z}''$ and vanishes in the direction of $\mathbf{Z}''$. The direction of polarization of this radiation (direction of $\mathbf{E}$) is the projection of $\mathbf{Z}''$ on the plane perpendicular to $\mathbf{R}$.

† See H. C. Brinkmann, *Zur Quantenmechanik der Multipolstrahlung*, Proefschrift, Utrecht 1932.

The total energy radiated per unit time S is given by integrating (25) over all angles

$$S = \frac{2}{3}\frac{Z''^2}{c^3}. \tag{26}$$

The simplest model of a light source is a harmonic oscillator, i.e. a single charge bound elastically to a centre of force and moving with simple harmonic motion of frequency ν along the x-axis. In this case we can put

$$\mathbf{Z} = e\mathbf{x} = e\mathbf{x}_0 \cos \nu t, \qquad \mathbf{Z}'' = -e\nu^2\mathbf{x}.$$

The radiation emitted from this oscillator is then monochromatic with the same frequency ν. The time average value of (26) becomes

$$S = \frac{2}{3}\frac{e^2}{c^3}\nu^4\overline{x^2} = \frac{1}{3}\frac{e^2}{c^3}\nu^4 x_0^2. \tag{27}$$

The energy emitted per unit time is therefore proportional to ν^4.

4. Reaction of the field, line breadth

In § 3 we have seen that a moving point charge in general emits radiation. The energy radiated per unit time for a slowly moving particle is given by § 3 eq. (26). According to the equations for the energy balance of the field (§ 1 eq. (22)), this energy has to be contributed by the forces keeping the charge in motion. Consequently, the kinetic energy of the particle must decrease with the time. It would not, therefore, be correct to determine the particle's motion only from the external forces acting on it (e.g. the quasi-elastic forces of an oscillator), since the motion is also influenced by the radiation emitted. In order to give a correct account of the conservation of energy we have to consider the *reaction* of the field produced by the charge on its own motion.

We shall find, however, that in general the reaction force of the radiation is small compared with other forces. This makes it possible to consider the reaction effect as a small correction, so that in a first approximation we may assume the motion of the particle to be determined by the external forces only.

We shall calculate this reaction in two different ways. The first way is purely phenomenological: we assume that in the first approximation the motion of the particle is not altered by the radiation emitted and that the latter is given by § 3 eq. (26). We then add as a second approximation in the equation of motion another small term (damping-force) which is determined so that it just preserves the energy balance.

The second method is more profound and, from the point of view of the general theory, more consistent. We shall calculate directly the force which the field produced by the charge exerts on the charge itself. The expression for the reaction force obtained in this way is the same as that obtained from the energy balance. However, this method leads to further results which are of prime importance in the discussion of the whole radiation theory.

1. *First way: the energy balance.* For simplicity we confine ourselves to the non-relativistic case of a slow particle. According to § 3 (26) and (20) the energy radiated per unit time by an accelerated particle is then given by

$$S = \tfrac{2}{3}e^2\dot{\mathbf{v}}^2/c^3. \tag{1}$$

In the first approximation the equation of motion of the particle is

$$\mathbf{K} = m\dot{\mathbf{v}}, \tag{2}$$

where $\mathbf{K}$ represents the external force. To this equation (2) we add another term $\mathbf{K}_s$ ('self-force') which is to take account of the energy loss (1):

$$\mathbf{K}+\mathbf{K}_s = m\dot{\mathbf{v}}. \tag{3}$$

In order to simplify the energy balance (§ 1 eq. (22)) we shall assume that for two times t_1, t_2 the state of motion of the particle is the same at t_2 as at t_1. Then the energy of the field U is also the same at t_2 as at t_1. The work performed by the force $\mathbf{K}_s$ during this time interval must then be equal to the total energy radiated

$$\int_{t_1}^{t_2} (\mathbf{K}_s\,\mathbf{v})\,dt = -\frac{2}{3}\frac{e^2}{c^3}\int_{t_1}^{t_2} \dot{\mathbf{v}}^2\,dt. \tag{4}$$

Under the above condition, this equation for $\mathbf{K}_s$ can be solved. Integrating (4) by parts we obtain

$$\int_{t_1}^{t_2} (\mathbf{K}_s\,\mathbf{v})\,dt = +\frac{2}{3}\frac{e^2}{c^3}\int_{t_1}^{t_2} (\mathbf{v}\ddot{\mathbf{v}})\,dt. \tag{5}$$

(5) is satisfied if we put for each instant of time:

$$\mathbf{K}_s = \frac{2}{3}\frac{e^2}{c^3}\ddot{\mathbf{v}}. \tag{6}$$

(6) gives the desired reaction force. It is proportional to the *time derivative of the acceleration*. For a particle moving (to a first approximation) with simple harmonic motion one can of course replace $\ddot{\mathbf{v}}$ by $-\nu^2\mathbf{v}$.

It can readily be proved that the force (6) also gives a correct account of the conservation of momentum and angular momentum.

This deduction is only correct so long as the reaction force (6) is small compared with the other forces. For a harmonic oscillator this is always the case as long as the frequency is not too high. Putting $\mathbf{x} = \mathbf{x}_0 \cos \nu_0 t$ the quasi-elastic force is $\nu_0^2 m\mathbf{x}$ and our condition gives

$$m\nu_0^2 \gg \frac{e^2}{c^3}\nu_0^3, \tag{7}$$

or introducing the wave-length λ,

$$\lambda = \frac{c}{\nu_0} \gg \frac{e^2}{mc^2} \equiv r_0. \tag{7'}$$

r_0 represents, as we shall see in subsection 3, the classical electronic radius and is of the order of 10^{-13} cm. Thus (7) holds for all wave-lengths large compared with r_0.

(7) is satisfied for all optical frequencies and also for the γ-rays of nuclear physics. It is not satisfied for the γ-rays of cosmic radiation, but here we are already far in the realm of quantum theory.

2. *Second way: the self-force.*† We shall now find directly the force which the field produced by the charge exerts on the charge itself. If the density of the charge is given by a function ρ and the field produced by this charge is given by $\mathbf{E}_s$, $\mathbf{H}_s$, the self-force is given by § 1 eq. (4)

$$\mathbf{K}^s = \int \rho\, d\tau\left(\mathbf{E}_s + \frac{1}{c}[\mathbf{v}\mathbf{H}_s]\right). \tag{8}$$

(We write the index s on the top because (8) will not be quite identical with (6).) For the field we may not of course insert the field of a point charge since it is the field inside the particle with which we are dealing. We must first assume a certain charge distribution; the transition to a point charge can be made in the final result.

(8) can be evaluated in the following way. We consider two charge elements de and de' at a distance $\mathbf{r}$ and determine the force which the field produced by de exerts on the element de'. For this field we may take the retarded field of a point charge de' as calculated in § 3. The total self-force is then obtained by integrating over all charge elements de as well as over all de'. The magnetic field will not give any contribution to the self-force, and we shall omit it in the following considerations.

For the calculation we make the following simplifying assumptions:

(1) The charge distribution is *rigid* (for small velocities; for high velocities we have to assume a Lorentz contraction). For a given instant of time all charge elements will then have the same velocity and acceleration.

(2) The charge distribution is to be spherical, with a finite extension of the order of magnitude of the characteristic radius $\bar{r}_0$ (electronic

† Cf. H. A. Lorentz, *Theory of the Electron* (Leipzig, 1916).

radius). We shall discuss in subsection 3 in what sense $\bar{r}_0$ can afterwards be put equal to zero.

Care must be taken in applying the formulae § 3 (10) for the field produced by de. These formulae give the field of a point charge at a *fixed* point P which is held at rest. But in our case, the charge element de', the field within which is under consideration, is rigidly connected with the charge element de which produces the field, both having the same velocity at any instant. The formulae § 3 (10) can, therefore, only be applied to our problem if we put the velocity $\mathbf{v}$ of the charge equal to zero at the time t at which the self-force is to be calculated, i.e. for a particle at rest: thus,

$$\mathbf{v}(t) = 0. \tag{9}$$

This does not, however, mean that the velocity vanishes at the retarded time $t' = t - r/c$. (9) does not represent any real restriction of our results, since the self-force of a moving particle can always be obtained by a Lorentz transformation.

The electric field produced by de at the position of de' at the time t is given by § 3 eq. (10 a)

$$d\mathbf{E}_s(t) = \frac{de'}{s^3}\left\{\frac{1}{c^2}\left[\mathbf{r}\left[\mathbf{r}+\frac{\mathbf{v}(t')}{c}r, \dot{\mathbf{v}}(t')\right]\right] - \left(1-\frac{v^2(t')}{c^2}\right)\left(\mathbf{r}+\frac{\mathbf{v}(t')}{c}r\right)\right\}, \tag{10}$$

$$s = r + \left(\frac{\mathbf{v}(t')}{c}\mathbf{r}\right), \qquad \mathbf{v}(t) = 0, \qquad t' = t - \frac{r(t')}{c}.$$

Also $\mathbf{r} \equiv \mathbf{r}(t')$ and $r \equiv r(t')$ have at all points to be taken at the retarded time t'.

(3) *The motion of the particle varies so slowly*, that the change of its acceleration within the time which light needs to pass the charge distribution is smaller than the acceleration itself, or:

$$\frac{\bar{r}_0}{c}\ddot{\mathbf{v}} \ll \dot{\mathbf{v}}. \tag{11}$$

If $\bar{r}_0 > 0$, (11) is not only a condition which the external forces must satisfy, but also a restriction on the self-force itself, but if $\bar{r}_0 \to 0$, (11) is always satisfied.

Assuming the condition (11), the effect of retardation in equation (10) becomes a small correction. We can then develop any function of t' (such as $\mathbf{r}(t')$, $\mathbf{v}(t')$) in a power series of $t - t' = r(t')/c$. Using the fact that $\mathbf{v}(t) = 0$, we have, for example,

$$\mathbf{r}(t') = \mathbf{r}(t) + \frac{1}{2}\frac{r^2(t')}{c^2}\dot{\mathbf{v}}(t) - \frac{1}{6}\frac{r^3(t')}{c^3}\ddot{\mathbf{v}}(t) + \dots. \tag{12}$$

The coefficients on the right are still $r(t')$. This retarded distance $r(t')$ is by no means equal to the distance of the two charge elements for the particle at rest. Even if $\mathbf{v} = 0$ at the time t, the particle was in motion (and accelerated) at the retarded time t'. To express the coefficients in (12) by $r(t)$ we take the square and re-insert the series on the right. This gives first $r(t')$ expressed by quantities to be taken at t only. We now write r for $r(t)$, etc. Then

$$r(t') = r+\frac{r}{2c^2}(\mathbf{r}\dot{\mathbf{v}})+\frac{r}{8c^4}(r^2\dot{\mathbf{v}}^2+3(\mathbf{r}\dot{\mathbf{v}})^2)-\frac{r^2}{6c^3}(\mathbf{r}\ddot{\mathbf{v}})+\dots. \tag{13 a}$$

Substituting this in the series (12) we obtain the expansion for $\mathbf{r}(t')$

$$\mathbf{r}(t') = \mathbf{r}+\frac{r^2}{2c^2}\dot{\mathbf{v}}+\frac{r^2}{2c^4}(\mathbf{r}\dot{\mathbf{v}})\dot{\mathbf{v}}-\frac{r^3}{6c^3}\ddot{\mathbf{v}}+\dots. \tag{13 b}$$

Similarly $\mathbf{v}$ and $\dot{\mathbf{v}}$ are expanded

$$\mathbf{v}(t') = -\frac{r}{c}\dot{\mathbf{v}}-\frac{r}{2c^3}(\mathbf{r}\dot{\mathbf{v}})\dot{\mathbf{v}}+\frac{r^2}{2c^2}\ddot{\mathbf{v}}+\dots, \tag{13 c}$$

$$\dot{\mathbf{v}}(t') = \dot{\mathbf{v}}-\frac{r}{c}\ddot{\mathbf{v}}+\dots \tag{13 d}$$

and the quantity s occurring in (10)

$$s \equiv r(t')+\frac{1}{c}(\mathbf{r}(t')\mathbf{v}(t')) = r-\frac{r}{2c^2}(\mathbf{r}\dot{\mathbf{v}})-\frac{r}{8c^4}((\mathbf{r}\dot{\mathbf{v}})^2+3r^2\dot{\mathbf{v}}^2)+\frac{r^2}{3c^3}(\mathbf{r}\ddot{\mathbf{v}})+\dots.$$

All these expansions are power series in r. We carry the expansion of $d\mathbf{E}_s$ (10) up to terms independent of r; the higher terms will vanish later in the transition to a point particle $\bar{r}_0 \to 0$. The result is greatly simplified by the observation that we have afterwards to integrate over the charge distribution. Since this is spherically symmetrical, terms of odd order in the vector $\mathbf{r}$ vanish. (This averaging must, of course, not be carried out over $\mathbf{r}(t')$.) Omitting terms of odd order in $\mathbf{r}$ we obtain

$$d\mathbf{E}_s(t) = de'\left[-\frac{\mathbf{r}(\mathbf{r}\dot{\mathbf{v}})}{2r^3c^2}-\frac{\dot{\mathbf{v}}}{2c^2r}+\frac{2\ddot{\mathbf{v}}}{3c^3}\right]. \tag{14}$$

Integrating (14) over all charge elements de, de' we obtain the self-force

$$\mathbf{K}^s = \mathbf{K}_0+\mathbf{K}_s, \tag{15}$$

where

$$\mathbf{K}_0 = -\frac{2}{3}\frac{\dot{\mathbf{v}}}{c^2}\int\frac{dede'}{r} = -\frac{4}{3}\frac{\dot{\mathbf{v}}}{c^2}\mu_0, \tag{16 a}$$

$$\mathbf{K}_s = \frac{2}{3}\frac{\ddot{\mathbf{v}}}{c^3}\int dede' = \frac{2}{3}\frac{e^2}{c^3}\ddot{\mathbf{v}}. \tag{16 b}$$

We discuss first the second term $\mathbf{K}_s$. This is *independent of the charge distribution* and identical with the self-force deduced in subsection 1 eq. (6) from the energy balance. Thus, the damping force (6) which simply takes account of the energy balance, represents the second approximation—due to the *retardation inside the particle*—to the self-force.

Comparing (16b) with the condition (11), which had to be imposed on the motion of the particle, this condition becomes

$$\mathbf{K}_s \ll \frac{2}{3}\frac{e^2}{\bar{r}_0 c^2}\dot{\mathbf{v}} = \frac{2}{3}m\dot{\mathbf{v}}\frac{r_0}{\bar{r}_0}. \tag{17}$$

We shall later put $\bar{r}_0$ equal to zero. The conditions (11), (17) are then satisfied automatically. The condition (7) used in the derivation of the previous subsection is then only required if the reaction force is to be considered as a small effect (which is the case in all applications) but not as a matter of principle.

If we take the expansion (12) to the next power of r/c, we obtain further terms of $\mathbf{K}_s$, the next one being of the order

$$\mathbf{K}' \sim \frac{\dddot{\mathbf{v}}}{c^4}\int r\,de\,de' \sim \frac{e^2\dddot{\mathbf{v}}}{c^4}\bar{r}_0.$$

This term depends upon the structure of the electron (the same is true for all higher terms). It is, however, proportional to $\bar{r}_0$ and vanishes for a point electron.

For a slowly moving point electron $\mathbf{K}_s$ is the only contribution to the reaction force produced by the field of the electron (apart from the $\mathbf{K}_0$ discussed below). This is not so in quantum theory where further important contributions to the reaction will arise. These will be treated in Chapter VI.

It is not difficult to generalize $\mathbf{K}_s$ for fast particles with relativistic velocities. We only give the result:

$$\mathbf{K}_s = \frac{2}{3}\frac{e^2}{c^3}\frac{1}{1-\beta^2}\left\{\ddot{\mathbf{v}} + \frac{\dot{\mathbf{v}}(\mathbf{v}\ddot{\mathbf{v}})}{c^2(1-\beta^2)} + \frac{3\dot{\mathbf{v}}(\mathbf{v}\dot{\mathbf{v}})}{c^2(1-\beta^2)} + \frac{3\mathbf{v}(\mathbf{v}\dot{\mathbf{v}})^2}{c^4(1-\beta^2)^2}\right\}$$

$$= \frac{2}{3}\frac{e^2}{\mu c^4}\sqrt{(1-\beta^2)}\left\{\frac{d^2\mathbf{u}}{ds^2} - \frac{1}{\mu^2}\mathbf{u}\sum_{\nu=1}^{4}\left(\frac{du_\nu}{ds}\right)^2\right\} \tag{18}$$

where $ds = c\sqrt{(1-\beta^2)}\,dt$ is the element of the proper time and $\mathbf{u}$ the kinetic momentum (cf. § 2). The equation of motion for the particle is, if $\mathbf{K}^e$ is the force from external fields, etc.,

$$\frac{1}{c}\frac{d\mathbf{u}}{dt} = \mathbf{K}_s + \mathbf{K}^e. \tag{18'}$$

(18) will not be used in this book, because for fast particles quantum effects must always be taken into account.

3. *Self-energy*. The first term of the self-force (16 a) is proportional to the *acceleration*. It therefore takes the same form as the *inertia force* $m\dot{\mathbf{v}}$. The factor $\frac{1}{2}\int dede'/r = \mu_0$ represents the electrostatic *self-energy* or the energy contained in the static field of the particle. The latter depends upon the structure of the particle, and becomes *infinite* for a point charge. For a charge with the extension $\bar{r}_0$ it is of the order of magnitude

$$\mu_0 = \frac{e^2}{\bar{r}_0}. \tag{19}$$

This term cannot be distinguished from the inertia term by any method. Since we know nothing about the nature of the inertial mass m, we could just as well take the two terms together and assume that the self-force $\mathbf{K}_0$ is contained in the definition of the mass m. In Abraham's theory it was even assumed that no 'non-electromagnetic' mass exists (see § 2.4). $\mathbf{K}_0$ is put equal to the *whole* observed inertia $m\dot{\mathbf{v}}$. μ_0 is then of the order $\mu = mc^2$ and

$$\bar{r}_0 \sim r_0 \equiv e^2/mc^2. \tag{20}$$

For this reason the universal constant r_0 is called the classical electron radius

$$r_0 = 2{\cdot}818\times 10^{-13}\text{ cm}. \tag{21}$$

Abraham's procedure is, however, faced by a great difficulty. A moving particle has a 'self-momentum' as well as a self-energy, and this should also be included in the momentum of the particle. But this is not possible since we have seen in § 2 that energy and momentum of the field of a particle do not form a 4-vector and do not have the relativistic transformation properties of a particle. One could only overcome these difficulties by introducing non-electromagnetic (i.e. mechanical) internal stresses which compensate the wrong transformation properties of the self-field of the particle, but this means giving up the idea of a purely electromagnetic nature of the inertia of the particle.† The necessity for non-electromagnetic forces in the structure of a particle with finite extension is also clear from the fact that a finite charge distribution cannot be stable under purely electromagnetic forces.

For point particles the additional difficulty occurs that the *self-energy* μ_0 *diverges*. The two difficulties have not been overcome satisfactorily yet, although we shall see that the problem takes a less severe form in quantum theory (§ 29 and appendix 7). It will be seen that the correct

† H. Poincaré, *Rend. di Palermo*, **21** (1906), 129.

transformation properties of the self-field can be enforced although the mathematical procedure used for this purpose is not entirely free from objections. Moreover, the divergence of the self-energy is of a much milder kind, only $\sim \log \bar{r}_0$, and $\bar{r}_0$ can even be defined as a relativistic invariant. So even for an exceedingly small $\bar{r}_0$ the self-energy will be quite small compared with the mechanical rest energy mc^2.

In the present preliminary stage of the theory the procedure to be adopted is this: we know from experiment that the electron has a finite mass and that its energy and momentum form a 4-vector. In this 4-vector the contribution from the self-field must be included. Therefore, if the *observed* mass of the particle is used to describe it, the *contribution from the self-field must be omitted* altogether. This is what we shall do throughout in this book. The difficulty is then pushed to an unobservable realm and does not prevent the application of the theory to observable phenomena.

4. *Line breadth.* Let us consider now the effect which the reaction force (16 b) has on the motion of the particle. The change in the motion due to this reaction force will of course also affect the radiation emitted by the particle.

We consider a linear harmonic oscillator as a simple model for a light source. If we neglect the reaction force, the oscillator will vibrate for an infinitely long time. But because of the damping force the amplitude of the oscillator will decrease. The equation of motion is

$$m\ddot{x} = -m\nu_0^2 x + \frac{2}{3}\frac{e^2}{c^3}\dddot{x}. \tag{22}$$

We assume that (7) is fulfilled. Then the reaction force is small and we can, as a first approximation for $\dddot{x}$, insert the value for the undamped motion $-\nu_0^2 \dot{x}$. Thus (22) becomes

$$\ddot{x} = -\nu_0^2 x - \gamma\dot{x}, \tag{23}$$

$$\gamma = \frac{2}{3}\frac{e^2\nu_0^2}{mc^3} = \frac{2}{3}\frac{\nu_0^2}{c}r_0 \ll \nu_0. \tag{24}$$

The solution of (23) is approximately $(\gamma \ll \nu_0)$†

$$x = x_0 e^{-\gamma t/2} e^{-i\nu_0 t}. \tag{25}$$

The energy of the oscillator averaged over one period is

$$W = \tfrac{1}{2}m(\dot{x}^2 + \nu_0^2 x^2) = W_0 e^{-\gamma t}. \tag{26}$$

† If we write for a classical wave an exponential function, the real part has always to be understood.

Thus the energy decreases exponentially, $1/\gamma$ representing the time taken to decrease in the ratio $e:1$. We therefore call $1/\gamma$ the *lifetime* of the oscillator. The condition (24) ($\gamma \ll \nu_0$) expresses the fact that this time is long compared with one period; otherwise the motion would not be even approximately periodic. γ is only a function of the frequency and does not depend upon the amplitude of the oscillator.

The light which is emitted by such an oscillator has an amplitude which is proportional to $\ddot{x}$, i.e. to $-\nu_0^2 x$. It decreases with t, therefore, in the same way as the amplitude of the oscillator, so that (for $t > 0$)

$$\mathbf{E} = \mathbf{E}_0 e^{-\gamma t/2} e^{-i\nu_0 t}. \tag{27}$$

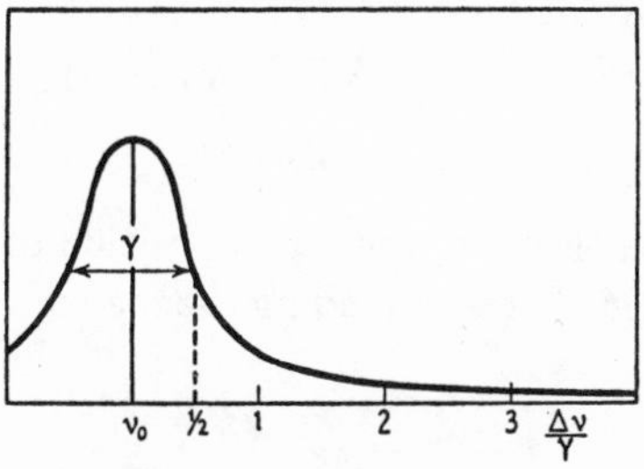

FIG. 2. Natural line breadth.

(27) no longer represents a monochromatic wave but a wave with a certain intensity distribution $I(\nu)$. To obtain this intensity distribution we develop (27) in a Fourier series:

$$\mathbf{E} = \int_{-\infty}^{+\infty} \mathbf{E}(\nu) e^{-i\nu t}\, d\nu, \qquad \mathbf{E}(\nu) = \frac{1}{2\pi}\mathbf{E}_0 \int_0^{\infty} e^{-i(\nu_0-\nu)t} e^{-\gamma t/2}\, dt$$

or

$$\mathbf{E}(\nu) = \frac{1}{2\pi}\mathbf{E}_0 \frac{1}{i(\nu_0-\nu)+\gamma/2}.$$

The intensity distribution becomes

$$I(\nu) \simeq |E(\nu)|^2 = I_0 \frac{\gamma}{2\pi} \frac{1}{(\nu-\nu_0)^2+\gamma^2/4}. \tag{28}$$

The factor I_0 has been chosen so that the total intensity $\int I(\nu)\, d\nu$ is equal to I_0. The line represented by equation (28) has a maximum intensity for the frequency ν_0, i.e. for the frequency of the undamped oscillator.† For $\nu_0-\nu = \gamma/2$ the intensity is half of the maximum intensity. We therefore call γ also the *breadth at half maximum* (see Fig. 2).

† The *exact* solution of (22) leads to a very small shift of the maximum, of order γ^2/ν_0.

The breadth of the line is equal to the reciprocal *lifetime*. If we express the line breadth in wave-length units we have according to (24)†

$$\Delta(2\pi\lambda) = 2\pi c\frac{\Delta\nu}{\nu_0^2} = 2\pi\frac{c\gamma}{\nu_0^2} = \frac{4\pi}{3}r_0 = 1{\cdot}18\times10^{-4}\,\text{Å.U.} \tag{29}$$

The line breadth is then independent of the frequency and equal to the electronic radius r_0 (besides a numerical factor) which is a universal constant.

In the quantum theory we shall treat the problem of the line breadth in § 18.

5. Scattering, absorption

1. *Scattering by free electrons.* A light wave with the frequency ν and the field strength

$$\mathbf{E} = \mathbf{E}_0\,e^{-i\nu t}$$

may fall upon a free electron, the mean position of which is at rest. If we neglect relativistic effects (including the magnetic force) the equation of motion becomes

$$m\ddot{\mathbf{x}} = e\mathbf{E}_0\,e^{-i\nu t}+\frac{2}{3}\frac{e^2}{c^3}\dddot{\mathbf{x}} \tag{1}$$

with the solution

$$\mathbf{x} = -\frac{e\mathbf{E}_0}{m\nu^2}\,e^{-i\nu t}\frac{1}{1+i\kappa}, \qquad \kappa = \frac{2}{3}\frac{e^2}{mc^3}\nu. \tag{2}$$

The electron therefore performs a vibration with the same frequency as the incident wave. This gives rise to the emission of a secondary wave of the same frequency. The time average value of the intensity, according to § 3 eq. (25), at a distance R from the electron is

$$I = |\mathbf{S}| = \frac{e^2\overline{\ddot{x}^2}\sin^2\Theta}{4\pi R^2c^3} = \frac{e^4\sin^2\Theta\,E_0^2}{8\pi R^2m^2c^3}\frac{1}{1+\kappa^2} \tag{3}$$

where Θ is the angle between the direction of observation ($\mathbf{R}$) and the direction of polarization of the incident wave $\mathbf{E}_0$. Introducing the intensity of the primary radiation $I_0 = cE_0^2/8\pi$ we have

$$I = \phi(\Theta)\frac{I_0}{R^2}, \qquad \phi(\Theta) = r_0^2\sin^2\Theta\frac{1}{1+\kappa^2} \tag{4}$$

where r_0 is the classical electronic radius and ϕ a quantity with the dimensions cm.² which we call the *cross-section* for *scattering*.

† We denote the wave-length by $2\pi\lambda$.

If the primary wave is unpolarized we have to average over Θ and obtain

$$\overline{\sin^2\Theta} = \tfrac{1}{2}(1+\cos^2\theta),$$

$$I = \frac{I_0}{R^2}\tfrac{1}{2}r_0^2(1+\cos^2\theta)\frac{1}{1+\kappa^2}, \quad \text{or} \quad \phi(\theta) = \frac{1}{2}\frac{r_0^2(1+\cos^2\theta)}{1+\kappa^2}, \tag{4'}$$

where θ represents the angle of scattering.

The damping effect is contained in the quantity κ. This is of the order r_0/λ and is exceedingly small for all wave-lengths down to those of the γ-rays of nuclear physics. κ would be appreciable only for cosmic ray wave-lengths and, for the shortest wave-lengths, even becomes large compared with unity. One might think that in this region the damping would have a decisive influence on the cross-section. However, in this region both relativistic and quantum effects are important and it is necessary to treat the damping also quantum-mechanically and relativistically. It will be shown in § 33 that the *damping is then small for all wave-lengths*.

The total scattered radiation is obtained from (4′) by integrating over all angles and the total cross-section becomes (neglecting κ)

$$\phi = \frac{8\pi}{3}r_0^2 = 6{\cdot}65\times10^{-25}\ \text{cm.}^2 \tag{5}$$

Thus the cross-section for scattering by a free electron is a *universal constant* and independent of the primary frequency.

The formulae (4) and (5) were first deduced by J. J. Thomson. We shall see that they also hold in the quantum theory as long as relativistic effects can be neglected.

2. *Scattering by an oscillator*. As a second case we shall consider the scattering of light of frequency ν by an elastically bound electron with the frequency ν_0. It is important here to take into account the damping force also. If the electron performs only the free vibration but no forced vibration, due to the light wave, its motion will be periodic with the frequency ν_0. Therefore, we can write for the damping force

$$\frac{2}{3}\frac{e^2}{c^3}\dddot{\mathbf{x}} = -m\gamma\dot{\mathbf{x}}, \qquad \gamma = \frac{2}{3}\frac{e^2\nu_0^2}{mc^3} = \frac{2}{3}r_0\frac{\nu_0^2}{c}. \tag{6}$$

The equation of motion then becomes

$$\ddot{\mathbf{x}}+\gamma\dot{\mathbf{x}}+\nu_0^2\mathbf{x} = \frac{e}{m}\mathbf{E}_0 e^{-i\nu t} \tag{7}$$

with the solution

$$\mathbf{x} = \mathbf{x}_0 e^{-i\nu t}, \qquad \mathbf{x}_0 = \frac{\mathbf{E}_0 e/m}{\nu_0^2-\nu^2-i\nu\gamma}. \tag{8}$$

The intensity of the radiation scattered per unit time and cm.^{2} in a distance R becomes (for $\ddot{x}$ we have to insert the real part):

$$I = \frac{e^2 \sin^2\theta}{4\pi R^2 c^3}\ddot{x}^2 = \frac{c}{4\pi R^2} E_0^2 r_0^2 \frac{\nu^4 \sin^2\theta}{(\nu_0^2-\nu^2)^2+\nu^2\gamma^2}\cos^2(\nu t-\delta), \tag{9}$$

$$\tan\delta = \frac{\nu\gamma}{\nu_0^2-\nu^2}.$$

The scattered radiation shows a phase displacement δ which, however, is only appreciable in the neighbourhood of $\nu = \nu_0$. Taking the average over one period and introducing the intensity of the primary radiation we obtain

$$I = \frac{I_0}{R^2}\phi(\theta), \qquad \phi(\theta) = r_0^2 \frac{\nu^4 \sin^2\theta}{(\nu_0^2-\nu^2)^2+\nu^2\gamma^2}. \tag{10}$$

The cross-section ϕ for the total radiation scattered is obtained by integration over the sphere, giving

$$\phi = \frac{8\pi}{3} r_0^2 \frac{\nu^4}{(\nu_0^2-\nu^2)^2+\nu^2\gamma^2}. \tag{11}$$

(11) represents the well-known dispersion formula. For $\nu_0 \to 0$ and $\gamma \ll \nu$ we obtain again equation (5) for the scattering by a free electron. For frequencies far away from the resonance frequency $\gamma^2\nu^2$ may be neglected. In the neighbourhood of $\nu = \nu_0$ (11) becomes very large and we have the case of *resonance fluorescence*. We may then put $\nu \sim \nu_0$:

$$\phi = \frac{2\pi}{3} r_0^2 \frac{\nu^2}{(\nu_0-\nu)^2+\gamma^2/4}. \tag{12}$$

In the quantum theory we shall meet formulae which represent a simple generalization of (11) and (12) (see §§ 19, 20).

3. *Absorption.* Finally, we shall consider the energy transfer from the incident wave to the oscillator. In the case of resonance, which is particularly interesting, we shall obtain a definite result only if we assume that the primary radiation has, in the neighbourhood of ν_0, a continuous *intensity distribution* $I_0(\nu)\,d\nu$, being the energy per cm.^{2} sec., say. Since we shall discuss the shape (breadth) of the absorption line in the quantum theory, we shall calculate here only the total absorption and can therefore neglect the damping γ. For a single Fourier component ν the equation of motion of the oscillator is

$$\ddot{\mathbf{x}}+\nu_0^2\mathbf{x} = \frac{e}{m}\mathbf{E}(\nu)\cos(\nu t+\delta_\nu), \tag{13}$$

where δ_ν is the phase of a single wave. We assume that these phases are distributed at random.

As for a single frequency ν the energy transfer will depend essentially upon the phase difference of the oscillator and the wave, we must take into account also the free vibration of the oscillator. We choose a solution of (13) so that for $t = 0$ only the free vibration is excited. This solution is

$$\mathbf{x} = \frac{e}{m}\mathbf{E}(\nu)\frac{1}{\nu_0^2-\nu^2}[\cos(\nu t+\delta_\nu)-\cos(\nu_0 t+\delta_\nu)]+\mathbf{b}\sin(\nu_0 t+\theta), \tag{14}$$

where $\mathbf{b}$ represents the amplitude and θ the phase of the oscillator at $t = 0$. The energy transfer per unit time (and per frequency interval) to the oscillator is equal to the work performed by the light wave ν

$$\epsilon_\nu = e(\dot{\mathbf{x}}\mathbf{E}(\nu))\cos(\nu t+\delta_\nu). \tag{15}$$

If we then integrate (15) over a time τ containing an integral number of periods $1/\nu$ the term $\cos(\nu t+\delta_\nu)$ of (14) vanishes and we obtain

$$\int_0^\tau \epsilon_\nu\,dt = \frac{e^2E^2(\nu)}{m}\frac{\nu_0}{\nu_0^2-\nu^2}\int_0^\tau dt\,\sin(\nu_0 t+\delta_\nu)\cos(\nu t+\delta_\nu)+$$

$$+e(\mathbf{E}(\nu)\mathbf{b})\nu_0\int_0^\tau dt\,\cos(\nu_0 t+\theta)\cos(\nu t+\delta_\nu). \tag{16}$$

This integral depends upon the phases and may even assume negative values. For certain phases, therefore, the oscillator transfers energy to the light wave (*induced emission of light*). But since the phases δ_ν were distributed at random, we may take the average over δ_ν. Then the last term of (16) vanishes, the first term is always positive and becomes

$$\int_0^\tau \bar{\epsilon}_\nu\,dt = \frac{e^2E^2(\nu)}{2m}\frac{\nu_0}{\nu_0^2-\nu^2}\frac{1-\cos(\nu_0-\nu)\tau}{\nu_0-\nu}. \tag{17}$$

This energy transfer is large only in the neighbourhood of the resonance frequency $\nu = \nu_0$. We may therefore put $\nu \sim \nu_0$. (17) represents the contribution of one single wave ν. We have now to integrate it over a certain frequency interval taking into account the fact that the intensity distribution was

$$I_0(\nu)\,d\nu = \frac{c}{4\pi}E^2(\nu)\overline{\cos^2(\nu t+\delta_\nu)}\,d\nu = \frac{c}{8\pi}E^2(\nu)\,d\nu.$$

In the region where (17) is appreciable we can assume $I_0(\nu)$ to be constant ($= I_0(\nu_0)$). The integration over ν can be extended from 0 to ∞, since (17) has a very strong maximum at $\nu = \nu_0$. If $\nu_0 \tau \gg 1$, the integral occurring in (17) is of the form

$$\int_{-\infty}^{+\infty} \frac{1-\cos x}{x^2}\, dx = \pi, \tag{18}$$

when $x = (\nu_0 - \nu)\tau$. Thus the energy absorbed per unit time becomes

$$S = \frac{1}{\tau} \int_0^\infty \int_0^\tau \bar{\epsilon}_\nu \, dt d\nu = \frac{2\pi^2 e^2}{mc} I_0(\nu_0). \tag{19}$$

The energy transferred to the oscillator is on the average *proportional to the time* τ, *and to the incident intensity* at the resonance frequency ν_0. Apart from that it is independent of ν_0.

The quantum theory will lead to a very similar formula (§ 17).

6. The field as a superposition of plane waves. Hamiltonian form of the field equations

As a preliminary to the extension of the theory, which is necessary to take account of quantum phenomena, it is useful to express it in another form. Since the quantum theory of a particle is based essentially on the canonical form of classical dynamics it will be convenient for our purpose to express also the classical *radiation theory* in a *canonical form*. In fact one can easily write the whole system of field equations § 1 (2), (5) in such a way that they appear as Hamiltonian equations with a single Hamiltonian function depending on the coordinates of the particles and on some other variables which describe the field.

1. *The pure radiation field.* We consider first a pure radiation field (light waves). According to § 1.3, one can derive it from the vector potential **A** only, normalizing the scalar potential so that $\phi = 0$. **A** will then satisfy the equations

$$\nabla^2 \mathbf{A} - \frac{1}{c^2}\ddot{\mathbf{A}} = 0 \tag{1 a}$$

$$\operatorname{div} \mathbf{A} = 0. \tag{1 b}$$

A is a function defined at all points of space and time. If therefore we wish to describe **A** in terms of canonical variables, the number of such variables must necessarily be infinite. It is possible, however, to choose an *enumerable* set. For this purpose we assume that the whole radiation field is enclosed in a certain volume, for instance, a cube of volume L^3, and that it has to satisfy some boundary conditions on the surface of this volume. In order to obtain running waves as well as standing waves, we shall postulate as boundary conditions that **A** and its derivatives have the same values on two opposite planes of the volume, i.e. that

$$\mathbf{A} \text{ is periodic on the surface.} \tag{2}$$

L is to be considered large compared with the dimensions of the material system. The physical behaviour of the system will not then depend upon L. For convenience we shall, as a rule, put $L = 1$.

With the boundary condition (2) the general solution of (1) can be represented as a series of orthogonal 'eigenwaves':

$$\mathbf{A} = \sum_\lambda q_\lambda(t)\mathbf{A}_\lambda(\mathbf{r}), \tag{3}$$

where $\mathbf{A}_\lambda$ depends only upon the space coordinates and q_λ only upon the time. $\mathbf{A}_\lambda$ has to satisfy the boundary condition (2). The $\mathbf{A}_\lambda$ will satisfy the wave equation

$$\nabla^2\mathbf{A}_\lambda + \frac{\nu_\lambda^2}{c^2}\mathbf{A}_\lambda = 0 \tag{4a}$$

with

$$\operatorname{div}\mathbf{A}_\lambda = 0, \tag{4b}$$

$$\mathbf{A}_\lambda \text{ periodic in } L. \tag{4c}$$

Then q_λ satisfies the equation for a harmonic oscillator

$$\ddot{q}_\lambda + \nu_\lambda^2 q_\lambda = 0, \tag{5}$$

The solutions of (4) represent an infinite set of orthogonal waves which we shall normalize in the following way:

$$\int (\mathbf{A}_\lambda \mathbf{A}_\mu)\, d\tau = 4\pi c^2 \delta_{\lambda\mu}. \tag{6}$$

The $\mathbf{A}_\lambda$ can, for example, be chosen to be cos and sin functions

$$\sqrt{(8\pi c^2)}\mathbf{e}_\lambda \cos(\boldsymbol{\varkappa}_\lambda \mathbf{r}), \qquad \sqrt{(8\pi c^2)}\mathbf{e}_\lambda \sin(\boldsymbol{\varkappa}_\lambda \mathbf{r}), \qquad |\kappa_\lambda| = \nu_\lambda/c,$$

where $\boldsymbol{\varkappa}_\lambda$ gives the direction of propagation, and the unit vector $\mathbf{e}_\lambda$ the direction of polarization, which according to (4 b) is always perpendicular to $\boldsymbol{\varkappa}_\lambda$. $\boldsymbol{\varkappa}_\lambda$ can assume a discrete set of values only, viz.

$$\kappa_{\lambda x} = \frac{2\pi}{L} n_{\lambda x}, \qquad \kappa_{\lambda y} = \frac{2\pi}{L} n_{\lambda y}, \qquad \kappa_{\lambda z} = \frac{2\pi}{L} n_{\lambda z}, \tag{7}$$

where the $n_{\lambda x}, n_{\lambda y},\ldots$ are positive integers. A change of sign of $\boldsymbol{\varkappa}_\lambda$ does not give a new $\mathbf{A}_\lambda$. For each wave with given $\boldsymbol{\varkappa}$ two independent directions of polarization can be chosen arbitrarily. We could also take two circular polarized waves. Also, instead of cos and sin, suitable linear combinations could be chosen.

In the general formulae (3)–(6) we denote different polarizations and the sin and cos functions by different indices λ.

Since the $\mathbf{A}_\lambda$ are given functions in space the field is characterized by the amplitudes q_λ. The field equations have been replaced by equations of the type (5). These can be written very simply as canonical equations; the Hamiltonian for an oscillator is

$$H_\lambda = \tfrac{1}{2}(p_\lambda^2+\nu_\lambda^2 q_\lambda^2), \tag{8a}$$

and the Hamiltonian equations

$$\frac{\partial H_\lambda}{\partial q_\lambda} = -\dot{p}_\lambda, \qquad \frac{\partial H_\lambda}{\partial p_\lambda} = \dot{q}_\lambda = p_\lambda \tag{8b}$$

are identical with (5).

The total field is described by an infinite set of canonical variables q_λ, p_λ, and a total Hamiltonian

$$H = \sum_\lambda H_\lambda. \tag{9}$$

In this way the field is represented as a system of independent oscillators. In classical dynamics H_λ would represent the energy of the oscillator. The same is true here. We show that the total energy of the radiation field is equal to the sum of the energies of all oscillators

$$U = \frac{1}{8\pi}\int (E^2+H^2)\,d\tau = \sum_\lambda H_\lambda. \tag{10}$$

The field strengths are given by

$$\begin{aligned} \mathbf{E} &= -\frac{1}{c}\dot{\mathbf{A}} = -\frac{1}{c}\sum_\lambda \dot{q}_\lambda \mathbf{A}_\lambda = -\frac{1}{c}\sum_\lambda p_\lambda \mathbf{A}_\lambda, \\ \mathbf{H} &= \operatorname{curl}\mathbf{A} = \sum_\lambda q_\lambda \operatorname{curl}\mathbf{A}_\lambda. \end{aligned} \tag{11}$$

(11) has to be inserted in the expression for U. We then obtain integrals of the form $\int (\mathbf{A}_\lambda \mathbf{A}_\mu)\,d\tau$ and $\int (\operatorname{curl}\mathbf{A}_\lambda \operatorname{curl}\mathbf{A}_\mu)\,d\tau$. For the first one we can apply the orthogonality relation (6). The second one can be transformed as follows:

$$\int (\operatorname{curl}\mathbf{A}_\lambda \operatorname{curl}\mathbf{A}_\mu)\,d\tau = \oint d\sigma[\mathbf{A}_\lambda \operatorname{curl}\mathbf{A}_\mu]_n + \int (\mathbf{A}_\lambda \operatorname{curlcurl}\mathbf{A}_\mu)\,d\tau.$$

The surface integral vanishes because of the boundary condition (4c). Furthermore, curlcurl = graddiv$-\nabla^2$. Then, according to (4a), (4b),

$$\int (\operatorname{curl}\mathbf{A}_\lambda \operatorname{curl}\mathbf{A}_\mu)\,d\tau = \frac{\nu_\lambda^2}{c^2}\int (\mathbf{A}_\lambda \mathbf{A}_\mu)\,d\tau. \tag{12}$$

Making use now of the orthogonality equation (6) we obtain for the field energy

$$U = \tfrac{1}{2}\sum (\dot{q}_\lambda^2+\nu_\lambda^2 q_\lambda^2) = \sum H_\lambda. \tag{13}$$

Thus the energy of the field is equal to the sum of the energies of all the oscillators.

For applications to quantum theory, it is more convenient not to represent the field by cos and sin waves but by (complex) exponential functions. Since the potential is real one can represent it by a series

$$\mathbf{A} = \sum_\lambda (q_\lambda(t)\mathbf{A}_\lambda + q^*_\lambda(t)\mathbf{A}^*_\lambda), \tag{14}$$

where the amplitude $q_\lambda(t)$ is now complex.

The solution of (4) and (5) can be written as follows:

$$\mathbf{A}_\lambda = \mathbf{e}_\lambda\sqrt{(4\pi c^2)}e^{i(\boldsymbol{\kappa}_\lambda \mathbf{r})}, \qquad |\kappa_\lambda| = \nu_\lambda/c, \tag{15a}$$

$$q_\lambda = |q_\lambda|e^{-i\nu_\lambda t}. \tag{15b}$$

$q_\lambda \mathbf{A}_\lambda$ represents a wave travelling in the direction of $+\boldsymbol{\kappa}_\lambda$. $\boldsymbol{\kappa}_\lambda$ can again assume the values (7), where the $n_{\lambda x}$ are now *positive or negative* integers. Waves with opposite directions $\boldsymbol{\kappa}_\lambda$ and $-\boldsymbol{\kappa}_\lambda$ are denoted by a different λ. The $\mathbf{A}_\lambda$ are orthogonal in the following sense:

$$\int (\mathbf{A}_\lambda \mathbf{A}^*_\mu)\, d\tau = \int (\mathbf{A}_\lambda \mathbf{A}_{-\mu})\, d\tau = 4\pi c^2 \delta_{\lambda\mu}, \tag{16}$$

where $\mathbf{A}_{-\mu}$ is the wave with the propagation vector $-\boldsymbol{\kappa}_\mu$ ($\mathbf{e}_{-\lambda} = \mathbf{e}_\lambda$).

In this representation the q_λ and q^*_λ are not canonical, but one can introduce new canonical variables (which are real)

$$Q_\lambda = q_\lambda + q^*_\lambda, \tag{17}$$

$$P_\lambda = -i\nu_\lambda(q_\lambda - q^*_\lambda) = \dot{Q}_\lambda.$$

The field equation (5), which holds for both q_λ and its conjugate complex, can be deduced from the Hamiltonian

$$H_\lambda = 2\nu^2_\lambda q_\lambda q^*_\lambda = \tfrac{1}{2}(P^2_\lambda + \nu^2_\lambda Q^2_\lambda). \tag{18}$$

The Hamiltonian equations are

$$\frac{\partial H_\lambda}{\partial Q_\lambda} = -\dot{P}_\lambda, \qquad \frac{\partial H_\lambda}{\partial P_\lambda} = \dot{Q}_\lambda. \tag{19}$$

In the same way as before one can show that $\sum H_\lambda$ represents the total field energy U.†

Finally, we determine the number of radiation oscillators contained in the volume L^3 with a given polarization, a given direction of propagation

† The connexion between the Q_λ, P_λ and the q_λ, p_λ (8) is this: If we denote the coefficients of cos and sin (for the same $\boldsymbol{\kappa}_\lambda$) by $q_{1\lambda}$ and $q_{2\lambda}$ respectively then

$$\sqrt{2}\,q_{1\lambda} = Q_\lambda + Q_{-\lambda}; \qquad \sqrt{2}\,q_{2\lambda} = -(P_\lambda - P_{-\lambda})/\nu_\lambda$$

$$\sqrt{2}\,p_{1\lambda} = P_\lambda + P_{-\lambda}; \qquad \sqrt{2}\,p_{2\lambda} = \nu_\lambda(Q_\lambda - Q_{-\lambda}).$$

This is a canonical transformation between the $q_{1\lambda},\ldots, p_{2\lambda}$ and the $Q_\lambda,\ldots, P_{-\lambda}$. ($Q_{-\lambda}$ belongs to $-\boldsymbol{\kappa}_\lambda$.)

(within an element of the solid angle $d\Omega$), and a given frequency between ν and $\nu+d\nu$. According to (7), this number must be equal to the volume element in n-space ($n_{\lambda x}$ being integers), at least if we assume that the wave-length c/ν is small compared with L. Since ν is given by

$$\nu_\lambda^2 = \left(\frac{2\pi c}{L}\right)^2 (n_{\lambda x}^2+n_{\lambda y}^2+n_{\lambda z}^2) \tag{20}$$

the volume element in n-space is given by

$$\rho_\nu \, d\nu d\Omega \, L^3 = n^2 \, dn d\Omega = \nu^2 \, d\nu d\Omega \, L^3/(2\pi c)^3.$$

This expression is proportional to the volume L^3. It is actually independent of the shape of the enclosure. The number of radiation oscillators per unit volume is therefore equal to

$$\rho_\nu \, d\nu d\Omega = \frac{\nu^2 \, d\nu d\Omega}{(2\pi c)^3} = \frac{1}{(2\pi)^3} \, d\kappa_x \, d\kappa_y \, d\kappa_z \equiv \frac{1}{(2\pi)^3} \, d^3\kappa. \tag{21}$$

We call ρ_ν the *density function* for the light waves.

2. *Hamiltonian of a particle.* As a next step we consider the relativistic equation of motion of a charged particle in a given field (§ 2 eq. (35)). To write this equation in Hamiltonian form, it is only necessary to use the fact that the Hamiltonian represents the total energy E of the particle. In § 2 eq. (37) we saw that E represents the fourth component of a 4-vector p_μ. Now E must be expressed as a function of the canonical coordinates and momenta. For a cartesian system of coordinates the canonical momenta are identical with the ordinary momenta. Then equation (38) in § 2 gives the required relation between $p_4 = iE$ and the momenta $p_1 = p_x$, etc. This equation can be written as follows:

$$H \equiv E = e\phi+\sqrt{\{\mu^2+(\mathbf{p}-e\mathbf{A})^2\}}, \qquad \mu = mc^2. \tag{22}$$

We are thus led to expect that (22) will represent the correct Hamiltonian. In fact, we obtain (using energy units for the momenta),

$$\frac{\partial H}{\partial p_x} = \frac{1}{c}\dot{q}_x \equiv \frac{v_x}{c} = \frac{p_x-eA_x}{\sqrt{\{\mu^2+(\mathbf{p}-e\mathbf{A})^2\}}}, \tag{23 a}$$

$$\frac{\partial H}{\partial x} = -\frac{\dot{p}_x}{c} = e\frac{\partial \phi}{\partial x}-\frac{e}{c}\left(v_x\frac{\partial A_x}{\partial x}+v_y\frac{\partial A_y}{\partial x}+v_z\frac{\partial A_z}{\partial x}\right). \tag{23 b}$$

On the other hand, the total derivative of A_x, with respect to t, is given by

$$\frac{dA_x}{dt} = \frac{\partial A_x}{\partial t}+v_x\frac{\partial A_x}{\partial x}+v_y\frac{\partial A_x}{\partial y}+v_z\frac{\partial A_x}{\partial z}. \tag{24}$$

Adding (24) to (23 b) we obtain with the definition of the potentials given in § 1 (7 a) and (7 b)

$$\frac{1}{c}\frac{d}{dt}(p_x-eA_x) = e\Big(E_x+\frac{1}{c}[\mathbf{vH}]_x\Big). \tag{25}$$

(25) is identical with the equation of motion § 2 eq. (35). Since $p_\mu - eA_\mu = u_\mu$ represents the 4-vector of the kinetic momentum, (23 a) gives the correct connexion between u_1 and v_x,

$$u_1 = \frac{mcv_x}{\sqrt{(1-\beta^2)}}, \qquad \beta = \frac{v}{c}.$$

Thus (22) represents the correct Hamiltonian for a particle. It contains the interaction of the particle with the field. Just as in the expression for the Lorentz force, the field which must be inserted in (22) is the external field produced by magnets, condensers, light sources, etc., as well as the field produced by the charge itself. The latter gives the reaction of the field to the moving charge. (See § 4.)

If the momentum is small compared with the rest-energy μ we may take for (22) the non-relativistic approximation (omitting the constant term μ)

$$H = e\phi + \frac{(\mathbf{p}-e\mathbf{A})^2}{2\mu}. \qquad \text{N.R. (26)}$$

This is the ordinary non-relativistic energy function for a particle in a field having the potentials ϕ and $\mathbf{A}$. In a pure electrostatic field ($\mathbf{A} = 0$) the second term reduces to the kinetic energy $p^2/2\mu$.

3. *General system of particles and field.*† Hitherto we have transformed to the Hamiltonian form the equations of motion of a particle and of a field consisting purely of light waves. As our last step, we must now consider a general system consisting of a particle and a field of any type In order to obtain a relativistic theory also for the interaction of two particles, we shall assume there to be several particles of charges e_k. Each particle is then described by a set of canonical variables, q_k, p_k say, with Hamiltonian

$$H_k = e_k\phi(k)+\sqrt{[\mu_k^2+\{\mathbf{p}_k-e_k\mathbf{A}(k)\}^2]}, \tag{27}$$

where $\phi(k)$, $\mathbf{A}(k)$ represent the field at the position of the kth particle.

† For the following sections see: E. Fermi, *Rev. Mod. Physics*, **4** (1932), 131, or H. Weyl, *Gruppentheorie und Quantenmechanik*, 2nd edition, Leipzig 1933. Also H. A. Kramers, *Hand- und Jahrb. Chem. Phys.* Leipzig 1938, **1**, chap. 8.

The total Hamiltonian for all particles becomes

$$H = \sum H_k, \tag{28 a}$$

$$\frac{1}{c}\dot{p}_k = -\frac{\partial H}{\partial q_k}, \qquad \frac{1}{c}\dot{q}_k = \frac{\partial H}{\partial p_k}. \tag{28 b}$$

In (28 a) no interaction between the particles has as yet been assumed; it will be seen that the latter is contained in the Hamiltonian for the field.

The field which has to be inserted in (27) consists of the external field produced by charges which do not belong to the system considered, and the field produced by all particles including the kth particle itself. It is, however, convenient to separate the external field ϕ^e, $\mathbf{A}^e$. Since this occurs in (27) only as a given potential energy (and potential momentum) for each particle and does not depend upon the position of the particles we may, for the following discussion, take the terms $\mathbf{A}^e$, ϕ^e together with p_k, H_k, simply writing p_k, H_k instead of $p_k - e\mathbf{A}^e$, $H_k - e\phi^e$. In the final result we can easily insert again the correct field.

We first use *Lorentz gauge*. Then the field satisfies the equations

$$\nabla^2\mathbf{A} - \frac{1}{c^2}\ddot{\mathbf{A}} = -\frac{4\pi}{c}\rho\mathbf{v}, \tag{29 a}$$

$$\nabla^2\phi - \frac{1}{c^2}\ddot{\phi} = -4\pi\rho, \tag{29 b}$$

$$\operatorname{div}\mathbf{A} + \frac{1}{c}\dot{\phi} = 0. \tag{29 c}$$

We have to express these equations in canonical form. For this purpose we assume again that the field is enclosed in a volume L^3 and that *all potentials* (and their derivatives) satisfy the boundary condition:

$$\mathbf{A},\ \phi \text{ periodic on the surface.} \tag{30}$$

We then develop the potentials into a Fourier series as before.

The special case of a transverse field for which $\operatorname{div}\mathbf{A} = \phi = 0$ has been treated in subsection 1. For the general case $\operatorname{div}\mathbf{A} \neq 0$ we notice that every vector-field can be divided into two parts, of which the first has a divergence equal to zero and the second is the gradient of a scalar field. Thus

$$\mathbf{A} = \mathbf{A}_1 + \mathbf{A}_2, \qquad \operatorname{div}\mathbf{A}_1 = 0, \qquad \mathbf{A}_2 \sim \operatorname{grad}\psi. \tag{31}$$

$\mathbf{A}_1$ is identical with the transverse field

$$\mathbf{A}_1 = \sum_\lambda q_\lambda \mathbf{A}_\lambda.$$

$\mathbf{A}_2$ can be developed in a similar way

$$\mathbf{A}_2 = \sum_\sigma q_\sigma(t)\mathbf{A}_\sigma, \tag{32}$$

where the $\mathbf{A}_\sigma$ satisfy the wave equation and boundary condition

$$\nabla^2\mathbf{A}_\sigma + \frac{\nu_\sigma^2}{c^2}\mathbf{A}_\sigma = 0, \qquad \mathbf{A}_\sigma \text{ periodic.} \tag{33 a}$$

According to (31) $\mathbf{A}_\sigma$ can be represented as the gradient of a scalar function

$$\mathbf{A}_\sigma = \frac{c}{\nu_\sigma}\operatorname{grad}\psi_\sigma, \qquad \operatorname{curl}\mathbf{A}_\sigma = 0, \tag{33 b}$$

where the ψ_σ obviously satisfy the same equation (33 a). The factor c/ν_σ has been added for reasons of normalization. The $\mathbf{A}_\sigma$ also represent a set of orthogonal waves which satisfy the wave equation and boundary condition. We shall now prove that they are also orthogonal to all transverse waves $\mathbf{A}_\lambda$. For this purpose we apply a general formula of vector calculus,

$$\int d\tau\,[(\operatorname{curl}\mathbf{a}\operatorname{curl}\mathbf{b}) + \operatorname{div}\mathbf{a}\operatorname{div}\mathbf{b} + (\mathbf{a}\nabla^2\mathbf{b})] = \oint d\sigma\,\{[\mathbf{a}\operatorname{curl}\mathbf{b}]_n + \mathbf{a}_n\operatorname{div}\mathbf{b}\}$$

(n = component normal to the surface). Inserting $\mathbf{b} = \mathbf{A}_\lambda$, $\mathbf{a} = \mathbf{A}_\sigma$, the surface integral vanishes, because of the boundary conditions. The first and second terms on the left-hand side vanish by (4 b) and (33 b) respectively. Thus

$$\int (\mathbf{A}_\sigma\nabla^2\mathbf{A}_\lambda)\,d\tau = -\frac{\nu_\lambda^2}{c^2}\int (\mathbf{A}_\sigma\mathbf{A}_\lambda)\,d\tau = 0. \tag{34}$$

The $\mathbf{A}_\lambda$ and $\mathbf{A}_\sigma$ represent a *complete* set of orthogonal waves, satisfying the wave equation and boundary condition (but not $\operatorname{div}\mathbf{A} = 0$).

In contrast to the transverse waves $\mathbf{A}_\lambda$, the $\mathbf{A}_\sigma$ represent *longitudinal waves*.

In the same way one can develop the scalar potential

$$\phi = \sum_\sigma q_{0\sigma}(t)\phi_\sigma, \tag{35 a}$$

$$\nabla^2\phi_\sigma + \frac{\nu_\sigma^2}{c^2}\phi_\sigma = 0, \qquad \phi_\sigma \text{ periodic.} \tag{35 b}$$

These ϕ_σ must be identical with the scalar functions ψ_σ introduced in (33 b) since they satisfy the same equation and boundary condition. Thus we can write

$$\mathbf{A}_\sigma = \frac{c}{\nu_\sigma}\operatorname{grad}\phi_\sigma. \tag{36}$$

Both ϕ_σ and $\mathbf{A}_\sigma$ are then normalized in the same way

$$\int \phi_\sigma \phi_\rho \, d\tau = \int (\mathbf{A}_\sigma \mathbf{A}_\rho) \, d\tau = 4\pi c^2 \delta_{\rho\sigma}. \tag{37}$$

The coefficients $q_{0\sigma}(t)$ and $q_\sigma(t)$ in (35 a) and (32) are not independent, since the Lorentz equation (29 c) has to be satisfied. Inserting (32) and (35) into (29 c) we obtain the important equation

$$\nu_\sigma q_\sigma(t) = \dot{q}_{0\sigma}(t) \tag{38}$$

to be satisfied for all times t.

The differential equations which the amplitudes q_λ, q_σ, $q_{0\sigma}$ have to satisfy can easily be obtained. They will not be identical with the equations of a harmonic oscillator, since the field must now satisfy the inhomogeneous wave equations (29). We insert (3), (32), (35 a) into (29), multiply by $\mathbf{A}_\lambda$, $\mathbf{A}_\sigma$, ϕ_σ respectively, and integrate over space. Assuming that all charges are point charges we obtain

$$\ddot{q}_\lambda + \nu_\lambda^2 q_\lambda = \frac{1}{c} \sum_k e_k (\mathbf{v}_k \mathbf{A}_\lambda(k)), \tag{39 a}$$

$$\ddot{q}_\sigma + \nu_\sigma^2 q_\sigma = \frac{1}{c} \sum_k e_k (\mathbf{v}_k \mathbf{A}_\sigma(k)), \tag{39 b}$$

$$\ddot{q}_{0\sigma} + \nu_\sigma^2 q_{0\sigma} = \sum_k e_k \phi_\sigma(k), \tag{39 c}$$

where $\mathbf{A}_\lambda(k)$ represents the value of $\mathbf{A}_\lambda$ at the position of the kth particle. Equations (39) represent a *forced vibration* of an oscillator, the *force being due to the presence of charged particles.*

The relation (38) can be expressed as an *initial condition* for the solutions of equations (39). Differentiating (39 c) with respect to the time, we obtain, according to (36),

$$\dddot{q}_{0\sigma} + \nu_\sigma^2 \dot{q}_{0\sigma} = \frac{d}{dt} \sum_k e_k \phi_\sigma(k) = \sum_k e_k (\mathbf{v}_k \operatorname{grad} \phi_\sigma(k))$$

$$= \frac{\nu_\sigma}{c} \sum_k e_k (\mathbf{v}_k \mathbf{A}_\sigma(k)),$$

and inserting (39 b)

$$\left(\frac{d^2}{dt^2} + \nu_\sigma^2\right)(q_\sigma \nu_\sigma - \dot{q}_{0\sigma}) = 0. \tag{40}$$

Therefore (38) is always satisfied if, at the time $t = 0$,

$$\nu_\sigma q_\sigma = \dot{q}_{0\sigma} \quad \text{and} \quad \nu_\sigma \dot{q}_\sigma = \ddot{q}_{0\sigma}. \tag{41}$$

Going back to the expansions of **A** and ϕ, and using (39 c) for $\ddot{q}_{0\sigma}$, the initial conditions are

$$\operatorname{div}\mathbf{A}+\frac{1}{c}\dot{\phi}=0, \qquad \operatorname{div}\mathbf{E}=4\pi\rho, \qquad \text{for } t=0. \tag{41'}$$

If these *two* conditions hold at $t = 0$, (29 c) is always satisfied.

If we consider only solutions which satisfy the initial conditions (41) we may consider all oscillators q_σ and $q_{0\sigma}$ as *independent*.

The differential equations (39) can easily be written in canonical form. The forces on the right-hand side are due to the particles and can therefore only be obtained from a term in the Hamiltonian which depends upon the variables of the particles. This term will be simply the Hamiltonian $\sum H_k$ of the particles (equation (27)) which also depends upon the field and therefore contributes additional terms to the equations describing the field. Thus, (27) contains the interaction between field and particle.

For each oscillator (39) we shall have a Hamiltonian of type $\frac{1}{2}(p^2+\nu^2q^2)$. The longitudinal and scalar waves are represented by two oscillators q_σ, $q_{0\sigma}$. But since the *scalar potential* ϕ occurs in (27) with the *opposite sign to the vector potential* **A** we must take the Hamiltonian for the $q_{0\sigma}$ with a *minus sign*. This minus sign will prove to be of great importance. Thus for the longitudinal and scalar waves we assume a Hamiltonian

$$H_\sigma = \tfrac{1}{2}(p_\sigma^2+\nu_\sigma^2 q_\sigma^2)-\tfrac{1}{2}(p_{0\sigma}^2+\nu_\sigma^2 q_{0\sigma}^2) \tag{42}$$

where $p_{0\sigma}$ represents the canonical conjugate momentum to $q_{0\sigma}$.

For the entire system comprised by the particles and the field we shall expect a Hamiltonian

$$H = \underset{\substack{k\\ \text{particles}}}{\sum} H_k + \underset{\substack{\lambda\\ \text{transverse}\\ \text{waves}}}{\sum} H_\lambda + \underset{\substack{\sigma\\ \text{longitudinal}\\ \text{and scalar}\\ \text{waves}}}{\sum} H_\sigma \tag{43}$$

with the canonical equations:

$$\frac{\partial H}{\partial q_k} = -\frac{1}{c}\dot{p}_k \qquad \frac{\partial H}{\partial p_k} = \frac{1}{c}\dot{q}_k \quad \text{(particles)} \tag{44 a}$$

$$\frac{\partial H}{\partial q_\lambda} = -\dot{p}_\lambda \qquad \frac{\partial H}{\partial p_\lambda} = \dot{q}_\lambda \quad \text{(transverse waves)} \tag{44 b}$$

$$\frac{\partial H}{\partial q_\sigma} = -\dot{p}_\sigma \qquad \frac{\partial H}{\partial p_\sigma} = \dot{q}_\sigma \quad \text{(longitudinal waves)} \tag{44 c}$$

$$\frac{\partial H}{\partial q_{0\sigma}} = -\dot{p}_{0\sigma} \qquad \frac{\partial H}{\partial p_{0\sigma}} = \dot{q}_{0\sigma} \quad \text{(scalar waves).} \tag{44 d}$$

In fact (44) is entirely equivalent to the field equations (39) and the equation of motion of the particles (25). We may, for instance, prove that (39 c) is identical with (44 d). $q_{0\sigma}, p_{0\sigma}$ occur in H_σ and H_k. According to (27), (35 a), and (42), we have

$$\frac{\partial H_k}{\partial q_{0\sigma}} = e_k \phi_\sigma(k), \qquad \frac{\partial H_\sigma}{\partial q_{0\sigma}} = -\nu_\sigma^2 q_{0\sigma},$$

$$\frac{\partial H_k}{\partial p_{0\sigma}} = 0, \qquad \frac{\partial H_\sigma}{\partial p_{0\sigma}} = -p_{0\sigma}.$$

Thus (44 d) yields

$$-\dot{p}_{0\sigma} = \ddot{q}_{0\sigma} = -\nu_\sigma^2 q_{0\sigma} + \sum_k e_k \phi_\sigma(k),$$

which is identical with (39 c). The Lorentz condition (41) can also be written in canonical variables (as initial condition)

$$p_{0\sigma} = -\nu_\sigma q_\sigma, \qquad \nu_\sigma p_\sigma = -\nu_\sigma^2 q_{0\sigma} + \sum_k e_k \phi_\sigma(k). \tag{44 e}$$

We have now represented the whole system of field equations in canonical form. The physical significance of the Hamiltonian is quite clear: the first term represents the energy of the particles (kinetic+potential); the second term the energy of the light waves (see subsection 1); the third term would represent the energy of the 'longitudinal and scalar waves'. They occur, of course, only in the presence of particles and have a very simple significance. They represent the *Coulomb interaction* between the particles. This will be shown in the following subsection.

The above canonical representation of Maxwell's equations and the equations of motion was based on the Fourier decomposition of the field. It is also possible to regard $\mathbf{A}(\mathbf{r},t)$, $\phi(\mathbf{r},t)$ directly as canonical variables in each space point and develop a canonical formalism without expanding the field. This will be shown in connexion with quantum theory in § 13.

4. *Coulomb gauge.* Instead of the Lorentz gauge used in the previous subsection we can also use the Coulomb gauge

$$\operatorname{div} \mathbf{A} = 0 \tag{45}$$

with the field equations § 1 eq. (13)

$$-\Box\mathbf{A} + \frac{1}{c}\operatorname{grad}\dot{\phi} = \frac{4\pi}{c}\rho\mathbf{v}, \tag{46 a}$$

$$\nabla^2\phi = -4\pi\rho. \tag{46 b}$$

The solution of (46 b) is given immediately by the *static potential*, i.e. for point charges

$$\phi(x) = \sum_k \frac{e_k}{r_{kx}} \tag{47}$$

where r_{kx} is the distance between the position of the charge e_k and the point x where ϕ is considered. In this gauge no longitudinal field occurs at all. The scalar field is the *static field* (as if the charges were at rest).

We can expand $\mathbf{A}$ and ϕ again, as before. Naturally, the expansion of $\mathbf{A}$ contains only transverse waves $q_\lambda \mathbf{A}_\lambda$; q_σ does not occur. To distinguish the Coulomb gauge from the Lorentz gauge we shall use dashes to denote all canonical variables in this gauge (including q'_k, p'_k, for the particles).

Treating (46) as before, we obtain from (46 a), by multiplying alternatively by $\mathbf{A}_\lambda$ and $\mathbf{A}_\sigma$ and using (36):

$$\ddot{q}'_\lambda + \nu_\lambda^2 q'_\lambda = \frac{1}{c}\sum_k e_k(\mathbf{v}_k\,\mathbf{A}_\lambda(k)), \tag{48 a}$$

$$\nu_\sigma\,\dot{q}'_{0\sigma} = \frac{1}{c}\sum_k e_k(\mathbf{v}_k\,\mathbf{A}_\sigma(k)) \tag{48 b}$$

and from (46 b) (multiplying by ϕ_σ and using (35 b))

$$\nu_\sigma^2 q'_{0\sigma} = \sum_k e_k\,\phi_\sigma(k). \tag{48 c}$$

Now (48 b) is a *consequence* of (48 c), as is seen by differentiation, using

$$\dot{\phi}_\sigma(k) = (\mathbf{v}_k\,\mathrm{grad}\,\phi_\sigma) = \frac{\nu_\sigma}{c}(\mathbf{v}_k\,\mathbf{A}_\sigma(k)).$$

(48 b) may hence be disregarded. Moreover, the scalar field ϕ at the time t is, by (47), a well-defined function of the particle coordinates (and x) at the same time t.† Since, according to (48 c), $q'_{0\sigma}$ depends on ϕ at the position of the particles only, $q'_{0\sigma}$ at any t is a unique function of the particle coordinates q'_k at the same t only, and is not an independent canonical variable. We can therefore regard ϕ as a mere definition of a certain function of the q'_k. In the Hamiltonian formulation of (48), therefore, only the canonical pairs q'_k, p'_k and q'_λ, p'_λ occur. The Hamiltonian is easily found:

$$H' = \sum_k H'_k + \sum_\lambda H'_\lambda + \tfrac{1}{2}\sum_{i,k}\frac{e_i e_k}{r_{ik}}, \qquad r_{ik} = |\mathbf{q}'_i - \mathbf{q}'_k|, \tag{49}$$

$$H'_\lambda = \tfrac{1}{2}(p'^2_\lambda + \nu_\lambda^2 q'^2_\lambda), \tag{49 a}$$

$$H'_k = \Big\{\mu_k^2 + \Big(\mathbf{p}'_k - e_k\sum_\lambda q'_\lambda\,\mathbf{A}_\lambda(k)\Big)^2\Big\}^{\frac{1}{2}}. \tag{49 b}$$

The part due to the scalar field gives rise to the *static* Coulomb interaction between all particles. H'_k differs essentially from the H_k (27), not

† This is not the case for the retarded potentials which depend on the coordinates of the particles at some previous time, in a very involved way.

merely by the dashes on the variables p, q. The $\mathbf{A}$ occurring in H'_k is transverse, whereas in H_k the transverse and the longitudinal fields appear. The canonical equations of motion

$$\frac{\partial H'}{\partial q'_k} = -\frac{1}{c}\dot{p}'_k, \qquad \frac{\partial H'}{\partial p'_k} = \frac{1}{c}\dot{q}'_k; \qquad \frac{\partial H'}{\partial q'_\lambda} = -\dot{p}'_\lambda, \qquad \frac{\partial H'}{\partial p'_\lambda} = \dot{q}'_\lambda \tag{50}$$

lead, on the one hand, to (48 a) and, on the other hand, to the correct equation of motion (25), where $\mathbf{E} = -\operatorname{grad}\phi - (1/c)\dot{\mathbf{A}}$ and ϕ is the function of the particle coordinates q'_k defined by (47), to be used at the position of the particle whose equation of motion is considered.

It would seem as if in this gauge the interaction of two particles were only the instantaneous Coulomb interaction and not the retarded interaction. However, this is not the case. The effect of retardation is contained in the part of the Hamiltonian which depends on the transverse waves; it appears in the present gauge as a mutual emission of light waves between the particles (compare the example in § 24).

In the Coulomb term (49) there are also terms of the form e_k^2/r_{kk}. They represent the infinite self-energy of a point charge. We met this difficulty in § 4. According to what was said there we have to omit these terms or rather to assume that they are already included in the rest energy of the particle μ_k. Therefore the term in (49) has to be understood as

$$H'_s = \frac{1}{2}\sum_{i\neq k}\frac{e_i e_k}{r_{ik}}. \tag{49'}$$

The relationship between the Lorentz and Coulomb gauges can be formulated as a gauge transformation, as is immediately evident from § 1. On the other hand, it can be formulated as a *canonical transformation* from the variables $q_k, p_k; q_\sigma, p_\sigma; q_\lambda, p_\lambda; q_{0\sigma}, p_{0\sigma}$; to the new set of variables $q'_k, p'_k; q'_\lambda, p'_\lambda$; the number of canonical pairs being thus *reduced*. This will now be shown.

A canonical transformation is best effected with the help of a generating function $\Omega(q, p')$ depending on the old coordinates q and the new momenta p' (writing q, p' for any number of pairs that may occur).

Then the old momenta and the new coordinates are defined by

$$p = \frac{\partial\Omega}{\partial q}, \qquad q' = \frac{\partial\Omega}{\partial p'}. \tag{51}$$

These relations enable us to express q and p as functions of q', p'

$$q = q(q', p'), \qquad p = p(q', p')$$

(by solving (51) with respect to q, p). The new Hamiltonian is then (if Ω does not depend explicitly on time):

$$H'(q',p') = H(q(q',p'),p(q',p'))$$

and q', p' satisfy the Hamiltonian equations automatically with H' as Hamiltonian.

In our case the canonical transformation will be of a somewhat singular type, in that the number of canonical pairs will be reduced. Certain variables p' will not occur in Ω or in the new Hamiltonian. This fact is due to the existence of the Lorentz condition. We put

$$\Omega = \sum_\sigma q_\sigma\Big(\sum_k \frac{e_k\,\phi_\sigma(k)}{\nu_\sigma} - q_{0\sigma}\,\nu_\sigma\Big) + \frac{1}{c}\sum_k q_k\,p'_k + \sum_\lambda q_\lambda\,p'_\lambda. \tag{52}$$

No p'_σ or $p'_{0\sigma}$ occurs. Applying (51) to all four pairs of variables we find:

$$p_\lambda = \frac{\partial\Omega}{\partial q_\lambda} = p'_\lambda, \qquad q'_\lambda = \frac{\partial\Omega}{\partial p'_\lambda} = q_\lambda. \tag{53 a}$$

The variables of the transverse field are left unchanged.

$$p_\sigma = \frac{\partial\Omega}{\partial q_\sigma} = \sum_k \frac{e_k\,\phi_\sigma(k)}{\nu_\sigma} - q_{0\sigma}\,\nu_\sigma, \qquad q'_\sigma = \frac{\partial\Omega}{\partial p'_\sigma} = 0. \tag{53 b}$$

The first equation is identical with the second Lorentz condition (44 e). $q'_\sigma = 0$ means that after the canonical transformation the longitudinal field vanishes.

$$p_{0\sigma} = \frac{\partial\Omega}{\partial q_{0\sigma}} = -q_\sigma\,\nu_\sigma, \qquad q'_{0\sigma} = \frac{\partial\Omega}{\partial p'_{0\sigma}} = 0. \tag{53 c}$$

The first equation is identical with the first Lorentz condition (44 e). $q'_{0\sigma} = 0$ means that no scalar field exists either after the canonical transformation. Finally:

$$\frac{1}{c}\mathbf{p}_k = \frac{\partial\Omega}{\partial \mathbf{q}_k} = \frac{1}{c}\mathbf{p}'_k + e_k\sum_\sigma q_\sigma \frac{1}{\nu_\sigma}\operatorname{grad}\phi_\sigma(k) = \frac{1}{c}(\mathbf{p}'_k + e_k\,\mathbf{A}_{\text{long}}(k)),$$

$$\mathbf{q}'_k = c\frac{\partial\Omega}{\partial \mathbf{p}'_k} = \mathbf{q}_k. \tag{53 d}$$

The coordinates of the particles are also left unchanged. The momenta of the particles are, however, changed. $\mathbf{p}'_k$ is the kinetic momentum of a particle moving in the longitudinal field $\mathbf{A}_{\text{long}}$ alone. This is just what will be necessary to change $H_k \rightarrow H'_k$.

We now find the new Hamiltonian by expressing all the q, p in H in terms of the q', p'. Owing to the singular character of the transformation this seems at first impossible because (53) does not permit us to express

$q_\sigma, p_\sigma, q_{0\sigma}, p_{0\sigma}$ by the dashed variables. Instead relations between the old variables themselves appear. These are just the Lorentz condition. These relations will enable us to obtain the new Hamiltonian H' as a function of those new variables which now alone occur, namely, $q'_k, p'_k, q'_\lambda, p'_\lambda$. The old Hamiltonian can be rewritten as

$$H = \sum_k \bar{H}_k + H_s + \sum_\lambda H_\lambda, \qquad \bar{H}_k = H_k - e\phi = \sqrt{\{\mu_k^2 + (\mathbf{p}_k - e_k \mathbf{A})^2\}}, \tag{54 a}$$

$$H_s = \sum_{k,\sigma} e_k q_{0\sigma} \phi_\sigma + \tfrac{1}{2} \sum_\sigma (p_\sigma^2 + \nu_\sigma^2 q_\sigma^2) - \tfrac{1}{2} \sum_\sigma (p_{0\sigma}^2 + \nu_\sigma^2 q_{0\sigma}^2). \tag{54 b}$$

By (53 a), H_λ is unchanged. Next express $\bar{H}_k$ as a function of p'_k instead of p_k. (53 d) shows that the change consists in that $\mathbf{p}_k - e_k \mathbf{A} = \mathbf{p}'_k - e_k \mathbf{A}_{\text{tr}}$, where $\mathbf{A}_{\text{tr}}$ is the transverse part of $\mathbf{A}$ only. $\bar{H}_k$ is therefore just the Hamiltonian of the particle in Coulomb gauge (49 b)

$$\bar{H}_k(q_k, p_k) = H'_k(q'_k, p'_k). \tag{55}$$

Finally, the part H_s is to be rewritten using (53 b, c). Nearly all terms cancel each other and we are left with

$$H_s = \frac{1}{2} \sum_{i,k,\sigma} e_i e_k \frac{\phi_\sigma(i)\phi_\sigma(k)}{\nu_\sigma^2},$$

a function of the coordinates of the particles alone. The value of H_s is readily found. H_s satisfies, as a function of q_i, the Poisson equation

$$\nabla_i^2 H_s = -\frac{1}{c^2} \sum_{\sigma,k} e_i e_k \phi_\sigma(i)\phi_\sigma(k) = -4\pi \sum_k e_i e_k \delta(\mathbf{q}_i - \mathbf{q}_k), \tag{56}$$

where $\delta(\mathbf{q}_i - \mathbf{q}_k) = 0$ except when the two particles have the same position. The last equation expresses a general property of all orthogonal systems. (See the definition of the δ-function in § 8.) The solution is

$$H_s = \frac{1}{2} \sum_{i,k} \frac{e_i e_k}{r_{ik}}. \tag{57}$$

Thus the new Hamiltonian is

$$H' = \sum_k H'_k + \sum_\lambda H'_\lambda + H_s \tag{58}$$

which, by (55), (57) is identical with (49).

The motion of the particle is now governed by the transverse field $\mathbf{A}_{\text{tr}}$ and the position of the other particles, as occurring in H_s, alone. Thus we see that the Hamiltonian for Coulomb gauge arises from that for Lorentz gauge by a canonical transformation.

Both gauges have their particular advantages and disadvantages. The Coulomb gauge is simpler, only physically really important variables occur and there is no extra condition connecting different types of variables. It has, however, the disadvantage of not appearing in relativistically invariant form. Whilst the theory as a whole is, of course, invariant, the Coulomb gauge is not an invariant concept and the potentials have to be regauged when going over to a different Lorentz frame. The Lorentz gauge has the great advantage of being covariant and no regauging is required during a Lorentz transformation. It is, however, more complicated, because there are more sets of variables occurring, which are partly linked together by an auxiliary condition. This causes some complication in quantum theory. Also the negative sign of the Hamiltonian for the scalar field complicates matters further. Once, however, these complications are dealt with in a general way (§ 10) the use of the Lorentz gauge is of great advantage for problems for which it is desirable to exhibit their Lorentz invariance explicitly. For the simpler problems of radiation theory we shall use the Coulomb gauge. In the following chapters both gauges will be developed in quantum theoretical form.

II

QUANTUM THEORY OF THE PURE RADIATION FIELD

7. Quantization of the radiation field

1. *Introduction.* The classical theory as developed in Chapter I is only correct in so far as one can neglect all effects which arise from the finite value of Planck's action constant h. Before we introduce this new constant into the theory we emphasize some of the experimental and theoretical facts from which the necessity for a quantization of the field becomes evident. Historically, it was in the theory of radiation itself that a departure from the classical theory first became necessary. In the problem of radiation in thermal equilibrium with a black body the classical theory leads to the well-known 'ultra-violet difficulty', as the density of energy in the form of short waves diverges. To avoid this difficulty Planck (and later Einstein) assumed that the energy of a monochromatic wave with the frequency ν could only assume values which are an integral multiple of a certain unit proportional to the frequency

$$E = n\hbar\nu. \tag{1}$$

n being an integer, the number of light quanta or photons. $2\pi\hbar = h$ is the universal Planck's constant. This assumption (1) leads to a correct formula for the radiation from a black body. It means, of course, that the classical theory, which is incompatible with such an assumption, has to a large extent to be abandoned.

The quantization (1) of a monochromatic wave, together with the law of conservation of energy, led to Bohr's well-known *frequency condition.* The latter, as well as the photon-picture of light waves, is in agreement with innumerable experiments of atomic and nuclear physics.

According to (1) a beam of light (X-rays, γ-rays) consists, on the one hand, of a number of photons. On the other hand, it shows the diffraction phenomena characteristic for its classical wave nature.

It is this *dual nature of light*, according to which it appears *as waves and as particles*, which necessitates the quantization of light waves for its description.

The necessity for a quantization of the electromagnetic field is shown also from another more theoretical consideration: if we assume the theory of *quantum mechanics* to be correct for particles we find the quantization of the field to be logically connected with that theory. The quantum properties of a particle are contained in the uncertainty relation for the position and momentum

$$\Delta q \Delta p \sim \hbar c \tag{2}$$

(energy units for the momentum). This relation could be disproved immediately if the classical theory were valid for a beam of light. For if this were the case one could measure the position of the particle *exactly* by a convergent beam of light (Heisenberg's γ-ray microscope) without transferring an appreciable amount of momentum to the particle, since the momentum of the lightbeam could be made as small as one likes. Thus, if p was measured before, one could obtain knowledge of position and momentum exceeding the limits given by equation (2). In order that (2) should hold notwithstanding the possibility of this experiment, it is necessary that there should be for the *light beam*, an *uncertainty relation similar to* (2). This will be seen (subsection 4) to be the case if the light wave is quantized. The light beam will then have a minimum momentum which is uncertain by an amount Δp if it has a shape and frequency suitable for measuring the position of a particle with an accuracy Δq. This momentum will be transferred to the particle in a way outside the control of the experimenter, and the uncertainty relation (2) still holds after the measurement of the position of the particle.†

In the development of the formalism of quantum electrodynamics it is easiest to let ourselves be guided by the *formal* analogy between classical mechanics and classical electrodynamics pointed out in § 6. In that section we represented the field by a set of canonical variables and the field equations were put in the form of Hamiltonian equations. The quantum of action h can then be introduced in the same way as in ordinary quantum mechanics.

2. *Quantization of the pure radiation field.*‡ We consider a pure radiation field which can be formed by superposition of transverse waves only. In Coulomb gauge no other field exists. This field can be derived from

† For a detailed discussion of this experiment see N. Bohr, *Atomtheorie und Naturbeschreibung*, Berlin 1931, or W. Heisenberg, *Die physikalischen Prinzipien der Quantentheorie*, Leipzig 1930.

‡ The quantum theory of the pure radiation field was first developed by P. A. M. Dirac, *Proc. Roy. Soc.* A, **114** (1927), 243, 710, and P. Jordan and W. Pauli, *Zs. f. Phys.* **47** (1928), 151. The general theory of quantum electrodynamics is due to W. Heisenberg and W. Pauli, ibid. **56** (1929), 1; **59** (1930), 169. For the quantization of the longitudinal and scalar field and the transition to Coulomb gauge see also E. Fermi, *Rev. Mod. Phys.* **4** (1931), 131. General accounts: W. Pauli, *Handb. d. Phys.* XXIV. 1. G. Wentzel, *Quantentheorie der Wellenfelder*, Edwards Bros. 1946. H. A. Kramers, *Hand- und Jahrb. d. chem. Phys.* vol. 1, Leipzig 1938.

the vector potential **A** which, according to § 6 eq. (14), may be written as a series of plane waves (we use the complex representation)

$$\left.\begin{aligned} \mathbf{A} = \sum_\lambda (q_\lambda \mathbf{A}_\lambda + q_\lambda^* \mathbf{A}_\lambda^*), \qquad \operatorname{div} \mathbf{A}_\lambda = 0 \\ \mathbf{A}_\lambda = \sqrt{(4\pi c^2)}\mathbf{e}_\lambda e^{i(\boldsymbol{\kappa}_\lambda \mathbf{r})} \end{aligned}\right\}. \tag{3}$$

Introducing the canonical variables

$$Q_\lambda = q_\lambda + q_\lambda^*, \qquad P_\lambda = -i\nu_\lambda(q_\lambda - q_\lambda^*), \tag{4}$$

the energy of a single wave is given by

$$H_\lambda = \tfrac{1}{2}(P_\lambda^2 + \nu_\lambda^2 Q_\lambda^2). \tag{5}$$

When the radiation theory is expressed in this form, it is quite obvious how the quantum of action h is to be introduced. By exact analogy with the ordinary quantum theory we have to consider the *canonical variables of each radiation oscillator as non-commutable operators* satisfying the *commutation relations*:

$$\left.\begin{aligned} [P_\lambda Q_\lambda] \equiv P_\lambda Q_\lambda - Q_\lambda P_\lambda = -i\hbar \\ [P_\lambda Q_\mu] = [P_\lambda P_\mu] = [Q_\lambda Q_\mu] = 0 \end{aligned}\right\}. \tag{6}$$

The result of this quantization for the Hamiltonian (5) is given by the well-known wave mechanical treatment of a harmonic oscillator. The eigenvalues of the energy of such an oscillator are given by

$$E_\lambda = (n_\lambda + \tfrac{1}{2})\hbar\nu_\lambda, \tag{7}$$

where n_λ is an integer. The amplitude Q_λ can be represented as a hermitian matrix (formed with the time-independent eigenfunctions of the oscillator) for each λ:

$$Q_{n,n+1} = Q_{n+1,n}^* = \sqrt{\left(\frac{\hbar(n+1)}{2\nu}\right)}, \tag{8}$$

$$Q_{n,n'} = 0 \qquad \text{if } n' \neq n \pm 1.$$

Thus Q_λ only has matrix elements for those transitions in which the quantum number n_λ of the λth radiation oscillator increases or decreases by one. According to (4) the complex amplitudes q, q^* which are not hermitian, can be represented as follows

$$\left.\begin{aligned} q_{n,n+1} &= \sqrt{\left(\frac{\hbar(n+1)}{2\nu}\right)} \\ q_{n+1,n}^* &= \sqrt{\left(\frac{\hbar(n+1)}{2\nu}\right)} \\ q_{n+1,n} &= q_{n,n+1}^* = 0 \end{aligned}\right\}. \tag{9}$$

q^* is the matrix hermitian conjugate to q, but we retain the notation q^*.

Above, we have put the volume $L^3 = 1$, otherwise all amplitudes q, q^* have a factor $L^{-\frac{3}{2}}$.

According to (4), (6), (9) the q's satisfy the commutation relation

$$q_\lambda q_\mu^* - q_\mu^* q_\lambda = \frac{\hbar}{2\nu_\lambda}\delta_{\lambda\mu} \tag{10}$$

($\delta_{\lambda\mu} = 0$, 1 according to whether the oscillators λ, μ are different or identical).

Each radiation oscillator has an energy which is an integral multiple of $\hbar\nu_\lambda$, in agreement with the original hypothesis of Planck (1). It seems, however, as though each oscillator had a *zero point energy* $\frac{1}{2}\hbar\nu_\lambda$ even in its lowest state $n_\lambda = 0$. Since the number of radiation oscillators, for a given volume, is infinite, this conclusion leads us to ascribe to the vacuum an infinite zero point energy. This difficulty is, however, purely formal. This method of making the transition from the classical theory to the quantum theory is not unique, since the q, q^* are non-commutable quantities. The Hamiltonian (5) can also be written in terms of the q's

$$H_\lambda = \nu_\lambda^2(q_\lambda q_\lambda^* + q_\lambda^* q_\lambda). \tag{11}$$

But (11) may equally well be written with the order of q_λ^* and q_λ interchanged in one of its terms without disturbing the correspondence with the classical theory. We may, for instance, write instead of (11)

$$H_\lambda = 2\nu_\lambda^2 q_\lambda^* q_\lambda = \tfrac{1}{2}(P_\lambda^2 + \nu_\lambda^2 Q_\lambda^2) - \tfrac{1}{2}\hbar\nu_\lambda. \tag{12}$$

Hence the Hamiltonian (12) has the eigenvalues

$$E_\lambda = n_\lambda \hbar\nu_\lambda, \qquad \frac{\hbar}{2\nu_\lambda} n_\lambda = q_\lambda^* q_\lambda \tag{13}$$

and the zero point energy has disappeared.

A state of the radiation field is now described by the numbers n_λ for all radiation oscillators.

In classical theory the amplitudes Q_λ or q_λ, q_λ^* depend on time. In quantum theory they are replaced by the *time-independent* operators, just as in the Schrödinger equation $p_x(t)$ becomes $-i\hbar\,\partial/\partial x$. The time dependence of any phenomenon is expressed by the time variation of the wave function. Since classically $\dot{Q}_\lambda = P_\lambda$, $\dot{P}_\lambda = -\nu_\lambda^2 Q_\lambda$ we see from (4) that the time derivatives of q_λ, q_λ^* go over into operators:

$$\dot{q}_\lambda \text{ becomes } -i\nu_\lambda q_\lambda, \qquad \dot{q}_\lambda^* \text{ becomes } +i\nu_\lambda q_\lambda^*. \tag{14}$$

Hence it follows that the *field strengths* are *represented* by the *operators* (using (3)):

$$\left.\begin{aligned}\mathbf{E} &= -\frac{1}{c}\dot{\mathbf{A}} = +\frac{i}{c}\sum_\lambda \nu_\lambda(q_\lambda \mathbf{A}_\lambda - q_\lambda^* \mathbf{A}_\lambda^*) \\ \mathbf{H} &= \operatorname{curl}\mathbf{A} = i\sum_\lambda (q_\lambda[\boldsymbol{\varkappa}_\lambda \mathbf{A}_\lambda] - q_\lambda^*[\boldsymbol{\varkappa}_\lambda \mathbf{A}_\lambda^*])\end{aligned}\right\}. \tag{15}$$

The Hamiltonian is then

$$H = \frac{1}{8\pi}\int (E^2+H^2)\,d\tau = \sum_\lambda H_\lambda = \sum_\lambda n_\lambda \hbar\nu_\lambda. \tag{15'}$$

The representation in which the amplitudes are represented by time-independent operators will be called the Schrödinger representation, to distinguish it from a different representation introduced below.

We now consider the momentum of the field, which is classically defined by

$$\mathbf{G} = \frac{1}{4\pi}\int [\mathbf{EH}]\,d\tau.$$

G can also be represented as a sum

$$\mathbf{G} = \sum_\lambda \mathbf{G}_\lambda, \qquad \mathbf{G}_\lambda = \frac{1}{4\pi}\int [\mathbf{E}_\lambda \mathbf{H}_\lambda]\,d\tau \tag{16}$$

where $\mathbf{G}_\lambda$ is the momentum of a plane wave.

Using the normalization of $\mathbf{A}_\lambda$ (§ 6 eq. (16)) and $(\boldsymbol{\varkappa}_\lambda \mathbf{e}_\lambda) = 0$, we obtain

$$\mathbf{G}_\lambda = 2\nu_\lambda c\boldsymbol{\varkappa}_\lambda q_\lambda^* q_\lambda, \qquad \kappa_\lambda = \nu_\lambda/c, \tag{17}$$

where $\boldsymbol{\varkappa}_\lambda$ is the vector with the direction of the wave and the absolute value of the reciprocal of the wave-length. In (17) again we have chosen the order of the q, q^* so that no zero point momentum occurs. (17) is identical with the energy function (12) apart from a numerical factor. The momentum therefore commutes with the energy, its eigenvalues being

$$\mathbf{G}_\lambda = c\boldsymbol{\varkappa}_\lambda n_\lambda \hbar = n_\lambda \mathbf{k}_\lambda, \qquad |\mathbf{k}_\lambda| = \hbar\nu_\lambda, \tag{18}$$

where $\mathbf{k}_\lambda$ is the vector with the direction of propagation and the value $\hbar\nu_\lambda$.

Thus the energy and momentum of a light wave are integral multiples of a unit $\mathbf{k}_\lambda$. Furthermore we have seen in § 2 that they transform like a 4-vector under a Lorentz transformation. In its energy and momentum properties a plane wave behaves therefore exactly as a *beam of n free particles each with energy* $\hbar\nu$ *and momentum* $\mathbf{k}$ ($k = \hbar\nu$). These particles are called *light quanta* or *photons*. The rest energy of a light quantum is, according to (13) and (18), equal to zero

$$G_\lambda^2 - E_\lambda^2 = 0. \tag{19}$$

We shall see later that, for the interaction of a light quantum with a free electron for instance, energy and momentum are conserved.

On the other hand, it will be seen that the quantized wave still has classical *wave properties* showing interference phenomena, etc.

The transformation properties of a light quantum under a Lorentz transformation are the same as those deduced in § 2 for the total momentum and energy of a particle. From § 2 eq. (44) we obtain for a Lorentz system moving in the x-direction

$$k'_x = (k_x - \beta k)\gamma, \qquad k'_y = k_y, \qquad k' = (k - \beta k_x)\gamma, \tag{20}$$
$$\gamma = 1/\sqrt{(1-\beta^2)},$$

or if we denote the angle between k and x by θ and write (20) in terms of the frequencies

$$\nu' = \nu \frac{1-\beta\cos\theta}{\sqrt{(1-\beta^2)}}, \qquad \cos\theta' = \frac{\cos\theta - \beta}{1-\beta\cos\theta}. \tag{21}$$

The first equation represents the well-known formula for the *Doppler effect*. The second, which shows that the direction of the light quantum in a moving system of coordinates is different from that in the system at rest, gives the *aberration*. Both effects are of course *classical* (but relativistic) and can also easily be deduced from the transformation formulae in § 2.

For each plane wave λ we have chosen a certain direction of polarization $\mathbf{e}_\lambda$. Therefore each light quantum also has a given polarization. Instead of the linear polarizations chosen originally, we could have expanded into circularly polarized plane waves. We would have obtained circularly polarized photons. Going over from one set of independent polarizations (q_1, q_2) to another set involves a linear transformation of the two independent amplitudes q_1, q_2 (for the same $\boldsymbol{\varkappa}$). This can be expressed as a canonical transformation between the operators q_1, q_2 in a rather trivial way.

We have expanded the field into *plane* waves. We could have chosen also an expansion, for example, into spherical or cylindrical waves and would have obtained photons represented by such waves. This representation is suitable for a discussion of the *angular momentum* of light. The transition from plane to spherical waves involves a linear transformation of waves with different directions $\boldsymbol{\varkappa}$, and is again expressible as a canonical transformation. We discuss the angular momentum of light in appendix 1.

The physical content of the quantum theory of transverse waves developed in this section is included essentially in Planck's assumption (1). Our theory is simply a consistent formalism erected on the basis of Planck's original assumption (1).

The 'dual nature' of light, as a wave and as a beam of free particles, which results from this quantization is analogous to that of a beam of free electrons, which have the nature of particles and of de Broglie waves. This analogy was extraordinarily fruitful in the development of the quantum theory, but it should not now be overstressed. The existence of a discrete set of light quanta is only a result of the quantization. The corresponding classical theory is essentially a *field theory* since if we make $\hbar \to 0$ the light quanta have no further existence; whereas for a beam of electrons the wave properties are due to the quantization and the classical theory is essentially a particle theory. The particle properties of the light quanta are comprised by the above-mentioned energy and momentum relations. But there is no indication that, for instance, the idea of the 'position of a light quantum' (or the 'probability for the position') has any simple physical meaning.

To ensure that in the transition to the classical theory quantum electrodynamics should go over into a field theory it is essential that the light quanta should satisfy the *Einstein–Bose statistics*. This is evident, since the light quanta occur in the theory only as quantum numbers attached to the radiation oscillators. Two light quanta cannot therefore be distinguished from each other. Furthermore, the number of quanta attached to each oscillator is not limited. Considering the radiation oscillators as 'quantum cells' a state of the total radiation field is described by the number of indistinguishable particles per quantum cell. In statistical mechanics these are exactly the variables by which a microscopic state of an Einstein–Bose assembly is defined. Applying the usual statistical methods Planck's distribution law is obtained.

If the light quanta satisfied the Fermi–Dirac statistics, i.e. if each radiation oscillator contained not more than one quantum, one could never obtain a field theory by making the transition to the classical theory. For if this were possible the intensity even of a radio wave could not be greater than $\hbar\nu$ and would decrease with increasing wave-length. Thus long waves could hardly exist at all. The principle of superposition, which is characteristic for the classical field theory, would not be valid because by superposing two waves with equal wave-lengths and equal phases we can obtain a wave with the same wave-length but higher intensity.

Thus, a classical field theory cannot exist for a Fermi–Dirac assembly;

the latter can only behave classically as a system of particles. (Compare also the remarks in § 12, end of subsection 3.)

3. *The state vector of the radiation field.* After quantization the field quantities **A**, and hence **E** and **H** have become operators which must act on a wave function or state vector Ψ.† This state vector obeys the general Schrödinger equation

$$i\hbar\dot{\Psi} = H\Psi, \tag{22}$$

where H is the Hamiltonian of the system. For a pure radiation field this is given by (15′), $H = \sum H_\lambda$. We are evidently dealing with an infinite number of degrees of freedom, each radiation oscillator representing one degree of freedom. If the Hamiltonian is (15′), there is no interaction between the radiation oscillators, and an eigenstate of H must be the product $\Psi^{(1)}\Psi^{(2)}...\Psi^{(\lambda)}...$, $\Psi^{(\lambda)}$ being a normalized eigenstate of H_λ. Since the eigenvalues of H_λ are $E_\lambda = n_\lambda \hbar\nu_\lambda$, we can characterize the various states $\Psi^{(\lambda)}$ by the number of photons n_λ, $\Psi^{(\lambda)} = \Psi_{n_\lambda}$, $n_\lambda = 0, 1, 2,...$

$$H_\lambda \Psi_{n_\lambda} = n_\lambda \hbar\nu_\lambda \Psi_{n_\lambda}. \tag{23}$$

A general solution of (22) can then certainly be represented by

$$\Psi(t) = \sum_{n_1...n_\lambda...} c_{n_1...n_\lambda...}(t)\Psi_{n_1}\Psi_{n_2}...\Psi_{n_\lambda}.... \tag{24}$$

$|c_{...n_\lambda...}(t)|^2$ is the probability for finding n_1 photons of type 1, n_λ photons of type λ, etc. The eigensolution of (22) with a total energy $E = \sum_\lambda E_\lambda$ is

$$\Psi(t) = \prod_\lambda \Psi_{n_\lambda}(t) = \prod_\lambda e^{-E_\lambda t/\hbar}\Psi_{n_\lambda} = e^{-iEt/\hbar} \prod_\lambda \Psi_{n_\lambda}. \tag{25}$$

We write $\Psi_{n_\lambda}(t)$ for the eigenfunction *with* the time exponential.

If the field interacts with particles, H will have contributions describing the interaction (Chapter III). The expansion (24) will then still be possible (at least when Coulomb gauge is used and the field consists of transverse waves only) but the coefficients will vary with time in a more complicated way.

So far we have not said what are the variables on which Ψ_{n_λ} depends. This is irrelevant. One might choose the amplitudes Q_λ, in which case Ψ_{n_λ} would be the well-known hermitian orthogonal functions. However, Q is hardly an observable quantity and is, at any rate, of little importance. In any case, whichever variables are chosen, the Ψ_n are orthogonal and normalized $(\Psi_n^* \Psi_{n'}) = \delta_{nn'}$ for each λ. The bracket indicates an integra-

† We shall henceforth prefer the name state vector for Ψ to distinguish it from the wave function ψ of a particle satisfying the Dirac equation, which will later (§ 12) also be subjected to a process of second quantization. Ψ generally describes the state of a system of any number of photons, electrons, etc.

tion over the variables on which Ψ depends. The matrix elements of the *operators* q, q^* formed with Ψ_{n_λ} are given by eq. (9) (for each λ):

$$\begin{aligned} q_{n,n+1} &\equiv (\Psi_n^* q_{\text{op}} \Psi_{n+1}) = \sqrt{\frac{\hbar}{2\nu}}\sqrt{(n+1)}, \\ q^*_{n+1,n} &\equiv (\Psi_{n+1}^* q^*_{\text{op}} \Psi_n) = \sqrt{\frac{\hbar}{2\nu}}\sqrt{(n+1)} \end{aligned} \tag{26}$$

where the subscript *op* has been added to emphasize the role of q as an operator acting on Ψ. The matrix elements of q_{op}, q^*_{op}, formed with the time-dependent eigenfunction $\Psi_{n_\lambda}(t)$ are evidently

$$(\Psi_n^*(t) q_{\text{op}} \Psi_{n+1}(t)) = q_{n,n+1}\, e^{-i\nu t}, \qquad (\Psi_{n+1}^*(t) q^*_{\text{op}} \Psi_n(t)) = q^*_{n+1,n}\, e^{+i\nu t}. \tag{26$'$}$$

The effect of q_{op} acting on Ψ can now be seen immediately from (26) and the orthogonality of Ψ. Denoting the variables on which Ψ depends by x (whatever quantity we may choose for them) we have

$$\begin{aligned} \sqrt{\left(\frac{2\nu}{\hbar}\right)} q_{\text{op}} \Psi_n(x) &= \sqrt{n}\, \Psi_{n-1}(x), \\ \sqrt{\left(\frac{2\nu}{\hbar}\right)} q^*_{\text{op}} \Psi_n(x) &= \sqrt{(n+1)}\, \Psi_{n+1}(x); \end{aligned} \tag{27}$$

the same arguments occur on both sides.

Thus the effect of q_{op}, q^*_{op} acting on Ψ_n is to produce the functions Ψ_{n-1}, Ψ_{n+1} respectively with the same arguments. The state described by Ψ_{n-1} is that which has one photon (for the λ considered) less than Ψ_n. Therefore q_λ describes the *absorption of a photon* λ, and q^*_λ is the *emission operator*. q and q^* have matrix elements for absorption and emission of a single photon respectively and no others (read indices from right to left!).

Apart from the Schrödinger representation (where the operators are time independent and Ψ is time dependent) one can use a representation where the time dependence is transferred to the operators and Ψ is independent of time. This is the *Born–Heisenberg representation* used by these authors in their matrix formulation of quantum mechanics. Later, when we consider the interaction between light and particles, we shall frequently use an intermediate representation, called the *interaction representation*, where the time dependence is carried partly by the operators and partly by the state vector. For a pure radiation field the interaction and Born–Heisenberg representations are identical. Since the latter will not be used in this book, we use the name interaction representation here also.

The transition to interaction representation can be described by a trivial canonical transformation. We denote all quantities in the new representation by dashes. Then

$$\Psi' = e^{iHt/\hbar}\Psi, \quad \text{or} \quad \Psi'_{n_\lambda} = e^{in_\lambda \nu_\lambda t}\Psi_{n_\lambda}(t); \tag{28}$$

Ψ' is time independent:

$$i\hbar\dot{\Psi}' = i\hbar e^{iHt/\hbar}\dot{\Psi} - e^{iHt/\hbar}H\Psi = 0 \tag{29}$$

by (22). The eigensolution Ψ'_{n_λ} for a particular oscillator is then identical with the eigenfunction Ψ_{n_λ} without the time factors.

The new operators q'_{op}, $q^{*\prime}_{\text{op}}$, which will depend explicitly on time, are such that the matrix elements of q', q'^* formed with the new time-independent wave function Ψ' are the same as those of q, q^* formed with the old time-dependent wave functions $\Psi_n(t)$ given by (26′)

$$(\Psi'^*_n q'_{\text{op}} \Psi'_{n+1}) = (\Psi^*_n(t) q_{\text{op}} \Psi_{n+1}(t)),$$

or

$$q' = e^{iHt/\hbar}qe^{-iHt/\hbar}, \qquad q'^* = e^{iHt/\hbar}q^*e^{-iHt/\hbar}, \tag{30}$$

$$q'_{n,n+1} = q_{n,n+1}e^{-i\nu t}, \qquad q'^*_{n+1,n} = q^*_{n+1,n}e^{+i\nu t}$$

(for each λ). The commutation relations between q', q'^* are unchanged

$$[q'^*_\lambda q'_\lambda] = -\frac{\hbar}{2\nu_\lambda},$$

q', q'^* are now time dependent and the derivatives are

$$\dot{q}'_{\text{op}} = -i\nu q'_{\text{op}}, \qquad \dot{q}'^*_{\text{op}} = +i\nu q'^*_{\text{op}}, \tag{31}$$

the same as are satisfied by the classical quantities, but (31) are, of course, operator equations. Accordingly, the vector potential can now be written

$$\begin{aligned}\mathbf{A}'(t) &= \sqrt{(4\pi c^2)}\sum_\lambda (q'_\lambda \mathbf{A}_\lambda + q'^*_\lambda \mathbf{A}^*_\lambda)\\ &= \sqrt{(4\pi c^2)}\sum_\lambda (q_\lambda e^{-i\nu_\lambda t}\mathbf{A}_\lambda + q^*_\lambda e^{+i\nu_\lambda t}\mathbf{A}^*_\lambda).\end{aligned} \tag{32}$$

$\mathbf{A}'(t)$ depends explicitly on time, and the electric field strength is obtained directly by the operator relation $\mathbf{E} = -(1/c)\dot{\mathbf{A}}$, giving again (15) with q, q^* replaced by q', q'^*. In Born–Heisenberg representation the operators satisfy the same differential equations as are satisfied by the corresponding classical quantities.

The change to interaction representation is quite trivial but has the following advantage. If we later consider the interaction with particles, H will be supplemented by this interaction. If we use the present representation, inserting for H in (30) only the Hamiltonian of the free radiation field (and that of the non-interacting particles) Ψ' will again depend on time, but the time variation will only be due to the interaction, and the trivial exponential time dependence of the free fields is removed. This is very suitable for the treatment of the interaction as a perturbation.

4. *Light quanta, phases, and similar questions.* The quantities describing the radiation field, such as field strengths, number of light quanta, etc., have no definite numerical values but are quantum mechanical quantities which in general *do not commute*. Any two of those quantities will satisfy a certain commutation relation which will determine their behaviour. The commutation relations of the field strengths will be treated in detail in § 9. Here we consider briefly those relations in which the *number of light quanta* is involved.

The number of light quanta of the λth radiation oscillator is represented as a matrix

$$n_\lambda = \frac{2\nu_\lambda}{\hbar} q_\lambda^* q_\lambda. \tag{33}$$

From every quantum mechanical commutation relation a corresponding uncertainty relation can always be deduced. If two physical quantities A and B satisfy the equation

$$AB-BA = C \tag{34}$$

(C being a number), A and B satisfy the uncertainty relation

$$\Delta A \Delta B \geqslant |C|, \tag{35}$$

which has the following meaning: if the values of A and B have been determined approximately, and if there is an uncertainty ΔA in our knowledge of A, the uncertainty in our knowledge of B must be greater than $C/\Delta A$. Every experimental attempt to exceed the limits to our knowledge given by (35) by an exact measurement of, say, first A and then B fails because of the interaction between the measuring apparatus and the system. n_λ obviously does not commute with the field strength **E** or **H**, eq. (15). If n has a given value, the electric field strength has no definite magnitude but will fluctuate about a certain average value. This is the case even if no light quanta at all are present ($n = 0$). Although the average value of **E** is then equal to zero, **E** will show certain fluctuations about this value.† These zero point fluctuations of the electric field will, for instance, give rise to a certain self-energy of a free electron in empty space (see § 29).

† The zero point fluctuations of **E** have no direct connexion with the zero point energy (subsection 2) which is of purely formal character.

Instead of the electric field strength **E** we can introduce the *phase* ϕ of the wave, putting (for each λ)

$$q = \sqrt{\left(\frac{\hbar}{2\nu}\right)} e^{i\phi}\sqrt{n}, \qquad q^* = \sqrt{\left(\frac{\hbar}{2\nu}\right)}\sqrt{n}e^{-i\phi}.$$

For ϕ we then obtain the equation (from (10))

$$e^{i\phi}n - ne^{i\phi} = e^{i\phi}. \tag{36}$$

(36) is satisfied if ϕ and n satisfy the commutation relation†

$$\phi n - n\phi = -i \tag{37}$$

and the uncertainty relation

$$\Delta n \Delta\phi \geqslant 1. \tag{37'}$$

Thus the number of light quanta n and the phase ϕ (multiplied by $\hbar$) are canonically conjugate. From (37′) it follows that if *the number of light quanta of a wave are given, the phase of this wave is entirely undetermined* and vice versa. If for two waves the phase difference is given (but not the absolute phase) the total number of light quanta may be determined, but it is uncertain to which wave they belong.

We can now show that for a quantized light wave an uncertainty relation (2) is valid which was postulated in subsection 1. This uncertainty relation, however, does not refer to the position and momentum of a light quantum, since the idea of position of a light quantum has no definite meaning. Equation (2) expresses the following fact: if a beam of light is such as to give an image of a point (electron) in the x-direction, say, with an accuracy Δx, the x-component of the momentum of this light beam is uncertain by an amount $\Delta G_x \simeq c\hbar/\Delta x$.

According to classical optics an image of a point can be formed by a monochromatic convergent beam of light with a solid angle of aperture θ, say, and a wave-length λ. Owing to the diffraction, the focus has, however, a finite extension in the x-direction given by the formula

$$\Delta x = \frac{\lambda}{\sin\theta}. \tag{38}$$

(38) represents also the extension of the image, i.e. the inaccuracy of the measurement of the position of the electron.

A convergent beam of light can be obtained by superposition of plane waves with the same wave-length but various directions of propagation $\boldsymbol{\kappa}$.

† This can be shown in the following way: by repeated application of (37) we can easily deduce the equation

$$\phi^k n - n\phi^k = -ik\phi^{k-1}.$$

Developing the exponential function $\exp(i\phi) = \sum i\phi^k/k!$, (36) can be proved immediately.

These plane waves have, however, to be superposed with given phase differences, otherwise the beam has no definite focus. From (37′) it follows then that whilst the total number of light quanta may be known, the attribution to the individual plane waves is quite uncertain. Since the momentum **G** of the beam is given directly by the number of light quanta of each plane wave, **G** is not determined either.

The exact quantum mechanical representation of such a convergent beam of light is rather complicated. We can, however, easily obtain the uncertainty in **G** by the following consideration: supposing the total number of light quanta of the beam is equal to 1 it is uncertain to which plane wave this quantum belongs, i.e. its *direction* is not determined. It can have any direction within the angle of aperture θ. The inaccuracy of G_x is therefore given by

$$\Delta G_x \simeq k \sin\theta = \hbar\nu \sin\theta. \tag{39}$$

From (38) and (39) we obtain, in fact, the uncertainty relation

$$\Delta G_x \Delta x \simeq \hbar c \tag{40}$$

which was postulated in subsection 1.

We see that the uncertainty relation (40) is a simple consequence of the fact that the momentum cannot be smaller than $\hbar\nu$.

8. δ, Δ and related functions

In quantum mechanics, and even more so in quantum electrodynamics, the δ-function introduced by Dirac has become an important mathematical tool which has helped to simplify many deductions and calculations. In the quantum theory of radiation and the relativistic theory of the electron the relativistic generalizations of the δ-function are of particular importance. We devote this section to a mathematical survey of these singular functions as far as they will be needed.†

1. *The* $\delta(x)$, $\delta(\mathbf{r})$, $\mathscr{P}/x$, $\zeta(x)$*-functions.* The δ-function is an operational quantity which is meaningful only when it occurs under an integral. If here equations occur which contain the 'bare δ-function' these are always to be understood as such operational relations.

We define

$$\delta(x) = \frac{1}{2\pi} \lim_{K\to\infty} \int_{-K}^{+K} e^{i\kappa x}\, d\kappa = \frac{1}{\pi} \lim_{K\to\infty} \int_{0}^{K} \cos\kappa x\, d\kappa = \frac{1}{\pi} \lim_{K\to\infty} \frac{\sin Kx}{x}. \tag{1}$$

The limit $K \to \infty$, of course, does not exist in itself. If, however, the right-hand side of (1) is multiplied by a function which is regular at $x = 0$

† For an exact mathematical treatment of the δ-function see L. Schwartz, *Théorie des distributions*, Paris (1951).

and integrated with respect to x, over any interval covering the point $x = 0$, the limiting process can be carried out after the integration and exists. We obtain

$$\int_{-a}^{+b} \delta(x)f(x)\,dx = \frac{1}{\pi}\lim_{K\to\infty}\int_{-aK}^{bK} f\left(\frac{y}{K}\right)\frac{\sin y}{y}\,dy = f(0). \tag{2}$$

Thus, $\delta(x)$ can be visualized as a 'function' which vanishes for $x \neq 0$, but has a strong singularity at $x = 0$ such that

$$\int_{-a}^{+b} \delta(x)\,dx = 1 \tag{2'}$$

(put $f(x) = 1$). (1) is always to be understood in this sense. $\delta(x)$ has no meaning when multiplied by a function $f(x)$ which is itself singular at $x = 0$. Expressions such as $\delta(x)/x$ are essentially ambiguous and acquire a definite meaning only after further prescriptions are applied.

$\delta(x)$ can be represented in many different ways. Every complete set of orthogonal wave functions $u_n(x)$ provides a representation of $\delta(x)$ through the completeness theorem

$$\sum_n u_n^*(x)u_n(x') = \delta(x-x') \tag{3}$$

where n may be a discrete or continuous (or mixed) index. It appears from (1) that $\delta(x)$ is an even function of (x), that is

$$\delta(-x) = \delta(x). \tag{4}$$

In the sense (2) we can also state: $x\,\delta(x) = 0$. Further we see, from the defining property (1), that

$$\delta(cx) = \frac{1}{|c|}\,\delta(x), \tag{5}$$

$$\delta\big((x-x_1)(x-x_2)\big) = \frac{\delta(x-x_1)+\delta(x-x_2)}{|x_1-x_2|}. \tag{6}$$

(6) follows simply from the fact that contributions to the integral (2) arise only from points where the arguments of δ vanish, and from (5).

Similarly

$$\delta(x-x_1)\delta(x-x_2) = \delta(x-x_1)\delta(x_1-x_2) = \delta(x-x_2)\delta(x_1-x_2). \tag{6'}$$

$\delta(x)$ can also be defined by contour integration in the complex plane. Let C be a closed path containing the point $x = 0$ but no singularity of $f(x)$. Then

$$f(0) = \frac{1}{2\pi i}\oint_C \frac{f(x)}{x}\,dx,$$

and we can write
$$\delta(x) = \frac{1}{2\pi i}\frac{1}{x}\bigg|_C. \tag{7}$$

The symbol C means that the subsequent integration over x should be carried out over C. C may be taken as a very small circle.

The derivative $\delta'(x)$ can also be defined:

$$\delta'(x) = \frac{1}{\pi}\lim_{K\to\infty}\left(\frac{K\cos Kx}{x} - \frac{\sin Kx}{x^2}\right), \tag{8}$$

to be understood in the same sense as (2). $\delta'(x)$ when multiplied by $f(x)$ has the simple property that follows from partial integration and the fact that $\delta(x) = 0$ for $x \neq 0$,

$$\int \delta'(x)f(x)\,dx = -f'(0). \tag{9}$$

$\delta'(x)$ is obviously an odd function of x.

Since $\delta(x)$ is an even function of x the integration starting from $x = 0$ yields half the contribution of the interval $-a...b$. (This can be verified directly from (2).) Thus

$$\int_0^x \delta(\xi)\,d\xi = \tfrac{1}{2}\epsilon(x), \qquad \epsilon(x) = \begin{cases} +1\ (x > 0) \\ -1\ (x < 0) \end{cases} \tag{10}$$

$$\epsilon'(x) = 2\delta(x), \qquad \epsilon^2(x) = 1. \tag{10'}$$

The 3-dimensional δ-function $\delta(\mathbf{r})$ is defined by

$$\delta(\mathbf{r}) \equiv \delta(x)\delta(y)\delta(z) = \frac{1}{(2\pi)^3}\int e^{i(\boldsymbol{\kappa}\mathbf{r})}\,d^3\kappa \tag{11}$$

where $d^3\kappa = d\kappa_x\,d\kappa_y\,d\kappa_z$ and the integration is over the whole κ space.

$\delta(\mathbf{r})$ has the property

$$\int \delta(\mathbf{r})f(\mathbf{r})\,d\tau = f(0) \tag{11'}$$

where the integration is over a certain volume including $\mathbf{r} = 0$. If $f(\mathbf{r})$ is continuous at $\mathbf{r} = 0$, obviously one can also write

$$\delta(\mathbf{r}) = \frac{1}{2\pi}\frac{\delta(r)}{r^2} = -\frac{1}{2\pi}\frac{\delta'(r)}{r}. \tag{12}$$

Both expressions (12) have the same effect as $\delta(\mathbf{r})$ after volume integration. Here the division by r^2 and r is permitted because the volume element $d\tau$ contains r^2. In (12) the factor $1/2\pi$ and not $1/4\pi$ occurs because the integration is from $r = 0$ to a finite limit.

In addition to $\delta(x)$ defined by (1) the function†

$$\zeta(x) = -i \lim_{K\to\infty} \int_0^K e^{i\kappa x}\, d\kappa = \lim_{K\to\infty} \frac{1-e^{iKx}}{x}$$

$$= \lim_{K\to\infty}\left(\frac{1-\cos Kx}{x} - i\frac{\sin Kx}{x}\right) \equiv \frac{\mathscr{P}}{x} - i\pi\,\delta(x) \quad (13)$$

will play an important role.

The function

$$\frac{\mathscr{P}}{x} \equiv \lim_{K\to\infty} \frac{1-\cos Kx}{x} = \lim_{K\to\infty} \int_0^K \sin \kappa x\, d\kappa = \frac{1}{2i}\int_{-\infty}^{+\infty} e^{i\kappa x}\epsilon(\kappa)\, d\kappa \quad (14)$$

is called the principal value of $1/x$. It behaves like $1/x$ everywhere where $x \neq 0$ (the rapidly oscillating $\cos Kx$ does not contribute to any integration over x), but for $x = 0$, $\mathscr{P}/x$ vanishes. When multiplied by $f(x)$ and integrated over x this has the same effect as if a small interval $-\epsilon...+\epsilon$ covering the point $x = 0$ symmetrically, were left out of the range of integration:

$$\int_{-a}^{+b} f(x)\frac{\mathscr{P}}{x}\, dx = \int_{-a}^{-\epsilon} \frac{f(x)}{x}\, dx + \int_{+\epsilon}^{b} \frac{f(x)}{x}\, dx \quad (\epsilon \to 0). \quad (15)$$

That the interval lies symmetrically round $x = 0$ follows from the fact that $\mathscr{P}/x$ is an odd function of x. The complex conjugate of $\zeta(x)$ is:

$$\zeta^*(x) = +i \lim_{K\to\infty} \int_0^K e^{-i\kappa x}\, d\kappa = \frac{\mathscr{P}}{x} + i\pi\,\delta(x) = -\zeta(-x). \quad (16)$$

The functions $\delta(x)$, $\zeta(x)$, $\zeta^*(x)$ can be represented in many different ways. In the complex plane the three functions can be defined as $1/x$ each taken by a different path of integration. $\mathscr{P}/x$ is obviously to be taken along the real axis leaving out the interval $-\epsilon...+\epsilon$. For $\zeta(x)$, $\zeta^*(x)$ we note that the contour integral taken over a small semicircle round $x = 0$ in the upper half of the complex plane (in the anti-clockwise sense) has the value $+i\pi$. Thus, by (13) and (16), the functions $\mathscr{P}/x$, ζ, ζ^*, $\delta(x)$ are defined by the paths of integration shown in Fig. 3. By shifting the whole path of integration into a straight line slightly above or below

† $-\frac{1}{2\pi i}\zeta(x)$ is usually denoted by $\delta_-(x)$ in the literature and $\frac{1}{2\pi i}\zeta^*(x) \equiv \delta_+(x)$. Since in this book $2\pi i\delta_+$ will occur more frequently than δ_+ we found it convenient to introduce the notation ζ, ζ^*.

the real axis and changing the variable x to $x \pm i\sigma$, say, we also obtain the representations:

$$\zeta(x) = \lim_{\sigma\to 0}\frac{1}{x+i\sigma}, \qquad \zeta^*(x) = \lim_{\sigma\to 0}\frac{1}{x-i\sigma}, \tag{17 a}$$

$$\frac{\mathscr{P}}{x} = \lim_{\sigma\to 0}\frac{x}{x^2+\sigma^2}, \qquad \delta(x) = \frac{1}{\pi}\lim_{\sigma\to 0}\frac{\sigma}{x^2+\sigma^2}, \tag{17 b}$$

and by differentiation

$$\delta'(x) = -\frac{2}{\pi}\lim_{\sigma\to 0}\frac{\sigma x}{(x^2+\sigma^2)^2}. \tag{17 c}$$

The integration is now along the *real* axis. The limit $\sigma \to 0$ is, of course, always to be taken after the integration.

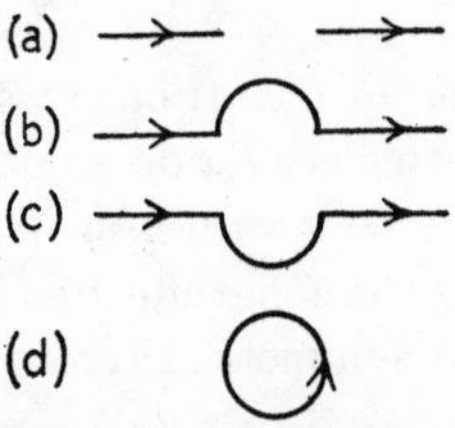

FIG. 3. Paths of integration for $1/x$ defining the following functions: (*a*) $\mathscr{P}/x$; (*b*) $\zeta(x)$; (*c*) $\zeta^*(x)$; (*d*) $2\pi i\delta(x)$.

Evidently
$$x\zeta(x) = x\zeta^*(x) = 1. \tag{17'}$$

Further properties of $\zeta(x)$ and $\zeta^*(x)$ will be given in eq. (40) and § 16 eq. (4′). The product

$$\frac{\mathscr{P}}{x}\delta(x) = \frac{1}{\pi}\lim_{\sigma,\sigma'\to 0}\frac{\sigma x}{(x^2+\sigma^2)(x^2+\sigma'^2)}$$

depends on which of the limits σ or σ' are dealt with first, and is ambiguous. If, however, we add the condition that both limits are to be carried out simultaneously with $\sigma = \sigma'$, we obtain

$$\frac{\mathscr{P}}{x}\delta(x) = -\tfrac{1}{2}\delta'(x). \tag{18}$$

A further representation of $\delta(x)$ which will occur in the following sections is

$$\delta(x) = \frac{1}{\pi}\lim_{K\to\infty}\frac{1}{K}\frac{1-\cos Kx}{x^2}. \tag{19}$$

Its δ-function property can easily be verified in the same way as in (2) and

$$\int_{\infty}^{+\infty}\frac{1-\cos y}{y^2}\,dy = \pi.$$

The representation (19) is also suggested by the fact that, according to (17),

$$\delta(x) = \frac{1}{\pi}\sigma|\zeta_\sigma(x)|^2,$$

where the index σ indicates that for ζ the 'σ-representation' (17) is used. Using instead the K-representation (13), we expect a similar relation with σ replaced by a multiple of $1/K$. In fact

$$\delta(x) = \lim_{K\to\infty}\frac{1}{\pi}\frac{1}{2K}|\zeta_K|^2,$$

by (13) and (19).

An obvious relativistic generalization of the 3-dimensional δ-function $\delta(\mathbf{r})$ is

$$\delta^4(x) \equiv \delta^4(x_\mu) \equiv \delta(\mathbf{r})\delta(x_0), \quad x_0 = ct \qquad (x_4 = ix_0). \tag{20}$$

$\delta^4(x)$ has, for instance, the property

$$\int f(x_\mu)\delta^4(x_\mu - x'_\mu)\,d^4x = f(x'_\mu), \qquad d^4x \equiv d\tau dx_0$$

and can be represented by

$$\delta^4(x) = \frac{1}{(2\pi)^4}\int\!\!\int\!\!\int\!\!\int e^{i\kappa_\mu x_\mu}\,d^4\kappa,$$

$$d^4\kappa = d^3\kappa\,d\kappa_0, \qquad \kappa_4 = i\kappa_0 \tag{21}$$

where κ_μ is a 4-vector. The integration extends over all values of $\kappa_x \ldots \kappa_0$. Henceforth we agree that a sum 1...4 is to be taken over any relativistic index that occurs twice.

2. *The relativistic Δ-function.* Apart from the trivial relativistic generalization (21) there is another, and more important, way of generalizing the δ-function in a relativistic way. In (21) the integration is carried out over all four variables $\boldsymbol{\kappa}$, κ_0 independently. If κ_μ is the 4-vector denoting the energy momentum of a particle or a photon (apart from a factor $\hbar c$) a relation exists between $\boldsymbol{\kappa}$ and κ_0, namely,

$$\kappa_\mu^2 = \boldsymbol{\kappa}^2 - \kappa_0^2 = -\eta^2, \qquad \eta = \frac{mc}{\hbar} \tag{22}$$

where m is the rest mass of the particle. We obtain a different function if we restrict the integration in (21) to values of κ_μ satisfying (22). In fact we obtain two different functions, depending on which of the two values $\pm\sqrt{(\kappa^2+\eta^2)}$ is taken by κ_0. As the sign of κ_0 is a relativistic invariant, we obtain two invariant functions. It will be more convenient to add and subtract the contributions for $\kappa_0 \gtrless 0$. The restriction of the domain of integration can best be carried out in a relativistic manner by adding the factor $\delta(\kappa_\mu^2+\eta^2)$ under the integral. According to (6), we can split

$$\delta(\kappa_\mu^2+\eta^2) = \frac{1}{2E}\{\delta(E-\kappa_0)+\delta(E+\kappa_0)\}, \qquad E = +\sqrt{(\kappa^2+\eta^2)}. \qquad (23)$$

We can also add a further factor $\epsilon(\kappa_0)$ (defined by (10)) that is also relativistically invariant. (The sign of a fourth component of a vector cannot change in a proper Lorentz transformation.) Then

$$\epsilon(\kappa_0)\delta(\kappa_\mu^2+\eta^2) = \frac{1}{2E}\{\delta(E-\kappa_0)-\delta(E+\kappa_0)\}. \qquad (24)$$

This follows from the observation that $\delta(E+\kappa_0) \neq 0$ only when $\kappa_0 = -E$.

We shall first consider the case where κ_μ is a null vector, $\eta = 0$, $E = |\boldsymbol{\kappa}|$, and therefore refers to photons. The Δ-functions thus obtained will play a fundamental role in the commutation relations of the electromagnetic field quantities. In the following subsection we shall consider the case $\eta \neq 0$ and the functions thus obtained will occur in the theory of particles with finite mass. We thus obtain the two functions

$$\Delta(\mathbf{r},t) = \frac{i}{(2\pi)^3}\iiiint d^4\kappa\, e^{i\kappa_\mu x_\mu}\epsilon(\kappa_0)\delta(\kappa_\mu^2), \qquad (25\text{ a})$$

$$\Delta_1(\mathbf{r},t) = \frac{1}{(2\pi)^3}\iiiint d^4\kappa\, e^{i\kappa_\mu x_\mu}\delta(\kappa_\mu^2). \qquad (25\text{ b})$$

We have left out a factor $1/2\pi$ because the integration is really only 3-dimensional. A factor i has been added in Δ so that Δ is real.

The integration can easily be performed. We get at once

$$\Delta(\mathbf{r},t) = \frac{1}{(2\pi)^3}\int d^3\kappa\, e^{i(\boldsymbol{\kappa}\mathbf{r})}\frac{\sin\kappa x_0}{\kappa} = \frac{1}{2\pi^2 r}\int_0^\infty d\kappa\, \sin\kappa r\,.\,\sin\kappa x_0, \qquad (26\text{ a})$$

$$\Delta_1(\mathbf{r},t) = \frac{1}{(2\pi)^3}\int d^3\kappa\, e^{i(\boldsymbol{\kappa}\mathbf{r})}\frac{\cos\kappa x_0}{\kappa} = \frac{1}{2\pi^2 r}\int_0^\infty d\kappa\, \sin\kappa r\,.\,\cos\kappa x_0. \qquad (26\text{ b})$$

For the further integration we split up the sin and cos again into exponentials $\exp(\pm i\kappa(r\pm x_0))$, and obtain with the help of (1) and (14)

$$\Delta(\mathbf{r},t) = \frac{1}{4\pi r}\{\delta(r-x_0)-\delta(r+x_0)\} = \frac{1}{2\pi}\epsilon(x_0)\delta(x_\mu^2), \qquad (27\text{ a})$$

$$\Delta_1(\mathbf{r},t) = \frac{1}{4\pi^2 r}\left\{\frac{\mathscr{P}}{r-x_0}+\frac{\mathscr{P}}{r+x_0}\right\} = \frac{1}{2\pi^2}\frac{\mathscr{P}}{x_\mu^2}. \qquad (27\text{ b})$$

Both Δ and Δ_1 are singular on the light cone $x_0 = \pm r$. Δ has a δ-type singularity, Δ_1 a $\mathscr{P}/x$ type singularity.

$\Delta(\mathbf{r},t)$ vanishes *everywhere* except on the light cone. This property is not shared by Δ_1. In particular $\Delta_1 \neq 0$ *outside* the light cone. This distinction will be important for the discussion of the measurability of the electromagnetic field strengths.

It follows immediately from (26), (11) that

$$\Delta(\mathbf{r},t=0) = 0, \qquad \frac{\partial\Delta(\mathbf{r},t)}{\partial t}\bigg|_{t=0} = c\,\delta(\mathbf{r}),$$
$$\Delta_1(\mathbf{r},t=0) = \frac{1}{2\pi^2}\frac{\mathscr{P}}{r^2}, \qquad \frac{\partial\Delta_1}{\partial t}\bigg|_{t=0} = 0. \tag{28}$$

The defining formulae (25) show that Δ, Δ_1 are the Fourier transforms of $\epsilon(\kappa_0)\delta(\kappa_\mu^2)$ and $\delta(\kappa_\mu^2)$ respectively. Note that Δ has the property of being its own Fourier transform. This is not so for Δ_1.

Both Δ and Δ_1 satisfy the wave equation $\Box\Delta = 0$, $\Box\Delta_1 = 0$. This is seen immediately from (25). The operator $\Box$ produces a factor κ_μ^2 under the integral and $\kappa_\mu^2\,\delta(\kappa_\mu^2) = 0$. Δ is an odd function of t, Δ_1 is even.

3. *The D, D_1-functions.* If κ_μ is the momentum vector of a particle with finite mass it satisfies $\kappa_\mu^2 = -\eta^2$. We then obtain a generalization of (25) by replacing $\delta(\kappa_\mu^2)$ by $\delta(\kappa_\mu^2+\eta^2)$. Thus we define:†

$$D(\mathbf{r},t) = \frac{i}{(2\pi)^3}\iiiint d^4\kappa\; e^{i\kappa_\mu x_\mu}\epsilon(\kappa_0)\delta(\kappa_\mu^2+\eta^2), \tag{29 a}$$

$$D_1(\mathbf{r},t) = \frac{1}{(2\pi)^3}\iiiint d^4\kappa\; e^{i\kappa_\mu x_\mu}\delta(\kappa_\mu^2+\eta^2). \tag{29 b}$$

The possible values of κ_0 are now $\pm\sqrt{(\kappa^2+\eta^2)}$. Thus

$$D(\mathbf{r},t) = \frac{1}{(2\pi)^3}\int d^3\kappa\; e^{i(\boldsymbol{\kappa}\mathbf{r})}\frac{\sin\sqrt{(\kappa^2+\eta^2)}x_0}{\sqrt{(\kappa^2+\eta^2)}}, \tag{30 a}$$

$$D_1(\mathbf{r},t) = \frac{1}{(2\pi)^3}\int d^3\kappa\; e^{i(\boldsymbol{\kappa}\mathbf{r})}\frac{\cos\sqrt{(\kappa^2+\eta^2)}x_0}{\sqrt{(\kappa^2+\eta^2)}}. \tag{30 b}$$

The square roots are always to be taken with the positive sign.

Both functions evidently satisfy the differential equations

$$\Box D - \eta^2 D = 0, \qquad \Box D_1 - \eta^2 D_1 = 0. \tag{31}$$

From (30) we see that

$$D(\mathbf{r},t=0) = 0, \qquad \frac{\partial D}{\partial t}\bigg|_{t=0} = c\,\delta(\mathbf{r}), \qquad \frac{\partial D_1}{\partial t}\bigg|_{t=0} = 0. \tag{32}$$

These relations are the same as for the Δ, Δ_1-functions.

† In part of the recent literature the notations Δ, D are interchanged.

The singularities of D, D_1 arise solely from infinitely large values of κ. But for large values of $|\boldsymbol{\kappa}|$, η can be neglected. It follows that the strongest singularities of D and D_1 are precisely the same as for Δ, Δ_1 and we can write

$$D(\mathbf{r},t) = \frac{1}{2\pi}\epsilon(x_0)\delta(x_\mu^2)+\tilde{D}(\mathbf{r},t), \tag{33 a}$$

$$D_1(\mathbf{r},t) = \frac{1}{2\pi^2}\frac{\mathscr{P}}{x_\mu^2}+\tilde{D}_1(\mathbf{r},t) \tag{33 b}$$

where $\tilde{D}$, $\tilde{D}_1$ have at most lesser singularities for all values of $\mathbf{r}$, t, and are, of course, real. We shall see that $\tilde{D}$ is finite (or zero) and $\tilde{D}_1$ still has a logarithmic singularity on the light cone. In particular it should be noted that the singular parts of (33) are independent of η.

The explicit integration of (33) presents no difficulty and can be reduced to well-known integral formulae for the Hankel functions. The result is:

$$\tilde{D}(\mathbf{r},t) = -\frac{\eta}{4\pi}\epsilon(x_0)\mathscr{R}\left\{\frac{H_1^{(1)}(i\eta\sqrt{(r^2-x_0^2)})}{i\sqrt{(r^2-x_0^2)}}\right\}, \tag{34 a}$$

$$\tilde{D}_1(\mathbf{r},t) = -\frac{\eta}{4\pi}\mathscr{R}\left\{\frac{H_1^{(1)}(i\eta\sqrt{(r^2-x_0^2)})}{\sqrt{(r^2-x_0^2)}}\right\}-\frac{1}{2\pi^2}\frac{\mathscr{P}}{x_\mu^2}. \tag{34 b}$$

$\mathscr{R}$ denotes the real part. Since $H_1^{(1)}(iy)$ (y real) is real, and $H_1^{(1)}(y)$ (y real) has both real and imaginary parts the following properties result:

(i) *Outside the light cone* $r > x_0$, $\tilde{D}(\mathbf{r},t) = 0$, and therefore $D(\mathbf{r},t) = 0$ also. At the same time we see that $D_1(\mathbf{r},t) \neq 0$ outside the light cone, as was in fact already the case for Δ_1. For $|\eta\sqrt{(r^2-x_0^2)}| \gg 1$, D_1 decreases exponentially.

(ii) *Inside the light cone* $r < x_0$, the argument of $H_1^{(1)}$ is real and neither D nor D_1 vanish. As regards D, this is in contrast to the Δ-function which vanishes for $r < x_0$ also. For $|\eta\sqrt{(r^2-x_0^2)}| \gg 1$, $\tilde{D}$ and $\tilde{D}_1$ tend to zero like $1/(x_0^2-r^2)$. This follows from the asymptotic formulae for $H_1^{(1)}$.

(iii) *Near the light cone* we use the expansion (neglecting terms $\sim y^2$)

$$H_1^{(1)}(y) = -\frac{2i}{\pi}\left(\frac{1}{y}-\frac{y}{2}\log\frac{\gamma y}{2i}+\frac{y}{4}\right) \qquad (\gamma = 0{\cdot}577...),$$

and obtain:

$$\tilde{D} = \begin{cases} 0, \text{ when } r > |x_0| \\ -\dfrac{\eta^2}{8\pi}\epsilon(x_0), \text{ when } r < |x_0| \end{cases} \tag{35 a}$$

$$\tilde{D}_1 = \frac{\eta^2}{4\pi^2}\log\frac{\eta\gamma}{2}\sqrt{|r^2-x_0^2|}+\text{finite parts}. \tag{35 b}$$

We see that D_1 has an additional logarithmic singularity and D a jump, on the light cone. Both vanish for $\eta = 0$.

4. *The D_2-function.* Besides the D and D_1-functions the following linear combination is important:†

$$D_2 \equiv D_1 - i\epsilon(x_0)D. \tag{36}$$

The Fourier representation of D_2 cannot be read off immediately because $\epsilon(x_0)D$ has not been represented yet as a Fourier integral. To obtain this we form

$$\int d^4x\ \epsilon(x_0)De^{-i\kappa'_\mu x_\mu} = \frac{i}{(2\pi)^3}\int d^4x \int d^4\kappa\ e^{i(\kappa_\mu-\kappa'_\mu)x_\mu}\epsilon(x_0)\epsilon(\kappa_0)\delta(\kappa_\mu^2+\eta^2)$$

$$= 2\int d^4\kappa\ \epsilon(\kappa_0)\delta(\kappa_\mu^2+\eta^2)\delta(\boldsymbol{\kappa}-\boldsymbol{\kappa}')\frac{\mathscr{P}}{\kappa_0-\kappa_0'} = 2\frac{\mathscr{P}}{\kappa_\mu'^2+\eta^2}$$

using (14) for the integration over x_0. Hence it appears, after transforming back, that

$$\epsilon(x_0)D(\mathbf{r},t) = \frac{1}{(2\pi)^3\pi}\int d^4\kappa\ e^{i\kappa_\mu x_\mu}\frac{\mathscr{P}}{\kappa_\mu^2+\eta^2}. \tag{37}$$

Combining this with (29 b) we find

$$D_2(\mathbf{r},t) = -\frac{i}{(2\pi)^3\pi}\int e^{i\kappa_\mu x_\mu}\zeta^*(\kappa_\mu^2+\eta^2)\ d^4\kappa. \tag{38}$$

The complex conjugate of D_2 ($D_2^* = D_1 + i\epsilon(x_0)D$) would be represented by $\zeta(\kappa_\mu^2+\eta^2)$ but this will not be needed. Nor shall we require the Fourier representation of $\epsilon(x_0)D_1$, which could also easily be obtained.

Differentiating the integral (38) we see with the help of (17′) and (21) that

$$\Box D_2 - \eta^2 D_2 = 2i\delta^4(x). \tag{38′}$$

When $\eta = 0$ the function D_2 reduces to $\Delta_2 = \Delta_1 - i\epsilon\Delta$. This is given explicitly by (using the definition (13)):

$$\Delta_2(\mathbf{r},t) = \frac{1}{2\pi^2}\zeta(x_\mu^2). \tag{39}$$

So we see that $\zeta(x_\mu^2)$ is the Fourier transform of its complex conjugate $\zeta^*(\kappa_\mu^2)$. It might be useful to summarize the various Fourier transforms in tabular form, for the case $\eta = 0$. When $\eta \neq 0$ further contributions with at most a logarithmic singularity are to be added in x-space (or κ-space when the reverse transformation is considered). The common factor $1/(2\pi)^3$ occurring everywhere in κ-space is left out.

† This function is often denoted by D_C in the European and by D_F in the American literature.

x-space	κ-space (times$(2\pi)^3$)
$\Delta = \frac{1}{2\pi}\epsilon(x_0)\delta(x_\mu^2)$	$i\epsilon(\kappa_0)\delta(\kappa_\mu^2)$
$\Delta_1 = \frac{1}{2\pi^2}\frac{\mathscr{P}}{x_\mu^2}$	$\delta(\kappa_\mu^2)$
$\pi\epsilon(x_0)\Delta = \frac{1}{2}\delta(x_\mu^2)$	$\frac{\mathscr{P}}{\kappa_\mu^2}$
$\pi\Delta_2 = \frac{1}{2\pi}\zeta(x_\mu^2)$	$-i\zeta^*(\kappa_\mu^2)$
$\pi\Delta_2^* = \frac{1}{2\pi}\zeta^*(x_\mu^2)$	$+i\zeta(\kappa_\mu^2)$

The singular functions discussed here from the mathematical point of view all have a good physical significance. For example, the first term of the Δ-function $\delta(r-x_0)/r$ represents the retarded potential at r and at the time $t = x_0/c$ due to a flashlike event taking place at the origin at $t = 0$ and lasting an infinitely short time. (Insert in § 1 eq. (14 a)

$$\rho(P',t') = \delta(\mathbf{r}')\delta(t'), \qquad t' = t-r_{PP'}/c.)$$

Similarly, $\delta(r+x_0)/r$ represents the 'advanced potential' which is the same as the retarded potential when the source and the space-time point where the potential is considered are interchanged. Thus we can write $\Delta \equiv \Delta_{\text{ret}}-\Delta_{\text{adv}}$.

The ζ and Δ_2-functions† are characteristic for the sequence of events which *follow* each other in time and are causally connected. This rests essentially on the capacity of the ζ-function of singling out a particular time direction in an analytical way. The following simple formula is an example

$$\int\limits_{-\infty}^{+\infty} dx\, \zeta(x)e^{-ixt} = \begin{cases} -2\pi i & \text{for } t > 0 \\ 0 & \text{for } t < 0. \end{cases} \tag{40}$$

This is different from zero only when $t > 0$. (40) is immediately verified by completing the path of integration through the lower/upper half-plane. Thus we shall always meet these functions in the theory of transition probabilities, collision processes, etc. (See in particular §§ 16, 28.)

9. Commutation and uncertainty relations of the field strengths

1. *Commutation relations of the field strengths in coordinate space.* In § 7 we have represented the vector potential $\mathbf{A}$ by a Fourier series whose

† The ζ-function was first introduced by P. A. M. Dirac (*The Principles of Quantum Mechanics*, Oxford University Press 1935) in the theory of collisions. For the role of the Δ_2-function in quantum electrodynamics see: E. C. G. Stückelberg and D. Rivier, *Helv. Phys. Acta*, **23** (1950), 215 (and previous papers, ibid.); R. P. Feynman, *Phys. Rev.* **76** (1949), 769; M. Fierz, *Helv. Phys. Acta*, **23** (1950), 731.

amplitudes have become non-commuting operators. If we now consider **A** as a function of the coordinates, it is clear that **A** *in each space point* is an operator and will not in general commute with **A** in another space point. Here the coordinates play the role of a parameter and not, like the coordinates of a particle, of a physical quantity which itself is an operator. If interaction representation is used (§ 7.3) **A** is also a function of t, and the values of **A** taken at two different space-time points are two physical quantities which do not in general commute.

The commutation relations of the potentials themselves can have no immediate physical significance because they must necessarily depend on the gauge used, although they are very important for the computational technique. They will be derived in appendix 2 for Coulomb gauge and in § 10 for Lorentz gauge. In this section we consider the commutation relations of the field strengths which are independent of the gauge and have immediate physical significance.

In interaction representation the field strengths are given by (see § 7 eqs. (15) and (32))

$$\mathbf{E} = i\sqrt{(4\pi)}\sum_\lambda \nu_\lambda \mathbf{e}_\lambda\{q_\lambda e^{i(\boldsymbol{\kappa}_\lambda \mathbf{r}-\nu_\lambda t)} - q_\lambda^* e^{-i(\boldsymbol{\kappa}_\lambda \mathbf{r}-\nu_\lambda t)}\}, \tag{1}$$

$$\mathbf{H} = i\sqrt{(4\pi c^2)}\sum_\lambda [\boldsymbol{\varkappa}_\lambda \mathbf{e}_\lambda]\{q_\lambda e^{i(\boldsymbol{\kappa}_\lambda \mathbf{r}-\nu_\lambda t)} - q_\lambda^* e^{-i(\boldsymbol{\kappa}_\lambda \mathbf{r}-\nu_\lambda t)}\} \tag{2}$$

with

$$[q_\lambda q_\lambda^*] = \frac{\hbar}{2\nu_\lambda}, \qquad [q_\lambda q_\mu^*] = 0. \tag{3}$$

Consider, for example, the commutator of two components of **H**, H_i and H_k, in two different space-time points. Denoting the point $\mathbf{r}, t$ by P, we obtain the commutation relations

$$[H_i(P_1)H_k(P_2)]$$
$$= -i4\pi c^2\hbar \sum_\lambda [\boldsymbol{\varkappa}_\lambda \mathbf{e}_\lambda]_i [\boldsymbol{\varkappa}_\lambda \mathbf{e}_\lambda]_k \frac{1}{\nu_\lambda} \sin[(\boldsymbol{\varkappa}_\lambda, \mathbf{r}_2 - \mathbf{r}_1) - \nu_\lambda(t_2 - t_1)]. \tag{4}$$

If we carry out the summation over the directions of polarization $\mathbf{e}_\lambda$ ($\mathbf{e}_\lambda$ is always perpendicular to $\boldsymbol{\varkappa}_\lambda$) and write $\mathbf{r}$ for $\mathbf{r}_2 - \mathbf{r}_1$ and t for $t_2 - t_1$, we obtain

$$[H_i(P_1)H_k(P_2)] = -4\pi i\hbar c \sum (\kappa^2\delta_{ik} - \kappa_i \kappa_k)\frac{\sin[(\boldsymbol{\varkappa}\mathbf{r}) - \kappa ct]}{\kappa}$$

$$= -4\pi i\hbar c\left(\frac{1}{c^2}\frac{\partial^2}{\partial t_1 \partial t_2}\delta_{ik} - \frac{\partial^2}{\partial x_{i1}\partial x_{k2}}\right)\sum \frac{\sin[(\boldsymbol{\varkappa}\mathbf{r}) - \kappa ct]}{\kappa}. \tag{5}$$

Here the summation $\sum$ has to be carried out over all directions and values of $\boldsymbol{\varkappa}$. This can be replaced by an integration, taking into account the fact that according to § 6 eq. (21) the number of waves per unit volume for which $\boldsymbol{\kappa}$ lies in the element $d\kappa_x\, d\kappa_y\, d\kappa_z$ is $d^3\kappa/(2\pi)^3$. Now the integral

$$\frac{1}{(2\pi)^3}\int d^3\kappa\,\frac{\sin[(\boldsymbol{\kappa}\mathbf{r})-\kappa ct]}{\kappa} = -\frac{1}{(2\pi)^3}\int \frac{e^{i(\boldsymbol{\kappa}\mathbf{r})}\sin\kappa ct}{\kappa}\,d^3\kappa = -\Delta(\mathbf{r},t) \quad (6)$$

is just the singular Δ-function discussed in § 8. By § 8 eq. (27 a), it is explicitly given by

$$\Delta = \frac{\delta(r-ct)-\delta(r+ct)}{4\pi r}. \quad (7)$$

The points at which Δ is different from zero form a double cone in 4-dimensional space. These points can be reached by a light signal emitted at the origin $r = 0$ at $t = 0$ $(r = +ct)$ and conversely a light signal emitted at the point $\mathbf{r}, t$ can reach the origin $r = 0$ at the time $t = 0$ $(r = -ct)$.

The Δ-function occurring in (6) contains as argument

$$|\mathbf{r}_2-\mathbf{r}_1|-c(t_2-t_1) \quad \text{and} \quad |\mathbf{r}_2-\mathbf{r}_1|+c(t_2-t_1). \quad (7')$$

It is therefore different from zero only if the two points of space time at which the field strengths are considered can be connected by a light signal. *Therefore the field strengths at two points of space time which cannot be connected by light signals commute with each other.*

The commutation relation (5) now becomes:

$$[H_i(P_1)H_k(P_2)] = 4\pi i\hbar c\left(\frac{1}{c^2}\frac{\partial^2}{\partial t_1\,\partial t_2}\delta_{ik}-\frac{\partial^2}{\partial x_{i1}\,\partial x_{k2}}\right)\Delta. \quad (8\,\text{a})$$

In the same way one can find the commutation relations for the other field strengths:

$$[E_i(P_1)E_k(P_2)] = [H_i(P_1)H_k(P_2)], \quad (8\,\text{b})$$

$$[E_i(P_1)H_i(P_2)] = 0, \quad (8\,\text{c})$$

$$[E_i(P_1)H_k(P_2)] = -4\pi i\hbar\frac{\partial^2}{\partial x_{l_2}\,\partial t_1}\Delta \quad (8\,\text{d})$$

$(i \neq k;\ i,\ k,\ l$ forming an even permutation of $x,\ y,\ z)$.

The relations (8) were first deduced by Jordan and Pauli.†

The universal constants occurring in the commutation relations (8) are c and $\hbar$. No constant referring to the atomic structure of matter (m or e) occurs. The non-commutability of the field strength which can be considered as characteristic for present quantum electrodynamics is therefore purely an effect of the union of quantum theory with classical electrodynamics and has no connexion with the problem of the elementary particles.

† P. Jordan and W. Pauli, *Zs. f. Phys.* **47** (1928), 151.

2. *Uncertainty relations for the field strengths.* From the quantum mechanical commutation relations (8) we can obtain corresponding uncertainty relations. If two physical quantities A and B satisfy the equation

$$AB-BA = C,$$

where C represents an ordinary number (not a matrix), A and B satisfy the uncertainty relation

$$\Delta A\Delta B \sim |C|.$$

The uncertainty relations deduced in this way from (8) are, however, not very physical. They represent relations for the field strengths at given points of space and time. But the only quantities which can be measured are the *average values* of the field strengths over certain *regions* of space and time. To obtain relations for these average values we integrate (8) over two regions of space and time $L_1^3 T_1$ and $L_2^3 T_2$, respectively for the two field strengths occurring in each equation. We shall denote these two regions by I_1 and I_2, and the average values of the field strengths by E_{xI_1} or $E_{xL_1T_1}$, etc. The result of the integration on the right-hand side will depend upon the relative position of these two regions, i.e. on which point of I_1 can be reached from I_2 by a light signal and vice versa. We shall confine ourselves to a few characteristic cases.

(*a*) Both time regions are identical: $T_1 = T_2$. According to (7) and (7′) Δ is antisymmetrical in the two times t_1 and t_2. (8 a) is symmetrical in the derivatives with respect to the two times. The time integral of the right-hand side of (8 a) over $T_1 = T_2$ therefore vanishes. Thus,

$$\Delta E_{iL_1T}\,\Delta E_{kL_2T} = \Delta H_{iL_1T}\,\Delta H_{kL_2T} = 0. \tag{9}$$

The average values of two components of the electric or magnetic field strength over the same time region but different space regions commute and can therefore be measured simultaneously.†

(*b*) Both space regions are identical: $L_1 = L_2$. Then the integral over the right-hand side of (8 d) vanishes and we have

$$\Delta E_{iLT_1}\,\Delta H_{kLT_2} = 0. \tag{10}$$

The average values of the electric field strength and the magnetic field strength over the same space region but different time regions commute and can therefore be measured simultaneously.

From (9) and (10) it follows, of course, that the average values of any two components of the field strengths over the same region of *space and time* can always be measured simultaneously.

† The expression 'simultaneous measurement of two quantities' is not used here in the sense 'measurement at the same time', but means that the reciprocal influence of the two measurements is taken into account.

(c) The two regions I_1 and I_2 are situated so that light signals emitted from some at least of the points of I_1 can reach I_2, but no light signal emitted from I_2 can reach any point of L_1^3 during the time T_1 (Fig. 4). Then all contributions arising from the second term of the Δ-function (7) vanish.

We consider the two cases of a simultaneous measurement of the x-component of the electric field strength E_x in I_1 and I_2 and of E_x in I_1 and H_y in I_2. The equations (8) give immediately the uncertainty relations

$$\Delta E_{xI_1}\Delta E_{xI_2} = \frac{\hbar c}{L_1^3 L_2^3 T_1 T_2}\int\limits_{L_1L_2T_1T_2}\left(\frac{1}{c^2}\frac{\partial^2}{\partial t_1\,\partial t_2}-\frac{\partial^2}{\partial x_1\,\partial x_2}\right)\frac{\delta[|\mathbf{r}_2-\mathbf{r}_1|-c(t_2-t_1)]}{|\mathbf{r}_2-\mathbf{r}_1|} \tag{11}$$

and $$\Delta E_{xI_1}\Delta H_{yI_2} = \frac{\hbar}{L_1^3 L_2^3 T_1 T_2}\int\limits_{L_1L_2T_2}\frac{\partial}{\partial z_2}\frac{\delta[|\mathbf{r}_2-\mathbf{r}_1|-c(t_2-t_1)]}{|\mathbf{r}_2-\mathbf{r}_1|}\Bigg|_{t_{10}}^{t_1'}. \tag{12}$$

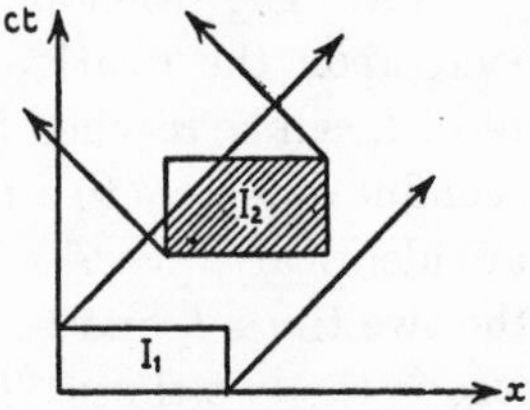

FIG. 4. Two regions of space and time. Light signals from I_1 can reach I_2, but no light signal from I_2 can reach I_1.

In (12) we have carried out the integration over T_1; t_{10} and t_1' represent the times of the beginning and the end of the interval T_1 respectively.

The right-hand sides of (11) and (12) can be interpreted physically in a simple way. This will be done in subsections 3 and 4. We first estimate the order of magnitude. We assume that $L_1 \sim L_2$, $T_1 \sim T_2$ and that the distance between the two space regions L_1 and L_2 is r.

Furthermore the region of I_2 which can be reached from I_1 by light signals shall be of the same order as I_2 itself (Fig. 4). The order of magnitude of the right-hand side of (12), for instance, depends on whether $L \gtrless cT$. One can easily find for the two cases

$$\begin{aligned}\Delta E_{xI_1}\Delta H_{yI_1} &\sim \hbar/r^2LT \qquad (L \gg cT)\\ &\ \hbar/r^2cT^2 \qquad (L \ll cT),\end{aligned} \tag{13}$$

and a similar expression for (11). Thus the two field strengths can be measured more exactly the larger the distance is between the two space regions, which is a very natural result.

(13) also gives the condition for the quantum properties of the field to be essential or for the classical theory to be applicable. The latter is the case if the field strengths are large compared with the term expressing the effect of non-commutability given by the right-hand side of (13). For field strengths of the order E we obtain (putting the distance of the two space regions of the order L)

$$E^2L^3cT \gg \hbar c \quad (L > cT). \tag{14}$$

Thus the typical quantum region is that of *weak fields*. For a light wave of frequency ν (14) simply expresses the condition that the number of light quanta n contained in L^3 must be large: Since $E^2L^3 = n\hbar\nu$ and since the time interval T must be chosen to be smaller than $1/\nu$ (otherwise the average value of E vanishes), we obtain from (14): $n \gg 1$.

3. *Measurement of the average value of a field strength.* For a critical understanding of the quantum mechanical formalism it is important to make sure that the uncertainty relations are consistent with the accuracy which can be best attained by *measurements*. The well-known discussion of the uncertainty of the position and momentum of an electron, for instance, shows that it is impossible to exceed the limits of accuracy given by the uncertainty relation $\Delta p \Delta q \sim \hbar$ by a direct measurement of the position of the electron when the momentum was known beforehand. As a counterpart to this ideal experiment we shall show that our uncertainty relations for the field strengths (11), (12) are consistent with the accuracy which can be attained by a simultaneous measurement of two field strengths.† For this purpose we must first consider the way in which a single field strength can be measured. The quantum mechanical formalism presumes that a single physical quantity such as the average value of the x-component of the electric field strength E_{xI} can be measured exactly. This assumption has, of course, first to be tested before we can interpret the uncertainty relations for two field strengths.

The simplest way to measure E_{xI} would be to take a *charged test body* of mass M covering the region L^3 and having a uniformly distributed charge ϵ. If we measure the momentum p_x‡ of this test body at the beginning t_0 and at the end t' of the time interval T, the average value of E_x is given by§

$$E_{xI} = \frac{p_{x0} - p_x'}{\epsilon T}. \tag{15}$$

† We follow in these considerations the paper of N. Bohr and L. Rosenfeld, *Det. Kgl. dansk. Vid. Selskab.* XII (1933), 8.

‡ In this section we use the usual notation for 'momentum'. (Dimension gm. cm. sec.$^{-1}$)

§ In this procedure it is assumed that the test body filling the region is a *rigid body*. It has been proved in the paper of Bohr and Rosenfeld (loc. cit.) that for this case the assumption that such rigid bodies actually exist is justified.

Since the right-hand sides of the uncertainty relations (11), (12) depend only upon the universal constants $\hbar$, c and upon the geometrical circumstances but not upon any quantity which refers to an *elementary particle* (e or m), the problem of the measurement of field strengths cannot have anything to do with the *atomic structure of matter*. Thus the test body may have any size and any charge, and we shall see in fact that the highest possible accuracy is reached with a *heavy* test body containing a large number of elementary charges. Therefore, the difficulties which are connected with elementary particles (infinite self-energy, etc.) do not play any role in our problem.

We have, however, to take into account that even a heavy test body has to satisfy the general laws of quantum mechanics especially the uncertainty relation for its position x and momentum p_x

$$\Delta x \Delta p_x = \hbar. \tag{16}$$

Therefore, if the momentum of the test body is measured at the beginning of the time interval t_0 with an accuracy Δp_x, we know the position of the test body during the whole time T with an accuracy of only $\hbar/\Delta p_x$.

Furthermore, if we measure the momentum at the time t_0 within a short interval Δt_0 ($\Delta t_0 \ll T$) a certain velocity v_{x0} will be transferred to the test body according to the well-known relation (E_0 = energy)

$$\Delta E_0 \Delta t_0 = v_{x0} \Delta p_{x0} \Delta t_0 = \hbar. \tag{17}$$

We shall assume that before p_{x0} is measured, the test body is in a fixed position exactly covering the region L^3. v_{x0} is then just the velocity which is necessary to displace the test body by Δx within the time Δt_0

$$v_{x0} = \frac{\hbar}{\Delta p_{x0} \Delta t_0} = \frac{\Delta x}{\Delta t_0}. \tag{18}$$

In contrast to the uncertainties of x, p_x the velocity v_{x0} is a *known* quantity since the intervals Δx, Δt_0, can be *chosen* (disregarding an uncertainty Δv_{x0} which is small if the mass M of the test body is large). In a similar way we measure the momentum after the time interval T at t' within a short interval $\Delta t'$ and then bring the test body back to its original position so that it again covers the region L^3 exactly.

These uncertainties (16), (17) will, as we shall see in subsection 4, give rise to the limitation of the accuracy of the simultaneous measurement of two field strengths, but they do not restrict the accuracy of the measurement of a single field strength. In this case all these uncertainties can be compensated.

The accuracy of the measurement of E_{xI} would first of all be restricted by the following facts:

(*a*) the inaccuracy Δp_x with which p_x is measured at the beginning and the end of the time interval T;

(*b*) by the fact that the test body does not cover the region L^3 exactly during the time T because of:

- (α) the acceleration exerted by the field itself;
- (β) the velocity v_{x0} which the test body has after t_0. v_{x0} even becomes large if Δt and Δp_x are small (according to (18));
- (γ) the unknown displacement Δx of the test body which is connected with the knowledge of p_x.

These inaccuracies can be compensated in the following way:

(*b*, α) The field does not change the position of the test body appreciably if the latter is sufficiently *heavy*.

(*b*, β) Since the velocity v_{x0} transferred to the test body by the measurement of the momentum during Δt_0 is known we can compensate for this effect by giving the test body a kick immediately after the measurement of the momentum, i.e. practically at the beginning of the time interval T. The same has to be done after the second measurement of the momentum p'_x.

(*b*, γ) and (*a*). The unknown displacement Δx can only be made small if the measurement of p_x is rather inaccurate. This does not, however, restrict the accuracy of the measurement of E_{xI} since we have another parameter available—the charge ϵ. The inaccuracy of the field strength is, according to (15), given by

$$\Delta E_{xI} = \frac{\Delta p_x}{\epsilon T} = \frac{\hbar}{\epsilon \Delta x T}. \tag{19}$$

Thus for any Δx, however small, E_{xI} can be measured with any desired accuracy if the *charge* ϵ of the test body is *sufficiently high*. If we want an accuracy high enough to test the quantum properties of the field ($\Delta E \ll E$) we see from our estimate (14) ($cT \sim L$, $E^2 \sim \hbar c/L^4$) and from (19) that we must have $\hbar c/\epsilon^2 \ll \Delta x^2/L^2 \ll 1$, that is that ϵ has to be large compared with the elementary charge e.

But, if ϵ is large, another difficulty arises: the field which is measured accurately in the way described above is not only the external field which we wish to measure but also the field $\mathscr{E}$ produced by the test body itself, and it is just for a large charge ϵ that $\mathscr{E}$ also becomes large. This difficulty exists, however, only in so far as we do not know (and cannot calculate) this field exactly because we do not know the position and motion of the test body exactly according to (16), (17).

As we shall see in the next subsection, it is just this inaccuracy $\Delta\mathscr{E}$ of the field produced by the test body which gives rise to the limitations of accuracy for the measurement of two field strengths. For the measurement of a single field strength the effect arising from the inaccuracy $\Delta\mathscr{E}$ can again be compensated.

We shall have to work out the field $\mathscr{E}$ produced by the test body in subsection 4. We see from the calculations (22)–(28) that the average value of $\mathscr{E}$ over the region L^3 is proportional to the displacement x

$$\mathscr{E}_{xI} = Fx, \qquad \Delta\mathscr{E}_{xI} = F\Delta x. \tag{20}$$

But then the force which the field B exerts on the test body itself can be compensated entirely if we subject the test body during the whole time interval T to another force of a purely mechanical nature which is also proportional to the displacement x from the original position L^3. Such a force can, for instance, be realized by a spring. If we choose the strength of the spring so that the force is equal to

$$K_x = -\epsilon Fx, \tag{21}$$

the force exerted by the field $\mathscr{E}$ is just compensated, whatever the—known or unknown—displacement of the test body is. The measurement of $\mathscr{E}$ carried out in this way then indicates the external field only. In principle there is no limit to the accuracy of this measurement.†

4. *Measurement of two field strengths.* Our next task is the physical interpretation of the uncertainty relations (11), (12). According to the general quantum mechanical interpretation of the uncertainty relations, we have to show that a measurement of the average value of E_x, say, over the space time region I_1 does not allow an accurate measurement of another field strength in the region I_2. The limitation of accuracy for the latter is due to the fact that the test body used for the measurement in I_1 produces in I_2 a field $\mathscr{E}$, $\mathscr{H}$, which is to some extent unknown. This unknown field is superposed on the field E, H in I_2 which we wish to measure and cannot be separated from it. It gives rise to an inaccuracy of the value of E, H in I_2 even if we measure the field in I_2 exactly.

To calculate this inaccuracy we have to work out the uncertainty of

† We have not mentioned two points which have to be taken into account for the measurement of E_x. (1) The *reaction* of the field $\mathscr{E}_x$ produced by the test body on the test body itself; (2) the fact that the field $\mathscr{E}_x$ which we have computed classically is also quantized. For the discussion of both points we refer to the paper of Bohr and Rosenfeld. Neither of them gives rise to any restriction on the accuracy of field measurements.

the field which is produced by the first test body. The sources of this field are the following:

(a) As we have seen in subsection 3, during the whole time T the test body is displaced by an unknown quantity $\sim \Delta x$. This gives rise to the field of an *electric dipole* with its moment $\epsilon \Delta x$ in the x-direction. This dipole moment is distributed uniformly over the space L_1^3 with a density $\epsilon \Delta x / L_1^3$. The uncertainty of the scalar potential at a point $\mathbf{r}_2, t_2$, produced by a volume element $d\tau_1$ at the time t_1, is therefore, if we take into account that the field travels with the velocity c,

$$\Delta \phi(\mathbf{r}_2, t_2)\, d\tau_1 = \frac{\epsilon \Delta x}{L_1^3} \frac{\partial}{\partial x_1} \frac{c\delta(r-ct)}{r}\, d\tau_1, \tag{22}$$

where we have denoted $|\mathbf{r}_2 - \mathbf{r}_1|$ by r and $t_2 - t_1$ by t. (The factor c in (22) comes from the definition of the δ-function $c \int \delta\, dt = 1$.)

(b) At the beginning t_{10} of the time interval T_1 the test body has a velocity v_{x0} for a short time Δt_{10} (v_{x0} is afterwards compensated). Thus the test body represents a *current density*

$$i_x\, d\tau_1 = \frac{\epsilon}{L_1^3} d\tau_1 \frac{\Delta x}{\Delta t_{10}} \quad (\text{at } t = t_{10}). \tag{23}$$

At the time t_1' (the end of the interval T_1) the test body again has a velocity $v_x' = \Delta x / \Delta t_1'$. But since we bring the test body *back* to its original position L_1^3 the time integral of the current density at t_1' has the same value but the opposite sign to (23). Assuming that Δt_{10}, $\Delta t_1'$ are infinitely small we can write for the total current density

$$i_x\, d\tau_1 = \frac{\epsilon \Delta x\, d\tau_1}{L_1^3} [\delta(t-t_{10}) - \delta(t-t_1')]. \tag{24}$$

This current density gives rise to a vector potential $\mathscr{A}$ at the space time point $\mathbf{r}_2, t_2$. Taking into account the retardation we have to insert $t_2 - |\mathbf{r}_2 - \mathbf{r}_1|/c$ for t in (24). Thus the unknown vector potential is equal to

$$\begin{aligned} \Delta \mathscr{A}_x(\mathbf{r}_2 t_2)\, d\tau_1 &= \frac{\epsilon \Delta x}{c L_1^3} d\tau_1\, c\, \frac{\delta[c(t_2-t_{10})-r] - \delta[c(t_2-t_1')-r]}{r} \\ &= -\frac{\epsilon \Delta x}{L_1^3} d\tau_1 \frac{\delta(r-ct)}{r}\bigg|_{t_{10}}^{t_1'}. \end{aligned} \tag{25}$$

From the potentials (22), (25) we obtain the uncertainties of the field strengths

$$\Delta \mathscr{E}_x = -\frac{\partial \Delta \phi}{\partial x_2} - \frac{1}{c} \frac{\partial \Delta \mathscr{A}_x}{\partial t_2},$$

$$\Delta \mathscr{H}_y = \frac{\partial \Delta \mathscr{A}_x}{\partial z_2}. \tag{26}$$

Integrating over all points of I_1 and taking the average over I_2, we obtain the total unknown contribution to the field in I_2 arising from the test body in I_1

$$\Delta\mathscr{E}_{xI_2} = \frac{\epsilon\Delta xc}{L_1^3 L_2^3 T_2} \int\limits_{L_1 L_2 T_2} \left\{\frac{1}{c^2}\frac{\partial}{\partial t_2}\frac{\delta(r-ct)}{r}\bigg|_{t_{10}}^{t_1'} - \frac{\partial^2}{\partial x_1 \partial x_2}\int\limits_{T_1}\frac{\delta(r-ct)}{r}\right\}, \tag{27}$$

$$\Delta\mathscr{H}_{yI_2} = \frac{\epsilon\Delta x}{L_1^3 L_2^3 T_2} \int\limits_{L_1 L_2 T_2} \frac{\partial}{\partial z_2}\frac{\delta(r-ct)}{r}\bigg|_{t_{10}}^{t_1'}. \tag{28}$$

The formulae (27), (28) give the inaccuracy of the field measurement in I_2. We can therefore write

$$\Delta\mathscr{E}_{xI_2} = \Delta E_{xI_2}.$$

Since we have assumed in subsection 2 that no light signal emitted from I_2 can reach the region I_1, no similar disturbance of the measurement in I_2 by the test body in I_1 can exist.

If we multiply (27), (28) by the inaccuracy of the field measurement in I_1 as given by (19)

$$\Delta E_{xI_1} = \frac{\hbar}{\epsilon\Delta x T_1}, \tag{29}$$

we obtain exactly the *uncertainty relations* (11), (12), deduced from the formalism. The charge ϵ and the unknown displacement Δx cancel.

If the two regions are so situated that light signals from I_2 can also reach I_1, the arrangement of the measurements which is necessary to reach the highest possible accuracy is more complicated. This case has been treated in the paper of Bohr and Rosenfeld (loc. cit.), where also all other details concerning the problem of field measurements have been carefully discussed.†

Thus we have proved that the consequences of the quantum electrodynamical formalism agree with the possibilities given by field measurements. This proof is analogous to the proof of the uncertainty relation for the position and momentum of an electron where the quantum properties of the *light beam* in a γ-ray microscope prevent an exact measurement (§ 7.4). In our case the quantum *mechanical* properties of the test body prevent an exact measurement of two *field strengths*. From this point of view it becomes clear that quantum electrodynamics and quantum mechanics form two inseparable parts of a single body of quantum theory, and neither is consistent without the other.

† For an extension of these considerations when pair creation is taken into account and for the question of the measurability of currents in quantum electrodynamics see N. Bohr and L. Rosenfeld, *Phys. Rev.* **78** (1950), 794 and E. Corinaldesi, Manchester Thesis (1951).

10. Quantization of the longitudinal and scalar fields

The quantization of the electromagnetic field carried out in the previous sections was based on the fact that Coulomb gauge can be used and the longitudinal and scalar fields replaced by the instantaneous Coulomb interaction between all charges (§ 6). **A** is then gauged so that div **A** $= 0$. This relation also holds when **A** is quantized as an *identity* which the operator **A** satisfies, as is evident from § 7 eq. (3). This method, though from the point of view of quantum theory by far the simplest, has the disadvantage of destroying the Lorentz invariant appearance of the theory, although the theory itself, of course, remains covariant. In the present preliminary state of the theory this is a little more than a mere defect in elegance. We shall see in Chapter VI that the theory is handicapped by certain ambiguities and we shall also see that a powerful guidance as to how these ambiguities should be removed is obtained from the demand of relativistic covariance of the result. For this purpose it is desirable to carry the covariant appearance of the theory as far as possible. It will also be seen that for the computation of more complicated radiation phenomena a symmetrical treatment of all four components of $\mathbf{A}_\mu$ is simpler than dividing the field into its transverse part and the Coulomb interaction.

1. *Expansions and commutation relations.* In this section we shall treat the quantization of the electromagnetic field using Lorentz gauge

$$\frac{\partial A_\alpha}{\partial x_\alpha} = 0 \tag{1}$$

(to avoid confusion with the index λ we now use α, β for the relativistic indices). This task is not as straightforward as it looks, and some fresh points of theory will arise. We expand all four components of A_α into a Fourier series as in § 6. Using interaction representation (i.e. adding the time factors) and the complex amplitudes q^*, q instead of the real q, p, the expansions can be written:

$$A_i = \sqrt{(4\pi c^2)} \sum_\lambda \{q_{i\lambda}^* e^{-i\kappa_\mu x_\mu} + q_{i\lambda} e^{i\kappa_\mu x_\mu}\}, \tag{2 a}$$

$$\phi \equiv A_4/i = \sqrt{(4\pi c^2)} \sum_\sigma \{q_{0\sigma}^* e^{-i\kappa_\mu x_\mu} + q_{0\sigma} e^{i\kappa_\mu x_\mu}\}, \tag{2 b}$$

$$\kappa_\mu x_\mu = (\boldsymbol{\varkappa}_\lambda \mathbf{r}) - \nu_\lambda t \quad (\text{or } (\boldsymbol{\varkappa}_\sigma \mathbf{r}) - \nu_\sigma t).$$

Here we have expanded the x, y, z components of **A**, and called the amplitudes $q_{i\lambda}$, instead of separating them into the two transverse components $\mathbf{e}_\lambda q_\lambda$ and the longitudinal component q_σ (§ 6).† λ does not now include the description of polarization. For the space parts $q_{i\lambda}$, the canonically conjugate variables were $Q_{i\lambda} = q^*_{i\lambda}+q_{i\lambda}$ and

$$P_{i\lambda} = -i\nu_\lambda(q_{i\lambda}-q^*_{i\lambda})$$

(§ 7 eq. (4)) and so, in quantum theory, the commutation relations are to be postulated (§ 7 eq. (10)):

$$[q_{i\lambda}\, q^*_{k\lambda'}] = \frac{\hbar}{2\nu_\lambda}\delta_{ik}\,\delta_{\lambda\lambda'}. \tag{3}$$

This is not so for the scalar field. It was shown in § 6 that the *scalar part* of the *Hamiltonian* is *negative* and consequently, in the variables $Q, P, \dot{Q}_{0\sigma} = -P_{0\sigma}$ (§ 6 eq. (42), (44 d)). Thus if $Q_{0\sigma} = q_{0\sigma}+q^*_{0\sigma}$, $P_{0\sigma}$ must be $+i\nu_\sigma(q_{0\sigma}-q^*_{0\sigma})$, with a change of sign. The commutation relations $[P_{0\sigma}\, Q_{0\sigma}] = -i\hbar$, which are valid for all canonical pairs, yield, again with a change of sign,

$$[q_{0\sigma}\, q^*_{0\sigma}] = -\frac{\hbar}{2\nu_\sigma} \tag{4}$$

and, omitting the zero point energy,

$$H_{\text{scalar}} = -\sum_\sigma 2\nu_\sigma^2\, q^*_{0\sigma}\, q_{0\sigma}. \tag{5}$$

To make the commutation relations look symmetrical in all four components we re-introduce the relativistic i and define

$$q_4 = +iq_0, \qquad q^*_4 = +iq^*_0. \tag{6}$$

q^*_4 is then not the complex conjugate of q_4, and if ordinary methods of quantization are applied, q^*_4 would not be the hermitian conjugate of q_4. q_4, q^*_4 are the expansion coefficients of the imaginary $A_4 = i\phi$. The commutation relations become $[q_4\, q^*_4] = +\hbar/2\nu$ and can be included in (3), writing now λ also for σ:

$$[q_{\alpha\lambda}\, q^*_{\beta\lambda}] = \frac{\hbar}{2\nu_\lambda}\delta_{\alpha\beta} \quad (\alpha, \beta = 1,\ldots, 4). \tag{7}$$

Thus the quantization of the longitudinal and scalar fields gives rise to two further types of photons, longitudinal and scalar. Their roles will become clear below. Altogether we have now four different 'polariza-

† If a normal coordinate system is chosen for a particular wave $\varkappa_\lambda$ such that the z-axis has the direction of $\varkappa_\lambda$, $q_{x\lambda}$ and $q_{y\lambda}$ describe the two transverse components and $q_{z\lambda}$ the longitudinal component. For any other choice of the coordinate system an orthogonal transformation of the normal components can be carried out. Since the commutation relations are the same for the amplitudes of all three normal components, the $q_{i\lambda}$, $q^*_{i\lambda}$ again satisfy the same commutation relations, (3). Thus one single coordinate system can be used for all partial waves λ. The factor $\delta_{ik}\,\delta_{\lambda\lambda'}$ in (3) is zero unless the polarizations as well as $\varkappa_\lambda$ are the same for the two factors of the commutator.

tions'. However, the change of sign in (4), or the fact that classically q_4, q_4^* are not complex conjugates of each other, will necessitate a different method of quantization for the scalar field.

The field A_α must now satisfy the Lorentz condition (1). Having used interaction representation we can differentiate the operators A_α directly (see § 7.3). Since (7) should hold for each space-time point $\mathbf{r}, t$, it should also hold for each partial wave and we would obtain from (2), (6)

$$\kappa_\alpha q_\alpha = 0, \qquad \kappa_\alpha q_\alpha^* = 0 \qquad (\kappa_\alpha \equiv \boldsymbol{\kappa},\, i\nu/c). \tag{8}$$

Now it is seen at once that (8) is incompatible with the commutation relations (7). Multiplying (7) by κ_β and summing over $\beta = 1,..., 4$, the right-hand side is $(\hbar/2\nu)\kappa_\alpha$ whereas, by (8), the left-hand side is zero. This means that, as soon as the operators q, q^* are subjected to the quantum conditions (7) the operators can no longer satisfy the Lorentz condition. The difficulty is particular to the Lorentz gauge, for we have seen that in Coulomb gauge the condition $\operatorname{div} \mathbf{A} = 0$, is satisfied by the *operators* $\mathbf{A}$ identically.

A way out of this difficulty was first indicated by Fermi† (we shall have to modify his procedure below): consider the set of *all* quantum states described by state vectors Ψ which satisfy the wave equation, and let the operators $\mathbf{A}_\alpha$ occurring in the Hamiltonian satisfy $\Box A_\alpha = 0$ and the commutation relations (7), but *not* the Lorentz condition (1). The solutions Ψ will then in general not describe a state which, classically, corresponds to a solution of Maxwell's equations. The latter reduce to $\Box A_\alpha = 0$ only when also $\partial A_\alpha/\partial x_\alpha = 0$. (In particular the equations $\operatorname{div} \mathbf{E} = 0$ and $\operatorname{curl} \mathbf{H} - (1/c)\dot{\mathbf{E}} = 0$ would be violated.) On the other hand, there may be *special* solutions, say, which have the property that in the classical limit, Maxwell's equations are satisfied. This is the case if we select amongst the whole set of Ψ's solutions for which

$$\frac{\partial A_\alpha}{\partial x_\alpha}\Psi = 0, \tag{9}$$

i.e. for which the eigenvalue of the operator $\partial A_\alpha/\partial x_\alpha$ is zero. If we succeed in finding such a set of solutions it is clear that the expectation value of $\partial A_\alpha/\partial x_\alpha$ formed with these wave functions, i.e. $(\Psi^* \, \partial A_\alpha/\partial x_\alpha \, \Psi)$ vanishes. Then one half of Maxwell's equations is satisfied in the same sense,

$$(\Psi^* \operatorname{div} \mathbf{E}\, \Psi) = 0, \quad \left(\Psi^* \left(\operatorname{curl} \mathbf{H} - \frac{1}{c}\dot{\mathbf{E}}\right)\Psi\right) = 0$$

† E. Fermi, *Rev. Mod. Phys.* **4** (1932), 131.

whereas the other half holds in the form of operator *identities*. However, (9) also leads to inconsistencies.‡

A consistent way of handling the situation has been shown by Gupta and Bleuler§ and we follow their procedure in the following:

Two questions are to be settled: (i) The handling of the sign in the commutation relations for q_0, or the question of reality of ϕ, and (ii) a consistent way of formulating the Lorentz condition. It will be seen below that by applying, what will be called the quantization by *indefinite metric*, it will be possible to regard q_4 and q_4^* as hermitian conjugates and A_4 as hermitian, thus to interpret q_4, like q_i, as absorption and q_4^* as emission operators, and yet obtain purely imaginary expectation values for A_4 and, as is necessary, real expectation values for ϕ. This is quite in contrast to ordinary quantum theory where a hermitian operator always has real expectation values. When this is achieved, the Lorentz condition will easily be formulated in a consistent way.

2. *Quantization by indefinite metric.* In ordinary wave mechanics the wave functions form an orthogonal set ψ_i, normalized to $(\psi_i^* \psi_k) = \delta_{ik}$. The expectation value of an operator Q in a state ψ_i is $\langle Q \rangle = (\psi_i^* Q\psi_k)$ and if Q is hermitian $\langle Q \rangle$ is real. Similarly the eigenvalues q of Q satisfying $Q\psi = q\psi$ are real.

This is now by no means the most general formalism of quantum mechanics. A generalization (due originally to Dirac‖ and used for a very different purpose) is the following: Suppose η is an operator, which we can assume to be hermitian and also to satisfy (this is the simplest generalization)

$$\eta^2 = 1, \qquad \eta^\dagger = \eta. \tag{10}$$

So, when η is written in form of a diagonal matrix, it has diagonal elements ± 1. Instead of the norm $(\psi^* \psi)$ we now require that the norm should be:

$$(\psi_i^* \eta\psi_k) = \pm\delta_{ik} \equiv N_i \delta_{ik}, \qquad N_i^2 = 1. \tag{10'}$$

‡ The procedure which is followed in a large part of the literature is now the following: The Lorentz condition is stated in the form (9). To cope with the change of sign in the commutation relations and to preserve the reality of ϕ the roles of emission and absorption are reversed for q_0 and q_0^* (q_0 emission, q_0^* absorption). q_0, q_0^* are considered as hermitian conjugates, ϕ as hermitian, A_4 antihermitian. Although with suitable handling correct results can be obtained in this way the method is not consistent. For example, a normalizable state vector, satisfying (9), does not exist at all. For a criticism of this procedure see: F. J. Belinfante, *Phys. Rev.* **76** (1949), 226; S. T. Ma, ibid. **75** (1949), 535. Compare also F. Coester and J. M. Jauch, ibid. **78** (1950), 149.

§ S. N. Gupta, *Proc. Phys. Soc.* **63** (1950), 681; K. Bleuler, *Helv. Phys. Acta*, **23** (1950), 567.

‖ P. A. M. Dirac, *Comm. Dublin Inst. Advanced Studies* A, No. 1 (1943). See also W. Pauli, *Rev. Mod. Phys.* **15** (1943), 175.

We call η the metric operator. The hermitian conjugate $Q^\dagger$ of Q is, as usual, defined by $(g^*Qf)^* = (f^*Q^\dagger g)$. If Q is hermitian, $Q^\dagger = Q$.

All matrix elements and expectation values must now be formed in this way. If Q is an operator (not necessarily hermitian) its matrix element and expectation value will be defined as

$$Q_{ik} = N_i(\psi_i^* \eta Q\psi_k), \tag{11}$$

$$\langle Q\rangle_i = (\psi_i^* \eta Q\psi_i) = N_i Q_{ii}. \tag{11'}$$

(10′) replaces the orthogonality relation. In addition there will be a completeness relation (replacing the usual $\sum_i \psi_i^*(x)\psi_i(x') = \delta(x-x')$) namely,

$$\sum_i N_i\psi_i^*(x)\eta\psi_i(x') = \delta(x-x'). \tag{12}$$

It is then seen immediately that Q_{ik} defined by (11) follows the rules of matrix multiplication: $(QP)_{ik} = \sum_l Q_{il}P_{lk}$. Also, if Q is hermitian, $Q_{ik}^* = Q_{ki}$, provided we assume η diagonal, $\eta\psi_i = N_i\psi_i$. Since ψ is now normalized to ± 1, it is clear that $(\psi_i^* \eta\psi_i)$ cannot be interpreted as a probability. This, however, will not give rise to difficulties.

Suppose Q is hermitian and anticommutes with η

$$\eta Q = -Q\eta.$$

In the usual theory Q would have real expectation values. With the present method this is quite different. $\langle Q\rangle$ is now given by (11′) and thus

$$\langle Q\rangle_i^* = (\psi_i^* \eta Q\psi_i)^* = (\psi_i^* Q^\dagger\eta^\dagger\psi_i) = (\psi_i^* Q\eta\psi_i) = -\langle Q\rangle_i.$$

In this case $\langle Q\rangle_i$ is purely imaginary. Alternatively, if Q is anti-hermitian ($Q^\dagger = -Q$) and anticommutes with η, $\langle Q\rangle$ will be real. For operators Q, which anticommute with η, η is equivalent to a factor i.

It is now quite clear how we can apply the above method of quantization to the scalar field to obtain correct reality properties for A_4 and ϕ. We consider first a single scalar radiation oscillator.

Let A_4 (not ϕ) be hermitian. This means q_4 and q_4^* have the same reality properties as the space parts q_i and q_i^* and are hermitian conjugates. The metric operator η is now required to anticommute with q_4 and q_4^*:

$$\eta q_4 = -q_4\eta, \qquad \eta q_4^* = -q_4^*\eta. \tag{13}$$

It follows immediately that the expectation value of A_4, $\langle A_4\rangle$ is purely imaginary, and therefore that of $\phi = -iA_4$ is real, as it should be. This permits us to treat q_4 on the same footing as q_i, i.e. to consider q_4 as absorption, q_4^* as emission operators, and to call $(2\nu/\hbar)q_4^* q_4 = n_4$ the number of 'scalar photons'. These behave in all respects, except in the use of the indefinite metric, like the spatial photons n_1, n_2, n_3. For a definite description we use the number of scalar photons n_4 as characterizing the state. Acting on a state vector Ψ_{n_4}, the operators q_4, q_4^* have the properties (see § 7 eq. (27))

$$\begin{aligned} q_4 \Psi_{n_4+1} &= \sqrt{\frac{\hbar}{2\nu}}\sqrt{(n_4+1)}\Psi_{n_4}, \\ q_4^* \Psi_{n_4} &= \sqrt{\frac{\hbar}{2\nu}}\sqrt{(n_4+1)}\Psi_{n_4+1}. \end{aligned} \tag{14}$$

The metric operator, which anticommutes with q_4 and q_4^*, is

$$\eta = (-1)^{n_4}. \tag{15}$$

Since q_4 and q_4^* change n_4 by 1, η necessarily incurs a minus sign when moved from the left of q_4 to the right. η is diagonal in this representation. The normalization of Ψ_{n_4} is then

$$(\Psi_{n_4}^* \eta \Psi_{n_4'}) \equiv (-1)^{n_4}\delta_{n_4 n_4'} \equiv N_{n_4}\delta_{n_4 n_4'}. \tag{16}$$

The matrix elements of q_4 and q_4^* are obtained immediately from (11), (10′), (14):

$$\begin{aligned} q_{4 n_4, n_4+1} &= \sqrt{\frac{\hbar}{2\nu}}\sqrt{(n_4+1)}, \\ q^*_{4 n_4+1, n_4} &= \sqrt{\frac{\hbar}{2\nu}}\sqrt{(n_4+1)}. \end{aligned} \tag{14'}$$

They are precisely the same as for the space parts.

The extension to the case of all radiation oscillators is trivial. η will be the direct product of factors, one for each radiation oscillator, but, of course, for the spatial oscillators the factors are all unity. Thus η has the properties

$$\eta A_i = A_i \eta, \qquad \eta A_4 = -A_4 \eta. \tag{17}$$

Ψ is a product, as in § 7.3, and the norm of Ψ is

$$\Big(\prod_\lambda \Psi^*_{n_{1\lambda}\ldots n_{4\lambda}} \eta \prod_\lambda \Psi_{n_{1\lambda}\ldots n_{4\lambda}}\Big) \equiv N = (-1)^{n_4}, \tag{18}$$

$$n_4 = \sum_\lambda n_{4\lambda},$$

where n_4 is now the total number of scalar photons. For each oscillator the matrix elements of q_α, q_α^*, no matter whether $\alpha = 1, 2, 3$, or 4, are given by

$$q_{\alpha n_\alpha, n_\alpha+1} = q^*_{\alpha n_\alpha+1, n_\alpha} = \sqrt{\frac{\hbar}{2\nu}}\sqrt{(n_\alpha+1)}. \tag{19}$$

It is seen that no asymmetry in the interpretation of q_α, q_α^* as emission or absorption operators is required. The asymmetry in space and time components is moved to the metric operator but does not appear explicitly in the commutation relations, matrix elements, hermiticity conditions, or interpretation.

3. *The Lorentz condition.* Our next task is to introduce the Lorentz condition. We must certainly require that only such state vectors Ψ_L are to be admitted as physically possible, for which the expectation value of $\partial A_\alpha/\partial x_\alpha$ vanishes. In the present theory the condition takes the new form

$$\left(\Psi_L^* \eta \frac{\partial A_\alpha}{\partial x_\alpha} \Psi_L\right) = 0. \tag{20}$$

(20) must be valid for each space point $\mathbf{r}$ and *each time t*. For a smooth transition to classical theory a little more than (20) is actually required. It is necessary that the expectation value of the energy-momentum tensor should also go over into the classical value. Since the former is quadratic in the field strengths it is also necessary that the expectation value of $(\partial A_\alpha/\partial x_\alpha)^2$ should vanish,† i.e.

$$\left(\Psi_L^* \eta \left(\frac{\partial A_\alpha}{\partial x_\alpha}\right)^2 \Psi_L\right) = 0. \tag{21}$$

We shall see below that (21) will be satisfied.

We again consider first a radiation oscillator with fixed wave number $\boldsymbol{\kappa}$. It is convenient now to choose the coordinate system so that $\boldsymbol{\kappa}$ lies in the z-direction. Then

$$\frac{\partial A_1}{\partial x_1} + \frac{\partial A_2}{\partial x_2} = 0$$

automatically and we are only concerned with the third and fourth components. If we suppress the arguments n_1, n_2 (transverse photons) our condition is

$$\left(\Psi_L^* \eta \left(\frac{\partial A_3}{\partial x_3} + \frac{\partial A_4}{\partial x_4}\right) \Psi_L\right) = 0. \tag{22}$$

(22) consists of two terms, one arising from the q's and proportional to $e^{i(\boldsymbol{\kappa}\mathbf{r})-i\nu t}$ and the other arising from the q^*'s and proportional to $e^{-i(\boldsymbol{\kappa}\mathbf{r})+i\nu t}$. Evidently both have to vanish separately, because (22) must be valid for all values of $\mathbf{r}, t$. Thus (22) reduces to ($\kappa_4 = i\kappa_3$):

$$\begin{aligned} (\Psi_L^* \eta (q_3 + i q_4) \Psi_L) &= 0, \\ (\Psi_L^* \eta (q_3^* + i q_4^*) \Psi_L) &= 0. \end{aligned} \tag{23}$$

The second equation (23) is equivalent to the first, because q^* is the hermitian conjugate of q, and η commutes with q_3 but anticommutes with q_4, so that the second equation can also be written

$$(\Psi_L^* (q_3^* - i q_4^*) \eta \Psi_L) = 0$$

which is the complex conjugate of the first.

† This is discussed in detail by F. J. Belinfante, *Physica*, **12** (1946), 17.

To satisfy the first equation (22) it is sufficient to demand†

$$(q_3+iq_4)\Psi_L = 0. \tag{24}$$

Ψ_L can certainly be represented as a superposition of normalized eigenfunctions with sharp photon numbers n_3, n_4, called $\Psi_{n_3n_4}$,

$$\Psi_L = \sum_{n_3,n_4} c_{n_3n_4}\Psi_{n_3n_4}, \tag{25}$$

$$(\Psi^*_{n_3n_4}\,\eta\Psi_{n_3'n_4'}) = (-1)^{n_4}\delta_{n_3n_3'}\,\delta_{n_4n_4'}. \tag{25'}$$

Using the operator properties (14) (and the same for q_3) we obtain from (24)

$$\sqrt{(n_3+1)}c_{n_3+1,n_4}+i\sqrt{(n_4+1)}c_{n_3,n_4+1} = 0 \tag{26}$$

where, of course, $c = 0$ when any index is negative. We obtain a set of possible Ψ_L by choosing linear combinations with the same total number n_3+n_4:

$$\begin{aligned}
\Psi_L^{(0)} &= \Psi_{00}\\
\Psi_L^{(1)} &= \Psi_{10}+i\Psi_{01}\\
\Psi_L^{(2)} &= \Psi_{20}+i\sqrt{2}\,\Psi_{11}-\Psi_{02}\\
&\ldots\ldots\ldots\\
\Psi_L^{(n)} &= \Psi_{n0}+\ldots+i^r\sqrt{\binom{n}{r}}\Psi_{n-r,r}+\ldots+i^n\Psi_{0n}.
\end{aligned} \tag{27}$$

For a general coordinate system ($\boldsymbol{\kappa}$ not in 3-direction), the combinations are more complicated. These new wave functions are not normalized, for reasons seen below. We call them for brevity the Lorentz set.

If we split up A_α into (compare subsection 6)

$$\begin{aligned}
A_\alpha &= A_\alpha^- + A_\alpha^+,\\
A_\alpha^- &= \sqrt{(4\pi c^2)}\sum_\lambda q_{\alpha\lambda}e^{i\kappa_\mu x_\mu},\\
A_\alpha^+ &= \sqrt{(4\pi c^2)}\sum_\lambda q^*_{\alpha\lambda}e^{-i\kappa_\mu x_\mu}
\end{aligned}$$

and similarly the Lorentz operator

$$\frac{\partial A_\alpha}{\partial x_\alpha} \equiv L = L^- + L^+,$$

(24) is seen to mean

$$L^-\Psi_L = 0. \tag{28}$$

This is our general Lorentz condition in quantum theory for a pure radiation field (instead of (9)). In this form the condition is, of course, independent of the choice of the coordinate system.

† Alternatively, (23) would be satisfied if $(q_3^*+iq_4^*)\Psi_L = 0$. This, however, cannot be satisfied because there cannot be any state from which the emission of photons is forbidden. This is also essentially the reason why (9) cannot be satisfied.

The hermitian conjugate of (28) is $\Psi^* L^{-\dagger} = 0$, $L^{-\dagger} \sim q_3^* - q_4^*$. On multiplication by η from the right, this is equivalent to

$$\Psi_L^* \eta L^+ = 0 \tag{29}$$

from which we again obtain (20), viz.

$$(\Psi_L^* \eta (L^+ + L^-)\Psi_L) \equiv (\Psi_L^* \eta L \Psi_L) = 0.$$

Also $(\Psi_L^* \eta L^2 \Psi_L) = (\Psi_L^* \eta L^- L^+ \Psi_L)$, on account of (28), (29). Now the explicit representation (23) shows that L^- commutes with L^+, $[q_3 q_3^*]$ being compensated by $-[q_4 q_4^*]$, so that‡

$$(\Psi_L^* \eta L^2 \Psi_L) = (\Psi^* \eta L^+ L^- \Psi_L) = 0 \tag{30}$$

in agreement with the requirement (21).

(24) does not contain the time because both q_3, q_4 and Ψ_L are independent of t (except for the time exponentials which are irrelevant). We may therefore regard (24) or (28) as *initial conditions* to be satisfied at $t = t_0$, say. It follows then that the same equations and (20) hold for all times. Ψ_L will still depend on the number of transverse photons which are not affected by the Lorentz condition. Naturally, the number of transverse photons must be the same in all terms of the same $\Psi_L^{(n)}$.

The new combinations $\Psi_L^{(n)}$ have a very remarkable property. It is seen immediately from (27) and (16) that the *norm of all* $\Psi_L^{(n)}$ *vanishes*, *except for* $n = 0$, and for $n = 0$ the norm is positive. Besides, the $\Psi_L^{(n)}$ are orthogonal to each other. Thus we can normalize

$$(\Psi_L^{(0)*} \eta \Psi_L^{(0)}) = 1 \tag{31 a}$$

and

$$(\Psi_L^{(n)*} \eta \Psi_L^{(n')}) = 0 \quad (n \text{ or } n' \neq 0). \tag{31 b}$$

Negative norms do not occur in the Lorentz set Ψ_L. This makes it again possible to apply the usual probability interpretation to physically observable quantities, e.g. the number of transverse photons. A physical state will consist of a linear combination of $\Psi_L^{(n)}$'s which we shall now require to *include* $\Psi_L^{(0)}$. Such a state vector can be normalized to $+1$. $\Psi_L^{(0)}$ itself will consist in general of various terms, with different numbers of transverse photons, and the coefficients are the probabilities for a specified number of such photons to occur. Any admixture of states $\Psi_L^{(n)}$, $n \neq 0$, does not alter these probabilities. The longitudinal and scalar photons are not observable (or, otherwise expressed, the probabilities for their occurrence are zero).

‡ It can be concluded in the same way that the expectation value of L^n vanishes. In ordinary quantum mechanics it would follow then that $L\Psi = 0$ and that L has the sharp value 0. This conclusion cannot be drawn here and $L^+\Psi \neq 0$.

The state vectors $\Psi_L^{(0)}$ and $\Psi_L^{(0)}+\sum_{n\neq 0} c_n \Psi_L^{(n)}$ describe the same physical state and for physically observable quantities the 'admixtures' $\Psi_L^{(n)}$ are irrelevant. Yet they are an essential part of the theory and cannot simply be omitted from the calculation, as will be shown presently.

4. *Gauge invariance.* We show now that the 'admixtures' with vanishing norm are closely connected with the various gauges still possible within the Lorentz gauge. Classically, the potentials A_α can still be changed without violating (1), by adding a 4-gradient:

$$A_\alpha \rightarrow A_\alpha + \frac{\partial \chi}{\partial x_\alpha}, \qquad \Box \chi = 0. \tag{32}$$

We show now: if the state vector $\Psi_L^{(0)}$ is changed by adding an arbitrary admixture $\sum_{n\neq 0} c_n \Psi_L^{(n)}$, the expectation values of A_α suffer a gauge transformation (32). For this purpose we first form the expectation value of the field strengths $f_{\alpha\beta}$, which are gauge invariant. Let

$$\Psi_L = \Psi_L^{(0)} + \sum_{n\neq 0} c_n \Psi_L^{(n)} \tag{33}$$

be a general state vector. Then

$$\langle f_{\alpha\beta}\rangle = \left(\Psi_L^* \eta \left(\frac{\partial A_\beta}{\partial x_\alpha} - \frac{\partial A_\alpha}{\partial x_\beta}\right)\Psi_L\right). \tag{34}$$

Using the expansions for A_α (2) it is immediately verified that $f_{\alpha\beta}$ commutes with L^-

$$[f_{\alpha\beta} L^-] = 0. \tag{35}$$

Since $L^-\Psi_L = 0$ it follows that also

$$L^- f_{\alpha\beta} \Psi_L = 0. \tag{36}$$

Now Ψ_L comprises *all* state vectors which, when acted upon by L^-, vanish. It follows then from (36) that $f_{\alpha\beta}$ acting on Ψ_L produces a state vector belonging to the Lorentz set:

$$f_{\alpha\beta} \Psi_L = \Psi'_L.$$

Moreover, $f_{\alpha\beta}$ when acting on $\Psi_L^{(n)}$ $(n \neq 0)$ produces a superposition of $\Psi_L^{(m)}$'s which does not include $\Psi_L^{(0)}$:

$$f_{\alpha\beta} \Psi_L^{(n)} = \sum_m b_{nm} \Psi_L^{(m)}, \qquad b_{n0} = 0 \quad (n \neq 0). \tag{37}$$

To see this consider again a partial wave with $\boldsymbol{\kappa}$ in the 3-direction ($\kappa_1 = \kappa_2 = 0$). f_{ik}, f_{14}, f_{24} (i, k spatial) affect the number of transverse photons only ($f_{ik} \sim q_i \kappa_k - q_k \kappa_i$, and since $\kappa_1 = \kappa_2 = 0$, only operators q_1, q_2 occur but not q_3, q_4; similarly in $f_{14} \sim q_1 \kappa_4 - q_4 \kappa_1$). f_{34} is, because of $i\kappa_3 = \kappa_4$, essentially the operator $-i(L^- + L^+)$. $L^-\Psi_L^{(n)}$ vanishes.

L^+ acting on any state vector only increases the number of photons (of any description). Thus $f_{\alpha\beta}$ acting on $\Psi_L^{(n)}$ cannot produce $\Psi_L^{(0)}$.

Inserting (33) into (34) and using (37) and (31 b) we see that

$$\langle f_{\alpha\beta}\rangle_L = (\Psi_L^{(0)*}\eta f_{\alpha\beta}\Psi_L^{(0)}) = (\Psi_L^{(0)*}f_{\alpha\beta}\Psi_L^{(0)}). \tag{38}$$

In other words: The expectation value of the field strengths and therefore of any gauge-invariant operator is independent of the admixtures $\Psi_L^{(n)}$ ($n \neq 0$), and can be formed with the wave function $\Psi_L^{(0)}$ only, which contains a zero number of longitudinal and scalar photons. For the expectation values of gauge-invariant quantities the admixtures are irrelevant. η is also equal to unity in an expression involving $\Psi_L^{(0)}$ only.

Consider next the expectation value of the potentials

$$\langle A_\alpha\rangle_L = (\Psi_L^* \eta A_\alpha \Psi_L). \tag{39}$$

Let us split up $\langle A_\alpha\rangle_L$ into the part arising from $\Psi_L^{(0)}$ and the rest

$$\langle A_\alpha\rangle_L = \langle A_\alpha\rangle^0 + \langle A_\alpha\rangle'. \tag{40}$$

Since the coordinates $\mathbf{r}$, t are parameters not operators, any derivative of the expectation value with respect to the coordinates equals the expectation value of the derivative. Then (38) can be stated as

$$\langle f_{\alpha\beta}\rangle = \frac{\partial\langle A_\beta\rangle^0}{\partial x_\alpha} - \frac{\partial\langle A_\alpha\rangle^0}{\partial x_\beta} \quad \text{or} \quad \frac{\partial\langle A_\beta\rangle'}{\partial x_\alpha} - \frac{\partial\langle A_\alpha\rangle'}{\partial x_\beta} = 0. \tag{41}$$

From (41) it follows that $\langle A_\alpha\rangle'$ can be represented as the 4-gradient of a scalar

$$\langle A_\alpha\rangle' = \frac{\partial\chi}{\partial x_\alpha},$$

and thus

$$\langle A_\alpha\rangle_L = \langle A_\alpha\rangle^0 + \frac{\partial\chi}{\partial x_\alpha}. \tag{42}$$

Of course χ, like A_α itself, satisfies $\Box\chi = 0$. Thus $\langle A_\alpha\rangle_L$ differs from $\langle A_\alpha\rangle^0$ only by a change of gauge. An admixture of $\Psi_L^{(n)}$ to $\Psi_L^{(0)}$ means a gauge transformation of the expectation values of the potentials.

The admixtures $\Psi_L^{(n)}$, to a state vector, are as essential a part of the formalism as are the potentials themselves and their various gauges. The role of the longitudinal and scalar fields is still more important in the actual computation of radiation phenomena, when charges are present. They then replace the physical effects of the Coulomb interaction as was shown in § 6, and will occur in the so-called virtual or intermediate states. A simple example (interaction of two electrons) will be given in § 24, and further examples in Chapter VI. It will be seen that the metric operator nowhere occurs explicitly and that all that is needed are the matrix elements (19). They are formally identical for all four components. For a logically consistent formulation of the theory, however, the use of the indefinite metric is essential.

The Lorentz invariance of the above method of quantization and of the Lorentz condition can easily be proved.† We can refrain from a detailed exposition, because the covariance of the theory will be evident at a later stage (see §§ 13, 28, etc.), where η will not occur explicitly and the results are formulated in a covariant way.

Thus the laws of quantum electrodynamics for a pure field in Lorentz gauge can be summarized as:

$$\Box A_\alpha = 0, \qquad \frac{\partial A_\alpha^-}{\partial x_\alpha}\Psi_L = 0,$$
$$[q_{\alpha\lambda}\, q^*_{\beta\lambda'}] = \frac{\hbar}{2\nu_\lambda}\delta_{\alpha\beta}\,\delta_{\lambda\lambda'} \tag{43}$$

for all four components, and the statement that A_α (including A_4) is hermitian. The generalization for the case where charges are present will be treated in § 13 and appendix 3.

5. *Four-dimensional Fourier expansion. Commutation relations of* A_α. This and the following subsections are concerned with further purely formal developments which will be helpful for the computational technique.

The four components of the potential A_α are now all treated on an equal footing. They satisfy the same hermiticity conditions (A_α hermitian) and the interpretation is the same (q_α absorption operator). This makes it possible to formulate all relations satisfied by A_α in a covariant manner. In particular it will be useful to express A_α by a 4-dimensional Fourier expansion.

The expansion (2) can be written as a Fourier integral in $\boldsymbol{\kappa}$ space instead of a sum over the partial waves $\boldsymbol{\kappa}_\lambda$. The number of waves in the element $d^3\kappa$ is $d^3\kappa/(2\pi)^3$. Thus:‡

$$A_\alpha = \sqrt{(4\pi c^2)} \int \frac{d^3\kappa}{(2\pi)^{\frac{3}{2}}} \{q_\alpha(\boldsymbol{\kappa})e^{i\kappa_\mu x_\mu} + q^*_\alpha(\boldsymbol{\kappa})e^{-i\kappa_\mu x_\mu}\}. \tag{44}$$

For discrete waves, q_λ, q^*_λ commute when $\lambda \neq \lambda'$. With $\boldsymbol{\kappa}$ now continuous, we must replace the factor $\delta_{\lambda\lambda'}$ by $\delta(\boldsymbol{\kappa}-\boldsymbol{\kappa}')$ in the commutation relations, showing that $q(\boldsymbol{\kappa})$ and $q^*(\boldsymbol{\kappa}')$ commute except when $\boldsymbol{\kappa} = \boldsymbol{\kappa}'$. The numerical factor must be such that when the commutator is

† See K. Bleuler, loc. cit. η replaces the relativistic i which is invariant.

‡ It is convenient to define the $q(\boldsymbol{\kappa})$ with a factor $(2\pi)^{\frac{3}{2}}$ in the denominator rather than $(2\pi)^3$. This avoids the occurrence of factors $(2\pi)^3$ in the commutation relations.

integrated over $\boldsymbol{\kappa}'$ $\left(\int d^3\kappa'/(2\pi)^3\right)$ the result is the same as when (7) is summed over the partial waves. We get

$$[q_\alpha(\boldsymbol{\kappa})q_\beta^*(\boldsymbol{\kappa}')] = \frac{\hbar}{2c\kappa}\delta_{\alpha\beta}\,\delta(\boldsymbol{\kappa}-\boldsymbol{\kappa}') \quad (\kappa \equiv |\boldsymbol{\kappa}| = \nu/c). \tag{45}$$

We can now go over to a 4-dimensional Fourier representation by allowing the integration to be carried out independently over all *values* of the 4-vector κ_μ. The Fourier components must then be such that they are different from zero only when $\kappa_0 = \pm|\boldsymbol{\kappa}|$. The two terms in (44) differ by the sign of $\kappa_0 x_0$ in the exponent. The sign of $\boldsymbol{\kappa}$ is not significant because an integration over $\boldsymbol{\kappa}$ including the directions is performed. We can thus also write (44) as:

$$A_\alpha(\mathbf{r},t) = \sqrt{(4\pi c^2)}\int \frac{d^4\kappa}{(2\pi)^{\frac{3}{2}}} A_\alpha(\kappa_\mu)e^{i\kappa_\mu x_\mu} \tag{46}$$

where $A_\alpha(\kappa_\mu)$ is a function of the four components κ_μ. By comparison with (44) the Fourier components $A_\alpha(\kappa_\mu)$ can now be identified with the $q_\alpha(\boldsymbol{\kappa})$, $q_\alpha^*(\boldsymbol{\kappa})$ as follows:

$$A_\alpha(\kappa_\mu) = q_\alpha(\boldsymbol{\kappa})\delta(\kappa_0-\kappa)+q_\alpha^*(-\boldsymbol{\kappa})\delta(\kappa_0+\kappa). \tag{47}$$

Inserting (47) into (46) the integration over κ_0 can be performed and leads back to (44). $A_\alpha(\kappa_\mu)$ and $A_\alpha(\kappa'_\mu)$ do not, in general, commute. It will be convenient to form $[A_\alpha(\kappa_\mu)A_\beta(-\kappa'_\mu)]$ (reversing the sign of the *whole* 4-vector κ_μ in the second factor), because it is really these quantities which do not commute when $\kappa_\mu = \kappa'_\mu$.

From (45) and (47) we obtain:

$$\begin{aligned}
&[A_\alpha(\kappa_\mu)A_\beta(-\kappa'_\mu)] \\
&\quad = [q_\alpha(\boldsymbol{\kappa})q_\beta^*(\boldsymbol{\kappa}')]\delta(\kappa_0-\kappa)\delta(\kappa'-\kappa'_0)+[q_\alpha^*(-\boldsymbol{\kappa})q_\beta(-\boldsymbol{\kappa}')]\delta(\kappa+\kappa_0)\delta(\kappa'+\kappa'_0) \\
&\quad = \frac{\hbar}{c}\delta_{\alpha\beta}\,\delta^4(\kappa_\mu-\kappa'_\mu)\frac{\delta(\kappa-\kappa_0)-\delta(\kappa+\kappa_0)}{2\kappa} \\
&\quad = \frac{\hbar}{c}\delta_{\alpha\beta}\,\delta^4(\kappa_\mu-\kappa'_\mu)\epsilon(\kappa_0)\delta(\kappa_\mu^2),
\end{aligned} \tag{48}$$

using § 8 eqs. (6′), (24) (and $\kappa = \kappa'$).

In § 9 the commutation relations of the field strengths were given in coordinate space. It is convenient to have the corresponding relations for A_α also. These necessarily depend on the gauge used but take a very simple form in Lorentz gauge. (In Coulomb gauge they are given in appendix 2.) The commutator of $A_\alpha(\mathbf{r}_2, t_2)$ and $A_\beta(\mathbf{r}_1, t_1)$ follows from (46) and (48):

$$[A_\alpha(\mathbf{r}_2, t_2)A_\beta(\mathbf{r}_1, t_1)] = \frac{4\pi c^2}{(2\pi)^3}\int d^4\kappa \int d^4\kappa'[A_\alpha(\kappa_\mu)A_\beta(-\kappa'_\mu)]e^{i\kappa_\mu x_{\mu 2}-\kappa'_\mu x_{\mu 1}}$$

$$= 4\pi\hbar c\,\delta_{\alpha\beta}\int \frac{d^4\kappa}{(2\pi)^3}e^{i\kappa_\mu(x_{\mu 2}-x_{\mu 1})}\epsilon(\kappa_0)\delta(\kappa_\mu^2).$$

We have denoted the Fourier components for A_β by $A_\beta(-\kappa'_\mu)$, which is possible because the integration is over the whole 4-dimensional κ'_μ-space. The integral on the right is the Fourier decomposition of the Δ-function, § 8 eq. (25a). Thus:

$$[A_\alpha(\mathbf{r}_2, t_2)A_\beta(\mathbf{r}_1, t_1)] = -4\pi i\hbar c\,\delta_{\alpha\beta}\Delta(\mathbf{r}_2-\mathbf{r}_1, t_2-t_1). \tag{49}$$

This is equivalent to the commutation relations (43). The covariance of (49) is evident. Putting $t_1 = t_2$ we also obtain from § 8 eq. (28)

$$[A_\alpha(\mathbf{r}_2, t)A_\beta(\mathbf{r}_1, t)] = 0,$$

$$\left[\frac{\partial}{\partial t}A_\alpha(\mathbf{r}_2, t),\, A_\beta(\mathbf{r}_1, t)\right] = -4\pi i\hbar c^2\delta_{\alpha\beta}\,\delta(\mathbf{r}_2-\mathbf{r}_1). \tag{49'}$$

The properties of the Δ-function were discussed in § 8. We see that (49) is *different from zero on the light cone* only when $|\mathbf{r}_2-\mathbf{r}_1| = c|t_2-t_1|$, i.e. if the *two points* $\mathbf{r}_2$, $\mathbf{r}_1$ *can be connected by a light signal* either travelling from $\mathbf{r}_2$ to $\mathbf{r}_1$ or vice versa. The significance of this fact for the measurement of field strengths was explained in § 9. By differentiating (49) with respect to $\mathbf{r}_2$, $\mathbf{r}_1$, t_2, t_1 we obtain the commutation relations of any two components of **E**, **H** given in § 9 eq. (8a–d).†

The commutation relations (49) are only valid if Lorentz gauge is used for A_α. Although this condition has not been used explicitly in the derivation of (49), the expansion (46) presupposes the validity of $\Box A_\alpha = 0$. This can also be seen directly from (49) since $\Box\Delta = 0$. $\Box A_\alpha = 0$ is, however, only equivalent to the Maxwell equations if the Lorentz condition is satisfied in the sense of subsection 3. On the other hand, it is interesting that the commutation relations hold whichever gauge is used within the framework of the Lorentz gauge. A restricted gauge transformation, leaving $\partial A_\alpha/\partial x_\alpha = 0$ unaltered, exhibits itself not by a change of the commutation relations but by the admixtures to the state vector as discussed in subsection 4.

† For this purpose note that $(\partial^2/c^2\partial t^2)\Delta = \nabla^2\Delta$ and $\partial\Delta/\partial x_1 = -\partial\Delta/\partial x_2$.

6. *Photon vacuum, expectation values.* It is, for some purposes, useful to separate the emission and absorption operators. This can be done in coordinate representation also. We define

$$A_\alpha^+ = \sqrt{(4\pi c^2)} \int \frac{d^4\kappa}{(2\pi)^{\frac{3}{2}}} A_\alpha^+(\kappa_\mu) e^{i\kappa_\mu x_\mu}, \qquad A_\alpha^+(\kappa_\mu) = q_\alpha^*(-\boldsymbol{\kappa})\delta(\kappa_0+\kappa);$$

$$A_\alpha^- = \sqrt{(4\pi c^2)} \int \frac{d^4\kappa}{(2\pi)^{\frac{3}{2}}} A_\alpha^-(\kappa_\mu) e^{i\kappa_\mu x_\mu}, \qquad A_\alpha^-(\kappa_\mu) = q_\alpha(\boldsymbol{\kappa})\delta(\kappa_0-\kappa). \tag{50}$$

Thus A^+ is the emission, A^- the absorption operator.† Going over to the complex conjugate and changing the sign of κ_μ under the integral we see that A_α^+ is the adjoint of A_α^-.

It follows immediately that

$$[A_\alpha^+ A_\beta^+] = [A_\alpha^- A_\beta^-] = 0 \tag{51}$$

because all q and q^* commute between themselves.

On the other hand, we get as in (48),

$$[A_\alpha^+(-\kappa'_\mu)A_\beta^-(\kappa_\mu)] = -\frac{\hbar}{2c\kappa}\,\delta_{\alpha\beta}\,\delta^4(\kappa_\mu-\kappa'_\mu)\delta(\kappa_0-\kappa)$$

$$= -\frac{\hbar}{c}\,\delta_{\alpha\beta}\,\delta^4(\kappa_\mu-\kappa'_\mu)\frac{1+\epsilon(\kappa_0)}{2}\,\delta(\kappa_\mu^2) \tag{52}$$

(see § 8 eqs. (23), (24)). In coordinate representation we find from (50), (52)

$$[A_\alpha^+(\mathbf{r}_2,t_2)A_\beta^-(\mathbf{r}_1,t_1)] = -4\pi\hbar c\,\delta_{\alpha\beta}\int \frac{d^4\kappa}{(2\pi)^3}\,\frac{1+\epsilon(\kappa_0)}{2}\,\delta(\kappa_\mu^2)e^{i\kappa_\mu(x_{\mu 1}-x_{\mu 2})}$$

$$= -4\pi\hbar c\,\delta_{\alpha\beta}\,\tfrac{1}{2}\{\Delta_1(\mathbf{r}_2-\mathbf{r}_1,t_2-t_1)+i\Delta(\mathbf{r}_2-\mathbf{r}_1,t_2-t_1)\}. \tag{52'}$$

The right-hand side is a linear combination of the Δ and Δ_1 functions (§ 8 eq. (25)). (Note that Δ_1 and Δ depend on $|\mathbf{r}_2-\mathbf{r}_1|$ and are even and odd functions of t respectively.)

$A^+(\mathbf{r},t)$, $A^-(\mathbf{r},t)$ may be interpreted as the emission and absorption operators, acting at one particular point $\mathbf{r},t$ of space time.

The operators q, q^*, A^+, A^-, etc., act on the state vector Ψ of the field, which may be regarded as depending on the number of photons of each description ($\alpha = 1,\ldots, 4$).

A particularly important state is Ψ_{0000} with the numbers of all four types of photons zero. This state will be called the *photon vacuum*. Some explanation is needed as to what is meant by the absence of longitudinal

† In the recent literature the notation A^+, A^- is usually the other way round. We have reversed the notation because + is certainly more suggestive of emission, and − of absorption.

and scalar photons. We have seen (subsection 4) that states differing by the number of longitudinal and scalar photons (provided that they belong to the Lorentz set Ψ_L) are physically equivalent and differ only by the gauge of the expectation value of A. Thus the states

$$\Psi_L^{(0)} + \sum_{n\neq 0} c_n \Psi_L^{(n)} \text{ (eq. (27))},$$

with the number of transverse photons $= 0$, might equally well be called the photon vacuum. They can be transformed into Ψ_{0000} by a gauge transformation. The situation is different when charges are present. We can then still consider a state with no transverse photons, but owing to the static field of the charges no state will exist with neither longitudinal nor scalar photons. Nevertheless, the vacuum state as defined by the absence of longitudinal and scalar photons has some significance in many cases, namely if the interaction between the particles as well as that between the particles and the radiation field can be treated as a perturbation. This will be the case for collisions between free particles. We return to these questions in § 13.

Let the photon-vacuum be defined by the absence of photons of any kind. It is clear then that the absorption operator acting on $\Psi_0 \equiv \Psi_{0000}$ must produce a zero result†

$$A_\alpha^- \Psi_0 = 0. \tag{53}$$

Since the adjoint of A^- is A^+, we have $\Psi_0^* A_\alpha^+ = 0$, and since η commutes with A_i^+ and anticommutes with A_4^+ also

$$\Psi_0^* \eta A_\alpha^+ = 0. \tag{53'}$$

Since Ψ_0 belongs to the Lorentz set, it is evident that (53), (53′) is compatible with the Lorentz condition (28).

We now consider the expectation values of certain field quantities in the particular state Ψ_0, i.e. for the photon vacuum. These expectation values will be denoted by an index 0, $\langle Q\rangle_0 = (\Psi_0^* \eta Q \Psi_0)$. It follows from (53), (53′) that

$$\langle A_\alpha^- \rangle_0 = \langle A_\alpha^+ \rangle_0 = \langle A_\alpha \rangle_0 = 0. \tag{54}$$

The expectation values of products A^+A^+ and A^-A^- are also zero. The same is true for a product A^+A^-, when A^- stands on the right. The only product of two factors A which does not vanish is A^-A^+.

† This simple form of the vacuum condition is made possible only by the use of the indefinite metric for the quantization of A_4. Otherwise A_4^- must be interpreted as emission operator, and $A_4^- \Psi \neq 0$, for any state. Compare footnote ‡ on p. 90.

To evaluate $\langle A^- A^+ \rangle_0$ we observe that $\langle A^+ A^- \rangle_0 = 0$, and hence

$$\langle A^- A^+ \rangle_0 = \langle [A^- A^+] \rangle_0.$$

The commutator $[A_\alpha^- A_\beta^+]$ is a c-number, not an operator.

Since the norm $(\Psi_0^* \, \eta \Psi_0) = 1$, we find from (52)

$$\begin{aligned} \langle A_\beta^- (\kappa_\mu) A_\alpha^+ (-\kappa_\mu') \rangle_0 &= \frac{\hbar}{c} \delta_{\alpha\beta} \, \delta^4(\kappa_\mu - \kappa_\mu') \frac{1+\epsilon(\kappa_0)}{2} \delta(\kappa_\mu^2) \\ &= \langle A_\beta(\kappa_\mu) A_\alpha(-\kappa_\mu') \rangle_0. \end{aligned} \tag{55}$$

The last equation is true because

$$\langle A^+ A^+ \rangle_0 = \langle A^+ A^- \rangle_0 = 0.$$

In coordinate space we have, from (52′), after interchanging $t_1 \rightleftarrows t_2$,

$$\langle A_\beta^- (\mathbf{r}_2, t_2) A^+(\mathbf{r}_1, t_1) \rangle_0 = 4\pi\hbar c \, \delta_{\alpha\beta} \, \tfrac{1}{2}(\Delta_1 - i\Delta)(\mathbf{r}_2 - \mathbf{r}_1, t_2 - t_1). \tag{56}$$

In the same way one could obtain the expectation values of products with more than two factors: By commuting them one can move all A^- to the right and all A^+ to the left, producing zero results, and one is left with the commutators themselves which are c-numbers.

III

THE ELECTRON FIELD AND ITS INTERACTION WITH RADIATION

11. The relativistic wave equation of the electron

1. *Dirac's equation.* We assume that the reader is acquainted with the elementary theory of the Dirac equation. This section is intended merely to summarize the facts that are needed in this book and to fix the notation.

The relativistic wave equation for an electron, interacting with an electromagnetic field, is

$$i\hbar\dot{\psi} = H\psi = \{(\boldsymbol{\alpha}, \mathbf{p}-e\mathbf{A})+\beta\mu+e\phi\}\psi, \tag{1}$$

$$\mathbf{p} = c\times\text{momentum}, \qquad \mu = mc^2, \qquad p_x = \frac{\hbar c}{i}\frac{\partial}{\partial x}.$$

$\boldsymbol{\alpha}$ and β are 4-row, 4-column matrices satisfying the relations:

$$\alpha_x\beta+\beta\alpha_x = 0, \qquad \alpha_x\alpha_y+\alpha_y\alpha_x = 0,$$
$$\alpha_x^2 = \beta^2 = 0, \text{ etc.} \tag{2}$$

$\boldsymbol{\alpha}$, β can, for instance, be represented in the form

$$\boldsymbol{\alpha} = \begin{pmatrix} 0 & \boldsymbol{\sigma} \\ \boldsymbol{\sigma} & 0 \end{pmatrix}, \qquad \beta = \begin{pmatrix} 1 & 0 \\ 0 & -1 \end{pmatrix} \tag{2'}$$

where 1, 0, σ are still 2-row, 2-column matrices, and $\boldsymbol{\sigma}$ are the Pauli spin matrices, satisfying

$$\sigma_x\sigma_y = -\sigma_y\sigma_x = i\sigma_z, \qquad \sigma_x^2 = 1. \tag{3}$$

A representation is

$$\sigma_x = \begin{pmatrix} 0 & 1 \\ 1 & 0 \end{pmatrix}, \qquad \sigma_y = \begin{pmatrix} 0 & -i \\ i & 0 \end{pmatrix}, \qquad \sigma_z = \begin{pmatrix} 1 & 0 \\ 0 & -1 \end{pmatrix}. \tag{3'}$$

The mechanical spin is $\quad \dfrac{\hbar}{2}\boldsymbol{\sigma} \equiv \dfrac{\hbar}{2}\begin{pmatrix} \boldsymbol{\sigma} & 0 \\ 0 & \boldsymbol{\sigma} \end{pmatrix}.$

Accordingly, ψ consists of four components, ψ_ρ, $\rho = 1,...,4$. ρ must be regarded as a discontinuous variable, on which ψ depends (not as an index denoting an eigenstate). ψ_ρ can be regarded as a column matrix, and a product $\alpha_x\psi$ can be regarded as a matrix product

$$(\alpha_x\psi)_\rho \equiv \sum_{\rho'} \alpha_{x\rho\rho'}\psi_{\rho'},$$

the result being again a column matrix,

There is an equation adjoint to (1). It is obtained by going over to the complex conjugate. If ψ^* is the complex conjugate to ψ, ψ^* must be regarded as a row matrix with operators $\boldsymbol{\alpha}$, etc., standing behind ψ^*. For instance,

$$(\psi^*\alpha_x)_\rho = \sum_{\rho'} \psi^*_{\rho'}\alpha_{x\rho'\rho}.$$

In fact ψ^* is the matrix adjoint to ψ.

To restore the symmetry in the wave equation it is convenient to introduce instead of ψ^*

$$\psi^\dagger = i\psi^*\beta. \tag{4}$$

The notation $\psi^\dagger$ is usual, although $\psi^\dagger$ is not the adjoint to ψ. Then the adjoint wave equation takes the form

$$-i\hbar\dot{\psi}^\dagger = \psi^\dagger\{(\boldsymbol{\alpha}, \mathbf{p}_{\text{op}}+e\mathbf{A})+\beta\mu+e\phi\}. \tag{5}$$

Here $\mathbf{p}_{\text{op}}$ is understood to act backwards on $\psi^\dagger$,

$$\psi^\dagger p_{x\text{op}} = \frac{\hbar c}{i}\frac{\partial\psi^\dagger}{\partial x},$$

which we have indicated by the index *op*, but it is convenient to write $\mathbf{p}_{\text{op}}$ also to the right of $\psi^\dagger$.

To exhibit the relativistic covariance of (1), (5) one can rewrite the equations by introducing the matrices γ_μ

$$\gamma_4 = \beta, \qquad \gamma_i = i\alpha_i\beta = -i\beta\alpha_i \quad (i = 1, 2, 3) \tag{6}$$

with

$$\gamma_\mu\gamma_\nu+\gamma_\nu\gamma_\mu = 2\delta_{\mu\nu}. \tag{6'}$$

Introducing further

$$p_\mu = \frac{\hbar c}{i}\frac{\partial}{\partial x_\mu} \qquad \left(p_4 = \frac{\hbar c}{i}\frac{\partial}{\partial x_4} = -\hbar\frac{\partial}{\partial t}\right)$$

which is the energy-momentum 4-vector, and $A_\mu = \mathbf{A}, i\phi$, (1) and (5) assume the form

$$\begin{aligned}\gamma_\mu(p_\mu-eA_\mu)\psi &= i\mu\psi,\\ \psi^\dagger\gamma_\mu(p_{\mu\text{op}}+eA_\mu) &= -i\mu\psi^\dagger.\end{aligned} \tag{7}$$

The γ_μ can be regarded, in a certain sense, as a matrix 4-vector. At any rate the quantities

$$i_\mu = ec(\psi^\dagger\gamma_\mu\psi) \tag{8}$$

or

$$\begin{aligned}i_k &= +ec(\psi^\dagger\gamma_k\psi) = ec(\psi^*\alpha_k\psi),\\ \rho &= \frac{1}{ic}i_4 = -ie(\psi^\dagger\beta\psi) = +e(\psi^*\psi)\end{aligned} \tag{8'}$$

form a 4-vector. $e = -|e|$ is the electronic charge. Moreover, we can see immediately from (7) that this 4-vector is divergence-free:

$$\frac{\partial i_\mu}{\partial x_\mu} = 0. \tag{9}$$

The physical interpretation of this 4-vector is that of the density of current and charge. A comparison with the classical expression, § 2 eq. (14), suggests that the operator $\boldsymbol{\alpha}$ has taken over the role of the particle velocity. This connexion is also corroborated by a comparison of the operator H with the classical Hamiltonian. The latter is, by § 6 eq. (22),

$$H = e\phi + [\mu^2 + (\mathbf{p} - e\mathbf{A})^2]^{\frac{1}{2}} = \frac{1}{c}(\mathbf{v}, \mathbf{p} - e\mathbf{A}) + \mu\sqrt{(1 - v^2/c^2)} + e\phi$$

where
$$\mathbf{v}/c = \frac{\mathbf{p} - e\mathbf{A}}{[\mu^2 + (\mathbf{p} - e\mathbf{A})^2]^{\frac{1}{2}}}$$

is the relativistic particle velocity. Comparing this with (1) we see that $\mathbf{v} \to \boldsymbol{\alpha} c$, $\sqrt{(1 - v^2/c^2)} \to \beta$. α_x has eigenvalues ± 1, and thus $\mathbf{v}_x$ appears to have eigenvalues $\pm c$. The explanation for this fact is that the electron carries out a fast irregular motion‡ ('Zitterbewegung')—which is responsible for the spin—whereas the mean velocity is given by the momentum $c\mathbf{p}/\mu$.

For a typical plane wave solution of a free electron we can put

$$\begin{gathered} \psi = u e^{i(\mathbf{pr})/\hbar c - iEt/\hbar}, \\ \psi^\dagger = u^\dagger e^{-i(\mathbf{pr})/\hbar c + iEt/\hbar} \quad (E^2 - \mathbf{p}^2 = \mu^2), \end{gathered} \tag{10}$$

where $\mathbf{p}$, E (or p_μ) are now numbers and u is a 4-component quantity, called a *spinor*, depending on p_μ but independent of $\mathbf{r}$, t. Then

$$\psi^\dagger p_{\mu op} = -p_\mu \psi^\dagger,$$

and (7) becomes:

$$\begin{gathered} \gamma_\mu p_\mu u = +i\mu u \qquad \text{or} \qquad Eu = [(\boldsymbol{\alpha}\mathbf{p}) + \beta\mu]u \\ u^\dagger \gamma_\mu p_\mu = +i\mu u^\dagger \qquad\qquad Eu^* = u^*[(\boldsymbol{\alpha}\mathbf{p}) + \beta\mu] \\ (p_\mu^2 = -\mu^2). \end{gathered} \tag{11}$$

Since u has four components there will also, for any given value of p_μ, be four different eigensolutions. These correspond (i) to two different directions of the spin and (ii) to the fact that in addition to solutions with positive energy those with *negative energy* occur also. The physical significance of the latter, the most important feature of the theory, will

‡ E. Schrödinger, *Sitz. Preuss. Akad.* XXIV, 1930.

be explained below. To see the type of solutions that exist, we consider the case of a free electron moving in z-direction, $p_x = p_y = 0$. Using the representation (2′), (3′) we obtain the following four solutions for u:

$$\begin{array}{c|cccc} & \multicolumn{2}{c}{E>0} & \multicolumn{2}{c}{E<0} \\ & \uparrow & \downarrow & \uparrow & \downarrow \\ \hline u_1 & 1 & 0 & -\dfrac{p_z}{\mu+|E|} & 0 \\ u_2 & 0 & 1 & 0 & +\dfrac{p_z}{\mu+|E|} \\ u_3 & +\dfrac{p_z}{\mu+E} & 0 & 1 & 0 \\ u_4 & 0 & -\dfrac{p_z}{\mu+E} & 0 & 1 \end{array} \left.\vphantom{\begin{array}{c}1\\1\\1\\1\\1\\1\end{array}}\right\} \left[1+\frac{p_z^2}{(\mu+|E|)^2}\right]^{-\frac{1}{2}} \tag{12}$$

$$(|E| = (\mu^2+p_z^2)^{\frac{1}{2}}).$$

To show the physical significance of these four solutions we apply the operators (i) σ_z and (ii) H on (12). We find immediately:

$$\sigma_z u = \pm u \quad \text{for} \uparrow\downarrow \text{ respectively,}$$

and $$Hu = \pm|E|u \quad \text{for } E \gtrless 0 \text{ respectively.}$$

Thus, for the solutions marked $\uparrow$, $\downarrow$ the spin is in the $\pm z$-direction respectively. (σ_x, σ_y are, of course, not sharp.) It is interesting that there are no solutions for which any other $\boldsymbol{\sigma}$-component is sharp except the one in the direction of motion. Similarly, the energy has $\pm$ values for the solutions marked $E \gtrless 0$. (12) is normalized to a unit volume.

The number of eigensolutions for a free electron in a volume L^3 with positive energy, given spin direction and momentum in the interval d^3p, is

$$d^3p\, L^3/(2\pi\hbar c)^3 = p^2\, dp d\Omega\, L^3/(2\pi\hbar c)^3 = pE\, dE d\Omega\, L^3/(2\pi\hbar c)^3.$$

The eigensolutions $\psi_{\rho n}(\mathbf{r})$ satisfy the usual orthogonality and completeness relations. Since ρ is a coordinate, the integration over the coordinates implies a summation over ρ also. Furthermore, the index n which denotes the eigensolution also includes a statement of whether $E \gtrless 0$ and of the spin direction.

Thus the orthogonality relation is:

$$\sum_\rho \int \psi^*_{\rho n}(\mathbf{r})\psi_{\rho n'}(\mathbf{r})\, d\tau \equiv \int (\psi^*_n \psi_{n'})\, d\tau = \delta_{nn'}. \tag{13}$$

$n \neq n'$ may also mean that either the spin directions or the signs of E are different. The completeness relation is

$$\sum_n \psi^*_{\rho n}(\mathbf{r})\psi_{\rho' n}(\mathbf{r}') = \delta_{\rho\rho'}\delta(\mathbf{r}-\mathbf{r}') \tag{13'}$$

(see § 8 eq. (3). For the discrete variable ρ the factor $\delta_{\rho\rho'}$ appears.)

Unless clarity demands it, we shall in future omit the index ρ and agree that a summation over ρ is to be carried out in products $(\psi^*...\psi)$, or $(\psi^\dagger...\psi)$ with some α, β operators between them.

The generalization to a system of several particles is quite trivial. Each particle has its own operators $\boldsymbol{\alpha}_k$, β_k, $\mathbf{p}_k$, etc. ψ depends on the variables of each particle, i.e. $\mathbf{r}_k$ and ρ_k which also have an index k. $\boldsymbol{\alpha}_k$ acts, of course, only on ρ_k but not on ρ_i. If there are n particles, $\boldsymbol{\alpha}_k$ can be regarded as a direct product of factors all of which are unity except the kth factor which is the $\boldsymbol{\alpha}$ represented by (2′). ψ has 4^n components altogether. ψ is antisymmetric in all particles.

2. *Spin summations.* We shall frequently find it necessary to carry out sums of free particle solutions over either the two spin directions as well as the two signs of energy or the two spin directions only with a given sign of E. These can be evaluated without explicit reference to the representations of α, β, or γ_μ. Let Q_1, Q_2 be two operators composed of matrices α, β. We denote the sum over all *four* states with given momentum $\mathbf{p}$ by $\sum^p$. Then the completeness theorem (13′) reduces to

$$\textstyle\sum^p u_\rho^* u_{\rho'} = \delta_{\rho\rho'}.\ddagger \tag{13″}$$

Hence, for example, we get immediately

$$\textstyle\sum^{p'} (u^*Q_1 u')(u'^*Q_2 u'') = (u^*Q_1 Q_2 u''), \tag{14a}$$

$$\textstyle\sum^p (u^*Qu) = \operatorname{Sp} Q \qquad \Big(\operatorname{Sp} Q \equiv \sum\limits_\rho Q_{\rho\rho}\Big), \tag{14b}$$

where $\operatorname{Sp} Q$ is the spur of the operator Q. The spurs can be evaluated without difficulty. It can be seen immediately from (2)–(3′) and (6′) that the spur of all products with an odd number of factors of either α_x or α_y, or β, or a particular γ_μ, vanishes.

Since $\operatorname{Sp} 1 = 4$ we have, for example,

$$\operatorname{Sp} \alpha_x \beta \alpha_x \beta = \operatorname{Sp} \alpha_x \alpha_y \alpha_x \alpha_y = \operatorname{Sp} \gamma_1 \gamma_2 \gamma_1 \gamma_2 = -4, \text{ etc.} \tag{15}$$

A cyclic permutation of the factors in an α, β, or γ product does not change the spur, nor is this the case if the order is reversed. So for example ($\mathbf{a}$, $\mathbf{b}$ fixed vectors)

$$\operatorname{Sp}(\boldsymbol{\alpha}\mathbf{a})Q(\boldsymbol{\alpha}\mathbf{a}) = \mathbf{a}^2 \operatorname{Sp} Q, \tag{15′}$$

$$\operatorname{Sp}(\boldsymbol{\alpha}\mathbf{a})(\boldsymbol{\alpha}\mathbf{b}) = \operatorname{Sp}(\boldsymbol{\alpha}\mathbf{b})(\boldsymbol{\alpha}\mathbf{a}) = +4(\mathbf{ab}). \tag{15″}$$

By S^p we denote a summation over the two spin directions only, keeping

‡ The four solutions u with a fixed value of $\mathbf{p}$ themselves form a complete orthogonal system. The sum over all states in (13′) is the sum Σ^p (giving $\delta_{\rho\rho'}$) and the sum over all values of $\mathbf{p}$ which involves the exp $i(\mathbf{pr})$ only gives $\delta(\mathbf{r}-\mathbf{r}')$.

the sign of E fixed. We add an index $+$, $-$ when $E \gtrless 0$, respectively. We then use the fact that

$$\begin{aligned} \frac{H+|E|}{2|E|}u &\equiv \frac{(\boldsymbol{\alpha}\mathbf{p})+\beta\mu+|E|}{2|E|}u = \begin{cases} u & (E>0) \\ 0 & (E<0) \end{cases} \\ \frac{-H+|E|}{2|E|}u &\equiv \frac{-(\boldsymbol{\alpha}\mathbf{p})-\beta\mu+|E|}{2|E|}u = \begin{cases} 0 & (E>0) \\ u & (E<0). \end{cases} \end{aligned} \tag{16}$$

Consequently, a sum $\mathsf{S}^p_\pm ... u$ can be transformed into

$$\sum^p \frac{|E|\pm H}{2|E|}u.$$

So we have

$$\mathsf{S}^{p'}_\pm(u^*Q_1 u')(u'^*Q_2 u'') = \left(u^*Q_1\frac{|E'|\pm H'}{2|E'|}Q_2 u''\right), \tag{17 a}$$

$$\mathsf{S}^p_\pm(u^*Qu) = \operatorname{Sp} Q\frac{|E|\pm H}{2|E|}. \tag{17 b}$$

Since $i(E+\boldsymbol{\alpha}\mathbf{p}+\beta\mu)\beta = \gamma_\mu p_\mu + i\mu$, (17) can also be written relativistically (applying (16) for u^*)

$$\mathsf{S}^p_\pm(u^\dagger Qu) = \sum^p \left(u^* i\frac{|E|\pm H}{2|E|}\beta Qu\right) = \pm\operatorname{Sp}\frac{(\gamma_\mu p_\mu+i\mu)Q}{2|p_0|}, \tag{18 a}$$

$$\mathsf{S}^{p'}_\pm(u^\dagger Q_1 u')(u'^\dagger Q_2 u'') = \pm\left(u^\dagger Q_1\frac{\gamma_\mu p'_\mu+i\mu}{2|p_0|}Q_2 u\right) \tag{18 b}$$

(note that for $E<0$, $p_4 = ip_0 = -i|E|$).

3. *Transition to non-relativistic case.* If the energy of an eigenstate is close to μ (not necessarily for a free electron) a transition to the non-relativistic wave equation can be made. If $E>0$ we see from (12) that the components u_3, u_4 are small compared with u_1, u_2 and are of order $p/\mu \sim v/c$. This is also so in the general case ϕ, $\mathbf{A} \neq 0$, and when $\mathbf{p}$ is still an operator, provided that we use the representation (2′) for $\boldsymbol{\alpha}$, β. We can then split the wave equation in two. We denote the two 'large components' ψ_1, ψ_2 by ψ_{I} (a 2-row matrix) and the small components by ψ_{II}. Then, for an eigenvalue E, (1) with (2′) reads:

$$(E-\mu)\psi_{\mathrm{I}} = e\phi\psi_{\mathrm{I}} + (\boldsymbol{\sigma}, \mathbf{p}-e\mathbf{A})\psi_{\mathrm{II}}, \tag{19 a}$$

$$(E+\mu)\psi_{\mathrm{II}} = e\phi\psi_{\mathrm{II}} + (\boldsymbol{\sigma}, \mathbf{p}-e\mathbf{A})\psi_{\mathrm{I}}. \tag{19 b}$$

$e\phi$ will be of the order of $E-\mu$ and therefore $\ll \mu$. We can then express ψ_{II} by ψ_{I}, with the help of (19 b). Neglecting orders $\sim p^2/\mu^2$, $E+\mu$ can be replaced by 2μ and, to the same order $e\phi\psi_{\mathrm{II}}$ can be neglected, thus

$$\psi_{\mathrm{II}} = \frac{1}{2\mu}(\boldsymbol{\sigma}, \mathbf{p}-e\mathbf{A})\psi_{\mathrm{I}},$$
$$\psi^*_{\mathrm{II}} = -\frac{1}{2\mu}\psi^*_{\mathrm{I}}(\boldsymbol{\sigma}, \mathbf{p}_{op}+e\mathbf{A}). \tag{20}$$

For solutions with momentum $\mathbf{p}$ (c-number) we can also write

$$\psi^*_{\mathrm{II}} = +\frac{1}{2\mu}\psi^*_{\mathrm{I}}(\boldsymbol{\sigma}, \mathbf{p}-e\mathbf{A}).$$

Inserting (20) into (19 a) we obtain a wave equation for ψ_{I} only

$$(E-\mu)\psi_{\mathrm{I}} = \left\{e\phi+\frac{1}{2\mu}(\mathbf{p}-e\mathbf{A})^2-\frac{e\hbar c}{2\mu}(\boldsymbol{\sigma}\mathbf{H})\right\}\psi_{\mathrm{I}}, \tag{21}$$

using (3) and $$[p_y A_x] = \frac{\hbar c}{i}\frac{\partial A_x}{\partial y}.$$

The last term of (21) is the well-known spin term with the magnetic moment $e\hbar/2mc$. (21) is correct up to and including terms of order v/c.

The charge and current density can be decomposed in the same way. It will be useful to do this for a transition element $(u^*_{p'}\boldsymbol{\alpha}u_p)$ where p, p' are different momenta. We obtain from (19) in the same approximation, for a component α_n, for example (for $\mathbf{A} = \phi = 0$):

$$(u^*_{p'}u_p) = (u^*_{p'\mathrm{I}}u_{p\mathrm{I}}),$$

$$(u^*_{p'}\alpha_n u_p) = \frac{1}{2\mu}\{(u^*_{p'\mathrm{I}}\sigma_n u_{p\mathrm{II}})+(u^*_{p'\mathrm{II}}\sigma_n u_{p\mathrm{I}})\}$$
$$= \frac{1}{2\mu}(u^*_{p'\mathrm{I}}[\sigma_n(\boldsymbol{\sigma}\mathbf{p})+(\boldsymbol{\sigma}\mathbf{p}')\sigma_n]u_{p\mathrm{I}})$$
$$= \frac{1}{2\mu}(u^*_{p'\mathrm{I}}(p_n+p'_n)u_{p\mathrm{I}})+\frac{i}{2\mu}(u^*_{p'\mathrm{I}}(\boldsymbol{\sigma},[\mathbf{p}'-\mathbf{p},\mathbf{n}])u_{p\mathrm{I}}) \tag{22}$$

where $\mathbf{n}$ is a unit vector in some direction. The first part of (22) is just the average velocity $(p+p')/2\mu$ in the two states, the second part is the current responsible for the spin. For negative energy states the decompositions are quite analogous.

4. *The hole theory.* We have seen that the relativistic wave equation has solutions for which the *energy is negative.* These negative energy states arise from the fact that the energy of a free particle is given by a square root

$$E = \pm\sqrt{(p^2+\mu^2)}.$$

Thus, for any given value of p the sign of E can be either positive or negative. The existence and properties of these negative energy states, if taken at their face value, would give rise to serious difficulties. In the classical theory, however, no difficulty arises, because one can define the energy to be the positive square root, and then it does not change with time.

This is not possible in the quantum theory. An external field (if it varies sufficiently rapidly) can cause *transitions* from a state of positive energy to a state of negative energy. The latter, therefore, cannot be excluded from the theory and, if the theory is not entirely wrong, must have some physical meaning.

The interpretation of these negative energy states has become clear through the discovery of particles which have all the properties of electrons but have a *positive charge*. These *positive electrons* or *positrons* can be created—together with a negative electron—by rapidly varying electromagnetic fields (γ-rays of high energies, collision of two fast particles, etc.).† Also, in company with a negative electron, they can be annihilated, emitting their rest energy $2mc^2$ in the form of light. A correct quantum electrodynamics must therefore give an account of the existence of these positive electrons, and of their creation and annihilation.

The connexion between the theoretical negative energy states and the observed positive electrons is given by Dirac's '*hole theory*'. This theory was put forward in order to avoid the difficulties of the negative energy states, two years before the positive electrons were discovered.

To avoid the conclusion that an electron can jump from a positive energy state into a negative energy one, we make two fundamental assumptions.‡

(1) All negative energy states with energies ranging from $-mc^2$ to $-\infty$ (for free electrons) are filled up with electrons. No electron can therefore jump into one of these occupied states. It is essential here that the electrons satisfy the exclusion principle.

(2) The electrons filling up the negative energy states do not produce an external field and do not give any contribution to the total charge, energy, and momentum of the system. The 'zero point' for the charge, energy, and momentum is represented by that electron distribution in which all negative energy states and no positive energy states are occupied. This state will be called the *electron vacuum*.

† First discovery: C. D. Anderson, *Phys. Rev.* **41** (1932), 405; **43** (1933), 491; **44** (1933), 406. Compare also P. M. S. Blackett, J. Chadwick, G. P. S. Occhialini, *Proc. Roy. Soc.* **144** (1934), 235.

‡ P. A. M. Dirac, *Quantum Mechanics*, 3rd ed., Oxford 1947, chap. xi.

In spite of the second of these assumptions, it is, however, assumed that an external field can act on the electrons in the negative energy states.

Consider what happens when one of the electrons with energy $E = -|E|$ and momentum $\mathbf{p}$ is removed. By the second assumption the whole system then has a charge, energy, and momentum different from zero:

$$E_+ = -E = |E|, \qquad e_+ = -e, \qquad \mathbf{p}_+ = -\mathbf{p}, \tag{23}$$

where e represents the charge of an ordinary electron. The *hole* in the distribution of the negative energy electrons has therefore a *positive charge*, *positive energy*, and a momentum and spin opposite to that corresponding to the negative energy state. It therefore behaves like an ordinary particle with electronic mass but with a positive charge. Energy and momentum are now connected by the equation

$$E_+ = +\sqrt{(p_+^2+\mu^2)} \tag{24}$$

with the positive sign of the square root. Thus positive electrons are represented as *holes* in the distribution of electrons filling up the negative energy states.

The creation and annihilation of a 'pair' of a positive and negative electron is interpreted in the following way: Starting from a state where no particle and hole is present, an external field acting on the electrons in the negative energy states may cause a transition of one of these electrons with energy E and momentum $\mathbf{p}$ to a state with positive energy E' and momentum $\mathbf{p}'$. We then have a pair present with energies and momenta:

$$\begin{aligned} \mathbf{p}_+ &= -\mathbf{p}, \qquad & E_+ &= -E = |E|, \\ \mathbf{p}_- &= \mathbf{p}', \qquad & E_- &= E'. \end{aligned} \tag{25}$$

The energy required to cause this transition must be larger than $2mc^2$:

$$E'-E = E_+ + E_- \geqslant 2mc^2. \tag{26}$$

On the other hand, if a pair is present initially, the negative electron can jump into the hole representing the positive electron and the pair is annihilated. By this process the energy (26) is given out as radiation, etc. In the creation and annihilation of pairs the *total number of particles* is *not conserved* whereas the *total charge* is *constant*. Processes in which negative energy electrons (or 'vacuum electrons') are taking part are essentially a many-body problem with as many particles as are involved in the process. *Antisymmetrical* wave functions must be used, antisymmetrical also with respect to an exchange of positive and negative energy electrons.

In processes where only a finite number of pairs are created or annihilated no use is actually made of the infinite number of electrons in negative energy states. The hole theory leads then simply to the following interpretation: A transition from a negative energy state to a positive energy state means the creation of a pair; the reverse transition means the annihilation of a pair.

The hole theory has far-reaching consequences that touch even upon our concept of a pure electromagnetic field. Consider what happens if an electromagnetic field, e.g. the Coulomb field of a nucleus, is superimposed on the electron vacuum. The negative energy states in the field are not the same as for free electrons. Therefore, if all negative energy states of the free electron were filled up, some of the negative energy states of the field would be vacant and some positive energy states occupied, in other words, a number of pairs would be present. The pairs act as dipoles (total charge zero) and so the vacuum becomes *polarized*. The polarizability of the vacuum is in some ways analogous to that of an inhomogeneous dielectric (because the Coulomb field is inhomogeneous) and is a conspicuous novel feature of the theory.

We shall treat the polarization of the vacuum in detail in § 32, but the qualitative consequences are easily seen: A constant field merely leads to a constant polarization. The polarizability is field independent and thus all charges are changed by a universal constant factor (the 'dielectric constant' of the vacuum). Since there is no way of experimenting in an 'ideal vacuum' where this polarizability is absent, this effect is unobservable in principle (although the polarizability turns out to be infinite owing to the infinite number of pairs which contribute). An inhomogeneous field creates, in addition, a non-uniform polarization whose effects are observable (and finite). For example, an additional light wave will be scattered by the polarization dipoles, i.e. in the Coulomb field. Similarly, two light waves will be scattered by each other (§ 32).

Phenomena of this nature can never be obtained from a linear theory for a pure electromagnetic field where the principle of superposition holds. The electromagnetic field no longer has an independent existence but is intimately connected with the 'electron field'. These departures from a pure Maxwell field are, in normal circumstances, very small and noticeable only for very strong fields or very high frequencies.

Another feature which we shall meet in the further development of the theory is what is called 'vacuum fluctuations'. In an electron vacuum the total charge is zero. This does not mean that the charge in a small volume element vanishes all the time. In fact there is a finite probability for observing a positive or negative electron (or several electrons) in any one space region at any time (compare also § 28.4).

The hole theory has turned out to be in excellent agreement with the experiments. Nevertheless, it must be admitted that the formulation of the theory, with the help of an infinite number of particles in negative energy states, is unsatisfactory. However, in the following section we shall see that the theory can be formulated directly in terms of positrons rather than negative energy electrons and much of the harshness of the picture will disappear.

12. Second quantization of the electron field

1. *Second quantization of a single electron wave.* Apart from treating a system of positrons and electrons as a many-body problem of particles with positive and negative energy there is an alternative method which introduces the properties of the positive electrons from the start. The method is to subject the ψ-function of the Dirac equation to a process of quantization similar to the quantization of the radiation field **A**. This method has the advantage that no 'infinite sea' of negative energy electrons occurs and only physically significant features are introduced. Moreover, it allows in a natural way for the possibility of creation and annihilation of particles. It has first been shown by Jordan and Wigner† that, instead of using an antisymmetrical wave function in a many-dimensional configuration space, one can also consider the ψ-function as a *field* like **A**, in 3-dimensional space only, while regarding it as a quantum-mechanical operator subject to certain quantum conditions. These quantum conditions must be such that the Pauli principle is fulfilled. The *particle* nature of the electrons arises then only through this process of 'second quantization'.

Every time-dependent solution of the Dirac equation can be expanded into a series of free electron solutions. This expansion will naturally include *both positive and negative energy* states. It will be essential to write these separately in the expansion. We denote the *normalized* 4-component amplitudes for the positive and negative energy states of a *free* electron with momentum $\mathbf{p}$, spin component s $(= \pm\frac{1}{2})$ by $u_{ps\rho}$ and $v_{ps\rho}$ respectively ($\rho = 1,..., 4$). These are, for example, the quantities § 11 eq. (12). It is convenient to reverse the signs of $\mathbf{p}$ and s for negative energy states, so that $\mathbf{p}, s$ denote the momentum and spin of the *positron*.

For a single partial wave $u_{ps\rho} \exp i(\mathbf{pr})/\hbar c$ is the eigensolution of an

† P. Jordan and E. Wigner, *Z. Phys.* **47** (1928), 631.

electron with $E > 0$, $v_{-p,-s,\rho}\exp(-i(\mathbf{pr})/\hbar c)$ that of an electron with $E < 0$, momentum $-\mathbf{p}$, and spin $-s$, so that $+\mathbf{p}$ is the momentum (s the spin) of the positron, if the negative energy electron is absent. In analogy to the expansion of the radiation field § 7 eq. (3), we can then write

$$\psi_\rho(t) = \sum_{p,s} \{a_{ps}(t)u_{ps\rho}\, e^{i(\mathbf{pr})/\hbar c} + b^*_{ps}(t)v_{-p-s\rho}\, e^{-i(\mathbf{pr})/\hbar c}\}. \tag{1 a}$$

Since ψ is, in contrast to $\mathbf{A}$, not real, b^* will not now be the complex conjugate to a. a_{ps} and b^*_{ps} are independent amplitudes depending on t. Unless clarity demands it we shall omit the indices s, ρ. There is a corresponding expansion for ψ^*:

$$\psi^* = \sum_{p,s} \{a^*_p(t)u^*_p\, e^{-i(\mathbf{pr})/\hbar c} + b_p(t)v^*_{-p}\, e^{i(\mathbf{pr})/\hbar c}\}. \tag{1 b}$$

We now go over to a process of quantization by making the amplitudes a, b, a^*, b^* into time-independent operators. At first it would seem plausible to do this by requiring $[a^*_p, a_p] = -1$, in analogy to § 7 eq. (10).† As in § 7 eq. (13) it would follow that $a^*_p a_p = n_p$ is the occupation number, i.e. the number of electrons in the eigenstate $\mathbf{p}$, s. However, it would also follow that n_p would have the eigenvalues 0, 1, 2,..., and this clearly contradicts the Pauli principle. Instead we must carry out the quantization in such a way that the number of particles in any given eigenstate is 0 or 1. This is effected if we require the *anticommutation* relation

$$a^*_p a_p + a_p a^*_p = 1, \tag{2}$$

with a plus sign instead of a minus sign. A matrix representation for a^*, a is easily found.‡ The simplest non-trivial representation must represent a^*, a by at least 2-dimensional matrices. Considering that a^* is, before quantization, the complex conjugate to a (and therefore after quantization must be the hermitian conjugate to a), and choosing a^*a diagonal, we find

$$a = \begin{pmatrix} 0 & 0 \\ 1 & 0 \end{pmatrix}, \qquad a^* = \begin{pmatrix} 0 & 1 \\ 0 & 0 \end{pmatrix}, \tag{3 a}$$

$$a^*a \equiv n = \begin{pmatrix} 1 & 0 \\ 0 & 0 \end{pmatrix}, \qquad aa^* = \begin{pmatrix} 0 & 0 \\ 0 & 1 \end{pmatrix} = 1-n. \tag{3 b}$$

That this is the only irreducible representation of (2) arises from the fact that (3) are formally nothing but linear combinations of the Pauli spin matrices and the irreducibility for the latter is well known.

† The factor $\hbar/2\nu$ does not occur here because u, v are already properly normalized.

‡ A trivial solution would be, of course, $a = a^* = 1/\sqrt{2}$, but this is useless because a, a^* would then be c-numbers.

From (3) it follows furthermore that

$$a^2 = a^{*2} = 0. \tag{4}$$

n has but two values, 0 and 1, and can thus be interpreted as the number of electrons in the state $\mathbf{p}$, s ($E > 0$), in accordance with the Pauli principle. Like the q, q^* for the radiation field, a, a^* will be the operators of 'absorption and emission' of an electron, i.e. of a change of the numbers of electrons present by ± 1. We shall have to introduce a *state vector* Ψ_n for the electron system, depending on the number n of electrons present (and the time), and a, a^* act on Ψ, much as in § 7 eq. (27).

There are but two states Ψ_1, Ψ_0, and we expect the action of the *operators* a_{op}, a^*_{op} on Ψ_n to be

$$\begin{aligned} a_{\text{op}}\,\Psi_1 &= \Psi_0, & a^*_{\text{op}}\,\Psi_0 &= \Psi_1, \\ a^*_{\text{op}}\,\Psi_1 = a_{\text{op}}\,\Psi_0 &= 0, & a^*_{\text{op}}\,a_{\text{op}}\,\Psi_n &= n\Psi_n. \end{aligned} \tag{5}$$

The matrix elements of a, a^* are then, indeed,

$$\begin{aligned} a_{01} &\equiv (\Psi_0^*\,a_{\text{op}}\,\Psi_1) = a^*_{10} \equiv (\Psi_1^*\,a^*_{\text{op}}\,\Psi_0) = 1, \\ a_{11} &= a_{10} = a_{00} = a^*_{11} = a^*_{01} = a^*_{00} = 0, \end{aligned} \tag{6}$$

which is identical with the matrix representation (3). Thus a^* and a play the roles of emission and absorption operators respectively, and the Pauli principle is satisfied.

Similarly, we require for the amplitudes b, b^*

$$bb^* + b^*b = 1. \tag{7}$$

Now, in accordance with the interpretation of the hole theory, we can interpret b as the absorption operator and b^* as the emission operator for a positive electron. For in the expansion of ψ (eq. (1)) the amplitude of the negative energy states is denoted by b^*, and this should describe the absorption of a negative energy electron which is the same as the emission of a positron. The matrix representation for b, b^* is the same as for a, a^*. Ψ depends now on the number of both positive and negative electrons, Ψ_{n^+,n^-}, and $a^*a = n^-$, $b^*b = n^+$.

2. *Set of electron waves.* The generalization to a set of many electron waves is not quite trivial. The state vector will depend now on the numbers of positive and negative electrons of each partial wave $\Psi_{\dots n_p^+ \dots n_p^- \dots}$. One might be tempted to regard a_p, etc., as a direct product of factors each referring to one partial wave (either for a positive or negative electron) and all being 2-row–2-column unit matrices except the factor referring to the partial wave p, which is given by (3a). For the time being we denote such direct products by $\bar{a}_p$, etc., whereas a_p, etc., denote

a single factor (2-dimensional matrix). It would follow then that $\bar{a}_{p'}\bar{a}_p^* - \bar{a}_p^*\bar{a}_{p'} = 0$, when $p \neq p'$. However, this cannot be correct. We may go over to the continuous spectrum and let $p \to p'$. The relation satisfied by $\bar{a}_{p'}$, $\bar{a}_p^*$ must then go over smoothly into (3a). This can only be the case if

$$\bar{a}_{ps}^*\bar{a}_{p's'}+\bar{a}_{p's'}\bar{a}_{ps}^* = \bar{b}_{ps}^*\bar{b}_{p's'}+\bar{b}_{p's'}\bar{b}_{ps}^* = \delta_{ss'}\delta_{pp'}. \tag{8}$$

For continuously varying p, p', $\delta_{pp'}$ will be replaced by $\delta(\mathbf{p}-\mathbf{p}')$ (see subsection 3). It can be seen that the different partial waves are not independent and we shall see that this is just the expression of the fact that in configuration space the wave function is antisymmetrical. To find a matrix representation satisfying (8) it is necessary to number the partial waves in some arbitrary order, 1, 2,..., λ,..., μ,..., etc. In this series all partial waves referring both to positive and negative electrons will occur. $\bar{a}_\lambda$ will then be a direct product of factors, one for each partial wave, in which not only the unit matrix 1_μ, etc., will occur but also the matrix

$$c = \begin{pmatrix} -1 & 0 \\ 0 & 1 \end{pmatrix}, \qquad c^2 = 1. \tag{9}$$

We put now ($\times$ means direct product)

$$\begin{aligned} \bar{a}_\lambda &= c_1\times c_2\times\ldots\times c_{\lambda-1}\times a_\lambda\times 1_{\lambda+1}\times 1_{\lambda+2}\times\ldots \\ \bar{a}_\lambda^* &= c_1\times c_2\times\ldots\times c_{\lambda-1}\times a_\lambda^*\times 1_{\lambda+1}\times 1_{\lambda+2}\times\ldots \end{aligned} \tag{10}$$

and similarly for b, b^*. By (3) and (9) c_λ anticommutes with a_λ and a_λ^*: $c_\lambda a_\lambda + a_\lambda c_\lambda = 0$. It follows immediately from (10) that

$$\bar{a}_\lambda\bar{a}_\mu^* + \bar{a}_\mu^*\bar{a}_\lambda = 0 \qquad (\mu \neq \lambda)$$

in agreement with (8). In addition

$$\bar{a}_\lambda\bar{a}_\mu + \bar{a}_\mu\bar{a}_\lambda = 0 \quad (\mu \neq \lambda \text{ or } \mu = \lambda). \tag{11}$$

We shall henceforth denote anticommutators like (11) by brackets {} and omit the bar again. The anticommutation relations of the a, a^*, b, b^* are then similar to those valid for the amplitudes of the radiation field, only all commutators [...] are replaced by anticommutators, viz:

$$\{a_{ps}^* a_{p's'}\} = \{b_{ps}^* b_{p's'}\} = \delta_{ss'}\delta_{pp'}, \tag{12a}$$

$$\{a_{ps}^* a_{p's'}^*\} = \{a_{ps} a_{p's'}\} = \{b_{ps} b_{p's'}\} = \{a_{ps} b_{p's'}^*\} = 0, \text{ etc.} \tag{12b}$$

(12a) are the only non-vanishing anticommutators.

Evidently a_{ps} commutes, in the ordinary sense, with a product $a_{p's'}^* a_{p's'}$:

$$[a_{ps}, a_{p's'}^* a_{p's'}] = 0 \quad (p \neq p' \text{ or } s \neq s').$$

Also
$$a_{ps}^2 = a_{ps}^{*2} = 0, \quad a_p^* a_p = n_p^-, \quad b_p^* b_p = n_p^+. \tag{12c}$$

The present formalism is equivalent to the use of antisymmetrical wave functions in configuration space. We shall make this plausible but refrain from a general proof which is rather cumbersome.‡

Consider two electrons in states p, p', the other states being unoccupied. The operator $a_p^* a_p a_{p'}^* a_{p'}$, acting on Ψ is then unity. Now an exchange of the two electrons can be effected by forming the operator

$$P = a_p^* a_{p'}^* a_p a_{p'},$$

i.e. (read from right to left): absorb electron p' first, then absorb electron p, then emit electron p', then emit electron p. Although here the electrons are not numbered, it should be plausible enough, that this operator, owing to the changed order of emission and absorption processes, describes the exchange of the two electrons. Using (12) we see that

$$P = -a_p^* a_p a_{p'}^* a_{p'} = -1, \quad \text{acting on } \Psi.$$

3. *Anticommutation relations for* ψ. We can go over to the continuous spectrum as in § 10, replacing the sums over p by integrals. The sum over the discrete spin still remains. The number of partial waves per unit volume in the interval d^3p with a given spin is $d^3p/(2\pi\hbar c)^3$. We first introduce interaction representation as in § 7, by affixing time factors to a, b, a^*, b^*:

$$\begin{aligned} a &\to ae^{-iEt/\hbar}, \qquad & a^* &\to a^*e^{iEt/\hbar}, \\ b &\to be^{-iEt/\hbar}, \qquad & b^* &\to b^*e^{+iEt/\hbar}. \end{aligned} \tag{13}$$

Here E is always positive, and, for the b's, means the energy of the positron. The expansions (1 a), (1 b) take the form (we use $\psi^\dagger = i\psi^*\beta$, $u^\dagger = iu^*\beta$ instead of ψ^*, u):

$$\begin{aligned} \psi_\rho(\mathbf{r},t) &= \sum_s \int \frac{d^3p}{(2\pi\hbar c)^{\frac{3}{2}}} \{a_{ps}\, u_{ps\rho}\, e^{i(\mathbf{pr})/\hbar c - iEt/\hbar} + b_{ps}^*\, v_{-p-s\rho}\, e^{-i(\mathbf{pr})/\hbar c + iEt/\hbar}\}, \\ \psi_\rho^\dagger(\mathbf{r},t) &= \sum_s \int \frac{d^3p}{(2\pi\hbar c)^{\frac{3}{2}}} \{a_{ps}^*\, u_{ps\rho}^\dagger\, e^{-i(\mathbf{pr})/\hbar c + iEt/\hbar} + b_{ps}\, v_{-p-s\rho}^\dagger\, e^{+i(\mathbf{pr})/\hbar c - iEt/\hbar}\}. \end{aligned} \tag{14}$$

The anticommutators of a_{ps}, etc., are given by (12), with the right-hand side replaced by $\delta_{ss'}\,\delta(\mathbf{p}-\mathbf{p}')$. (For the factor $(2\pi\hbar c)^{\frac{3}{2}}$ compare p. 98.) As a, a^*, etc., are now operators, ψ and ψ^* are likewise operators, rather like $A_\mu(\mathbf{r},t)$. There are, however, two points of difference: (i) ψ and ψ^* obey different commutation rules, in fact anticommutation rules. (ii) Before quantization ψ, $\psi^\dagger$ are complex, and therefore the operators ψ, $\psi^\dagger$ will not be hermitian.

‡ Jordan and Wigner, loc. cit.

Like $A_\mu(\mathbf{r},t)$, ψ and $\psi^\dagger$ are operators in each space-time point. Owing to the use of the interaction representation, ψ and $\psi^\dagger$ depend explicitly on time, which would not be the case in the Schrödinger representation (1). ψ is evidently an operator that gives rise to the absorption of an electron or the emission of a positron, i.e. to the subtraction of a negative charge. $\psi^\dagger$ gives rise to the addition of a negative charge.

We now form the anticommutators of ψ, $\psi^\dagger$ in two space-time points $\mathbf{r}_1, t_1$ and $\mathbf{r}_2, t_2$. Since all a's anticommute with themselves and with all b^*'s it follows that

$$\{\psi(\mathbf{r}_1,t_1)\psi(\mathbf{r}_2,t_2)\} = \{\psi^\dagger(\mathbf{r}_1,t_1)\psi^\dagger(\mathbf{r}_2,t_2)\} = 0. \tag{15}$$

Thus ψ anticommutes with ψ in any other space-time point. The only non-vanishing anticommutator is $\{\psi^\dagger(\mathbf{r}_1,t_1)\psi(\mathbf{r}_2,t_2)\}$. This follows from (12) and (14). Writing $(\mathbf{pr})-Ect = p_\mu x_\mu$, we obtain:

$$\{\psi^\dagger_{\rho'}(\mathbf{r}_1,t_1)\psi_\rho(\mathbf{r}_2,t_2)\}$$
$$= \int \frac{d^3p}{(2\pi\hbar c)^3}\Big[\sum_s (u^\dagger_{psp'}\, u_{psp})e^{ip_\mu(x_{\mu2}-x_{\mu1})/\hbar c} + \sum_s (v^\dagger_{-p-sp'}\, v_{-p-sp})e^{-ip_\mu(x_{\mu2}-x_{\mu1})/\hbar c}\Big]. \tag{16}$$

The sums $\sum_s$ in (16) can be evaluated by means of the operators § 11 eqs. (16), (18). These sums are to be extended over the spin directions only, keeping the sign of the energy fixed. Thus:

$$\sum_s (u^\dagger_{\rho'}\, u_\rho) \equiv \mathsf{S}^p_+ (u^\dagger_{\rho'}\, u_\rho) = \sum \Big(u^* \frac{\gamma_\mu p_\mu + i\mu}{2|p_0|}\Big)_{\rho'} u_\rho$$
$$= \Sigma^p \sum_{\rho''} u^*_{\rho''}\Big(\frac{\gamma_\mu p_\mu + i\mu}{2|p_0|}\Big)_{\rho''\rho'} u_\rho = \frac{1}{2|p_0|}(\gamma_\mu p_\mu + i\mu)_{\rho\rho'}. \tag{17}$$

In the last equation the completeness theorem, § 11 eq. (13″), is used. The index $\rho\rho'$ denotes the matrix element of the operators γ_μ.

Similarly:

$$\sum_s (v^\dagger_{-p\rho'}\, v_{-p\rho}) = -\frac{1}{2|p_0|}(-\gamma_\mu p_\mu + i\mu)_{\rho\rho'} \quad (p_0 = +|E|). \tag{18}$$

Here $-\gamma_\mu p_\mu$ occurs because we have v_{-p} on the left and we have defined p_0 to be positive.‡ Acting on the exponentials in (16) respectively

$$\sum_s (u^\dagger_{\rho'}\, u_\rho) = -\sum_s (v^\dagger_{\rho'}\, v_\rho) = \frac{1}{2|p_0|}\Big(\gamma_\mu \frac{\hbar c}{i}\frac{\partial}{\partial x_{\mu2}} + i\mu\Big)_{\rho\rho'}. \tag{19}$$

The integral in (16) then reduces to the D-function of § 8 eq. (30), if we reverse the sign of the integration variable $\mathbf{p}$ in the second term, write

‡ In contrast to § 11 eq. (18 a), where p_0 is the energy in a negative energy state.

$\exp(-iEt/\hbar)-\exp(iEt/\hbar) = -2i\sin Et/\hbar$ and use $E = +\sqrt{(\mu^2+p^2)}$. Finally we obtain:

$$\{\psi^\dagger_{\rho'}(\mathbf{r}_1,t_1)\psi_\rho(\mathbf{r}_2,t_2)\} = -\left(\gamma_\mu\frac{\partial}{\partial x_{\mu 2}}-\frac{\mu}{\hbar c}\right)_{\rho\rho'} D(\mathbf{r}_2-\mathbf{r}_1,t_2-t_1). \tag{20}$$

The relativistic invariance of these relations is evident.

As in the expansion of the electromagnetic field, it is often convenient to use a 4-dimensional expansion instead of the 3-dimensional Fourier expansion (14). We write, as in § 10.5,

$$\psi_\rho(x_\mu) = \sum_s \int \frac{d^4p}{(2\pi\hbar c)^{\frac{3}{2}}}\psi_\rho(p_\mu,s)e^{ip_\mu x_\mu/\hbar c}, \tag{21 a}$$

$$\psi^\dagger_\rho(x_\mu) = \sum_s \int \frac{d^4p}{(2\pi\hbar c)^{\frac{3}{2}}}\psi^\dagger_\rho(p_\mu,s)e^{-ip_\mu x_\mu/\hbar c},$$

$$d^4p = d^3p\,dp_0, \qquad p_0 = -ip_4. \tag{21 b}$$

The sign of p_μ is reversed for the second terms of (14) (p_0 is again negative for positrons). Comparison with (14) shows that

$$\begin{aligned}\psi_\rho(p_\mu,s) &= a_{ps}\,u_{ps\rho}\,\delta(p_0-E)+b^*_{-p-s}\,v_{ps\rho}\,\delta(p_0+E)\\ &\equiv \psi^-_\rho(p_\mu,s)+\psi^+_\rho(p_\mu,s),\end{aligned} \tag{22 a}$$

$$\begin{aligned}\psi^\dagger_\rho(p_\mu,s) &= a^*_{ps}\,u^\dagger_{ps\rho}\,\delta(p_0-E)+b_{-p-s}\,v_{ps\rho}\,\delta(p_0+E)\\ &\equiv \psi^{-\dagger}_\rho(p_\mu,s)+\psi^{+\dagger}_\rho(p_\mu,s),\end{aligned} \tag{22 b}$$

ψ^- and $\psi^{+\dagger}$ are absorption operators for negative and positive electrons respectively, ψ^+, $\psi^{-\dagger}$ are emission operators.

The anticommutators, e.g. $\{\psi^-(p_\mu,s)\psi^{-\dagger}(p'_\mu,s')\}$, are proportional to $\delta_{ss'}$. If we carry out the summation over the spins as above (eq. (17)), we obtain for the non-vanishing anticommutators:

$$\mathsf{S}\{\psi^{\mp\dagger}_{\rho'}(p'_\mu,s)\psi^\mp_\rho(p_\mu,s)\} = \pm T_{\rho\rho'}\frac{1}{2|E|}\delta(\mathbf{p}-\mathbf{p}')\delta(p_0\mp E)\delta(p'_0\mp E'),$$

$$T = \gamma_\mu p_\mu+i\mu, \qquad p_4 = ip_0 \tag{23}$$

where, by definition, $E = +\sqrt{(\mathbf{p}^2+\mu^2)}$. Owing to $\delta(\mathbf{p}-\mathbf{p}')$, $E = E'$ and $p_0 = p'_0 = \pm E$.

This can also be written, with the help of the function $\epsilon(x) = \pm 1$ when $x \gtrless 0$, and the relations § 8 eqs. (23), (24):

$$\mathsf{S}\{\psi^{\mp\dagger}_{\rho'}(p'_\mu,s)\psi^\mp_\rho(p_\mu,s)\} = \pm T_{\rho\rho'}\,\delta^4(p_\mu-p'_\mu)\,\frac{1\pm\epsilon(p_0)}{2}\,\delta(p^2_\alpha+\mu^2). \tag{24}$$

Finally, we have for the complete ψ's:

$$\mathsf{S}\{\psi^\dagger_{\rho'}(p'_\mu,s)\psi_\rho(p_\mu,s)\} = T_{\rho\rho'}\,\delta^4(p_\mu-p'_\mu)\epsilon(p_0)\delta(p^2_\alpha+\mu^2). \tag{25}$$

The Fourier transformation of this expression just leads back to (20).

The vacuum expectation values are also easily obtained. By vacuum is meant here the electron vacuum, i.e. the absence of positive and negative electrons. In any product $\psi^\dagger$, ψ contributions arise only if the emission operator stands to the right, and the absorption operator to the left. Thus (writing p for p_μ, s)

$$\langle \mathsf{S}\psi^\dagger_{\rho'}(p')\psi_\rho(p)\rangle_0 = \langle \mathsf{S}\psi^{+\dagger}_{\rho'}(p')\psi^+_\rho(p)\rangle_0 = \langle \mathsf{S}\{\psi^{+\dagger}_{\rho'}(p')\psi^+_\rho(p)\}\rangle_0$$

$$= -T_{\rho\rho'}\frac{1-\epsilon(p_0)}{2}\delta^4(p_\mu-p'_\mu)\delta(p_\alpha^2+\mu^2), \quad (26\text{a})$$

$$\langle \mathsf{S}\psi_\rho(p)\psi^\dagger_{\rho'}(p')\rangle_0 = \langle \mathsf{S}\psi^-_\rho(p)\psi^{-\dagger}_{\rho'}(p')\rangle_0$$

$$= +T_{\rho\rho'}\frac{1+\epsilon(p_0)}{2}\delta^4(p_\mu-p'_\mu)\delta(p_\alpha^2+\mu^2). \quad (26\text{b})$$

According to (15), (20), any two field quantities ψ or $\psi^\dagger$ taken at two different space-time points never commute, although they may anti-commute. The fact that we have obtained everywhere *anti*commutation relations is a direct expression for the Pauli principle. $\psi(\mathbf{r}_1, t_1)$ and $\psi(\mathbf{r}_2, t_2)$ can therefore never be measured simultaneously, however far away the two space points may be. This shows that ψ, in spite of the formal analogy with the quantization of the electromagnetic field, cannot be regarded as a measurable quantity like the electromagnetic field strengths. In the classical limit electrons behave as *particles* and the non-quantized ψ-field in 3-dimensional space has no physical existence. On the other hand, measurable field quantities can be constructed from ψ. An example is the current density, which is discussed below.

4. *The current and energy densities.* In Dirac's theory the current density is defined by

$$i_\mu = ec(\psi^\dagger\gamma_\mu\psi). \quad (27)$$

If ψ is subjected to second quantization i_μ is also an operator. (27), however, still includes as a contribution the current due to all the vacuum electrons in negative energy states. This is seen if we form the expectation value of the total current $I_\mu = \int i_\mu\, d\tau$ for a state with given occupation numbers n_p^+, n_p^-. Using the expansions (1) we find†

$$\langle I_\mu\rangle = ec\sum_{p,s}\{n_p^-(u_p^\dagger\gamma_\mu u_p)+(1-n_p^+)(v_{-p}^\dagger\gamma_\mu v_{-p})\}. \quad (28)$$

Now $e(u_p^\dagger\gamma_\mu u_p)$ is the current of a single negative electron with momentum $\mathbf{p}$, $e(v_{-p}^\dagger\gamma_\mu v_{-p})$ that of an electron in a negative energy state with

† I_μ itself has additional terms proportional to the operators of creation and annihilation of a pair with opposite momenta. The current is

$$I_k = \langle I_k\rangle + ec\sum_{p,s}\{a_p^* b_{-p}^*(u_p^*\alpha_k v_p)+b_{-p}a_p(v_p^*\alpha_k u_p)\}.$$

For the total charge I_4/ic these terms vanish. Evidently $\langle b_{-p}a_p\rangle = 0$.

momentum $-\mathbf{p}$ or minus the current of a positron with momentum $+\mathbf{p}$. ($e = -|e|$ is the charge of a negative electron.)

$$I'_\mu = ec \sum_{p,s} \{n_p^-(u_p^\dagger \gamma_\mu u_p) - n_p^+(v_{-p}^\dagger \gamma_\mu v_{-p})\} \tag{29}$$

is thus the current to be expected for a system of electrons and positrons, whereas (28) contains as an additional contribution the current of the vacuum electrons

$$I''_\mu = ec \sum_{p,s} (v_{-p}^\dagger \gamma_\mu v_{-p}). \tag{30}$$

This unwanted contribution can be removed in a very simple way. Before second quantization $\psi^\dagger$ and ψ were commuting functions of space and time. After applying second quantization these functions no longer commute and the order of the factors is at our disposal. We can use this fact and choose instead of an expression $\psi^\dagger ... \psi$ a linear combination of $\psi^\dagger ... \psi$ and $\psi ... \psi^\dagger$. In order that an expression of the form $(\psi ... \gamma ... \psi^\dagger)$ should make sense from the point of view of matrix multiplication in the space of the γ's we must insert the transposed matrices $\tilde{\gamma}_\mu$, instead of γ_μ, then, for example, $(u\tilde{\gamma}_\mu u^\dagger) = u_{\rho'} \tilde{\gamma}_{\mu\rho'\rho} u_\rho^\dagger = (u^\dagger \gamma_\mu u)$.

We now re-define the current density as follows:‡

$$i_\mu = +\frac{ec}{2}\{(\psi^\dagger \gamma_\mu \psi) - (\psi \tilde{\gamma}_\mu \psi^\dagger)\}. \tag{31}$$

If ψ and $\psi^\dagger$ anticommute we could reverse the order and the second term would be the same as the first. Proceeding now as above, the total current becomes

$$\langle I_\mu \rangle = \frac{ec}{2} \sum_{p,s} \{(a_p^* a_p - a_p a_p^*)(u_p^\dagger \gamma_\mu u_p) + (b_p b_p^* - b_p^* b_p)(v_{-p}^\dagger \gamma_\mu v_{-p})\}$$
$$= ec \sum_{p,s} \{n_p^-(u_p^\dagger \gamma_\mu u_p) - n_p^+(v_{-p}^\dagger \gamma_\mu v_{-p})\} = I'_\mu \tag{32}$$

where we have used the fact that for the same spin directions of the negative electron and positron $(u_p^\dagger \gamma_\mu u_p) = (v_{-p}^\dagger \gamma_\mu v_{-p})$, because, evidently, the currents of a single negative and positive electron are the same, apart from the sign of e. In (32) the contribution from the electrons in negative energy states has disappeared, and only those particles that really exist contribute to the current.

The current density $i_\mu(\mathbf{r}, t)$ in two different space-time points satisfies certain commutation relations (not anticommutation relations) which

‡ W. Heisenberg, *Z. Phys.* **90** (1934), 209; see also P. A. M. Dirac, *Proc. Camb. Phil. Soc.* **30** (1934), 150.

can easily be derived from (20) by straightforward calculation. One finds a relation of the type

$$[i_\mu(\mathbf{r}_1, t_1) i_\nu(\mathbf{r}_2, t_2)] = ...D(\mathbf{r}_2-\mathbf{r}_1, t_2-t_1). \tag{33}$$

The dots replace certain operators, still including one $\psi^\dagger$ and one ψ, γ-operators and differentiations of the D-function. What is important is the occurrence of the D-function (and its derivatives) as a common factor.‡ It has been shown in § 8 that the D-function vanishes if $\mathbf{r}_1, t_1$ lies outside the light cone of $\mathbf{r}_2, t_2$ or vice versa. Therefore the commutator vanishes unless the points in which i_μ and i_ν are considered can be connected by a signal travelling with velocity $\leqslant c$. If this is not the case it follows that $i_\mu(\mathbf{r}_1, t_1)$ and $i_\nu(\mathbf{r}_2, t_2)$ can be measured simultaneously. This is a necessary requirement if i_μ is to be regarded as a *measurable quantity* and if the principles of measurement are to be in accord with the principle of relativity. That $D \neq 0$ inside the light cone is connected with the fact that disturbances arising from the measurement of the current density can be carried by electrons which travel with velocity $\leqslant c$. The physically sound results obtained for the measurability of the current density hinge on the fact that in (20), and hence in (33), the D- and not the D_1-function occurs. One can discuss the measurements of the current density $i_\mu(\mathbf{r}, t)$ in a manner similar to our discussion of the measurements of the field strengths (§ 9). It turns out that the limitations imposed on the simultaneous measurement of i_μ in two different space-time regions conform with the uncertainty relations derived from the formalism. In particular this also applies to cases where pair creation must be taken into account. For details we refer to a paper by Bohr and Rosenfeld.§

In addition to the charge-current density there exist further quadratic expressions $\psi^\dagger$, ψ of physical importance. The Hamiltonian of a single free electron was $H = (\boldsymbol{\alpha}\mathbf{p})+\beta\mu$. Hence a Hamiltonian density

$$\mathscr{H} = \psi^*[(\boldsymbol{\alpha}\mathbf{p})+\beta\mu]\psi = \psi^\dagger((\boldsymbol{\gamma}\mathbf{p})-i\mu)\psi \tag{34}$$

can be defined for the quantized fields.

Using the fact that $((\boldsymbol{\alpha}\mathbf{p})+\beta\mu)u_p = E_p u_p$, $((\boldsymbol{\alpha}\mathbf{p})+\beta\mu)v_{-p} = -E_p v_{-p}$, and the normalization of the u's and v's the total Hamiltonian becomes

$$H = \int \mathscr{H}\, d\tau = \sum_{p,s} (n_p^- E_p + (n_p^+ - 1)E_p) \tag{35}$$

‡ Compare W. Pauli, *Ann. Inst. Henri Poincaré*, **6** (1936), 137; *Phys. Rev.* **58** (1940), 716. When $t_1 = t_2$, the commutation relation is given explicitly in appendix 3 eq. (12).

§ N. Bohr and L. Rosenfeld, *Phys. Rev.* **78** (1950), 794. Also E. Corinaldesi, Manchester Thesis (1951).

containing the contributions from negative electrons $n_p^- E_p$, positrons $n_p^+ E_p$, and an infinite contribution from the negative energy electrons. If we apply the same rule as for the current, however, the expression is only symmetrized and a zero point energy remains. Defining

$$\mathscr{H} = \tfrac{1}{2}\psi^*((\boldsymbol{\alpha}\mathbf{p})+\beta\mu)\psi - \tfrac{1}{2}\psi((\tilde{\boldsymbol{\alpha}}\mathbf{p})+\tilde{\beta}\mu)\psi^* \tag{36}$$

we get

$$H = \sum_{p,s} \{(n_p^- - \tfrac{1}{2}) + (n_p^+ - \tfrac{1}{2})\}E_p. \tag{37}$$

The zero point energy is quite analogous to that occurring for the radiation field, and has to be removed in the same way, namely by subtracting it from the total Hamiltonian.

H is diagonal simultaneously with n^+, n^-.

The Hamiltonian density $\mathscr{H}$ can be generalized to a complete energy-momentum density tensor. $\mathscr{H}$ is the 44-component of this tensor. The $4i$-components are essentially the density of momentum $(\psi^*\mathbf{p}\psi)$. This tensor is given in appendix 7.

13. Electrons interacting with radiation

1. *Hamiltonian of the complete system.* In classical theory (§ 6) Maxwell's equations as well as the equations of motion of the electrons can be derived from a Hamiltonian for the complete system which consists of (i) the Hamiltonian of the radiation field

$$H_{\text{rad}} = \frac{1}{8\pi}\int (E^2+H^2)\,d\tau \tag{1}$$

and (ii) the sum of the Hamiltonians of all particles k

$$H_{\text{el+int}} = \sum_k H_k \tag{2}$$

which also include the interaction of the particles with the electromagnetic field (and hence between themselves). In Chapter II the pure electromagnetic field was subjected to the quantum laws and in §§ 11, 12 we have translated H_k into quantum form, describing the electrons by Dirac's equation, either in $3N$-dimensional configuration space (N = number of particles) or by a 3-dimensional 'ψ-field' subjected to second quantization. The general problem of electrons (and positrons) interacting with the Maxwell field can now easily be formulated.

A particular advantage of Dirac's theory is its linearized form. This has the consequence that the Hamiltonian is also (in contrast to the classical Hamiltonian § 6 eq. (22)) linearly composed of two parts, the first is the Hamiltonian of the electrons in the absence of the radiation field, the second describes the interaction with the radiation field, and is linear in the potentials (see § 11 eq. (1)):

$$H_{\text{el+int}} = H_{\text{el}}+H_{\text{int}}.$$

The general state of the system will be described by a state vector Ψ, and its variation with time is determined in Schrödinger representation by

$$i\hbar\frac{\partial\Psi}{\partial t} = (H_{\text{rad}}+H_{\text{el}}+H_{\text{int}})\Psi. \tag{3}$$

The variables on which Ψ depends, as well as the explicit formulation of H, depend to some extent on the gauge used for the potentials (Coulomb or Lorentz gauge) and also on whether the electrons are described in configuration space or as a quantized field. We consider the two gauges separately.

(*a*) *Coulomb gauge.* $\operatorname{div}\mathbf{A} = \phi = 0$ hold as operator equations. H_{rad} consists of contributions from transverse waves only

$$H_{\text{rad}} = \sum_\lambda q_\lambda^* q_\lambda 2\nu_\lambda^2 = \sum_\lambda n_\lambda \hbar\nu_\lambda. \tag{4}$$

The q_λ satisfy, of course, the same commutation relations as for a pure field:

$$[q_\lambda q_{\lambda'}^*] = \frac{\hbar}{2\nu_\lambda}\delta_{\lambda\lambda'}.$$

If the electrons are described as particles in configuration space

$$H_{\text{el}} = \sum_k (\boldsymbol{\alpha}_k \mathbf{p}_k)+\beta_k \mu_k, \tag{5}$$

$$H_{\text{int}} = -\sum_k e_k(\boldsymbol{\alpha}_k \mathbf{A}(k))+\sum_{i>k}\frac{e_i e_k}{r_{ik}}, \tag{6}$$

where i, k refer to the individual electrons. For problems of atomic physics one would regard the Coulomb term of (6) as part of the unperturbed problem and include it in H_{el} rather than in H_{int}. Ψ depends on the coordinates of the electrons and the variables used to describe the radiation field and is, of course, antisymmetric in all electrons.

In the non-relativistic approximation $H_{\text{el}}+H_{\text{int}}$ goes over into § 11 eq. (21) and thus (neglecting the spin term which is already relativistic):

$$H_{\text{int}} = -\sum_k \left\{\frac{e_k}{\mu_k}(\mathbf{p}_k \mathbf{A}(k))-\frac{e_k^2}{2\mu_k}A^2(k)\right\}+\sum_{i>k}\frac{e_i e_k}{r_{ik}}. \tag{7}$$

If the electrons are described by second quantization the Hamiltonian is a *space integral* like § 12 eq. (35). i_μ is bilinear in ψ, ψ^*.†

$$H_{\text{el}} = \int \psi^*((\boldsymbol{\alpha}\mathbf{p})+\beta\mu)\psi\, d\tau, \qquad \rho = (\psi^*\psi), \text{ etc.} \tag{8}$$

It is clear that the Coulomb term in (6), $H^{(c)}$ say, must be replaced by the Coulomb interaction of the total charge-density with itself. Thus

$$H_{\text{int}} = -e \int \psi^*(\boldsymbol{\alpha}\mathbf{A})\psi\, d\tau + \tfrac{1}{2} \iint \frac{\rho(\mathbf{r})\rho(\mathbf{r}')}{|\mathbf{r}-\mathbf{r}'|}\, d\tau d\tau'. \tag{9}$$

After second quantization ψ^*, ψ are operators changing the numbers of negative and positive electrons. A state vector Ψ describing a state with fixed numbers of photons and electrons is $\Psi_{...n_\lambda,...,\overset{+}{n_p},...,\overset{-}{n_p}...}$, and a general state vector will be of the form

$$\sum c_{...n_\lambda...,...\overset{+}{n_p}...,...\overset{-}{n_p}...} \Psi_{...n_\lambda...,...\overset{+}{n_p}...,...\overset{-}{n_p}...}.$$

$|c|^2$ is then the probability for finding n_λ photons of type λ, n_p^+ positrons of type p, etc. The anticommutation relations of the expansion coefficients of ψ, ψ^* are given in § 12.

(*b*) *Lorentz gauge.* Classically, the potentials satisfy $\partial A_\alpha/\partial x_\alpha = 0$. The Hamiltonian of the electromagnetic field is (§ 10)

$$H_{\text{rad}} = \sum_\lambda \sum_\alpha q_{\alpha\lambda}^* q_{\alpha\lambda} 2\nu_\lambda^2, \tag{10}$$

with contributions from longitudinal and scalar partial waves also. After quantization

$$q_\alpha^* q_\alpha = \frac{\hbar}{2\nu} n_\alpha, \qquad [q_\alpha^* q_\beta] = -\frac{\hbar}{2\nu} \delta_{\alpha\beta}. \tag{11}$$

n_α is the number of photons including longitudinal and scalar photons with 'polarization' α. If the electrons are described by the quantized ψ-field, H_{el} is again given by (8). H_{int} contains now $e\phi - e(\boldsymbol{\alpha}\mathbf{A})$ (§ 11 eq. (1)) and can be written

$$H_{\text{int}} = -e \int (\psi^\dagger \gamma_\mu A_\mu \psi)\, d\tau = -\frac{1}{c} \int i_\mu A_\mu\, d\tau. \tag{12}$$

The state vector depends on the number of all four types of photons n_α.

The formulation of the Lorentz condition in the general case of a field interacting with electrons requires special considerations and will be given below. (12) is gauge-invariant, because $\partial i_\mu/\partial x_\mu = 0$.

It is a remarkable fact that the integrand of (12), i.e. the *Hamiltonian density* $\mathscr{H}_{\text{int}}$, is a *relativistic invariant*, although the Hamiltonian or part

† After second quantization we should actually put $\rho = \frac{1}{2}((\psi^*\psi)-(\psi\psi^*))$, etc., as in § 12.4.

of it has 'no obligation' to be so ($\mathscr{H}$ is generally the 44-component of a tensor). This peculiarity of quantum electrodynamics will be a great aid in the treatment of the radiative corrections (Chapter VI).

The amplitudes of the radiation field q_λ or q_α and those of the electron field a_p, b_p represent independent degrees of freedom as will also appear from subsection 3. Therefore q_λ, q_λ^* commute with a_p:

$$[q_\lambda a_p] = [q_\lambda^* a_p] = 0, \text{ etc.}$$

2. *Interaction representation. Lorentz condition.*† So far we have used Schrödinger representation where all operators A_α, ψ are time independent. In the previous sections frequent use was made of what was called interaction representation, where by a trivial transformation the time dependence was shifted from Ψ to the operators. In the case of a pure radiation field or a pure electron field the state vector became time independent. A similar transformation can be carried out in the general case of interacting fields. We shall, however, shift the time dependence to the operators only partially: Ψ will still depend on time but the time variation will be due to the interaction only, whereas the time dependence due to the Hamiltonian of the pure fields is carried by the operators A, ψ. They will depend on time in just the same way as would be classically the case for pure non-interacting fields (hence the name interaction representation).

Now let
$$H_0 = H_{\text{rad}} + H_{\text{el}}$$
be the Hamiltonian of the pure non-interacting fields. If the number of electrons and photons of all descriptions and in each state be given, H_0 is diagonal and equals the energy of the non-interacting system of photons and electrons (by (4) and § 12 eq. (35)). We then put (compare § 7.3):

$$\Psi' = e^{iH_0 t/\hbar}\Psi. \tag{13}$$

This implies a transformation for all operators

$$\begin{aligned} A' &\equiv A(t) = e^{iH_0 t/\hbar} A e^{-iH_0 t/\hbar}, \\ \psi' &\equiv \psi(t) = e^{iH_0 t/\hbar} \psi e^{-iH_0 t/\hbar}, \text{ etc.} \end{aligned} \tag{14}$$

Thus A', ψ' depend explicitly on time. Similar formulae hold for the expansion coefficients q, a, etc. For the emission and absorption operators q, q^*, a_p, etc., this merely has the effect of affixing the time factors $e^{\mp i\nu t}$, etc., as was done in §§ 7, 10, 12, for example: let q_λ be the absorption operator of a particular type of photon ν_λ. Then $H_0 = n_\lambda \hbar\nu_\lambda + E'$ where

† The use of interaction representation for radiation problems is old. It was first systematically investigated by S. Tomonaga, *Prog. Theor. Phys.* **1** (1946), 27; **2** (1947), 101, and subsequent papers.

E' is the energy of all other types of photons and electrons. q_λ commutes with E'‡ but not with $n_\lambda \hbar\nu_\lambda$. We get

$$q'_{n,n+1} = (e^{iH_0t/\hbar})_n q_{n,n+1}(e^{-iH_0t/\hbar})_{n+1} = q_{n,n+1}\, e^{-i\nu t}, \text{ etc.}$$

Since the total Hamiltonian is $H = H_0 + H_{\text{int}}$, Ψ' now satisfies the equation, according to (3) and (13),

$$i\hbar\dot{\Psi}' = -H_0\Psi' + i\hbar e^{iH_0t/\hbar}\dot{\Psi} = e^{iH_0t/\hbar}H_{\text{int}}\Psi$$

or

$$i\hbar\dot{\Psi}' = H'_{\text{int}}\Psi', \tag{15}$$

$$H'_{\text{int}} = e^{iH_0t/\hbar}H_{\text{int}}\, e^{-iH_0t/\hbar}.$$

Equation (15) will be the starting-point for the calculation of the transition probabilities. Interaction representation can, of course, be used for both Coulomb and Lorentz gauge. In Lorentz gauge the commutation relations between the A_α's take their simplest general form. H'_{int} differs from H_{int} merely in that all emission and absorption operators have the time exponentials. Thus the commutation relations of the time-dependent operators in interaction representation are the same as those for the pure fields (§ 10, eq. (49); § 12 eq. (20)):

$$[A_\alpha(\mathbf{r}_2, t_2)A_\beta(\mathbf{r}_1, t_1)] = -4\pi i\hbar c\delta_{\alpha\beta}\Delta(\mathbf{r}_2-\mathbf{r}_1, t_2-t_1),$$

$$[A_\alpha\psi_\rho] = [A_\alpha\psi^\dagger_\rho] = 0, \tag{16}$$

$$\{\psi^\dagger_{\rho'}(\mathbf{r}_1, t_1)\psi_\rho(\mathbf{r}_2, t_2)\} = -\left(\gamma_\mu\frac{\partial}{\partial x_{\mu 2}} - \frac{\mu}{\hbar c}\right)_{\rho\rho'} D(\mathbf{r}_2-\mathbf{r}_1, t_2-t_1). \tag{17}$$

A different representation still would be the Born–Heisenberg representation where the time dependence is transferred to the operators completely. This is effected by the transformation $\Psi'' = \exp(iHt/\hbar)\Psi$ where H is the total Hamiltonian. The commutation relations then take a form that is more complicated than (16), (17), and A''_α and ψ''_ρ taken at different times no longer commute. This representation, however, will not be used in this book.

In interaction representation where the operators Q' depend explicitly on time, the time derivative is, by (14),

$$\frac{\partial Q'}{\partial t} = \frac{i}{\hbar}[H_0\,Q']. \tag{14 a}$$

In addition one can define a total derivative dQ/dt (in all representations)

‡ Note that E' is quadratic in the electron operators a^*_p, a_p, etc.; an operator $a_{p'}$ also commutes with $a^*_p\, a_p$, although it anticommutes with each factor.

such that the time derivative of the expectation value of Q, viz. $\langle Q\rangle$ equals the expectation value of dQ/dt:

$$\frac{d}{dt}\langle Q\rangle = \frac{d}{dt}(\Psi^* Q\Psi) \equiv \left(\Psi^* \frac{dQ}{dt}\Psi\right)$$

(when indefinite metric is used replace Ψ^* by $\Psi^*\eta$). The time derivative of $\langle Q\rangle$ involves $\dot{\Psi}$, $\dot{\Psi}^*$ for which the wave equation is to be used: (eq. (3) in Schrödinger representation and eq. (15) in interaction representation. We readily find for the three representations mentioned

$$\frac{dQ}{dt} = \frac{i}{\hbar}[HQ], \qquad \frac{dQ'}{dt} = \frac{\partial Q'}{\partial t} + \frac{i}{\hbar}[H'_{\text{int}}\, Q'], \qquad \frac{dQ''}{dt} = \frac{\partial Q''}{\partial t}. \tag{14 b}$$

The values of $\langle Q\rangle$ and $d\langle Q\rangle/dt$ are, of course, independent of the representation.

The Lorentz condition (when Lorentz gauge is used) assumes a more general form when charges are present. For a pure field this condition is, according to § 10,

$$\left(\Psi_L^* \eta \frac{\partial A_\alpha}{\partial x_\alpha} \Psi_L\right) \equiv \left\langle \frac{\partial A_\alpha}{\partial x_\alpha} \right\rangle_L = 0, \quad \text{or} \quad \frac{\partial A_\alpha^-}{\partial x_\alpha} \Psi_L = 0, \tag{18}$$

where $\langle\,\rangle$ is the expectation value of the expression bracketed and Ψ_L is a restricted set of state vectors for which (18) is satisfied. (18) is formulated in interaction representation and Ψ_L is time independent. When charges are present Ψ_L depends on time. The condition replacing (18) is derived in appendix 3. It is shown there that

$$\left(\Psi_L^*(t)\eta \frac{\partial A_\alpha}{\partial x_\alpha}(t)\Psi_L(t)\right) = 0 \tag{19}$$

must hold for all t in order that Maxwell's equations should hold for the expectation values of the field strengths. As in § 6 eqs. (41), (41′), the validity of (19) is safeguarded if two initial conditions are satisfied, namely,

$$\left(\Psi_L^* \eta \frac{\partial A_\alpha}{\partial x_\alpha} \Psi_L\right)_{t_0} = 0, \tag{20 a}$$

$$\left(\Psi_L^* \eta \left[\frac{\partial}{\partial t}\frac{\partial A_\alpha}{\partial x_\alpha} + 4\pi c\rho\right] \Psi_L\right)_{t_0} = 0. \tag{20 b}$$

In interaction representation $\Box\phi = 0$ (operator equation) and thus

$$\frac{\partial}{\partial t}\frac{\partial A_\alpha}{\partial x_\alpha} = \operatorname{div}\dot{\mathbf{A}} + \frac{1}{c}\ddot{\phi} = -c\operatorname{div}\mathbf{E},$$

so the classical initial conditions ($\partial A_\alpha/\partial x_\alpha = 0$ and $\operatorname{div}\mathbf{E} = 4\pi\rho$) hold again as relations between expectation values. The second term in (20 b) expresses the fact that, where charges are present, a Coulomb field also exists and then longitudinal and scalar photons must be present and the state vector fulfils a different condition. For $\rho = 0$ both conditions (20) reduce to (18).

There is, however, a large class of problems for which the above modification of the Lorentz condition can be ignored, namely, for collisions between free particles and quanta, provided that we treat the interaction between them (including the Coulomb interaction) as a perturbation. Consider two particles approaching each other from infinity. At the initial time $t = -\infty$ they are infinitely far apart from each other. Their Coulomb field in any finite region of space vanishes; in particular the field of one particle vanishes at the position of the other. Thus the initial conditions reduce to (18) and we may choose as initial state at $t = -\infty$ a state which satisfies the Lorentz condition (18) for a pure field, e.g. the state Ψ_0 with no longitudinal and scalar photons. A long time after the collision, at $t = +\infty$ say, the particles will again be separated by an infinite distance and the state will necessarily fulfil (18) again and belong to the Lorentz set for a pure field. During the collision the full Lorentz condition (19) will be fulfilled, but this is automatically the case because (19) always holds if the conditions (20) are satisfied initially. We need not therefore impose any further restrictions at a later time.† Thus we can compute the transition probability by choosing an initial state without longitudinal and scalar photons, ignoring the Lorentz condition in the meantime, and by calculating only the probabilities of final states which have no longitudinal and scalar photons.‡ An example (collisions between two electrons) will be treated in § 24 where it will be shown explicitly that in this way the same results are obtained as if Coulomb gauge were used.

All this is only correct for an infinite time interval $-\infty ... +\infty$. The most important case where the Lorentz condition cannot be treated in this way is that of a *bound electron*. Transitions take place in a finite time interval and the Coulomb field of the electron is there all the time. In this case the full initial conditions (20) have to be used. Since the

† In the language of perturbation theory this means that longitudinal and scalar photons will occur as intermediate states but we need not impose any condition on them. For an exact proof of this statement see, for example, F. Coester and J. M. Jauch, *Phys. Rev.* **78** (1950), 149, 827.

‡ 'Admixtures' with longitudinal and scalar photons satisfying (18) may occur in the final state but these merely reflect the ambiguities due to the various gauges of the potentials (§ 10).

handling of these conditions is complicated we shall treat all problems involving bound electrons in Coulomb gauge (see Chap. V and § 34).

In cases where a permanent external field is applied, we may treat this field as a classical field not subjected to quantization (the Lorentz condition can then be used in the classical form) or treat the problem in Coulomb gauge.

3. *The canonical formalism.* The quantization of the electromagnetic field was based on the fact that, when the field is expanded into a Fourier series, canonical pairs of variables could be found for each partial wave. In the subsequent development it was shown that the field variables *at each space point* have to be considered as operators (or, in interaction representation, at each space-time point) and the commutation relations were derived from those of the Fourier amplitudes.

It will be of interest to see that the field equations and commutation laws can be derived in a more concise form, considering the potentials and ψ themselves at each *space point* as *separate variables*, by the same standard formalism that is used in dynamics leading from a classical Lagrangian to the establishment of a Hamiltonian and canonical pairs and hence to the commutation laws. The Lagrangian is chosen as a starting-point because of its relativistic invariance. We use Lorentz gauge and start with the classical Lagrange–Hamiltonian description, where also the electron field ψ is treated as a mere classical field in 3-dimensional space.†

We regard $A_\mu(\mathbf{r},t)$ and $A_\mu(\mathbf{r}',t)$ as different variables, $\dot{A}_\mu(\mathbf{r},t)$ as the velocity associated with $A_\mu(\mathbf{r},t)$ in the Lagrange–Hamilton sense. Similarly, $\psi(\mathbf{r},t)$ at each point will be independent variables for the electron field, $\dot{\psi}$ the associated velocities.

For a pure radiation field the Lagrangian is usually‡ given as

$$\int L\,dt = \int \mathscr{L}\,d\tau dt, \qquad \mathscr{L} = -\frac{1}{16\pi}f_{\mu\nu}^2 = -\frac{1}{16\pi}\left(\frac{\partial A_\mu}{\partial x_\nu}-\frac{\partial A_\nu}{\partial x_\mu}\right)^2, \quad (21)$$

where $f_{\mu\nu}$ are the field strengths. A certain complication arises from the fact that $\dot{A}_4$ does not occur in $\mathscr{L}$ and that therefore the momentum conjugate to A_4 cannot be found. The difficulty is connected with the Lorentz condition. There are various ways of avoiding the difficulty. A very simple way is the following. We regard the Lorentz condition and

† This is to be regarded as a purely formal procedure, and does not mean that a 'classical ψ' has any physical meaning. The profound difference between fields satisfying Bose–Einstein or Fermi–Dirac statistics has been repeatedly stressed (pp. 60, 121). The treatment of ψ is physically meaningful only after quantization.

‡ See, for example, G. Wentzel, *Quantentheorie der Wellenfelder*, Edwards Bros., 1946.

its time derivative as initial conditions, as in § 6 and subsection 2 above. We then use the Lorentz condition to rewrite the cross term in (21)

$$\int dt \int d\tau \sum_{\mu,\nu} \frac{\partial A_\mu}{\partial x_\nu}\frac{\partial A_\nu}{\partial x_\mu} = -\int dt \int d\tau \sum_{\mu,\nu} A_\nu \frac{\partial}{\partial x_\nu}\frac{\partial A_\mu}{\partial x_\mu} = 0 \qquad (22)$$

provided that A_ν vanishes at infinity, which we may well assume to be the case. Then the integrand in (21) can be replaced by $2(\partial A_\mu/\partial x_\nu)^2$, and now $\dot{A}_4$ also occurs.

We now state the classical Lagrangian of the complete system to be:‡

$$L = -\int d\tau\left\{\frac{1}{8\pi}\left(\frac{\partial A_\mu}{\partial x_\nu}\right)^2 + (\psi^\dagger(\gamma_\mu p_\mu - e\gamma_\mu A_\mu - i\mu)\psi)\right\}, \qquad p_\mu \equiv \frac{\hbar c}{i}\frac{\partial}{\partial x_\mu}. \qquad (23)$$

The fact that the space integral occurs here is an expression of the fact that the Lagrangian is a sum over all independent variables, and therefore in our case is a space integral. Evidently the integrand is Lorentz invariant, and the same is true for $\int L\,dt$ because the 4-dimensional volume element is also invariant. The momentum canonically associate to $A_\mu(\mathbf{r})$ is defined by $\partial\mathscr{L}/\partial\dot{A}_\mu(\mathbf{r})$, for a fixed value of $\mathbf{r}$. This differentiation singles out a particular value of $\mathbf{r}$ in the sum over the $\mathbf{r}$'s. We denote the canonical momentum of $A_\mu(\mathbf{r})$ by $B_\mu(\mathbf{r})$, that of $\psi_\rho(\mathbf{r})$ by $\chi_\rho(\mathbf{r})$. Then:

$$B_\mu = \frac{\partial\mathscr{L}}{\partial\dot{A}_\mu} = \frac{1}{4\pi c^2}\dot{A}_\mu,$$
$$\chi_\rho = \frac{\partial\mathscr{L}}{\partial\dot{\psi}_\rho} = \hbar(\psi^\dagger\gamma_4)_\rho = i\hbar\psi^*_\rho. \qquad (24)$$

Thus ψ^* is essentially the canonical conjugate to ψ.

The Hamiltonian is now given by $\mathscr{H} = \sum p\dot{q} - \mathscr{L}$ where the sum is to be carried out over all canonical pairs, which in our case again means an integration. Thus, expressing $\dot{A}_\mu$, $\dot{\psi}_\rho$ by B_μ, χ_ρ everywhere,

$$H = \int \mathscr{H}\,d\tau = \int d\tau(B_\mu\dot{A}_\mu + \chi\dot{\psi}) - \int \mathscr{L}\,d\tau$$
$$= \int d\tau\left\{\frac{1}{8\pi}[(4\pi c B_\mu)^2 + \mathrm{grad}^2 A_\mu] - \frac{i}{\hbar}(\chi[(\boldsymbol{\alpha}\mathbf{p}) + \beta\mu]\psi) - \right.$$
$$\left. - \frac{e}{\hbar}(\chi\gamma_4\gamma_\mu A_\mu\psi)\right\} = H_0 + H_{\text{int}}. \qquad (25)$$

The Hamiltonian equations of motion will lead, on the one hand, to the Maxwell equations and, on the other hand, to the Dirac equation for ψ.

‡ The electron part of L actually vanishes by virtue of the Dirac equation but this is irrelevant.

There is only one point to be considered: $\mathscr{H}$ depends not only on $A_\mu(\mathbf{r})$ but also on the space derivatives of $\partial A_\mu/\partial x_i$. The derivative $\partial\mathscr{H}/\partial A_\mu$ is then to be found as follows. Consider a variation of $\mathscr{H}$, varying

$$A_\mu \to A_\mu + \delta A_\mu.$$

Then
$$\delta\mathscr{H} = \frac{\partial\mathscr{H}}{\partial A_\mu(\mathbf{r})}\delta A_\mu(\mathbf{r}) + \frac{\partial\mathscr{H}}{\partial(\partial A_\mu/\partial x_i)}\frac{\partial}{\partial x_i}\delta A_\mu(\mathbf{r}).$$

In the second term the operations $\partial/\partial x_i$ and δ have been permuted. Since H is a space integral, the second term can be treated by partial integration and it appears that the derivative $\partial\mathscr{H}/\partial A_\mu$ is to be replaced by the variational derivative

$$\frac{\partial\mathscr{H}}{\partial A_\mu} \to \frac{\partial\mathscr{H}}{\partial A_\mu} - \sum_i \frac{\partial}{\partial x_i}\frac{\partial\mathscr{H}}{\partial(\partial A_\mu/\partial x_i)} \quad (i = 1,\ldots, 3).$$

The Hamiltonian equations are then

$$\dot{A}_\mu = \frac{\partial\mathscr{H}}{\partial B_\mu} - \sum_i \frac{\partial}{\partial x_i}\frac{\partial\mathscr{H}}{\partial(\partial B_\mu/\partial x_i)}, \qquad -\dot{B}_\mu = \frac{\partial\mathscr{H}}{\partial A_\mu} - \sum_i \frac{\partial}{\partial x_i}\frac{\partial\mathscr{H}}{\partial(\partial A_\mu/\partial x_i)}, \tag{26 a}$$

$$\dot{\psi} = \frac{\partial\mathscr{H}}{\partial \chi} - \sum_i \frac{\partial}{\partial x_i}\frac{\partial\mathscr{H}}{\partial(\partial \chi/\partial x_i)}, \qquad -\dot{\chi} = \frac{\partial\mathscr{H}}{\partial \psi} - \sum_i \frac{\partial}{\partial x_i}\frac{\partial\mathscr{H}}{\partial(\partial \psi/\partial x_i)}, \tag{26 b}$$

i.e.
$$\dot{A}_\mu = 4\pi c^2 B_\mu,$$
$$-\dot{B}_\mu = -\frac{1}{4\pi}\nabla^2 A_\mu - \frac{1}{\hbar}e(\chi\gamma_4\gamma_\mu\psi)$$

or, using (24):

$$\nabla^2 A_\mu - \frac{1}{c^2}\ddot{A}_\mu = -\frac{4\pi}{c}i_\mu, \qquad i_\mu = \frac{ec}{\hbar}(\chi\gamma_4\gamma_\mu\psi). \tag{27}$$

Furthermore,

$$i\hbar\dot{\psi} = \{\beta\mu + (\boldsymbol{\alpha}\mathbf{p}) - ie\gamma_4\gamma_\mu A_\mu\}\psi, \qquad p_x = \frac{\hbar c}{i}\frac{\partial}{\partial x},$$

$$-i\hbar\dot{\chi} = \chi\beta\mu - \frac{\hbar c}{i}\frac{\partial}{\partial x_i}\chi\alpha_i - ie\chi\gamma_4\gamma_\mu A_\mu. \tag{28}$$

These equations are identical with Dirac's equations for ψ and $\psi^\dagger$ (put $\chi = \hbar\psi^\dagger\gamma_4$ and use the relations between $\boldsymbol{\alpha}$ and γ_μ). (27) is the wave equation for A_μ, which is identical with Maxwell's equations, provided that the Lorentz condition is satisfied. This is now used as an initial condition. Since A_μ satisfies a second-order wave equation it is sufficient for $\partial A_\alpha/\partial x_\alpha = 0$ to be valid at all times if (compare the same reasoning in § 6 eq. (41) and appendix 3)

$$\frac{\partial A_\alpha}{\partial x_\alpha}\bigg|_{t=t_0} = 0, \qquad \frac{\partial}{\partial t}\frac{\partial A_\alpha}{\partial x_\alpha}\bigg|_{t=t_0} = 0. \tag{29}$$

In canonical variables (29) can be written (using (24) and (27) for B_4) as

$$\left.\begin{aligned} \sum_i \frac{\partial A_i}{\partial x_i} - i4\pi c B_4 = 0 \\ \sum_i \frac{\partial B_i}{\partial x_i} + \frac{1}{4\pi i c}\nabla^2 A_4 = -\frac{\rho}{c} \end{aligned}\right\} \quad (\text{at } t = t_0). \tag{30}$$

(Compare the same equations in Fourier decomposition, § 6 eq. (44e).) In canonical variables the field strengths are

$$H_{ik} = \frac{\partial A_k}{\partial x_i} - \frac{\partial A_i}{\partial x_k}, \qquad E_k = -4\pi c B_k + i\frac{\partial A_4}{\partial x_k}. \tag{31}$$

The second equation (30) is then identical with div $\mathbf{E} = 4\pi\rho$. H (eq. (25)) is identical with (8), (12), and the first term reduces to (10).

So far the formalism is 'classical' in the sense that A_μ, B_μ and ψ, χ have not yet been subjected to quantization. The quantum conditions are now quite obvious. We have only to take into account that $A_\mu(\mathbf{r})$ and $A_\mu(\mathbf{r}')$ are to be considered as two *different* variables. Thus we obtain the canonical relations:

$$\begin{aligned} [A_\mu(\mathbf{r}_1)B_\nu(\mathbf{r}_2)] &= i\hbar\delta(\mathbf{r}_2-\mathbf{r}_1)\,\delta_{\mu\nu}, \\ [A_\mu(\mathbf{r}_1)A_\nu(\mathbf{r}_2)] &= [B_\mu(\mathbf{r}_1)B_\nu(\mathbf{r}_2)] = 0. \end{aligned} \tag{32}$$

In the case of the ψ, χ field, the only formal difference is that the commutators are to be replaced by the anticommutators $\{...\}$:

$$\begin{aligned} \{\psi_\rho(\mathbf{r}_2)\chi_{\rho'}(\mathbf{r}_1)\} &= i\hbar\delta_{\rho\rho'}\,\delta(\mathbf{r}_2-\mathbf{r}_1), \\ \{\psi_\rho(\mathbf{r}_2)\psi_{\rho'}(\mathbf{r}_1)\} &= \{\chi_\rho(\mathbf{r}_2)\chi_{\rho'}(\mathbf{r}_1)\} = 0. \end{aligned} \tag{33}$$

In Schrödinger representation all these operators are time independent. In interaction representation (transform by (14)) the same commutation relations (32), (33) hold for the time-dependent operators if both operators are taken at the *same time*:

$$[A_\mu(\mathbf{r}_1,t)B_\nu(\mathbf{r}_2,t)] = i\hbar\,\delta(\mathbf{r}_2-\mathbf{r}_1)\,\delta_{\mu\nu}, \tag{34a}$$

$$\{\psi_\rho(\mathbf{r}_2,t)\chi_{\rho'}(\mathbf{r}_1,t)\} = i\hbar\,\delta(\mathbf{r}_2-\mathbf{r}_1)\,\delta_{\rho\rho'} \quad (\chi = i\hbar\psi^*). \tag{34b}$$

$$[A_\mu(\mathbf{r}_2,t)A_\nu(\mathbf{r}_1,t)] = \{\psi_\rho(\mathbf{r}_2,t)\psi_{\rho'}(\mathbf{r}_1,t)\} = [A_\mu(\mathbf{r}_2,t)\psi_\rho(\mathbf{r}_1,t)] = 0, \tag{34'}$$

etc. Moreover, it follows that

$$\frac{\partial A_\nu(\mathbf{r},t)}{\partial t} = \frac{i}{\hbar} e^{iH_0t/\hbar}[H_0 A_\mu(\mathbf{r})]e^{-iH_0t/\hbar}.$$

Inserting the expression (25) for H_0 and using the commutation relations (32) we obtain

$$\frac{\partial A_\nu(\mathbf{r},t)}{\partial t} = 4\pi c^2 B_\nu(\mathbf{r},t). \tag{35 a}$$

Similarly,

$$\frac{\partial \psi_\rho(\mathbf{r},t)}{\partial t} = \frac{i}{\hbar} e^{iH_0t/\hbar}[H_0 \psi_\rho]e^{-iH_0t/\hbar},$$

($[H_0 \psi]$ is the commutator, not anticommutator). Using the anti-commutation relations (33) and (25) one finds

$$i\hbar\frac{\partial \psi}{\partial t} = (\boldsymbol{\alpha}\mathbf{p}+\beta\mu)\psi(t). \tag{35 b}$$

Altogether, the time-dependent operators $A(t)$, $B(t)$, $\psi(t)$, $\chi(t)$ satisfy the same equations, in the role of *operator-equations*, as are satisfied by the *classical fields* in the *absence of the interaction*.

If we replace B_ν by $(\partial A_\nu/\partial t)/4\pi c^2$, χ by $i\hbar\psi^*$ it appears that (34), (34′) are specializations of the relations (16), (17), for $t_1 = t_2$, taking into account § 8 eqs. (28), (32). For (34′) and (34 a) this is evident. Putting $t_1 = t_2$ in (17) the only term which contributes is

$$\gamma_4 \frac{\partial}{\partial x_{42}} = -\frac{i}{c}\beta\frac{\partial}{\partial t_2},$$

and, as $\psi^\dagger = i\psi^*\beta$, (34 b) results.

The more general looking relations (16), (17) are not really more general than (32) and (33). If we expand the fields and determine the commutation relations so that (32), (33) are satisfied the more general relations (16), (17) follow, as in §§ 10, 12. This is also evident from the fact that in Schrödinger representation the canonical relations (32), (33) are the complete quantum conditions. The initial conditions (30) go over, in quantum theory, into the initial conditions for the expectation values (20). To satisfy the Pauli principle we had to demand anti-commutation rules for ψ. These do not go over into Poisson brackets in the classical limit, as is the case for commutators, which again shows that a classical ψ-field does not really exist.

The Hamiltonian density $\mathscr{H}$ can be generalized to a complete energy-momentum tensor which is given in appendix 7.

IV

METHODS OF SOLUTION

14. Elementary perturbation theory

1. *General considerations.* In the preceding sections the physical and mathematical basis for the treatment of the radiation field as well as the electrons was laid. We have now to develop methods by which we can derive physical results from the formalism. The quantum-mechanical behaviour is described by a wave function or state vector $\Psi(t)$ satisfying the wave equation

$$i\hbar\frac{\partial\Psi}{\partial t} = H\Psi.$$

H is the total Hamiltonian and consists of contributions from the radiation field, $H_{\rm rad}$, from the electrons, $H_{\rm el}$ (whether considered as a quantized field or not) and from their interaction with the electromagnetic field $H_{\rm int}$. It may be left open whether an external field or the Coulomb interaction between the particles is already included in $H_{\rm el}$ or in the interaction $H_{\rm int}$. Let $H_0 = H_{\rm rad}+H_{\rm el}$ be the energy of the non-interacting radiation field and the electrons. It was shown in § 13 that a slightly different representation can be used (interaction representation) whereby

$$i\hbar\frac{\partial\Psi'}{\partial t} = H'_{\rm int}\Psi', \qquad H'_{\rm int} = e^{iH_0t/\hbar}H_{\rm int}\,e^{-iH_0t/\hbar}. \tag{1}$$

H_0 and $H_{\rm int}$ have been given in § 13.1 for the two gauges used (Coulomb and Lorentz gauge). An eigenstate of $H_{\rm rad}$ is characterized by the number of photons n_λ of each description, $\Psi'_{...n_\lambda...}$. (In Lorentz gauge there are four types of photons $n_{\alpha\lambda}$, $\alpha = 1,..., 4$.) In addition, Ψ' depends on the variables describing the electron system. If we are dealing with only a limited number of electrons or positrons it is not necessary to apply the formalism of second quantization. An eigenstate of $H_{\rm el}$, a, say, depends then, for example, on the coordinates of the particles: $\psi_a(...r_k...)$. When second quantization is applied, Ψ' will depend on the numbers n_p^-, n_p^+ of negative and positive electrons in each specified state p. An eigenstate of H_0 is then $\Psi_{...n_\lambda...}\psi_a$ or, when second quantization is applied, $\Psi_{...n_\lambda...}\Psi_{...n_p^+...n_p^-...}$.

The electromagnetic field was described by the emission and absorption operators q_λ^*, q_λ, for example, and $H_{\rm int}$, being proportional to **A** (or A_α in Lorentz gauge) is linear in these operators.‡ Consequently, H'_{int} acting on $\Psi'_{\dots n\lambda\dots}$ changes the number of photons in the index by one unit. Similarly, if second quantization is applied, H'_{int} is bilinear in $\psi^\dagger$, ψ, and H'_{int} acting on Ψ' changes two of the electron numbers n_p^-, n_p^+ by one unit.

We denote an eigenstate of H_0 by Ψ_n for short, where n now includes a complete description of the whole system of radiation and electrons in the absence of their interaction. Let the energy of the system in state n be E_n. The exact solution of (1) can then be expanded

$$\Psi'(t) = \sum_n b_n(t)\Psi_n \tag{2}$$

and $|b_n(t)|^2$ will be the probability for the system to be in the unperturbed state n at the time t. We insert this into (1), multiply by Ψ_n^* and integrate over all variables on which Ψ_n depends (summation if these are discrete). We obtain

$$i\hbar\dot{b}_n(t) = \sum_n H'_{\text{int}\,n|m}\, b_m(t),$$

where $H'_{\text{int}\,n|m}$ is the matrix element of H'_{int} for the transition $m \to n$. The time dependence of H'_{int} is easily found from (1). Since H_0 is diagonal and equals E_n for the eigenstate Ψ_n, we can write

$$H'_{\text{int}\,n|m} = H_{n|m}\, e^{i(E_n-E_m)t/\hbar}$$

and

$$i\hbar\dot{b}_n(t) = \sum_m H_{n|m}\, e^{i(E_n-E_m)t/\hbar} b_m(t). \tag{3}$$

$H_{n|m}$ is independent of t and equals the matrix element of the time-independent H_{int}. It follows from (3) that the normalization of the b's remains constant:

$$\frac{d}{dt}\sum_n |b_n|^2 = 0. \tag{3'}$$

The matrix elements $H_{n|m}$ are different from zero if the number of photons in state m differs by 1 from that in state n. If the wave equation (3) is written out explicitly, it will be a very complicated set of an infinite number of simultaneous equations, one equation for each $b_{\dots n_\lambda\dots}$ with specified numbers n_λ on the left-hand side. On the right-hand side all the $b_{\dots n_\lambda \pm 1,\dots}$ occur. By the equation for $\dot{b}_{\dots n_\lambda \pm 1\dots}$ the b's with indices $\dots n_\lambda+2\dots$ and $\dots n_\lambda+1$, $n_\mu+1\dots$ will be drawn in, and are interconnected, etc. Such a set of equations is far too complicated to be solved exactly for any problem. In fact there is no radiation problem for which an exact solution has been found so far. We shall therefore be forced to apply perturbation theory of some kind, considering the interaction of electrons

‡ In non-relativistic approximation H_{int} also contains a term $\mathbf{A}^2$, and then q_λ^*, q_λ occur quadratically.

with radiation H'_{int} as a small perturbation. Fortunately, this is also quite justifiable. H'_{int} is proportional to e, and a solution expanded in a power series of e, if expressed in terms of a dimensionless quantity, must mean a power series in the fine structure constant $e^2/\hbar c = 1/137$. This is a small quantity. On the other hand, an expansion in a power series also implies some danger. It will be seen later that the theory leads to certain divergence difficulties and the possibility has to be kept in mind that these may be due to an impermissible expansion. However, we shall only be concerned with these difficulties in Chapter VI, where they will be discussed in detail, and it will be seen that they only concern unobservable features of the theory. For all problems of physical interest the application of perturbation theory is beyond doubt.

The mathematical situation then simplifies radically. For a large class of problems (mainly the problems treated in Chapter V) quite an elementary type of perturbation theory suffices which we shall develop first for the benefit of the reader who is only interested in the more elementary applications.

2. *Transition probability and energy change.* Most of the problems of interest in radiation theory are concerned with the calculation of transition probabilities, some are also concerned with the change of total energy due to the interaction with radiation. We shall confine ourselves to these two problems. At the time $t = 0$, say, we suppose the system to be in a definite unperturbed state. For example, we have an atomic electron in an excited state and no photon. Let this state be 0:

$$b_0(0) = 1, \qquad b_n(0) = 0 \quad (n \neq 0). \tag{4}$$

Through the action of the perturbation H'_{int}, this will change in course of time, and later some b_n's will be different from zero, for example, the probability amplitude for the atom in the ground state+a photon emitted. It is clear that, to a first approximation, only those b_n's for which the number of photons in n differs by unity from those in 0 will be different from zero. Thus to first approximation

$$i\hbar\dot{b}_n = H_{n|0}\, b_0\, e^{i(E_n - E_0)t/\hbar}, \tag{5a}$$

$$i\hbar\dot{b}_0 = \sum_n H_{0|n}\, b_n\, e^{i(E_0 - E_n)t/\hbar}. \tag{5b}$$

We can still go a step farther and insert (4) for b_0 in (5a). This will then be valid for comparatively short times t, when b_0 has not yet changed much, but this suffices to calculate the transition probability per unit time. Then the solution satisfying the initial condition (4) is

$$b_n(t) = H_{n|0}\frac{e^{i(E_n-E_0)t/\hbar}-1}{E_0-E_n}. \tag{6}$$

The probability for finding the system at the time t in the state n when it was in state 0 at $t = 0$ is thus

$$|b_n(t)|^2 = |H_{n|0}|^2\, 2\, \frac{1-\cos(E_n-E_0)t/\hbar}{(E_0-E_n)^2}. \tag{7}$$

We shall not be interested in times t which are *very* short, of the order of one period $\hbar/E_n$. Notwithstanding the fact that we have assumed t to be so small that b_0 does not vary much, i.e. *t small compared with the life-time* of the state 0, we may at the same time let $t \to \infty$ in (7), which means more precisely that $t \gg \hbar/E_n$ or $\hbar/E_0$. For $t \to \infty$ the expression containing t is a representation of the δ-function, multiplied by t, as shown in § 8 eq. (19). We have then

$$\frac{1}{t}\,|b_n(t)|^2 = \frac{2\pi}{\hbar}\,|H_{n|0}|^2\delta(E_n-E_0). \tag{8}$$

This has the following meaning: When $t \to \infty$, the expression

$$\frac{1-\cos(E_n-E_0)t/\hbar}{(E_0-E_n)^2 t}$$

vanishes except when $E_n = E_0$. In the latter case it becomes infinite, $\sim t$. The integral over E_n, however, covering the point E_0 is finite

$$\int dE_n \frac{1-\cos(E_n-E_0)t/\hbar}{(E_0-E_n)^2 t} = \frac{1}{\hbar}\int_{-\infty}^{+\infty} dy\frac{1-\cos y}{y^2} = \frac{\pi}{\hbar}.$$

This is expressed by the factor $\delta(E_n-E_0)$.

The left-hand side of (8) is the *transition probability per unit time*, for the transition from state 0 to n. The δ-factor shows that energy is conserved, and transitions only take place between states of equal unperturbed energy. It is assumed here explicitly that a time t exists so that $E_n t/\hbar \gg 1$, yet $t \ll$ lifetime of the initial state. It will be seen in Chapter V that this is always true. The fact is connected with the smallness of the damping effects (line breadth) as has already been pointed out in classical theory (§ 4).

Formula (8) with the δ-function on the right only has a definite meaning if an integration over either E_n or E_0 can be performed, i.e. if one of the two states belong to the continuous spectrum. This is the case if a photon is present in n or 0. We are then not interested in the probability for any definite state n, specified by the *exact* partial wave of the photon, etc., but only in transitions to all states with the momentum of the photon, etc., lying in certain infinitesimal intervals. We can multiply by the number of states $\rho_n\,dE_n$ with these specifications and the energy in the interval dE_n and integrate over E_n. Then

$$w_{n|0} = \frac{2\pi}{\hbar}|H_{n|0}|^2\rho_n. \tag{9}$$

This formula gives the desired transition probability per sec. to a first approximation and it will be used for the emission and absorption of light.

In many cases, however, (9) gives zero to a first approximation. Consider, for example, the scattering of a photon by an electron (whether free or bound). Here two occupation numbers n_λ have to change: The primary photon has to be absorbed and the secondary (scattered) photon to be emitted, whereas (9) gives rise to the change of one photon only. To describe this process we have to proceed to the next higher approximation and allow for two factors H to come in. In place of (5) we write

$$i\hbar\dot{b}_{n'} = H_{n'|0}\,b_0\,e^{i(E_{n'}-E_0)t/\hbar}, \tag{10 a}$$

$$i\hbar\dot{b}_{n} = \sum_{n'} H_{n|n'}\,b_{n'}\,e^{i(E_n-E_{n'})t/\hbar}. \tag{10 b}$$

The states n' differ from 0 by one photon, and n' again from n by one photon, therefore n differs from 0 by two photons as is required. Only those states n' that can bridge the transition from 0 to the wanted state n need be considered. We call these states n' *intermediate* or *virtual states*.

The initial condition (4) need not be satisfied for the virtual states. The $b_{n'}$ will be rapidly varying functions with small amplitude, and if we were to satisfy the initial condition, it would only be for a period of the order of $\hbar/E_n$ in any case. This is physically meaningless.

We solve (10) by again inserting $b_0 = 1$, and get

$$b_{n'} = \frac{H_{n'|0}}{E_0-E_{n'}}\,e^{i(E_{n'}-E_0)t/\hbar}, \tag{11}$$

$$i\hbar\dot{b}_n = \sum_{n'}\frac{H_{n|n'}H_{n'|0}}{E_0-E_{n'}}\,e^{i(E_n-E_0)t/\hbar}. \tag{12}$$

(12) is now identical with the problem (5 a), already solved, if we replace $H_{n|0}$ by the 'compound matrix element':

$$K_{n|0} = \sum_{n'}\frac{H_{n|n'}H_{n'|0}}{E_0-E_{n'}}. \tag{13}$$

Consequently the transition probability per unit time is

$$w_{n|0} = \frac{2\pi}{\hbar}\left|\sum_{n'} \frac{H_{n|n'}H_{n'|0}}{E_0-E_{n'}}\right|^2 \rho_n, \tag{14}$$

(9) is of the order e^2, (14) of the order e^4. We could continue in this way and consider more complicated radiation processes requiring still higher steps with more intermediate states. For example, when three photons are involved in a process, $w_{n|0}$ is again given by (9) but

$$K_{n|0} = \sum_{n',n''} \frac{H_{n|n'}H_{n'|n''}H_{n''|0}}{(E_0-E_{n'})(E_0-E_{n''})}. \tag{15}$$

In all cases, (9), (14), (15) only represent the transition probability to the first non-vanishing order and only when $E_0-E_{n'}$ cannot vanish.

In collision problems it is the *cross-section* rather than the transition probability per sec. which is of physical interest. Consider two particles in a volume L^3 colliding with opposite momenta. If the particles are represented by plane waves normalized to L^3 it is obvious that $w_{n|0} \to 0$ as $L^3 \to \infty$, for any state n. This can also be verified from the L^3-dependence of the matrix elements (see subsection 3) and ρ_n (§§ 6, 11). On the other hand, the chance for a specified collision $0 \to n$ can be represented by an area $\phi_{n|0}$ perpendicular to the line of motion, so that a collision is said to take place if the particles collide within this cross-section. The number of collisions taking place per sec. is then

$$\phi_{n|0}(v_1+v_2)/L^3 \quad \text{or} \quad \phi_{n|0} = \frac{L^3}{v_1+v_2} w_{n|0}. \tag{16}$$

v_1+v_2 is the relative velocity of the particles (this can also be larger than c!). $\phi_{n|0}$ is independent of L^3. Evidently the cross-section is invariant under Lorentz transformations along the line of motion.

The change in the energies of the unperturbed states can be considered in the same way. Since H'_{int} involves the emission or absorption of a photon, no diagonal elements of H'_{int} exist. The perturbation energy is at least of the second order in H'_{int}. Consider a state n with unperturbed energy E_n. We seek now a stationary solution, with total energy E, $E = E_n+\Delta E_n$, say. For b_n we have

$$i\hbar\dot{b}_n = \sum_{n'} H_{n|n'} b_{n'} e^{i(E_n-E_{n'})t/\hbar},$$

$$i\hbar\dot{b}_{n'} = H_{n'|n} b_n e^{i(E_{n'}-E_n)t/\hbar}.$$

It is sufficient to restrict the second equation to the one term b_n, because any other terms $b_{n''}$, say, would only give rise to higher approximations in ΔE_n. We find a stationary solution by putting

$$b_n(t) = c . e^{-i\Delta E_n t/\hbar}, \qquad b_{n'} = c . \frac{H_{n'|n}}{E_n - E_{n'} + \Delta E_n} e^{i(E_{n'} - E_n - \Delta E_n)t/\hbar}$$

and thus
$$\Delta E_n = \sum_{n'} \frac{H_{n|n'} H_{n'|n}}{E_n - E_{n'}} \tag{17}$$

omitting ΔE_n in the denominator. This is justified because ΔE_n is considered small compared with E_n. (18) is the usual formula for the second-order perturbation of the energy, and can be regarded as the diagonal element of (13). The change of energy of an electron, due to its interaction with radiation, or its *self-energy*, can be described as due to photons emitted in intermediate states n'. This radiative self-energy is in addition to that due to its Coulomb field already met with in § 4. Both contributions will be treated in quantum theory in detail in § 29.

If radiation problems are treated in non-relativistic approximation, H'_{int} consists of two parts, one linear in **A** the other $\sim$ $\mathbf{A}^2$, thus

$$H'_{\text{int}} = H^{(1)'} + H^{(2)'}.$$

$H^{(2)'}$ gives rise to a change of two photons directly. It is quite obvious that the formula (13) for second-order transitions is then to be replaced by

$$K_{n|0} \equiv \sum_{n'} \frac{H^{(1)}_{n|n'} H^{(1)}_{n'|0}}{E_0 - E_{n'}} + H^{(2)}_{n|0}. \tag{18}$$

The results of this section will be sufficient for the applications of the theory to radiative processes treated in the lowest non-vanishing order of approximation (whichever this lowest order may be), i.e. to nearly all applications of Chapter V. Only problems connected with the finite line breadth require an improvement of the above solutions (see § 18). A general theory of damping phenomena will be developed in § 16 and used in §§ 22, 34. A more general treatment of perturbation theory including higher-order corrections to radiative processes will be given in § 15 and Chapter VI.

3. *Matrix elements.* For the applications in Chapter V it will be convenient to work out explicitly the matrix elements of H_{int} which occur in the above perturbation theory. We consider here the electrons as particles in configuration space. In Coulomb gauge H_{int} is given by § 13 eq. (6) or, in non-relativistic approximation, (7). We consider only the part depending on the radiation field (the Coulomb interaction

may be included in H_{el}). For **A** we use the expansion § 7 eq. (3), viz.

$$\mathbf{A} = \sum_{\lambda} (q_\lambda \mathbf{A}_\lambda + q_\lambda^* \mathbf{A}_\lambda^*), \qquad \mathbf{A}_\lambda = \mathbf{e}_\lambda \sqrt{(4\pi c^2)} e^{i(\boldsymbol{\kappa}_\lambda \mathbf{r})}. \tag{19}$$

The matrix elements of q_λ and q_λ^* are given by § 7 eq. (9). For the relativistic interaction $H_{n|m}$ is different from zero only for a transition of the type

$$m \equiv b,..., n_\lambda,... \rightarrow n \equiv a,..., n_\lambda \pm 1..., \tag{20}$$

i.e. for the emission or absorption of one photon only and for any transition of the electron system $b \rightarrow a$.

If $\int$ denotes the integration over the space coordinates and summation over the index ρ of ψ we obtain for a single particle

$$H_{an_\lambda|bn_\lambda+1} = -e\sqrt{\left(\frac{2\pi\hbar^2c^2}{k_\lambda}\right)}\sqrt{(n_\lambda+1)} \int \psi_a^* \alpha_e e^{i(\boldsymbol{\kappa}_\lambda \mathbf{r})} \psi_b, \tag{21 a}$$

$$H_{an_\lambda+1|bn_\lambda} = -e\sqrt{\left(\frac{2\pi\hbar^2c^2}{k_\lambda}\right)}\sqrt{(n_\lambda+1)} \int \psi_a^* \alpha_e e^{-i(\boldsymbol{\kappa}_\lambda \mathbf{r})} \psi_b, \tag{21 b}$$

where α_e represents the component of the matrix vector $\boldsymbol{\alpha}$ in the direction of polarization of the light quantum.

For the transitions (20) the energy will not in general be conserved if either of the states in question is an intermediate state. The energy differences of the two states for the two transitions are

$$E_n - E_m = E_a - E_b \mp k_\lambda. \tag{22}$$

In the non-relativistic approximation, the first term of § 13 eq. (7), $H^{(1)}$, gives also matrix elements for transitions by which one quantum is emitted or absorbed. The second term is proportional to

$$A^2 = \sum_{\lambda,\mu} [q_\lambda q_\mu (\mathbf{A}_\lambda \mathbf{A}_\mu) + q_\lambda q_\mu^* (\mathbf{A}_\lambda \mathbf{A}_\mu^*) + q_\lambda^* q_\mu (\mathbf{A}_\lambda^* \mathbf{A}_\mu) + q_\lambda^* q_\mu^* (\mathbf{A}_\lambda^* \mathbf{A}_\mu^*)].$$

The matrix elements are different from zero if two quanta are emitted or absorbed (or one emitted, one absorbed). Thus (for one particle)

$$H^{(1)}_{an_\lambda|bn_\lambda+1} = -\frac{e}{\mu}\sqrt{\left(\frac{2\pi\hbar^2c^2}{k_\lambda}\right)}\sqrt{(n_\lambda+1)} \int \psi_a^* p_e e^{i(\boldsymbol{\kappa}_\lambda \mathbf{r})} \psi_b, \qquad \text{N.R. (23 a)}$$

$$H^{(1)}_{an_\lambda+1|bn_\lambda} = -\frac{e}{\mu}\sqrt{\left(\frac{2\pi\hbar^2c^2}{k_\lambda}\right)}\sqrt{(n_\lambda+1)} \int \psi_a^* p_e e^{-i(\boldsymbol{\kappa}_\lambda \mathbf{r})} \psi_b, \qquad \text{N.R. (23 b)}$$

where p_e represents the component of $\mathbf{p}$ in the direction of polarization,† and, for example,

$$H^{(2)}_{a,n_\lambda+1,n_\mu|b,n_\lambda,n_\mu+1} = \frac{e^2}{\mu}(\mathbf{e}_\lambda\,\mathbf{e}_\mu)\frac{2\pi\hbar^2c^2}{\sqrt{(k_\lambda k_\mu)}}\sqrt{\{(n_\lambda+1)(n_\mu+1)\}}\int \psi_a^*\,e^{i(\kappa_\mu-\kappa_\lambda,\mathbf{r})}\psi_b. \quad (24)$$

Finally, we consider the special but very important case of *free electrons*. The relativistic wave function for a free electron has been given in § 11. For two states ψ_a, ψ_b with the momenta $\mathbf{p}_a$, $\mathbf{p}_b$ we have

$$\psi_a = u_a\,e^{i(\mathbf{p}_a\mathbf{r})/\hbar c}, \qquad \psi_b = u_b\,e^{i(\mathbf{p}_b\mathbf{r})/\hbar c}.$$

Inserting these functions in the matrix elements (21) we obtain for the integral (putting $\boldsymbol{\kappa} = \mathbf{k}/\hbar c$)

$$\int = (u_a^*\,\alpha_e\,u_b)\int e^{i(\mathbf{p}_b-\mathbf{p}_a\pm\mathbf{k}_\lambda,\mathbf{r})/\hbar c}. \quad (25)$$

This integral is proportional to $\delta(\mathbf{p}_b-\mathbf{p}_a\pm\mathbf{k}_\lambda)$ and vanishes unless

$$\mathbf{p}_b-\mathbf{p}_a\pm\mathbf{k}_\lambda = 0, \quad (26)$$

for absorption and emission respectively.

Equation (26) expresses the law of *conservation of momentum*. For all interaction processes of a light quantum with a free particle, the momentum is therefore conserved.‡ On the other hand, the conservation of energy was a general result of the perturbation theory (subsection 2).

The matrix elements for transitions of a free electron for which the momentum is conserved are given by

$$H_{p_a n_\lambda+1|p_b n_\lambda} = H_{p_a n_\lambda|p_b n_\lambda+1} = -e\sqrt{\left(\frac{2\pi\hbar^2c^2}{k_\lambda}\right)}\sqrt{(n_\lambda+1)}(u_a^*\,\alpha_e\,u_b), \quad (27)$$

the wave functions of the plane waves being normalized for a unit volume $L^3 = 1$.

For the non-relativistic case the matrix elements are

$$H^{(1)}_{p_a,n_\lambda+1|p_b,n_\lambda} = H^{(1)}_{p_a n_\lambda|p_b n_\lambda+1} = -e\sqrt{\left(\frac{2\pi\hbar^2c^2}{k_\lambda}\right)}\sqrt{(n_\lambda+1)}\frac{p_e}{\mu} \quad \text{N.R. (28)}$$

and for transitions in which two quanta are involved, for instance,

$$H^{(2)}_{p_a,n_\lambda+1,n_\mu|p_b,n_\lambda,n_\mu+1} = \frac{e^2}{\mu}(\mathbf{e}_\lambda\,\mathbf{e}_\mu)\frac{2\pi\hbar^2c^2}{\sqrt{(k_\lambda k_\mu)}}\sqrt{\{(n_\lambda+1)(n_\mu+1)\}}. \quad \text{N.R. (28')}$$

† Note that it is irrelevant whether the operator p_e stands before or after $\exp(i\boldsymbol{\kappa}\mathbf{r})$ because $(\boldsymbol{\kappa}\mathbf{e}) = 0$.

‡ For a bound electron the momentum is not in general conserved since the nucleus can take any amount of momentum.

If Lorentz gauge is used all that changes is that four types of photons occur with occupation numbers n_α, $\alpha = 1,..., 4$. As has been shown in § 10, the matrix elements of $q_{\alpha\lambda}$ and $q^*_{\alpha\lambda}$ are precisely the same as for q_λ, q^*_λ. Thus in place of (27) we obtain for free particles:

$$H_{p_a, n_\alpha|p_b, n_\alpha+1} = -e\sqrt{\left(\frac{2\pi\hbar^2c^2}{k}\right)}\sqrt{(n_\alpha+1)}(u_a^\dagger\gamma_\alpha u_b) = H_{p_a, n_\alpha+1|p_b, n_\alpha} \quad (29)$$

(omitting the index λ). For bound particles the use of Lorentz gauge is impracticable because the Lorentz condition then takes a very complicated form.

In all the above formulae the volume L^3 in which the field is considered to be enclosed has been taken to be unity. Otherwise the expansion (19) has a factor $L^{-\frac{3}{2}}$, if $\mathbf{A}_\lambda$ is to be normalized to $4\pi c^2$. It follows that all matrix elements involving one photon (21), (23), (27), (28) are proportional to $L^{-\frac{3}{2}}$, and (24), (28′) to L^{-3}. The compound matrix elements (13), (15), etc., involving n photons are proportional to $L^{-3n/2}$.

15. General perturbation theory: free particles

For the more complicated applications of quantum electrodynamics it is necessary to develop the perturbation theory from a more general point of view which will make it possible to carry it to any desired order of approximation. It will be useful to distinguish between two classes of problems: (i) collisions between free particles and quanta, and (ii) processes involving bound atomic states. For the first class a number of interesting theorems and connexions will hold permitting a fairly general treatment whereas rather different considerations will have to be applied for the second class (§ 16).

In this section we consider collisions between free particles and quanta. We may also include the case when an external field $A_\mu^{(e)}$ is present, provided that this is considered as a perturbation causing transitions of the free particles (e.g. the scattering of an electron in a static field) and not as giving rise to the existence of discrete levels.

The characteristic simplifications in the case of free particles are the following. Before the collision the particles and quanta are so far apart from each other that their interaction vanishes. The total energy is that of the non-interacting particles and belongs to the continuous spectrum. Since energy is conserved during the collision, the exact total energy at any time coincides with the energy of the free *non-interacting particles* before the collision, and also with the final energy after the collision, when the interaction again vanishes. All these energies belong to the continuous spectrum. Thus, for collisions between free particles and quanta the *unperturbed energies before and after the collision are the same.*† This is not true for bound states where the unperturbed energy is in general different from the exact total energy. Moreover, the transition probability per unit time vanishes when the volume L^3 in which we consider the system to be enclosed is infinite. What is finite is the collision cross-section (see § 14 eq. (16)). It follows then that the 'line breadth' of the initial state vanishes and we need not be concerned here with questions of line breadth at all.

1. *Time-dependent canonical transformation.* We start from the wave equation in interaction representation § 13 eq. (15) (omitting the dash on Ψ'),

$$i\hbar\dot{\Psi} = (H(t)+H^{(e)}(t))\Psi \tag{1}$$

where $H(t) = H'_{\text{int}}$ is the interaction with the radiation field and $H^{(e)}(t)$ that with an external field (if present). $H(t)$ will be supplemented by a term describing the self-energies (subsection 2), but the following is independent of the explicit form of $H(t)$. Since interaction representation is used $H(t)$ and $H^{(e)}$ depend on t. In the relativistic theory of electrons $H(t)$ and $H^{(e)}$ are linear in e and we shall consider this as small to the first order. For non-relativistic problems $H(t)$ also contains a quadratic term $\sim e^2$ (see § 13 eq. (7)) which should be considered as of second order. If Coulomb gauge is used $H(t)$ includes the Coulomb interaction, $H^{(c)}$, which is also of second order $\sim e^2$. It will be very easy to generalize our results to include also terms of second order, but for the present we assume that $H(t)$ has only first-order terms.

Now consider the effect of the perturbation caused by $H(t)$ (leaving $H^{(e)}$ out of consideration). Let Ψ first be the unperturbed state vector of a *single* free particle or quantum. H, acting on Ψ, changes the number of photons by one and if second quantization is used for the electron field, changes the number of electrons also. The solution of (1), when written in a representation where the number of photons and electrons of each kind is given, will consist of a series with terms corresponding to any number of photons and electrons. A single free electron will no longer be a 'bare' electron but will be accompanied by photons (and electron pairs). Similarly a photon will be accompanied by electron pairs. These accompanying particles we call *virtual*. They are the particles which occur in intermediate states (see § 14). The classical analogue of this

† The 'unperturbed energy' of the free particles, however, includes the self-energies of the individual particles (see subsection 2).

virtual field is nothing but the field accompanying a moving particle in its neighbourhood. Thus to each unperturbed Ψ of a single particle there is associated a perturbed state vector, denoted by Ψ', which describes the same particle, but whilst Ψ is only the *zero*th order of approximation, Ψ' is the exact solution of (1) (if the perturbation is carried to an infinite order). Ψ' will be independent of time. The energy belonging to Ψ' is the perturbed energy which differs from the unperturbed energy by what we call the self-energy of the particle. This fact, however, will be taken into account at a later stage (subsection 2 and § 29). If now several particles or quanta are involved, a product state vector composed of such factors Ψ' (again called Ψ') will no longer be time independent. Transitions between the various states of the free particles will take place, giving rise, for instance, to the actual emission of a real photon, a scattering of the particles, etc. It is these transition probabilities which we wish to calculate. They only take place between states of the same energy. Thus our task is to determine Ψ' so that its time variation is due to such real transitions only. $\dot{\Psi}'$ will be determined by a new Hamiltonian K, which is different from zero only for transitions of the same energy.

Ψ' must be obtainable from Ψ by a canonical transformation which we write†

$$\Psi' = S^{-1}\Psi, \qquad S\Psi' = \Psi, \tag{2}$$

where $S(t)$ is a unitary operator (derivable from H) and obviously depends also on time. If we insert (2) in (1) we obtain a wave equation for Ψ':

$$i\hbar\dot{\Psi} = i\hbar(\dot{S}\Psi'+S\dot{\Psi}') = (H+H^{(e)})S\Psi'$$

or multiplying by S^{-1}:

$$\begin{aligned} i\hbar\dot{\Psi}' &= (S^{-1}H(t)S - i\hbar S^{-1}\dot{S} + S^{-1}H^{(e)}S)\Psi' \\ &\equiv (K(t)+K^{(e)})\Psi', \end{aligned} \tag{3}$$

$$K(t) = S^{-1}H(t)S - i\hbar S^{-1}\dot{S}, \qquad K^{(e)}(t) = S^{-1}H^{(e)}(t)S. \tag{3'}$$

As H was hermitian, K and $K^{(e)}$ will also be *hermitian*. So far S is perfectly arbitrary. The proper determination of S will be such that $K(t)$ has only elements for which the total energy is preserved. $\dot{\Psi}'$ ought to vanish for a single free particle but this will only be the case when proper regard is taken of the self-energy.

† A canonical transformation of this type was first used by F. Bloch and A. Nordsieck, *Phys. Rev.* **52** (1937), 54. The following version of perturbation theory does not follow a published pattern but uses elements from papers by F. J. Dyson (ref. on p. 162), T. Tati, and S. Tomonaga, *Prog. Theor. Phys.* **3** (1948), 391 (and subsequent papers); J. Schwinger, *Phys. Rev.* **74** (1948), 1439, **75** (1949), 651, **76** (1949), 790; and W. Heitler and S. T. Ma, *Phil. Mag.* **11** (1949), 651.

We introduce first a small but as yet finite energy interval ϵ and demand

$$K(t)_{n|m} = 0 \quad \text{for } |E_n - E_m| > \epsilon. \tag{4}$$

Later ϵ will tend to zero automatically. Until further notice (p. 155), the unperturbed energies can be taken for the E_n. We shall say that K only has elements on the 'energy shell', or is diagonal with respect to the total energy. For the evaluation of K we anticipate the limit $\epsilon \to 0$.

To determine S we expand S and K according to powers of e, noting that $H(t)$ is of first order:

$$S = 1+S_1+S_2+S_3+\dots,$$

$$S^{-1} = 1-S_1+S_1^2-S_2+S_1 S_2+S_2 S_1-S_1^3-S_3+\dots, \tag{5}$$

$$K(t) = K_1(t)+K_2(t)+K_3(t)+K_4(t)+\dots, \tag{6}$$

$S^{-1} = S^\dagger$ is determined so that $SS^{-1} = 1$. Writing (3′) in the form

$$SK = HS - i\hbar\dot{S}$$

we get the following set of equations:

$$K_1(t) = H(t) - i\hbar\dot{S}_1, \tag{7_1}$$

$$K_2(t) = H(t)S_1 - S_1 K_1(t) - i\hbar\dot{S}_2, \tag{7_2}$$

$$K_3(t) = H(t)S_2 - S_1 K_2(t) - S_2 K_1(t) - i\hbar\dot{S}_3, \tag{7_3}$$

$$K_4(t) = H(t)S_3 - S_1 K_3(t) - S_2 K_2(t) - S_3 K_1(t) - i\hbar\dot{S}_4. \tag{7_4}$$

We now consider the parts on and outside the energy shell separately (in the sense of eq. (4)). For the non-diagonal parts $K = 0$. The equations (7) then permit the determination of S_λ ($\lambda = 1, 2, 3,\dots$) in all orders in succession, by mere integrations. The diagonal parts do not determine K and S uniquely because only the combinations $K_\lambda + i\hbar\dot{S}_\lambda$ occur in (7). This quite corresponds to the physical situation. The system is highly degenerate and any linear transformation can be carried out between states of equal energy. This we do not wish to do. If we write

$$\Psi'_m = \sum_n S^{-1}_{m|n} \Psi_n,$$

and if we were to allow elements $S_{m|n}$ on the same energy shell to occur, Ψ' would be a linear combination of states Ψ between which *real* transitions could occur, for instance between states of an electron with different directions of the momentum. We therefore require that the elements of $S_{m|n}$ on the energy shell should vanish. Only terms $S_{m|m}$ which are diagonal with respect to *all* variables must be permitted to occur because these merely mean a renormalization of Ψ' which necessarily must occur if Ψ' is to be normalized to unity and therefore S is unitary. It is evident (and will be verified below) that $\dot{S}_{m|m} = 0$. In the following the index d denotes elements on the energy shell, n.d. means non-diagonal with respect to energy. The index D means diagonal with respect to all variables. So S only has elements $S_{\text{n.d.}}$ and S_D, and we can require

$$\dot{S}_d = \dot{S}_D = 0. \tag{8}$$

H, being linear in the electromagnetic field, only has matrix elements for emission and absorption of a single photon and, furthermore, for free particles momentum is conserved. Momentum and energy cannot be conserved simultaneously for a single emission act and consequently $H_d(t) = 0$. To simplify writing we shall write H for $H(t)$, H' for $H(t')$, $H^{(e)\prime}$ for $H^{(e)}(t')$, etc. We obtain now from (7_1)

$$K_1 = H_d = 0, \tag{9_1}$$

$$S_{1\,\text{n.d.}} = \frac{1}{i\hbar}\int^{t} H'\, dt'. \tag{10_1}$$

The lower limit of integration is an arbitrary constant which will be fixed later.

The S_D elements follow from the condition that S should be unitary. $S^\dagger = S^{-1}$, or, from (5),

$$S_1^\dagger + S_1 = 0, \qquad S_2^\dagger + S_2 = S_1^2, \qquad S_3 + S_3^\dagger = S_1 S_2 + S_2 S_1 - S_1^3.$$

For the diagonal elements we can assume without loss of generality that these are real (an imaginary contribution would merely mean an insignificant phase factor). Then

$$S_{1d} = S_{1D} = 0, \tag{11_1}$$

$$S_{2D} = \tfrac{1}{2}(S_1^2)_D, \tag{11_2}$$

$$S_{3D} = \tfrac{1}{2}(S_1 S_2 + S_2 S_1 - S_1^3)_D. \tag{11_3}$$

It also follows from (11_1) that $(S_1 K)_d = 0$, because $K = K_d$, and the n.d. elements of S_1 do not contribute to the d-product. Hence from (7_2), (11_2)

$$K_2(t) = \frac{1}{i\hbar}\int^{t} (HH')_d\, dt', \tag{9_2}$$

$$S_{2\,\text{n.d.}} = \frac{1}{(i\hbar)^2}\int^{t} dt' \int^{t'} dt''\, (H'H'')_{\text{n.d.}}, \tag{10_2}$$

$$S_{2D} = \tfrac{1}{2}(S_1^2)_D = \frac{1}{2}\frac{1}{(i\hbar)^2}\int^{t} dt' \int^{t} dt''\, (H'H'')_D. \tag{$10_2'$}$$

In (9_2) $H = H(t)$ is constant during the integration. The order of the factors HH' is, of course, essential. From (7_3) we obtain, using (9_2), (10_2), (11_2):

$$K_3(t) = \frac{1}{(i\hbar)^2} \int^t dt' \int^{t'} dt''\ (HH'H'')_d, \tag{9_3}$$

$$S_{3\,\text{n.d.}} = \frac{1}{(i\hbar)^3} \left\{ \int^t dt' \int^{t'} dt'' \int^{t''} dt'''\ (H'(H''H''')_{\text{n.d.}})_{\text{n.d.}} + \right.$$
$$\left. + \tfrac{1}{2}(i\hbar)^2 \int^t H'\, dt' (S_1^2)_D - \int^t dt' \int^{t'} dt'' \int^{t'} dt'''\ H''(H'H''')_d \right\}. \tag{10_3}$$

Carrying on up to K_4, S_{3D} will not be needed since it only occurs as factor of H in K_4 and $H_d = 0$. S_4 will not be needed either. Finally from (7_4):

$$K_4(t) = (HS_3)_d - S_{2D}\, K_2(t)$$
$$= \frac{1}{(i\hbar)^3} \left\{ \int^t dt' \int^{t'} dt'' \int^{t''} dt'''\ (HH'(H''H''')_{\text{n.d.}})_d - \right.$$
$$\left. - \int^t dt' \int^{t'} dt'' \int^{t'} dt'''\ (HH''(H'H''')_d)_d \right\} +$$
$$+ \tfrac{1}{2} K_2(t)(S_1^2)_D - \tfrac{1}{2}(S_1^2)_D\, K_2(t). \tag{9_4}$$

The lower limit of integration will be conveniently chosen to be some time $-T$, $T \to \infty$, say, in the remote past, before the collision. H, or any n.d.-product of H's, will be a fast periodic function of t

$$H_{n|m} \sim \exp i(E_n - E_m)t/\hbar,$$

and if we take the average over T, over an interval covering many periods, any integral of an n.d.-product vanishes at the lower limit. This is, of course, not true for the d-factors $(HH')_d$, etc., occurring in (9). Alternatively we could say that before the collision the interaction vanishes. The lower limits in all time integrations can then be chosen to be $-\infty$.† Likewise all n.d.-products vanish for $t = +\infty$.

The property that K has, of lying on the energy shell, is best brought out if we form the time integral $\int_{-\infty}^{+\infty} K\, dt$. In fact it will be seen (subsection 3) that this is all we need if we are only interested in the state a long time

† That the initial time can be chosen to be $-\infty$ hinges on the fact that the transition probability per sec. vanishes for $L^3 = \infty$. For problems involving bound states this simplification is not possible.

after the collision. In interaction representation, and using, for the time being, the unperturbed states for a representation, K can be written:

$$K_{n|m}(t) = K_{n|m}\, e^{i(E_n - E_m)t/\hbar}, \tag{12}$$

where $K_{n|m}$ is independent of t, and hence:

$$\bar{K}_{n|m} \equiv \frac{1}{2\pi\hbar} \int\limits_{-\infty}^{+\infty} K_{n|m}(t)\, dt = K_{n|m}\, \delta(E_n - E_m). \tag{12'}$$

By taking the time integrals the above formulae for K can be simplified further. First we see that

$$S_1(\infty) = \frac{1}{i\hbar} \int\limits_{-\infty}^{+\infty} H\, dt = 0.$$

Furthermore, since $H(-\infty) = 0$, it is seen from $(10_2')$ that†

$$S_{2D}(t) = \tfrac{1}{2}(S_1^2)_D$$

is independent of t. The second term of (9_4) can then be written, by partial integration

$$\frac{1}{(i\hbar)^3} \int\limits_{-\infty}^{+\infty} dt \int\limits_{-\infty}^{t} dt' \int\limits_{-\infty}^{t'} dt'' \int\limits_{-\infty}^{t'} dt'''\, (HH''(H'H''')_d)_d$$

$$= \int\limits_{-\infty}^{+\infty} dt\, \frac{dS_1}{dt}(t) \int\limits_{-\infty}^{t} dt'\, S_1(t')K_2(t')$$

$$= S_1(\infty) \int\limits_{-\infty}^{+\infty} S_1(t')K_2(t')\, dt' - \int\limits_{-\infty}^{+\infty} S_1^2(t)K_2(t)\, dt = -(S_1^2)_D \int\limits_{-\infty}^{+\infty} K_2(t)\, dt.$$

Thus we obtain for $\bar{K}$:

$$2\pi\hbar \bar{K}_1 = 0, \tag{13$_1$}$$

$$2\pi\hbar \bar{K}_2 \equiv \int\limits_{-\infty}^{+\infty} K_2(t)\, dt = \frac{1}{i\hbar} \int\limits_{-\infty}^{+\infty} dt \int\limits_{-\infty}^{t} dt'\, (HH')_d, \tag{13$_2$}$$

$$2\pi\hbar \bar{K}_3 = \frac{1}{(i\hbar)^2} \int\limits_{-\infty}^{+\infty} dt \int\limits_{-\infty}^{t} dt' \int\limits_{-\infty}^{t'} dt''(HH'H'')_d, \tag{13$_3$}$$

$$2\pi\hbar \bar{K}_4 = \frac{1}{(i\hbar)^3} \int\limits_{-\infty}^{+\infty} dt \int\limits_{-\infty}^{t} dt' \int\limits_{-\infty}^{t'} dt'' \int\limits_{-\infty}^{t''} dt'''\, (HH'(H''H''')_{\text{n.d.}})_d +$$
$$+ 2\pi\hbar\left(\tfrac{1}{2}(S_1^2)_D\, \bar{K}_2 + \tfrac{1}{2}\bar{K}_2 (S_1^2)_D\right)_d. \tag{13$_4$}$$

† The limit $t \to \infty$ must only be taken *after* the square is formed and

$$(S_1^2)_{t\to\infty} \neq (S_1(t = \infty))^2.$$

The same applies to all products, etc., occurring in what follows.

The terms $\sim \frac{1}{2}(S_1^2)_D$ arise from the renormalization of the new state vector Ψ' and will be called the renormalization terms. Apart from these $\bar{K}$ has the same structure in all orders.

By the above transformation transitions to virtual states are eliminated from the wave equation (3). The new Hamiltonian K gives rise to real transitions only.

Finally we have to consider the contribution from the external field $H^{(e)}$

$$K^{(e)}(t) = S^{-1}H^{(e)}(t)S.$$

We shall require $\bar{K}^{(e)}$ only up to the second order in H. To this order

$$K^{(e)} = H^{(e)}+H^{(e)}S_1-S_1 H^{(e)}+S_2^\dagger H^{(e)}+H^{(e)}S_2-S_1 H^{(e)}S_1$$
$$\equiv K_0^{(e)}+K_1^{(e)}+K_2^{(e)}.$$

Using the expressions for S_1 and S_2 and

$$S_{2\,\text{n.d.}}^\dagger = \frac{1}{(i\hbar)^2}\int^{t} dt' \int^{t'} dt''\, (H''H')_{\text{n.d.}}, \qquad S_{2D}^\dagger = S_{2D},$$

we get

$$2\pi\hbar\bar{K}_0^{(e)} = \int_{-\infty}^{+\infty} H^{(e)}\, dt, \tag{$13_0^{(e)}$}$$

$$2\pi\hbar\bar{K}_1^{(e)} = \frac{1}{i\hbar}\int_{-\infty}^{+\infty} dt \int_{-\infty}^{t} dt'\, (H^{(e)}H'-H'H^{(e)}),$$

$$2\pi\hbar\bar{K}_2^{(e)} = \frac{1}{(i\hbar)^2}\int_{-\infty}^{+\infty} dt \int_{-\infty}^{t} dt' \int_{-\infty}^{t'} dt''\, \{H^{(e)}(H'H'')_{\text{n.d.}}+(H''H')_{\text{n.d.}}\, H^{(e)}\}-$$

$$-\frac{1}{(i\hbar)^2}\int_{-\infty}^{+\infty} dt \int_{-\infty}^{t} dt' \int_{-\infty}^{t} dt''\, H'H^{(e)}H''+\frac{1}{2}\int_{-\infty}^{+\infty} ((S_1^2)_D H^{(e)}+H^{(e)}(S_1^2)_D)\, dt.$$

$\bar{K}_1^{(e)}$ and $\bar{K}_2^{(e)}$ can also be written more symmetrically. Changing the notation $t \rightleftarrows t'$ and then the order of integration in $\bar{K}_1^{(e)}$ and using the fact that

$$\int_{-\infty}^{+\infty} H\, dt = 0$$

we obtain $\quad -\int_{-\infty}^{+\infty} dt \int_{-\infty}^{t} dt'\, H'H^{(e)} = +\int_{-\infty}^{+\infty} dt \int_{-\infty}^{t} dt'\, HH^{(e)\prime}.$

$\bar{K}_2^{(e)}$ can be rewritten in a similar way. For any n.d.-product

$$\int_{t}^{\infty} dt'\, (...)_{\text{n.d.}} = -\int_{-\infty}^{t} dt'\, (...)_{\text{n.d.}}$$

because the integral $-\infty...+\infty$ vanishes. (This is not true for the terms $S_1^2 H^{(e)}$, $H^{(e)} S_1^2$.) Taking this into account we obtain after a simple calculation:

$$2\pi\hbar \bar{K}_1^{(e)} = \frac{1}{i\hbar} \int_{-\infty}^{+\infty} dt \int_{-\infty}^{t} dt' \, (HH^{(e)\prime} + H^{(e)} H'), \tag{$13_1^{(e)}$}$$

$$2\pi\hbar \bar{K}_2^{(e)} = \frac{1}{(i\hbar)^2} \int_{-\infty}^{+\infty} dt \int_{-\infty}^{t} dt' \int_{-\infty}^{t'} dt'' \, \{(HH')_{\text{n.d.}} H^{(e)\prime\prime} + HH^{(e)\prime} H'' +$$

$$+ H^{(e)}(H'H'')_{\text{n.d.}}\} + \frac{1}{2} \int_{-\infty}^{+\infty} dt \, \big((S_1^2)_D H^{(e)} + H^{(e)} (S_1^2)_D\big). \tag{$13_2^{(e)}$}$$

The structure of $(13^{(e)})$ is almost the same as that of (13). We have only to replace one factor H in (13) by $H^{(e)}$, alternately in each place. The further treatment of $(13^{(e)})$ is much the same as that of (13).

It was emphasized in § 13 that H_{int} is the space integral of a relativistic invariant if Lorentz gauge is used. It can be concluded that $\bar{K}$ is invariant: in the first place $d\tau dt$ is invariant and if all time integrations were extended to ∞ the fact would be obvious. However, the restriction of the time integration, e.g. in (13_3), is only such that $t > t' > t''$. Such a sequence in time also remains the same after a proper Lorentz transformation. Therefore $\bar{K}_2$, $\bar{K}_3$ are invariant. That the n.d.-restrictions and the terms $\sim S_1^2$ do not jeopardize the invariance can also easily be shown. We refrain from a detailed proof because the invariance of $\bar{K}$ will be evident in the further developments† (§§ 28, 30).

2. *Energy representation, self-energies.* We shall evaluate the expressions (13) and $(13^{(e)})$ in various ways for various purposes. A simple way is to use energy representation and to work out the matrix elements of K. These are then expressed in terms of matrix elements of H. We write

$$H(t)_{n|m} = H_{n|m} \, e^{i(E_n - E_m)t/\hbar},$$

denoting the time-independent matrix elements, as in § 14, by $H_{n|m}$, and using first the unperturbed states for a representation. To distinguish states on the same energy shell we denote these by A, B, C,..., whereas

† A formulation of quantum electrodynamics which exhibits its covariance at each stage of perturbation theory has been proposed by S. Tomonaga (*Progr. Theor. Phys.* **1** (1946), 27 and subsequent papers) and followed in a large part of the literature. In this theory t is replaced by a time-like surface $t(x, y, z)$. In spite of its intrinsic beauty the introduction of such supersurfaces is neither necessary for the covariance (see § 28) nor is it really useful because no significant result can depend on the shape of the surface which is arbitrary.

$n, m, k, \ldots$, denote states with $|E_n - E_A| > \epsilon$. The time integrals in (13) are worked out immediately. Lower limits of integration never give any contribution. From (13) we obtain, with the definition (12), (12′),

$$K_{1\,A|B} = 0, \tag{14$_1$}$$

$$K_{2\,A|B} = \frac{H_{A|n}\,H_{n|B}}{E_B - E_n}, \tag{14$_2$}$$

$$K_{3\,A|B} = \frac{H_{A|n}\,H_{n|m}\,H_{m|B}}{(E_B - E_n)(E_B - E_m)}, \tag{14$_3$}$$

$$K_{4\,A|B} = \frac{H_{A|n}\,H_{n|m}\,H_{m|k}\,H_{k|B}}{(E_B - E_n)(E_B - E_m)(E_B - E_k)}\bigg|_{|E_m - E_B| > \epsilon} - \\ -\frac{1}{2}\frac{H_{A|n}\,H_{n|A}\,H_{A|m}\,H_{m|B}}{(E_A - E_n)^2(E_B - E_m)} - \frac{1}{2}\frac{H_{A|n}\,H_{n|B}\,H_{B|m}\,H_{m|B}}{(E_B - E_n)(E_B - E_m)^2}. \tag{14$_4$}$$

In all the above expressions a summation is to be carried out over any index occurring twice. Similar formulae for $K^{(e)}$ could be obtained immediately from ($13^{(e)}$). In the last two terms of (14_4) eq. ($10_2'$) has been used. If any state n not on the energy shell has a continuously varying energy which may cover E_B the n.d.-restriction means that an interval ϵ is to be left out $(|E_n - E_B| > \epsilon)$ and that therefore all *denominators are principal values.*

Before we proceed farther an important modification must be made. K also contains elements which are diagonal in all variables $K_{A|A}$. Since K is the new Hamiltonian in the transformed wave equation (3) it is clear that $K_{A|A}$ is a contribution to the energy of the state A, the *self-energy* of A. Even for a single particle Ψ' would not vanish. The problem of the self-energies requires extensive investigation which will be carried out in Chapter VI. But even at this stage it is clear that the true and measured energies of a state A are not the unperturbed energies but the perturbed energies $\tilde{E}_A = E_A + K_{A|A}$. It will be more consistent to express K in terms of $\tilde{E}$ rather than the theoretical E. In order to express K by the true energies $\tilde{E}$ we go back to the definition of the interaction representation (§ 14 eq. (1)). Let $\delta_m H$ be a correction to the unperturbed Hamiltonian H_0 so that the eigenvalues of $H_0 + \delta_m H$ are the energies $\tilde{E}$. The operator $\delta_m H$ will be determined in § 29 and it will be seen that it can be described as a mere addition to the mass m of each particle and therefore has for a single free particle the form $\beta\,\delta\mu$.† We redefine now the interaction representation by

$$\Psi' = e^{i(H_0 + \delta_m H)t/\hbar}\Psi, \qquad H'_{\text{int}} = e^{i(H_0 + \delta_m H)t/\hbar} H_{\text{int}}\, e^{-i(H_0 + \delta_m H)t/\hbar}. \tag{15}$$

† To be added to the Hamiltonian $H_0 = (\boldsymbol{\alpha}\mathbf{p}) + \beta\mu$. $\delta\mu$ is a series $\delta\mu_2 + \delta\mu_4 + \ldots$.

Since Ψ satisfies the wave equation with the Hamiltonian

$$H_0+H_{\text{int}}+H^{(e)},$$

we obtain for Ψ' the wave equation (dropping the dash on Ψ again)

$$i\hbar\dot{\Psi} = (H(t)+H^{(e)}(t)-\delta_m H)\Psi \tag{16}$$

in place of (1). We see that we obtain an additional term in the Hamiltonian, namely $-\delta_m H$, which will be determined so that it just subtracts the self-energies. It should be observed that this is in no way a change of theory or an alteration of the Hamiltonian but a mere change of representation.

$\delta_m H$ should be embodied in the perturbation theory of subsection (1), replacing $H(t)$ by $H(t)-\delta_m H$ everywhere. Since a general treatment will be given in Chapter VI we confine ourselves here to the explicit formulae in energy representation. According to (15), the energies occurring now in the time exponentials and energy denominators are the eigenvalues of $H_0+\delta_m H$, i.e. the perturbed $\tilde{E}$.

We perform the canonical transformation as before, aiming at a new Hamiltonian $\tilde{K}$, which is different from zero on the energy shell only, and *energy shell now means equal energies* $\tilde{E}$ (within ϵ). Moreover, $\tilde{K}$ must not have any D-elements. We obtain

$$\tilde{K} = K+K^{(e)}+K^{(s)}, \qquad K^{(s)} = S^{-1}H^{(s)}S, \qquad H^{(s)} \equiv -\delta_m H. \tag{17}$$

Before we evaluate (17) in energy representation we note that we can also interpret the states A, n,... between which the matrix elements are taken as the *redefined* states Ψ' *with* their virtual accompaniments. $\tilde{K}$ is an operator function of the emission and absorption operators q, q^*, etc., and we are free to choose a matrix representation for these. So far we have used a representation where the number n of 'naked' particles (i.e. without their virtual fields) is diagonal and q absorbs such a naked particle. However, Ψ' is best expressed not by n (in which case Ψ' would generally be a series of terms) but by the number n' of physically real particles, i.e. with their virtual fields. As $\Psi' = S^{-1}\Psi$, n' and n are connected by a similar canonical transformation $n' = S^{-1}nS$. The numbers n' (and not n) are, of course, what is measured. Similarly, $q' = S^{-1}qS$ is the operator which absorbs one real particle. When n' is diagonal, q' is formally identical with the matrix expressions given in Chapters II, III if n is replaced by n', and the commutation relations are the same. If we now use the representation where n' is diagonal, the matrix elements of q', etc. (and also $\tilde{K}$), formed with the new state vectors Ψ' and expressed by n' are numerically the same as the matrix elements of q formed with the old state vectors Ψ and expressed by n,

$$(\Psi'^*_{n'} q' \Psi'_{m'}) = (\Psi^*_n q \Psi_m).$$

The same holds for $\tilde{K}(q)$, if q is replaced by q'. Hence, in the following, the states $A, n,...$ (we do not introduce a new notation) are the redefined states expressed by the number of real particles and the operators q, etc., act on these numbers. E_n, etc., are their energies.

The most important parts of $\delta_m H$ are diagonal simultaneously with H_0 or $H_0+\delta_m H$, and represent the self-energy in each state A. However, there are also certain non-diagonal elements of $\delta_m H$, as appears from the fact that β does not commute with $(\boldsymbol{\alpha}\mathbf{p})$ and therefore $\delta_m H$ not with $H_0+\delta_m H$. These are of relativistic nature and only n.d. elements exist (see § 29). For simplicity we disregard these contributions for the moment; it will be quite trivial to include them later (eq. (22)). Then S is the same transformation as before and K and $K^{(e)}$ in (17) are formally identical with (14), the difference being that all states are the redefined states with their energies $\tilde{E}$. We omit the $\sim$ sign on E and agree that in all formulae for $\tilde{K}$ the energies are the true energies.

Expanding S, and $H^{(s)} = H_2^{(s)}+H_4^{(s)}+...$, the additional term $K^{(s)}$ in (17) becomes

$$K^{(s)} = (H_2^{(s)}+H_4^{(s)})_D+\tfrac{1}{2}(S_1^2)_D H_{2D}^{(s)}+\tfrac{1}{2}H_{2D}^{(s)}(S_1^2)_D-S_1 H_{2D}^{(s)} S_1+\dots. \quad (18)$$

With the exception of the last term which has transition elements $B \to A$, all terms have D-elements only. $H^{(s)}$ is now to be determined so that $\tilde{K}_{A|A} = K_{A|A}+K_{A|A}^{(s)} = 0$:

$$\begin{aligned} H_{2A}^{(s)} &= -K_{2\,A|A}, \\ H_{4A}^{(s)} &= -K_{4\,A|A}-\tfrac{1}{2}(S_1^2)_A H_{2A}^{(s)}-\tfrac{1}{2}H_{2A}^{(s)}(S_1)_A^2+(S_1 H_{2D}^{(s)} S_1)_{A|A}. \end{aligned} \quad (19)$$

The transition element of $K^{(s)}$ becomes, using (19) and (10_1),

$$K_{A|B}^{(s)} = -\frac{H_{A|n} H_{n|m} H_{m|n} H_{n|B}}{(E_A-E_n)(E_B-E_n)(E_n-E_m)}, \quad (20)$$

(20) takes care of the self-energy in the intermediate state n and has to be added to (14_4) to give the new $\tilde{K}$. It can be included in the first term of (14_4) when $k = n$, by using

$$\frac{1}{(E_B-E_n)^2(E_B-E_m)}-\frac{1}{(E_B-E_n)^2(E_n-E_m)} = -\frac{1}{(E_B-E_n)(E_B-E_m)(E_n-E_m)}.$$

We find then for $\tilde{K}_{A|B}$, when $A \neq B$:

$$\tilde{K}_{4\,A|B} = \frac{H_{A|n}\,H_{n|m}\,H_{m|k}\,H_{k|B}}{(E_B-E_n)(E_B-E_m)(E_B-E_k)}\bigg|_{k\neq n} -$$

$$-\frac{H_{A|n}\,H_{n|m}\,H_{m|n}\,H_{n|B}}{(E_B-E_n)(E_B-E_m)(E_n-E_m)} - \frac{1}{2}\frac{H_{A|n}\,H_{n|A}\,H_{A|m}\,H_{m|B}}{(E_A-E_n)^2(E_B-E_m)} -$$

$$-\frac{1}{2}\frac{H_{A|n}\,H_{n|B}\,H_{B|m}\,H_{m|B}}{(E_B-E_n)(E_B-E_m)^2}. \quad (21_4)$$

For $\tilde{K}_1,\ldots,\tilde{K}_3$ we obtain formally the previous expressions $(14_{1,2,3})$

$$\tilde{K}_{1\,A|B} = 0, \qquad \tilde{K}_{2\,A|B} = K_{2\,A|B}, \qquad \tilde{K}_{3\,A|B} = K_{3\,A|B} \quad (21_{1,2,3})$$

although, of course, the energy denominators in all these formulae now have a different meaning.

In non-relativistic approximation H also has a term $H^{(2)}$ which is of second order. Also $H^{(s)}_{2\text{n.d.}}$ and, in Coulomb gauge $H^{(c)}$, are of second order.† The generalization of the above formulae for this case is quite obvious, without going through the calculation again. Each pair $(H_{n|m}\,H_{m|B})/(E_B-E_m)$ can be replaced by a factor $H^{(2)}_{n|B}$, if $H^{(2)}$ stands for any of these contributions. Thus, when $H = H^{(1)}+H^{(2)}$:

$$\tilde{K}_2 = \frac{H^{(1)}_{A|n}\,H^{(1)}_{n|B}}{E_B-E_n} + H^{(2)}_{A|B}, \quad (22_2)$$

$$\tilde{K}_3 = \frac{H^{(1)}_{A|n}}{E_B-E_n}\left(\frac{H^{(1)}_{n|m}\,H^{(1)}_{m|B}}{E_B-E_m} + H^{(2)}_{n|B}\right) + \frac{H^{(2)}_{A|m}\,H^{(1)}_{m|B}}{E_B-E_m}, \quad (22_3)$$

$$\tilde{K}_{4\,A|B} = \left(\frac{H^{(1)}_{A|n}\,H^{(1)}_{n|m}}{E_B-E_n} + H^{(2)}_{A|m}\right)\frac{1}{E_B-E_m}\left(\frac{H^{(1)}_{m|k}\,H^{(1)}_{k|B}}{E_B-E_k} + H^{(2)}_{m|B}\right)_{k\neq n} +$$

$$+\frac{H^{(1)}_{A|n}\,H^{(2)}_{n|k}\,H^{(1)}_{k|B}}{(E_B-E_n)(E_B-E_k)}\bigg|_{k\neq n} - \frac{H^{(1)}_{A|n}\,H^{(1)}_{n|m}\,H^{(1)}_{m|n}\,H^{(1)}_{n|B}}{(E_B-E_n)(E_B-E_m)(E_n-E_m)} -$$

$$-\frac{1}{2}\frac{H^{(1)}_{A|n}\,H^{(1)}_{n|A}}{(E_A-E_n)^2}\left(\frac{H^{(1)}_{A|m}\,H^{(1)}_{m|B}}{E_B-E_m} + H^{(2)}_{A|B}\right) -$$

$$-\frac{1}{2}\left(\frac{H^{(1)}_{A|n}\,H^{(1)}_{n|B}}{E_B-E_n} + H^{(2)}_{A|B}\right)\frac{H^{(1)}_{B|m}\,H^{(1)}_{m|B}}{(E_B-E_m)^2}, \quad (22_4)$$

(21), (22) is the generalization of the compound matrix elements of § 14.

Finally we add a remark concerning the renormalization terms with the factor $\frac{1}{2}$. We shall show that they can also be included in the main term (which has the same structure in all orders) provided that a properly

† With the inclusion of $H^{(s)}_{\text{n.d.}}$, (22) is relativistically exact.

defined limiting procedure is carried out for the denominators which then become singular. We illustrate this for (14_4). We write K_4 in the form

$$K_4 = \frac{H_{A|n}\,H_{n|m}\,H_{m|k}\,H_{k|B}}{(E_B-E_n)(E_B-E_m)(E_B-E_k)}. \tag{23}$$

We assume $E_A = E_B$, exactly. In (14_4) states $m = C$ (in particular also $m = A$ or B) on the energy shell of B would be excluded, but we now include these also. Then the denominator $\mathscr{P}/(E_B-E_m)$ becomes singular. To see what limiting value we can ascribe to (23) we replace E_B by a variable energy E and to fix the value of E we write

$$K_4 = \int dE\, \delta(E-E_B)K_4(E), \tag{24}$$

where $K_4(E)$ arises from (23) by replacing $E_B \to E$. We will show in § 16 and in appendix 4 that (24) is actually the precise form of K_4, when we start with finite line width and let the latter go to zero. Let now m be a state C on the same energy shell as B. It will be seen that only $C = B$ or A contribute. Thus $E_m = E_B$. This contribution, K_4', say, is

$$K_4' = \int dE\, \delta(E-E_B)\frac{\mathscr{P}}{E-E_B}f(E),$$

where $f(E)$ is non-singular for $E = E_B$. The function $\delta(x)\mathscr{P}/x$ is actually ambiguous. We can, however, ascribe to it a unique value, namely $-\frac{1}{2}\delta'(x)$, if the limiting procedures involved in $\delta(x)$ and $\mathscr{P}/x$ are carried out simultaneously† (cf. § 8 eq. (18)). With this understanding the part K_4' in question becomes

$$K_4' = -\tfrac{1}{2}\int dE\, \delta'(E-E_B)f(E) = +\frac{1}{2}\frac{\partial f}{\partial E}\bigg|_{E=E_B}$$

$$= -\frac{1}{2}\frac{H_{A|n}\,H_{n|C}\,H_{C|m}\,H_{m|B}}{(E_B-E_n)^2(E_B-E_m)} - \frac{1}{2}\frac{H_{A|n}\,H_{n|C}\,H_{C|m}\,H_{mB}}{(E_B-E_n)(E_B-E_m)^2}. \tag{25}$$

Furthermore, contributions arise only when C is identical with either A or B. This can be seen by considering the dependence on the volume L^3, when $L^3 \to \infty$. When $C = A$ (or B) the states n (or m) include virtual photons emitted and reabsorbed by A and these fill a continuous spectrum. The sum over n includes a density function which is proportional to L^3. If, however, $C \neq A$, B the states n, m can, on account of the conservation of momentum, only be isolated states and no such factor L^3 arises. It follows that in the limit $L^3 \to \infty$ only $C = A$ or B

† This will later be automatic (see § 30).

contribute.† When $C = A$ the second term of (25) has the self-energy of A as a factor. This cancels when the self-energy contribution is added and treated by the same limiting procedure. We are left with the contribution $C = A$ for the first term of (25) and $C = B$ for the second term. (25) is then identical with (14). The limiting process carried out above will occur in a similar form in the further development of the general theory in §§ 28, 30. If we understand that such a limiting process is to be carried out for singular contributions, we can use for K always the first terms in (13), (14) only without the renormalization terms. They have the same structure in all orders.

3. *Solution of the wave equation.* Our final task is to solve the wave equation (3)

$$i\hbar\dot{\Psi}' = \tilde{K}(t)\Psi'$$

(with a term $K^{(e)}$ added if necessary). K is now replaced by $\tilde{K}$ if care is taken of the self-energies. We omit the $\sim$ sign again. K has elements on the energy shell only, i.e. $K_{A|B}$, $|E_A - E_B| < \epsilon$. For the time being we must not yet let ϵ go to zero: this will turn out to be the case automatically when the solution is found. We desire a solution such that at the initial time $t = -\infty$ the system was in a definite state, O say, and the probabilities for all other states are then zero. We wish to know the probabilities for other states A, B,..., after the collision, say, at $t = +\infty$. For this purpose we put

$$\Psi' = \sum_A b_A(t)\Psi'_A, \qquad K_{A|B}(t) \equiv (\Psi'^*_A K \Psi'_B) = K_{A|B}\, e^{i(E_A - E_B)t/\hbar}, \tag{26}$$

where Ψ'_A is the normalized state vector of state A, including the accompanying virtual photons, etc. Hence

$$i\hbar\dot{b}_A = \sum_B K_{A|B}\, b_B\, e^{i(E_A - E_B)t/\hbar}. \tag{26'}$$

The solution involves essentially the function ζ defined in § 8:

$$\zeta(E) = \frac{1}{i\hbar}\int_0^\infty e^{iEt/\hbar}\,dt = \frac{\mathscr{P}}{E} - i\pi\delta(E), \tag{27}$$

$$E\zeta(E) = 1. \tag{27'}$$

To solve (26) we now put

$$b_A(t) = \delta_{AO} + U_{A|O}\,\zeta(E_O - E_A)e^{i(E_A - E_O)t/\hbar}, \tag{28}$$

where $U_{A|O}$ is an as yet unknown amplitude which is time independent

† For the volume dependence of K and the collision cross-section, see § 14. If, for example, $K_{A|B}$ refers to a scattering, the cross-section is $\phi \sim L^3\rho_A|K|^2$, $\rho \sim L^3$, and $K_{A|B}$ must be $\sim L^{-3}$ if ϕ is to be independent of L^3. This is the case for K_4 if the sum over n or m involves a factor L^3, because each H is $\sim L^{-\frac{3}{2}}$.

and $U_{O|O} = 0$. Using the integral representation (27), the following property of $\zeta(E)$ is proved:

$$\zeta(E)e^{-iEt/\hbar}\Big|_{t\to\pm\infty} = \frac{1}{i\hbar}\int_0^\infty e^{iE(t'-t)}\,dt'\Big|_{t\to\pm\infty}$$

$$= \frac{1}{i\hbar}\int_{\mp\infty}^{+\infty} e^{iEt'/\hbar}\,dt' = \begin{cases}-2\pi i\,\delta(E)\\ 0.\end{cases} \tag{29}$$

This shows immediately that (according to (28))

$$b_A(-\infty) = \delta_{AO}. \tag{30}$$

The initial conditions $b_A = 0$ when $A \neq O$ are therefore satisfied for the time $t = -\infty$.

To show that (28) satisfies the equation (26′) we insert it. On account of (27′) we get

$$i\hbar\dot{b}_A = U_{A|O}\,e^{i(E_A-E_O)t/\hbar}$$

and therefore

$$U_{A|O} = K_{A|O} + \sum_{B\neq O} K_{A|B}\,U_{B|O}\,\zeta(E_O - E_B). \tag{31}$$

This is an equation for $U_{A|O}$ which will be further reduced. Before we do this we calculate the transition probabilities. The probability for a state A is $|b_A(t)|^2$, and the transition probability per unit time is

$$\frac{d}{dt}|b_A(t)|^2 = \dot{b}_A b_A^* + \dot{b}_A^* b_A.$$

This will be independent of t. From (28) and (27), (27′) we obtain

$$\frac{d}{dt}|b_A|^2 \equiv w_{A|O} = \frac{i}{\hbar}|U_{A|O}|^2\{\zeta(E_O - E_A) - \zeta^*(E_O - E_A)\}$$

$$= \frac{2\pi}{\hbar}|U_{A|O}|^2\delta(E_A - E_O). \tag{32}$$

Thus we see that $|U_{A|O}|^2$ is essentially the transition probability and, moreover, that this is different from zero only when $E_A = E_O$. This means that we are permitted to let $\epsilon \to 0$ in the expressions for K. If we do this, (31) simplifies. The second term on the right is

$$\sum_B K_{A|B}\,U_{B|O}\,\zeta(E_O - E_B)$$

$$= \sum_B K_{A|B}\,U_{B|O}\frac{\mathscr{P}}{E_O - E_B} - i\pi K_{A|B}\,U_{B|O}\,\delta(E_O - E_B).$$

The summation over B includes an integration over E_B. If ϵ is sufficiently small $K_{A|B}$ and $U_{B|O}$ will be constant in the interval, and since $U_{A|O}$ is only wanted for $E_A = E_O$ we can write

$$\int \frac{\mathscr{P}}{E_O - E_B}\, dE_B = \int_{E_A-\epsilon}^{E_A+\epsilon} \frac{\mathscr{P}}{E_A - E_B}\, dE_B = 0.$$

Thus the equation for U reduces to

$$U_{A|O} = K_{A|O} - i\pi \sum_B K_{A|B}\,\delta(E_O - E_B) U_{B|O}. \tag{33}$$

This is an integral equation for $U_{A|O}$. It contains a summation over all states B, including in general integrations over angles, summations over spins and polarizations, etc. However, only states are included which lie on the energy shell. The integration over E_B can be performed immediately on account of the factor $\delta(E_O - E_B)$ and yields

$$U_{A|O} = K_{A|O} - i\pi K_{A|B}\,\rho_B\, U_{B|O}, \tag{33'}$$

where ρ_B is the number of states B per energy interval dE_B. In (33′) now $E_B = E_A = E_O$, i.e. all energies are equal.

K is given as a power series in e. If we wish to expand $U_{A|O}$ or $w_{A|O}$ we should consequently also obtain the solution of (33′) by expansion. This is, of course, possible but there is no need to do so. In fact (33′) is exact. It will appear that the second term of (33′) has quite a different physical meaning from the corresponding higher terms of K. We call the second term of (33′) the *damping term* because it will be responsible for what classically describes the radiation damping. On the other hand, the higher terms of K appear to be typical quantum effects and, moreover, are involved in ambiguities which require special treatment (Chapter VI). We shall study the situation for the example of the Compton effect in § 33.†

We may consider (33) as a matrix equation for U. The solution is found immediately. Defining, by analogy with $\bar{K}$, eq. (12′)

$$\bar{U}_{A|B} = U_{A|B}\,\delta(E_A - E_B),$$

we have, after multiplication of (33) by $\delta(E_A - E_O)$,

$$\bar{U} = \bar{K}(1 + i\pi\bar{K})^{-1}. \tag{34}$$

The occurrence of a denominator $1 + i\pi\bar{K}$ already indicates its significance as a damping factor.

† (33) was derived for special cases by: A. H. Wilson, *Proc. Camb. Phil. Soc.* **37** (1941), 301; A. Sokolow, *J. Phys. U.S.S.R.* **5** (1941), 231; W. Heitler, *Proc. Camb. Phil. Soc.* **37** (1941), 291; E. Gora, *Z. f. Phys.* **120** (1943), 121; for the general case by: W. Heitler and H. W. Peng, *Proc. Camb. Phil. Soc.* **38** (1942), 296. See also W. Pauli, *Meson Theory and Nuclear Forces*, New York 1946.

The $b_A(t)$ that satisfies the initial condition $b_A(-\infty) = \delta_{AO}$ can be regarded as the AO-element of a matrix $\mathscr{S}_{AO}(t)$. It is the matrix which transforms the state vector at $t = -\infty$, $\Psi''(-\infty) = \Psi''_O$, into the state vector $\Psi''(t)$,

$$\Psi''(t) = \mathscr{S}(t)\Psi''(-\infty). \tag{35}$$

The matrix $\mathscr{S} \equiv \mathscr{S}(+\infty)$ is called the S-matrix.† It is given by

$$\mathscr{S}_{AO} \equiv b_A(\infty) = \delta_{AO} - 2\pi i U_{A|O}\,\delta(E_A - E_O) = \left(\frac{1-i\pi\bar{K}}{1+i\pi\bar{K}}\right)_{A|O} \tag{35'}$$

according to (28), (29). $\mathscr{S}$ is evidently unitary. Its non-diagonal elements are essentially U and therefore also permit the calculation of the collision cross-sections. A slightly different definition is obtained from (33'),

$$\mathscr{S}' = \frac{1-i\pi K'}{1+i\pi K'}, \qquad K'_{A|B} \equiv \sqrt{\rho_A}\,K_{A|B}\sqrt{\rho_B}. \tag{35''}$$

For $\mathscr{S}(t)$ an expansion similar to that of K exists. By (35), $\mathscr{S}(t)$ also satisfies the wave equation (3). Thus, as is verified by straightforward differentiation,

$$\mathscr{S}(t) = 1 + \frac{1}{i\hbar}\int\limits_{-\infty}^{t} dt'\,K(t') + \frac{1}{(i\hbar)^2}\int\limits_{-\infty}^{t} dt' \int\limits_{-\infty}^{t'} dt''\,K(t')K(t'') + \ldots. \tag{36}$$

Inserting the expansion for K from subsection 1 we could obtain $\mathscr{S}$ in terms of H, but we shall not use this expansion.

We have based our perturbation theory on the expansion of $\bar{K}$ rather than $\mathscr{S}$ for three reasons: (i) for physical purposes it is desirable to isolate the damping effects and this is not possible if U or $\mathscr{S}$ is expanded; (ii) when (36) is used in higher orders it turns out to be singular on the energy shell and some special manipulation is needed. This is not the case for K; (iii) whilst $\mathscr{S}$ is unitary, this is not the case for the individual terms of the expansion (36). The expansion of $\bar{K}$ has the advantage that each of the terms is hermitian.

If we neglect the damping term,

$$|U_{A|B}|^2 = |K_{A|B}|^2 = |K_{B|A}|^2 = |U_{B|A}|^2.$$

† P. A. M. Dirac, *Principles of Quantum Mechanics*, Oxford (1947), p. 173; F. J. Dyson, *Phys. Rev.* **75** (1949), 486, 1736. There is an extensive literature on the properties of the $\mathscr{S}$-matrix. We quote the most important of the earlier papers: W. Heisenberg, *Z. Naturforschung*, **1** (1946), 608 (summary of previous papers); C. Møller, *Kgl. Dansk. Vid. Sels.* **23** (1945), No. 1, **22** (1946), No. 19. Various other forms of the expansion of K equivalent to the above have also been given: S. N. Gupta, *Proc. Camb. Phil. Soc.* **47** (1950), 454; T. Miyazima and N. Fukuda, *Prog. Theor. Phys.* **5** (1950), 849; J. Pirenne, *Phys. Rev.* **86** (1952), 395.

This is called the *principle of detailed balance*, frequently used in statistics. The damping term destroys the general validity of this principle, and only a weaker theorem holds which is derived in appendix 5.

16. General theory of damping phenomena

We now turn to problems involving discrete bound states. The treatment will be quite different since a number of the simplifications made in § 15 cannot be applied here. Consider an atom in a discrete excited level. The most essential differences are then: (i) there will be a *finite transition probability* per sec. into lower states accompanied by photon emission, etc., whereas for collisions between free particles this transition probability vanishes for $L^3 \to \infty$. Consequently, we cannot fix the initial condition for a time $t = -\infty$, but must fix it for a finite time, e.g. $t = 0$; (ii) the initial state (and in general all states) will have a *finite level width* Γ, whereas in the case of § 15, $\Gamma = 0$ (for $L^3 \to \infty$).

In the following we shall be concerned with an exact general theory of the *solutions* of a wave equation of type $i\hbar\dot{\Psi} = H(t)\Psi$ or equation (1) below, when discrete levels are involved. This theory (which appears as a generalization of § 15.3) has nothing to do with what the Hamiltonian of the interaction H is. For the most important applications (Chapter V), where only the first approximation to the line breadth, etc., is considered, we can use as a starting-point the unperturbed states. H is then the interaction Hamiltonian H_{int} and we can calculate transitions between such unperturbed states. However, from an exact point of view, in particular if we wish to calculate the higher approximations, this procedure should be modified.

A bound electron is, like a free electron, accompanied by its virtual field, i.e. the field in its neighbourhood, and its wave function contains admixtures with virtual photons, etc. It would be desirable to carry out a canonical transformation first, to redefine the atomic states such that these virtual states were included, as we did for free particles in § 15, and then to calculate the transitions between the states thus redefined. The transformation should be such that the redefined ground state of an atom is not further modified by the interaction with the field, whereas this interaction leads to real transitions only from excited states. The difficulties which occur in this programme when the line breadth is finite are explained at the end of this section and in § 34. In view of the most important applications we use in the following a notation and nomenclature that refers to the 'naked' unperturbed states with H_{int} as Hamiltonian. It should, however, be realized that the following theory remains the same if H_{int} is replaced by a transformed Hamiltonian K, say, in which case the probability amplitudes b are those of the redefined atomic states.

Applications of the theory will be made in § 20 (resonance fluorescence) and § 34 (line shift and radiative corrections to the line breadth). The simple problem of the line breadth in emission and absorption will be treated once more in an elementary way in § 18.

1. *General solutions.*† We start again from the set of equations giving the time variation of the probability amplitudes b_n for the states n in interaction representation (§ 14 eq. (3)):

$$i\hbar\dot{b}_n(t) = \sum_m H_{n|m} b_m(t) e^{i(E_n-E_m)t/\hbar}, \tag{1}$$

where H is the interaction energy between the charged particles and the radiation field. E_n may be regarded as the unperturbed energy H_0 of n. We may, however, also go over to a slightly changed representation, as in § 15.2, and understand that E_n *includes the self-energy of* n, and in this case H is the interaction energy minus the operator of the self-energy (subsection 3). The latter will be the more consistent representation. In this case H has also diagonal elements $H_{n|n}$. In relativistic problems also the Coulomb interaction has diagonal elements.

We desire a solution of (1), satisfying an initial condition such that at $t = 0$ the system is in a state O, say, and all other probability amplitudes are zero,

$$b_n(0) = 0, \qquad b_O(+0) = 1, \tag{2}$$

$t = +0$ means that t approaches zero from the positive side. (This must be indicated for b_O on account of its discontinuity, see below.)

In the following, the singular function $\zeta(x)$, defined in § 8, will be used extensively. Its main properties are (according to § 8.1, § 8 eq. (40), § 15 eq. (29))

$$\zeta(x) = -i\int_0^\infty e^{ixt}\,dt = \lim_{\sigma\to 0}\frac{1}{x+i\sigma} = \lim_{t\to\infty}\frac{1-e^{ixt}}{x} = \frac{\mathscr{P}}{x} - i\pi\,\delta(x), \tag{3}$$

$$x\zeta(x) = 1, \tag{3'}$$

$$\int_{-\infty}^{+\infty} \zeta(x)e^{ixt}\,dx = \begin{cases} 0, & \text{for } t > 0, \\ -2\pi i, & \text{for } t < 0, \end{cases} \tag{4}$$

$$\lim_{t\to\infty} \zeta(x)e^{\pm ixt} = \begin{cases} 0, \\ -2\pi i\,\delta(x). \end{cases} \tag{4'}$$

† For this section see: W. Heitler and S. T. Ma, *Proc. Roy. Ir. Ac.* **52** (1949), 109; E. Arnous and S. Zienau, *Helv. Phys. Acta,* **24** (1951), 279; also M. Schönberg, *Nuov. Cim.* **8** (1951), 817. A number of simplifications are due to K. Bleuler.

(1) and the solution $b_n(t)$ will be physically meaningful only when $t > 0$. For analytical reasons it will, however, be convenient to extend the solution to the negative time axis. We are quite free to choose the b_n's for negative t and our solution will be such that

$$b_n(t) = b_O(t) = 0, \quad \text{for } t < 0. \tag{5}$$

The wave equation (1) is then satisfied identically but the b's are normalized to zero for $t < 0$, instead of to unity. It follows then from (2) that b_O has a discontinuity at $t = 0$. It jumps from 0 at $t = -0$ to unity at $t = +0$. In the neighbourhood of $t \sim 0$, $\dot{b}_O$ will therefore be singular and behave like $\delta(t)$: $\dot{b}_O = \delta(t)$, for $t \sim 0$. By integration over a small interval of t covering $t = 0$ we just obtain the required jump

$$b_O(+0) - b_O(-0) = 1. \tag{2'}$$

$b_n(t)$, $n \neq O$, will be continuous at $t = 0$. We can extend the wave equation (1) to negative t's, including the point $t = 0$, by adding an inhomogeneous term which just accounts for the jump of b_O:

$$i\hbar \dot{b}_n = \sum_m H_{n|m} b_m e^{i(E_n - E_m)t/\hbar} + i\hbar\, \delta_{nO}\, \delta(t). \tag{6}$$

This wave equation is now valid for *all* t. It follows from (6) that b_n ($n \neq O$) is continuous at $t = 0$ whereas b_O makes the jump (2′). We desire a solution with $b_n = b_O = 0$ for $t < 0$; then (2) will be satisfied automatically.

To solve (6) we transform the $b(t)$ by a Fourier transformation:

$$b_n(t) = -\frac{1}{2\pi i} \int_{-\infty}^{+\infty} dE\; G_{n|O}(E) e^{i(E_n - E)t/\hbar} \tag{7}$$

and also

$$i\hbar\, \delta(t) = -\frac{1}{2\pi i} \int_{-\infty}^{+\infty} dE\; e^{i(E_O - E)t/\hbar}. \tag{7'}$$

E is a variable energy taking the place of the time variable t. Inserting this into (6) we see that the wave equation is satisfied if we put

$$(E - E_n) G_{n|O}(E) = \sum_m H_{n|m}\, G_{m|O}(E) + \delta_{nO}. \tag{8}$$

That this is also necessary will follow from the fact that the solution with the initial conditions (2) is unique.

To obtain an equation for $G_{n|O}$ we would have to divide by $E - E_n$. This division is not unique. Generally, when $xf(x) = g(x)$ and $g(x)$ is non-singular at $x = 0$, $f(x) = g(x)[1/x + \alpha\, \delta(x)]$ because $x\, \delta(x) = 0$, α is quite arbitrary. It will be seen that if the b's are to fulfil the initial conditions the result of the division by $E-E_n$ must just be a factor $\zeta(E-E_n)$ on the right-hand side of (8) (i.e. $\alpha = -i\pi$ and $1/x$ interpreted as principal value). For convenience we split off a factor $G_{O|O}$ and therefore put

$$G_{n|O}(E) = U_{n|O}(E)G_{O|O}(E)\zeta(E-E_n) \quad (n \neq O); \qquad U_{O|O} \equiv 0. \tag{9}$$

From (8) we then obtain the fundamental equation

$$U_{n|O}(E) = H_{n|O} + \sum_{m\neq O} H_{n|m}\,\zeta(E-E_m)U_{m|O}(E) \quad (n \neq O) \tag{10}$$

now always separating states $n = O$ and $n \neq O$.

Equation (10) is the equation from which U must be determined for every value of E. For $G_{O|O}$ we obtain from (8)

$$(E-E_O)G_{O|O}(E) = 1+G_{O|O}(E)\Big\{H_{O|O} + \sum_{m\neq O} H_{O|m}\,\zeta(E-E_m)U_{m|O}(E)\Big\}$$

or,

$$G_{O|O} = \frac{1}{E-E_O+\frac{1}{2}i\hbar\Gamma_{O|O}(E)}, \tag{11}$$

$$\tfrac{1}{2}\hbar\Gamma_{O|O}(E) = iH_{O|O}+i\sum_{m\neq O} H_{O|m}\,\zeta(E-E_m)U_{m|O}(E). \tag{12}$$

There is no singularity at $E = E_O$. $\Gamma_{O|O}$ has, in general, a real and an imaginary part, and we shall see that the real part, which is always positive, (taken at a particular point) describes the line breadth. When the equation (10) for U is solved and U known, Γ can be calculated from (12), and hence $G_{O|O}$ is known. We shall henceforth omit the index $O|O$ of Γ which always refers to the initial state.

Inserting (9), (11) into (7) we obtain for the amplitudes b_n, b_O

$$b_n(t) = -\frac{1}{2\pi i}\int_{-\infty}^{+\infty} dE\; U_{n|O}(E)\frac{\zeta(E-E_n)e^{i(E_n-E)t/\hbar}}{E-E_O+\frac{1}{2}i\hbar\Gamma(E)} \quad (n \neq O) \tag{13}$$

$$b_O(t) = -\frac{1}{2\pi i}\int_{-\infty}^{+\infty} dE\,\frac{e^{i(E_O-E)t/\hbar}}{E-E_O+\frac{1}{2}i\hbar\Gamma(E)} \tag{14}$$

where U is determined by (10) and Γ from U by (12).

We have now to prove that the initial conditions (2) are satisfied. So far we have shown that our solution fulfils the inhomogeneous wave equation (6), which means that b_n and b_O satisfy the homogeneous equation (1) for $t > 0$ and $t < 0$, and that b_O makes a jump of magnitude unity at $t = 0$, whereas b_n is continuous there. To show that (2) is true we have, for example, to show that $b_n = b_O = 0$ for $t = -0$. Since it is a general consequence of the *homogeneous equation* that the normalization of the b's is conserved

$$\frac{d}{dt}\Big(|b_O(t)|^2+\sum_{n\neq O}|b_n(t)|^2\Big)=0, \tag{15}$$

it is sufficient to show that $b_n = b_O = 0$ for $t = -\infty$. All b's must then necessarily vanish so long as the homogeneous equation holds, i.e. for all negative t's. The normalization then jumps by unity and remains unity for all positive t's.

For $t = -\infty$ it follows from (4′) that $\zeta(E-E_n)e^{i(E_n-E)t/\hbar} = 0$ and hence $b_n(-\infty) = 0$.† To show the same for b_O we make use of $x\,\zeta(x) = 1$ and write

$$b_O(t) = -\frac{1}{2\pi i}\int\limits_{-\infty}^{+\infty} dE\,\frac{(E-E_O)\zeta(E-E_O)}{E-E_O+\frac{1}{2}i\hbar\Gamma(E)}\,e^{i(E_O-E)t/\hbar}$$

$$= -\frac{1}{2\pi i}\int\limits_{-\infty}^{+\infty} dE\;\zeta(E-E_O)e^{i(E_O-E)t/\hbar}+$$

$$+\frac{\hbar}{4\pi}\int\limits_{-\infty}^{+\infty} dE\,\frac{\Gamma(E)\zeta(E-E_O)}{E-E_O+\frac{1}{2}i\hbar\Gamma(E)}\,e^{i(E_O-E)t/\hbar}$$

The first term vanishes for negative t by (4), the second term vanishes for $t = -\infty$ by (4′), as above. Hence also $b_O(-\infty) = 0$.

We have thus shown that the solution (13), (14) with U determined by (10) and Γ by (12) satisfy the homogeneous wave equation for $t > 0$, and the initial condition (2) for $t = +0$. The solution is thus unique. We see that it is thanks to the ζ-function that the initial conditions are satisfied.

It follows from the initial condition and (13) that

$$\int\limits_{-\infty}^{+\infty} dE\;U_{n|O}(E)G_{O|O}(E)\zeta(E-E_n) = 0. \tag{16}$$

This means that the sum of all residues (if any) of $U_{n|O}\,G_{O|O}/(E-E_n)$ in the upper half of the complex plane vanishes. (ζ is regular there.) We can use (16) to write $b_n(t)$ in another form.

If we subtract (16) from (13) $e^{i(E_n-E)t/\hbar}-1$ vanishes for $E = E_n$ and cancels the singularity of $\zeta(E-E_n)$. The ζ-factor can then be replaced

† Provided that $U(E)$ is non-singular for $E = E_n$. Unless unsuitable expansions are used, this is generally true, but (2) can also be proved if U is singular,

by $1/(E-E_n)$. Thus

$$b_n(t) = -\frac{1}{2\pi i}\int_{-\infty}^{+\infty} dE \frac{U_{n|O}(E)}{E-E_O+\frac{1}{2}i\hbar\Gamma(E)} \frac{e^{i(E_n-E)t/\hbar}-1}{E-E_n}, \tag{17}$$

which exhibits the initial condition.

Equations (10) and (12) can be comprised into a single matrix equation as the right-hand sides of these two equations are the non-diagonal and diagonal parts of the same quantity respectively. If E_n are the eigenvalues of the Hamiltonian H_0 we can write $[\zeta(E-H_0)H]_{n|m}$ for $\zeta(E-E_n)H_{n|m}$, etc. Writing ζ as short for $\zeta(E-H_0)$ we have

$$U = H+H\zeta U+\tfrac{1}{2}i\hbar\Gamma. \tag{10'}$$

Note that in this notation ζ does not commute with H. (10′) is a general matrix equation. In the representation where H_0 is diagonal, Γ is diagonal and U non-diagonal.

To obtain an idea of the physical significance of our solution we assume for a moment that Γ is independent of E. This is really the case for many simple problems to a very good approximation. Let

$$\Gamma = \mathscr{R}(\Gamma)+i\mathscr{I}(\Gamma).$$

b_O can then be evaluated immediately:

$$b_O = -\frac{1}{2\pi i}\int_{-\infty}^{+\infty} \frac{dE\, e^{i(E_O-E)t/\hbar}}{E-E_O-\frac{1}{2}\hbar\mathscr{I}(\Gamma)+\frac{1}{2}i\hbar\mathscr{R}(\Gamma)} = e^{-\frac{1}{2}i\mathscr{I}(\Gamma)t}e^{-\frac{1}{2}\mathscr{R}(\Gamma)t} \tag{18}$$

by contour integration through the lower half-plane. The exponential decrease shows that the real part of Γ, $\mathscr{R}(\Gamma)$, must be the *total transition probability per sec.* from the state O into all other states. On the other hand, the periodic factor shows that the imaginary part $\mathscr{I}(\Gamma)$ is a change of energy of the state O. This is what we have called the self-energy of the state O. $\mathscr{R}(\Gamma)$ will be responsible for the line breadth for transitions starting from O, and $\mathscr{I}(\Gamma)$ for a displacement of the line. When, however, a representation is used where the E_n, E_O include the self-energies, $\mathscr{I}(\Gamma)$ will vanish. (Compare subsection 3.) These results will be essentially corroborated by the exact theory. We shall also see that $U_{n|O}$ is responsible for the transition probability $O \to n$.

2. *Transition probabilities.* We now calculate the probabilities $|b_n(t)|^2$ for finding the system in a state n at the time t. To obtain $b_n(t)$ for any time requires the complete solution of (10) for each E. The physically most interesting cases are, however, only the following:

(i) The transition probability per unit time for the particular transition $O \to n$ is obtained for small times $\mathscr{R}(\Gamma)t \ll 1$. Notwithstanding this condition we are not interested in times comparable to the atomic period and we may allow $E_n t/\hbar \gg 1$ at the same time. This presupposes, of course, that the line breadth is small, $\hbar\mathscr{R}\Gamma \ll E_n$, a condition well satisfied for all atomic systems. This comes to the same as putting $\mathscr{R}(\Gamma)\to 0$, and afterwards $t\to\infty$.

(ii) The probability distribution of the various states n into which transitions can take place after a *long time* has elapsed so that at least some transitions from O have certainly taken place. This means $\Gamma t\to\infty$.

For large values of t the time-dependent factor in (17) becomes a representation of the ζ-function according to (3), and

$$b_n(\infty) = -\frac{1}{2\pi i}\int\limits_{-\infty}^{+\infty} dE\, U_{n|O}(E)\frac{\zeta(E_n-E)}{E-E_O+\frac{1}{2}i\hbar\Gamma(E)}. \tag{19}$$

The integral can be evaluated by comparing it with (16) and (11). (19) differs from (16) in that $\zeta(E-E_n)$ is replaced by $\zeta(E_n-E)$. We add (16) to (19) and remember that $\zeta(x)+\zeta(-x) = -2\pi i\,\delta(x)$. We find†

$$b_n(\infty) = \int\limits_{-\infty}^{+\infty} dE\, U_{n|O}(E)\frac{\delta(E-E_n)}{E-E_O+\frac{1}{2}i\hbar\Gamma(E)} = \frac{U_{n|O}(E_n)}{E_n-E_O+\frac{1}{2}i\hbar\Gamma(E_n)}. \tag{20}$$

We see that after long times only U and Γ at $E = E_n$ are needed. The probability distribution of the states n after a long time is therefore given by

$$|b_n(\infty)|^2 = \frac{|U_{n|O}(E_n)|^2}{(E_n-E_O-\Delta E)^2+\frac{1}{4}\hbar^2(\mathscr{R}\Gamma(E_n))^2}, \tag{20'}$$

where we have put

$$\tfrac{1}{2}\hbar\mathscr{I}(\Gamma(E_n)) \equiv \Delta E. \tag{21}$$

The formula already has the appearance of an 'intensity distribution with finite line breadth' $\mathscr{R}\Gamma(E_n)$ (compare the line shape of a classical oscillator § 4) with a maximum displaced by ΔE. However, in more complicated problems, U may also depend strongly on E_n (see § 20).

† This again is only true if U is non-singular for $E = E_n$. See footnote on p. 167. (20) follows also from (4′) and (13).

If we wish, we can also, as in § 15, call $b_n(\infty)$ the $\mathscr{S}$-matrix. One can expand $\mathscr{S}$, or even $b_n(t)$, into a series according to progressive powers of e (or H) by expanding both the numerator and the denominator of (20). After some calculation one finds:

$$\mathscr{S} = 1+\frac{1}{i\hbar}\int\limits_0^\infty H(t)\,dt+\frac{1}{(i\hbar)^2}\int\limits_0^\infty dt\int\limits_0^t dt'\,H(t)H(t')+\dots.$$

This is quite analogous to the expansion (36) in § 15; only the lower limit is replaced by 0 instead of $-\infty$ in accordance with the present initial condition. This expansion is, however, quite useless if we wish to study problems with finite line width.

To calculate the transition probability per unit time we have to perform the limiting process $\mathscr{R}(\Gamma) \to 0$ before letting $t \to \infty$. It will be seen below that $\mathscr{R}(\Gamma)$ is always positive. For $\mathscr{R}(\Gamma) \to 0$ the denominator in (13) is then a representation of the ζ-function. Thus,

$$b_n(t) = -\frac{1}{2\pi i} \int\limits_{-\infty}^{+\infty} dE\, U_{n|O}(E)\zeta_\Gamma(E-E'_O)\zeta_\sigma(E-E_n)e^{i(E_n-E)t/\hbar} \qquad (\mathscr{R}(\Gamma)\to 0), \quad (22)$$

$$E'_O \equiv E_O + \Delta E,$$

where we have indicated the imaginary parameters occurring in the representation of ζ by an index $(\zeta_\sigma(x) \equiv 1/(x+i\sigma))$. σ and Γ tend to 0 independently. We now have to find the limiting value of $\zeta_\Gamma(E-E'_O)\zeta_\sigma(E-E_n)\exp[i(E_n-E)t/\hbar]$. We cannot use (4′) directly because this holds only if no other singular factor depending on E occurs. We split up the product of the ζ-functions by partial fractions:

$$\zeta_\Gamma(E-E'_O)\zeta_\sigma(E-E_n) = [\zeta_\Gamma(E-E'_O)-\zeta_\sigma(E-E_n)]\zeta_{\sigma-\Gamma}(E'_O-E_n). \quad (23)$$

The common factor is a ζ-function only if in the limiting process σ is kept $> \Gamma$, but is a ζ^*-function if $\sigma < \Gamma$. The result is the same in both cases, so let $\sigma > \Gamma$. We now apply (4′) to both terms of (23) (inserted in (22)), separating off a factor $\exp[i(E_n-E'_O)t/\hbar]$ for the first term. The two terms then reduce to a $\delta(E-E'_O)$ and $\delta(E-E_n)$-function respectively. Thus

$$b_n(t\to\infty) = \zeta_{\sigma-\Gamma}(E'_O-E_n)\{U_{n|O}(E'_O)e^{i(E_n-E'_O)t/\hbar} - U_{n|O}(E_n)\}, \quad (24)$$

and hence also

$$\dot{b}_n(t\to\infty) = -\frac{i}{\hbar} U_{n|O}(E'_O)e^{i(E_n-E'_O)t/\hbar}, \quad (24')$$

using (3′).

We now form the transition probability per unit time,

$$w_{nO} = \frac{d}{dt}|b_n|^2 = \dot{b}_n b_n^* + \dot{b}_n^* b_n$$

and use (4′) again for $\zeta(E'_O-E_n)e^{-i(E_n-E'_O)t/\hbar}$ and its complex conjugate. The second term of (24) does not then contribute. We obtain

$$w_{nO} = +\frac{i}{\hbar}|U_{n|O}(E'_O)|^2\{\zeta(E'_O-E_n)-\zeta^*(E'_O-E_n)\}.$$

Since $\zeta(x)-\zeta^*(x) = -2\pi i\,\delta(x)$, this reduces to

$$w_{nO} = \frac{2\pi}{\hbar}|U_{n|O}(E_n)|^2\delta(E_n-E_O-\Delta E). \quad (25)$$

Thus $|U_{n|O}|^2$ is essentially the transition probability per unit time. Moreover, this is different from zero only when $E_n - E_O - \Delta E = 0$, i.e. if the energy, including any displacements of the energy levels caused by the interaction with radiation, is conserved.

While the analysis up to eq. (20) is *exact and general*, formula (25) hinges on the assumption that $\mathscr{R}(\Gamma) \ll E_n$, and is correct only in the limiting case $\mathscr{R}(\Gamma) \to 0$. For problems of atomic physics this is usually a very good approximation, but it should be kept in mind that the 'transition probability per unit time' is an *approximate* concept and $|b_n|^2$ is not exactly proportional to t in any time region.

Next we show that $\mathscr{R}(\Gamma)$ is the sum of all transition probabilities from O into all other states. Γ is given by (12), and for any fixed E we have (H is hermitian and $H_{O|O}$ therefore real)

$$\hbar\mathscr{R}(\Gamma(E)) = i \sum_{m \neq O} \{H_{O|m} U_{m|O} \zeta(E - E_m) + U^*_{m|O} H_{m|O} \zeta(E_m - E)\} \quad (26)$$

using $\zeta^*(x) = -\zeta(-x)$. On the other hand, $U_{m|O}(E)$ is given by (10). We multiply (10) by $U^*_{m|O}(E)\zeta(E_m - E)$ and the complex conjugate equation of (10) by $U_{m|O}(E)\zeta(E - E_m)$, sum over m, and add. Since $\zeta(x) + \zeta(-x) = -2\pi i\,\delta(x)$ we get

$$-2\pi i \sum_{m \neq O} |U_{m|O}(E)|^2 \delta(E - E_m)$$

$$= \sum_{m \neq O} \{H_{m|O} U^*_{m|O}(E)\zeta(E_m - E) + H_{O|m} U_{m|O}(E)\zeta(E - E_m)\} +$$

$$+ \sum_{m,n \neq O} \{H_{m|n} U_{n|O} U^*_{m|O} \zeta(E - E_n)\zeta(E_m - E) -$$

$$- H_{n|m} U^*_{n|O} U_{m|O} \zeta(E_n - E)\zeta(E - E_m)\}.$$

The double sum vanishes, as is seen by changing $m \rightleftarrows n$ in the second term. Comparing with (26) we see that

$$\mathscr{R}\Gamma(E) = \frac{2\pi}{\hbar} \sum_m |U_{m|O}(E)|^2 \delta(E - E_m). \quad (27)$$

This result is also exact. Evidently $\mathscr{R}\Gamma(E) > 0$. If we now put

$$E = E_O + \Delta E$$

(27) reduces to the sum of all transition probabilities

$$\mathscr{R}\Gamma(E_O + \Delta E) = \sum_n w_{nO}, \quad (28)$$

but this identification only holds when w_{nO} is given by (25).

The imaginary part of Γ becomes

$$\frac{\hbar}{2}\mathscr{I}(\Gamma(E)) = H_{O|O}+\tfrac{1}{2}\sum_{n\neq O}\{H_{O|n}\,\zeta(E-E_n)U_{n|O}+ +H_{n|O}\,\zeta^*(E-E_n)U^*_{n|O}(E)\}. \quad (29)$$

We shall consider this in subsection 3.

We see that all physical properties can be derived from $U_{n|O}(E)$, which is determined by the fundamental equation (10). If we wish to know the details of the time dependence of the various probability amplitudes this equation must be solved for each value of E. If, however, we only wish to know the probabilities for large t (not necessarily large Γt) it suffices to know U for $E = E_n$. This does not necessarily imply yet that $E_n = E_O+\Delta E$. If, in addition, $\Gamma t \ll 1$, the only values E_n which contribute are

$$E_n = E_O+\Delta E.$$

3. *Level displacements.* We have seen that the imaginary part of Γ plays the role of a displacement of the energy E_O (see eqs. (20′), (25)). It is, however, somewhat disconcerting that this displacement appears to be unsymmetrical in the initial and final states. One would expect E_n to be displaced also. This asymmetry occurs only as long as the states n, O,... are regarded as the 'naked' unperturbed states. Clearly it makes little sense to state that the atom was initially in the unperturbed state E_O when this level is actually displaced and has energy $\tilde{E}_O$, say. The symmetry can be restored quite easily by a mere change of representation exactly as in § 15.2.

$\mathscr{I}\Gamma$ at the point $E = E_O$ is nothing but the self-energy of the state O. If for U we use the first approximation which is H (according to (10)) and $\zeta(E-E_m)+\zeta^*(E-E_m) = 2\mathscr{P}/(E-E_m)$, we have

$$\tfrac{1}{2}\hbar\mathscr{I}\Gamma(E_O) = H_{O|O}+\sum_{m\neq O}\frac{H_{O|m}H_{m|O}}{E_O-E_m}. \quad (30)$$

We now introduce a matrix $-H^{(s)}$ which we assume to be diagonal simultaneously with H_0, the diagonal elements being the self-energy of each state. In contrast to § 15, $H^{(s)}$, which refers now to *bound states*, does not reduce to a mere change of mass. We use then for our representation not the eigenvalues of H_0 but those of $H_0-H^{(s)} = \tilde{H}_0$, say.†
Let the eigenvalues of $\tilde{H}_0$ be $\tilde{E}_n$, $\tilde{E}_O$, etc. Carrying out the transition to interaction representation as in § 15 eqs. (15), (16) we find:

(i) What has been denoted so far by E_n, etc., are the displaced energies

† For a precise formulation of this representation it is first necessary to account for the change of mass $\delta\mu$ in H_0. This will be carried out in § 34.

$\tilde{E}_O = E_O - H^{(s)}_{O|O}$; (ii) the Hamiltonian of the interaction is not H_{int} but $H_{\text{int}}+H^{(s)}$. $H^{(s)}$ is now determined by the condition that $\mathscr{I}\Gamma(\tilde{E}_O)$ should vanish (up to the second order in e). Then (30) becomes

$$\mathscr{I}\Gamma(\tilde{E}_O) = 0, \qquad -H^{(s)}_{O|O} = +\sum_{m\neq O} \frac{H_{\text{int}\,O|m}H_{\text{int}\,m|O}}{\tilde{E}_O - \tilde{E}_m} + H_{\text{int}\,O|O}. \quad (31)$$

This is the formula § 14 eq. (17) for the self-energy. (31) holds, of course, for any state, because every state can be an initial state. Only the Coulomb interaction contributes to $H_{\text{int}\,O|O}$ (if Coulomb gauge is used). Thus in the new representation $\mathscr{I}\Gamma(\tilde{E}_O)$ vanishes whereas in all formulae of this section the energies are the displaced energies, in particular also in (20), (25). The asymmetry between initial and final states has disappeared, at least inasmuch as $\mathscr{I}\Gamma(E)$ can be identified with $\mathscr{I}\Gamma(\tilde{E}_O)$. Strictly speaking this is not the case, not even if we consider the distribution of the final states after a long time. In the latter case $\mathscr{I}\Gamma(\tilde{E}_n)$ occurs and $\tilde{E}_n$ depends on the frequency of the emitted photon. The difference ΔE, in our new representation, is

$$\Delta E = \tfrac{1}{2}\hbar\mathscr{I}\Gamma(\tilde{E}_n) = \sum_{m\neq O}\left(\frac{H_{\text{int}\,O|m}H_{\text{int}\,m|O}}{\tilde{E}_n - \tilde{E}_m} - \frac{H_{\text{int}\,O|m}H_{\text{int}\,m|O}}{\tilde{E}_O - \tilde{E}_m}\right), \quad (31')$$

and this still occurs in the denominator of (20′) which reads now:

$$|b_n(\infty)|^2 = \frac{|U_{n|O}(\tilde{E}_n)|^2}{(\tilde{E}_n - \tilde{E}_O - \Delta E)^2 + \frac{1}{4}\hbar^2(\mathscr{R}\Gamma(\tilde{E}_n))^2}. \quad (31'')$$

Since $\Gamma(E)$ is a slowly varying function of E and E differs from E_O only by an amount of the line breadth, ΔE is exceedingly small. We consider it in § 34.4. (31″) is symmetrical in both states.

The level displacement (31) will be treated in § 34. It will be seen that it cannot be accounted for entirely by a change of mass (in contrast to the case of free electrons) but that it is a—very important—*observable effect*. It will also be seen that for allowed transitions it is usually much smaller than the line breadth.

The above way of restoring the symmetry in the level displacements is somewhat provisional. The self-energy of a state is due to the virtual field accompanying the particle. We should first include the virtual field in the definition of the atomic state, as explained at the beginning of this section, before calculating real transitions between different states. However, the separation of virtual photons from photons emitted in real transitions is not as clear-cut as in the case of free particles (§ 15). The level width extends in principle to infinity and a photon of any energy can occur as a virtual photon as well as a photon emitted in a real transition. Therefore the 'energy shell' of § 15 cannot be defined unambiguously except for the ground state which is sharp. In fact no exact definition of an isolated excited atomic state with a finite lifetime can be given at all. The emission of light must be treated in connexion with the *way in which the atom was excited* if the finer details of the emitted line are studied. These will depend to some extent on the *excitation conditions.*

The problem will arise when we consider the higher order (radiative) corrections to the line breadth in § 34.4. It will be seen that for this purpose a crude way of separating off the virtual field which reflects the excitation conditions suffices (utilizing the fact that the line breadth is very small compared with the level differences). The corrections to the line breadth will in fact be independent of the excitation conditions. In first approximation, of course, all line breadth phenomena can be computed by using the unperturbed states.

That a mere change of representation is not sufficient to eliminate the virtual field can also be seen as follows: according to elementary perturbation theory (§ 14 eq. (11)) the probability of virtual states is

$$\sum_n |b_n|^2 = \sum_n \frac{|H_{\text{int}\,O|n}|^2}{(E_O-E_n)^2} = -\tfrac{1}{2}\hbar\frac{\partial}{\partial E}\mathscr{I}\Gamma_{O|O}(E)\bigg|_{E=E_O} \tag{32}$$

in second approximation. A clear definition of the virtual states is certainly possible for the ground state of the atom which is sharp. Therefore (32) would vanish for the ground state if we had eliminated the virtual field properly (whereby H is replaced by the transformed Hamiltonian K). So far we have only made $\mathscr{I}\Gamma(E_O)$ vanish but not the derivative. This will also show itself when we use the results of this section to go over to the case of free particles (appendix 4). After the transformation used in § 34.4, (32) will be fulfilled for the ground state and also in the limiting case of free particles.

V

RADIATION PROCESSES IN FIRST APPROXIMATION

In this chapter we apply the theory of the preceding sections to calculate the probabilities for various processes in which light quanta are emitted, absorbed, scattered, or otherwise involved. As has been explained in Chapter IV the mathematical complication of the problem is such that it is necessary to apply an expansion in powers of the electronic charge e (or the fine structure constant $e^2/\hbar c$). For practical applications it is quite sufficient to use only the lowest order in e in which a non-vanishing result occurs, whereas the higher orders give but very small corrections. It will be seen that the lowest order of approximation is already in excellent agreement with the experiments. The treatment of the higher order corrections is by no means straightforward, and we devote a special chapter (VI) to them. There is, however, one class of problems which can be treated without difficulty, although they go, in some sense, beyond the lowest order of approximation, namely the damping effects. The damping phenomena correspond to the classical damping force proportional to $\dddot{x}$ (§ 4) and can be translated into quantum theory in a straightforward manner. They will for the most part be included in this chapter (§§ 18, 20). For the applications of this chapter the elementary perturbation theory of § 14 suffices, with easy generalizations for questions concerned with the line breadth. In § 20 the general theory of damping (§ 16) will be used.

17. Emission and absorption†

The emission and absorption of light by an atom can easily be understood by reference to the preceding theory. An atom and the radiation field form two quantum-mechanical systems with an interaction energy H_{int}. This interaction, regarded as a perturbation, will cause transitions of the unperturbed system (atom+radiation) in general consisting (i) of a transition of the atom from one quantum state to another, and (ii) of an emission or absorption of light quanta.

The radiation field has a continuous spectrum. If a light quantum $\mathbf{k}$ ($k = \hbar\nu$) is emitted or absorbed, we have the choice of assigning this

† The theory of emission and absorption of light was first developed by P. A. M. Dirac, *Proc. Roy. Soc.* A, **114** (1927), 243 and 710 (dispersion).

quantum to any one of a very great number of radiation oscillators (per unit volume L^3)

$$\rho_k\, dk = \frac{k^2\, dk d\Omega}{(2\pi\hbar c)^3}, \tag{1}$$

which all have the same frequency (within the interval dk), the same direction of propagation (within the element of the solid angle $d\Omega$), and the same polarization. Hence it follows, according to § 14, that a transition probability per unit time exists. Furthermore, if we neglect effects due to the line breadth, the *energy of the unperturbed system* is *conserved* for all transitions in which light quanta are emitted or absorbed.

The interaction between the atom and the radiation field can cause these radiative transitions even if, in the initial state, *no light quanta at all are present.* Supposing the atom to be excited in the initial state, then in the final state light quanta will be present. This process represents a *spontaneous emission of light.*

In § 14 we have seen that in the first approximation only *one light quantum* is emitted or absorbed. These transitions occur directly, without any intermediate states. The transition probabilities are then given by the matrix elements of H_{int} for the direct transition from the initial to the final state.

For the theory of emission and absorption we may confine ourselves to the non-relativistic approximation. Even for heavy atoms the energy of the K-shell is still much smaller than mc^2, and the relativistic correction, though appreciable for uranium and X-rays emitted in transitions to the K-shell, does not seriously affect the results. We may use, therefore, the interaction (§ 13 eq. (7))

$$H_{\text{int}} = -(e/\mu)(\mathbf{pA}). \tag{2}$$

The second term which is proportional to A^2 would give rise to transitions in which two quanta are involved; it can therefore be neglected. The matrix elements of (2) for the emission or absorption of a quantum $\mathbf{k}_\lambda$ are given by § 14 eq. (23)

$$H_{an_\lambda+1|bn_\lambda} = H^*_{bn_\lambda|an_\lambda+1} = -\frac{e}{\mu}\sqrt{\left(\frac{2\pi\hbar^2c^2}{k_\lambda}\right)}\sqrt{(n_\lambda+1)}\int \psi_a^* p_e\, e^{-i(\boldsymbol{\kappa}_\lambda \mathbf{r})}\psi_b, \tag{3}$$

where p_e denotes the component of $\mathbf{p}$ in the direction of polarization of $\mathbf{k}_\lambda$. For simplicity we confine ourselves to the case of one electron. If the atom contains several electrons the matrix element (3) must be written

$$e\int \psi_a^* p_e\, e^{i(\boldsymbol{\kappa}_\lambda \mathbf{r})}\psi_b \to \sum_k e_k \int \psi_a^* p_{ek}\, e^{-i(\boldsymbol{\kappa}_\lambda \mathbf{r}_k)}\psi_b. \tag{3'}$$

1. *Emission.* We calculate first the probability for the emission of light. We assume that there are two non-degenerate atomic states a, b with energies $E_b > E_a$. The conservation of energy states then that only light quanta with a frequency

$$k = \hbar\nu = E_b - E_a \tag{4}$$

can be emitted. Equation (4) represents *Bohr's* well-known *frequency relation.*

The transition probability per unit time is according to § 14 eq. (9)

$$w = \frac{2\pi}{\hbar}\rho_E|H|^2, \tag{5}$$

where ρ_E represents the number of final states with an energy between E and $E+dE$. For ρ_E in our case we have obviously to insert the number of radiation oscillators ρ_k. The energy of the final state is $k+E_a = E$ and therefore $dE = dk$, $\rho_E = \rho_k$. The formula (9) of § 14 has been obtained by taking the summation over all radiation oscillators with the same physical properties. For $|H|^2$ we have therefore to insert the average value of the square of the matrix element (3) over all these oscillators. Since this average value will then only depend upon the frequency, etc., but not upon the particular oscillator considered, we can replace n_λ by a quantity $\bar{n}_\nu$ denoting the *average number of light quanta per oscillator* having the frequency ν, the direction of propagation $\mathbf{k}$, etc., already present before the emission.

Inserting equations (1) and (3) in (5) we obtain the transition probability per unit time for the emission of a light quantum $\hbar\nu$ in a given direction:

$$w\,d\Omega = \frac{e^2}{\mu^2}\frac{\nu\,d\Omega}{2\pi\hbar c}|(p_e\,e^{-i(\boldsymbol{\kappa}\mathbf{r})})_{ab}|^2(\bar{n}_\nu+1). \tag{6}$$

In general one can assume that the wave-length of the light emitted, $1/\kappa$, is large compared with the dimensions of the atom. If E is the energy of the atom, the wave-length will be of the order of magnitude

$$\lambda \sim \hbar c/E. \tag{7}$$

On the other hand, the dimensions a of the atom are roughly

$$E \sim e^2/a \quad \text{or} \quad a \sim e^2/E. \tag{8}$$

Since $\hbar c/e^2 = 137$, λ will be large compared with a. We can then omit the factor $\exp(-i(\boldsymbol{\kappa}\mathbf{r}))$ in (6), since it is nearly constant in the region where ψ_a or ψ_b is different from zero. Putting $\mathbf{p}/\mu = \mathbf{v}/c$ and introducing the notation Θ for the angle between the direction of polarization and the vector $\mathbf{v}$, the transition probability becomes

$$w\,d\Omega = \frac{e^2\nu\,d\Omega}{2\pi\hbar c^3}|\mathbf{v}_{ab}|^2\cos^2\Theta(\bar{n}_\nu+1), \tag{9}$$

where $|\mathbf{v}_{ab}|^2 = v_{xab}^2+v_{yab}^2+v_{zab}^2$ and v_{xab} represents the matrix element of the x-component of v for the transition $b \to a$. Since, in quantum theory,

$$v_{xab} \equiv \dot{x}_{ab} = -i\nu x_{ab},$$

we obtain

$$w\,d\Omega = \frac{e^2\nu^3\,d\Omega}{2\pi\hbar c^3}|\mathbf{x}_{ab}|^2\cos^2\Theta(1+\bar{n}_\nu). \tag{10}$$

The probability of emission consists, according to (10), of two terms. The first term is independent of the intensity of radiation present before the emission. It gives rise to the *spontaneous emission* and is different from zero even if $\bar{n}_\nu = 0$. The second term is proportional to the intensity of radiation $\bar{n}_\nu$ of frequency ν present before the emission process. This term gives rise to a certain *induced emission of radiation*. The existence of such an induced emission was first postulated by Einstein, who has shown that it is necessary to account for the thermal equilibrium in a gas emitting and absorbing radiation. It can be deduced from the classical theory by means of the correspondence principle, since in § 5 we have seen that a light wave falling upon an oscillator not only causes an absorption of light but also, for certain phase differences between the oscillator and incident light wave, an emission of light. The analogous process in the quantum theory is given by the term proportional to $\bar{n}_\nu$.†

The total intensity radiated per unit time is obtained by multiplying (10) by $\hbar\nu$ and integrating over the angles. Taking the summation over the directions of polarization first, we obtain $\sin^2\theta$ instead of $\cos^2\Theta$, where θ represents the angle between the vector $\mathbf{x}$ (position of the electron related to the nucleus) and the direction of propagation $\mathbf{k}$. We obtain for the spontaneous emission:

$$S\,d\Omega = \frac{e^2\nu^4\,d\Omega}{2\pi c^3}|\mathbf{x}_{ab}|^2\sin^2\theta. \tag{11}$$

(11) gives the intensity emitted per unit time in the direction $\mathbf{k}$. The total intensity is given by integrating (11) over all angles

$$S = \frac{4}{3}\frac{e^2}{c^3}\nu^4|\mathbf{x}_{ab}|^2. \tag{12}$$

The formulae (11) and (12) are almost identical with the formulae obtained for an oscillator in the classical theory (§ 3 eqs. (25) and (27)).

† Cf. M. Planck, *Wärmestrahlung*, Leipzig 1923, pp. 145 ff.

We have only to replace the time average of the coordinate of the oscillator $\overline{x^2}$ by the *matrix element* of the same quantity for the transition $b \to a$

$$\overline{x^2} \to 2|\mathbf{x}_{ab}|^2. \tag{13}$$

(13) gives the well-known connexion between the classical quantities and the quantum theoretical quantities according to the correspondence principle.

If the atom contains several electrons, $\mathbf{x}_{ab}$ has to be equated to

$$e\mathbf{x}_{ab} \to \sum_k e_k \mathbf{x}_{kab}. \tag{14}$$

(14) is the total *dipole moment* of the atom. The radiation given by (12) is *dipole radiation* (cf. subsection 3) with the same intensity as would be emitted by a classical oscillator with the amplitude (13).

The order of magnitude of the transition probability per unit time will be according to (10), (7), (8) (putting $x_{ab} \sim a$)

$$w \sim \frac{e^2}{\hbar c^3}\nu^3 a^2 \sim \frac{1}{137}\left(\frac{\nu a}{c}\right)^2 \nu \sim \frac{\nu}{137^3}, \tag{15}$$

i.e. of the order 10^8 sec.$^{-1}$ for the optical region, 10^{11} for X-rays, and 10^{14} sec.$^{-1}$ for γ-rays. It is independent of the mass of the emitting particle, but not of course of the charge.

2. *Absorption.* The probability of the absorption of a light quantum can be obtained in the same way. We consider a light beam of intensity $I_0(\nu)\,d\nu$ (energy per cm.^{2} sec.) coming from a given direction within an element of the solid angle $d\Omega$. The light quantum can be absorbed from any of the radiation oscillators, with frequency in the interval $d\nu$. If in the initial state the average number of quanta per oscillator is $\bar{n}_\nu$, $I_0(\nu)$ is given by

$$I_0(\nu)\,d\nu = \bar{n}_\nu \hbar\nu c \frac{k^2\,dk d\Omega}{(2\pi\hbar c)^3} = \bar{n}_\nu \frac{\nu^3\,d\Omega d\nu}{(2\pi)^3 c^2}\hbar. \tag{16}$$

If we sum over all these radiation oscillators a transition probability per unit time exists again and is given by (5), ρ_E denoting the number of initial states. We obtain from (3) (transition from $\bar{n}_\nu$ to $\bar{n}_\nu - 1$)

$$w\,d\Omega = \frac{e^2}{\mu^2}\frac{\nu\,d\Omega}{2\pi\hbar c}|(p_e e^{i(\boldsymbol{\kappa}\mathbf{r})})_{ba}|^2 \bar{n}_\nu. \tag{17}$$

The probability of absorption is proportional to the intensity of the incident light beam as is to be expected. The coefficient is exactly the same as for the emission. The ratio of the two probabilities is therefore

$$\frac{w_{\text{emission}}}{w_{\text{absorption}}} = \frac{\bar{n}_\nu + 1}{\bar{n}_\nu}. \tag{18}$$

As is well known the ratio (18) is just that which is necessary to preserve the correct thermal equilibrium of the radiation with a gas.

For the same reason as in subsection 1 we can omit the factor $\exp i(\boldsymbol{\kappa}\mathbf{r})$ in (17). Averaging over all orientations of the atom (i.e. over the directions of $\mathbf{x}$, $\overline{\cos^2\Theta} = \frac{1}{3}$) relative to the incident beam and introducing the primary intensity $I_0(\nu)$ (16) instead of $\bar{n}_\nu$, we obtain for the energy absorbed per unit time:

$$S = \frac{4\pi^2}{3}\frac{e^2}{\hbar c}\nu|\mathbf{x}_{ba}|^2 I_0(\nu). \tag{19}$$

This formula corresponds to the formula (19) § 5 obtained for the absorption of a classical oscillator. For a 3-dimensional oscillator the quantum theory gives just

$$|\mathbf{x}_{ba}|^2 = \frac{3\hbar}{2m\nu}$$

and hence the classical formula deduced in § 5.

3. *Quadripole and magnetic dipole radiation.* In subsection 1 we have seen that the matrix element for the emission of light can in general be replaced by the matrix element $\mathbf{x}_{ab}$ of the dipole moment. For certain transitions $b \to a$, however, it may happen that the matrix elements of the dipole moment $\mathbf{x}_{ab}$ vanish. Those are classed as forbidden transitions. If, for a transition $b \to a$, $\mathbf{x}_{ab} = 0$, it is still possible that (3) does not vanish and that the transition occurs though in a higher approximation and with a smaller transition probability.

For a wave-length $\lambda \gg a$ we can expand the exponential function occurring in the matrix element of (3) in powers of the ratio a/λ:

$$e^{-i(\boldsymbol{\kappa}\mathbf{r})} = 1 - i(\boldsymbol{\kappa}\mathbf{r}) + \dots. \tag{20}$$

The matrix element (3) can then be developed in a similar way (inserting $-i\nu x_e/c$ for p_e/μ and denoting the displacement of the electron by $\mathbf{x}$ instead of $\mathbf{r}$)

$$(x_e e^{-i(\boldsymbol{\kappa}\mathbf{x})})_{ab} = x_{eab} - i\{x_e(\boldsymbol{\kappa}\mathbf{x})\}_{ab}. \tag{21}$$

For a forbidden transition, $\mathbf{x}_{ab} = 0$, the second term of (21) may then still give a certain transition probability. The intensity is, according to (6), given by

$$S\,d\Omega = \frac{e^2\nu^4\,d\Omega}{2\pi c^3}|\{\mathbf{x}(\boldsymbol{\kappa}\mathbf{x})\}_{ab}|^2\sin^2\theta. \tag{22}$$

The order of magnitude of (22) is smaller than for an allowed transition (11). Since $\kappa = 1/\lambda$ and $x \sim a$, the ratio of the intensity of a 'forbidden transition' to the intensity of an 'allowed transition' is of the order $(a/\lambda)^2$ (supposing, of course, that the matrix elements of (22) do not vanish).

The expansion (20) corresponds exactly to the expansion of the Hertzian vector $\mathbf{Z}$ in the classical theory (§ 3 eq. (22′)). The first term represents the dipole moment. The second term for a harmonically vibrating system of electrons can be written:

$$\mathbf{Z}_2 = -i\nu \sum e_k \mathbf{x}_k \frac{(\mathbf{x}_k \mathbf{R})}{Rc} = -i \sum_k e_k \mathbf{x}_k (\boldsymbol{\varkappa} \mathbf{x}_k). \tag{23}$$

($\mathbf{R}$ denotes the vector from the nucleus to the point of observation $\mathbf{R}/R = \boldsymbol{\varkappa}/\kappa$.) (23) is just the quantity occurring in (22). It represents the *electric quadripole* and *magnetic dipole moment* of the atom.

The expansion (21) can also be considered as an expansion by which the *retardation* between the different points of the atom is taken into account in successive degrees of approximation.

We see that, for the problem of emission and absorption of light, the quantum theory gives results which correspond in every detail to those of the classical theory, in the sense of Bohr's correspondence principle.

In a higher approximation it is also possible that two photons are emitted in the transition $b \to a$, $E_b - E_a = k_1 + k_2$. The probability is very small compared with that of single emission.† (Compare also § 23.)

18. Theory of the natural line breadth

In the classical treatment of the emission and absorption of light we have seen that the line emitted by an oscillator is not infinitely sharp. It has a certain *natural breadth* γ corresponding to an intensity distribution (§ 4 eq. (28))

$$I(\nu) = I_0 \frac{\gamma}{2\pi} \frac{1}{(\nu - \nu_0)^2 + \gamma^2/4}, \tag{1}$$

where ν_0 is the frequency of the oscillator. This natural breadth is due to the *damping force* of the emitted radiation on the oscillator (self-force of the electron). In the approximation where the natural breadth is taken into account the reaction force can also be derived simply from the law of conservation of energy. In this approximation, therefore, it has no connexion with the problem of the structure of the electron as is the case in all higher approximations.

In the quantum-mechanical formalism the radiation damping is included just as fully as in the classical theory. We may therefore expect that the quantum theory will also account for the natural breadth of a spectral line without difficulty.

In fact, the smallness of the interaction H_{int} between atom and radiation was not the only assumption made in our perturbation theory

† M. Mayer-Göppert, *Ann. Phys.* **9** (1931), 273.

(§ 14). (This assumption is of course fundamental.) We have solved the equations (5) § 14 for a time t which is small compared with the lifetime of the initial state, so that, up to the time t, the probability for a transition is very small. It is clear that this assumption makes a theory of the line breadth impossible, because the latter is due just to the gradual decrease (in the classical theory) of the amplitude of the oscillator or, in the quantum theory, to the falling off of the probability of the atom being in its initial state.

Weisskopf and Wigner† have given an improved solution of the equations of the perturbation theory which is also valid for times t comparable with the reciprocal of the transition probability.

1. *Atom with two states.* We consider first the simple case of an atom with two states a, b only ($E_b > E_a$). We return to the differential equations § 14 eq. (5). It is still sufficient to take only those states into account which can be reached directly from the initial state (first approximation in H). We assume that at the time $t = 0$ the atom is excited and that no light quantum is present. Then we may confine ourselves to the consideration of those states where the atom has jumped down to the lower state and one light quantum $\hbar\nu$ has been emitted with a frequency *nearly* equal to $E_b - E_a$. Denoting the probability amplitudes by b_{b0} and b_{a1_λ} we have:

$$i\hbar\dot{b}_{b0} = \sum_\lambda H_{b0|a1_\lambda} b_{a1_\lambda} e^{i(E_b-E_a-k_\lambda)t/\hbar}, \tag{2a}$$

$$i\hbar\dot{b}_{a1_\lambda} = H_{a1_\lambda|b0} b_{b0} e^{i(E_a-E_b+k_\lambda)t/\hbar}. \tag{2b}$$

The initial conditions are

$$b_{b0}(0) = 1, \qquad b_{a1_\lambda}(0) = 0. \tag{3}$$

We try to solve the equations (2) putting

$$b_{b0}(t) = e^{-\gamma t/2}, \tag{4}$$

i.e. we assume that the *probability of finding the atom in the excited state decreases exponentially* with a *lifetime* $1/\gamma$. (3) is obviously satisfied by (4).

Inserting (4) in (2b) we obtain the differential equation

$$i\hbar\dot{b}_{a1_\lambda} = H_{a1_\lambda|b0} e^{i(E_a-E_b+k_\lambda)t/\hbar-\gamma t/2} \tag{5}$$

with the solution

$$-b_{a1_\lambda} = H_{a1_\lambda|b0} \frac{e^{i(\nu_\lambda-\nu_0)t-\gamma t/2}-1}{\hbar(\nu_\lambda-\nu_0+i\gamma/2)} \tag{6}$$

where we have put $\qquad E_b - E_a = k_0 = \hbar\nu_0. \qquad (7)$

† V. Weisskopf and E. Wigner, *Zs. f. Phys.* **63** (1930), 54; ibid. **65** (1930), 18.

Finally we have to satisfy equation (2 a). Inserting (6) into (2 a) we obtain

$$-\frac{i\hbar\gamma}{2} = \sum_\lambda \frac{|H|^2[1-e^{[i(\nu_0-\nu_\lambda)+\gamma/2]t}]}{\hbar(\nu_0-\nu_\lambda-i\gamma/2)}. \tag{8}$$

The summation over the radiation oscillators λ on the right-hand side can be replaced by an integration over the frequencies ν. If $\rho_k\,dkd\Omega$ represents as before the number of radiation oscillators per unit volume with given physical properties, the right-hand side of (8) reduces to the integral

$$\int f(\nu)\frac{1-e^{[i(\nu_0-\nu)+\gamma/2]t}}{\nu_0-\nu-i\gamma/2}\,d\nu \tag{9}$$

with

$$f(\nu) = \int \rho_k|H|^2\,d\Omega,$$

where $\int d\Omega$ denotes integration over all directions of propagation, etc. If our solution is to be right, the integral (9) must be independent of t. We shall again be interested only in times t where $\nu_0 t \gg 1$, which still leaves it open whether γt is large or small. γ will turn out to be small compared with ν_0, an expression for the fact that the damping is small or that the lifetime is large compared with the frequency of the atom. γ can then be neglected in the integral (9). We divide the integrand into its real and imaginary parts ($f(\nu)$ is real):

$$\frac{1-e^{i(\nu_0-\nu)t}}{\nu_0-\nu} = \frac{1-\cos(\nu_0-\nu)t}{\nu_0-\nu} - i\frac{\sin(\nu_0-\nu)t}{\nu_0-\nu}. \tag{10}$$

For $\nu_0 t \gg 1$, the first term is what was called in § 8 the principal value of $1/(\nu_0-\nu)$. The cos, being a very rapidly varying function, gives no contribution to any integral over ν, except when $\nu_0-\nu = 0$, and then the function is zero. The second term is just the δ-function. Owing to the rapid variation of the sin it gives no contribution to the integral except for $\nu_0-\nu = 0$, when

$$\frac{\sin(\nu_0-\nu)t}{\nu_0-\nu} \to t$$

becomes infinite. The integral over ν is finite and has the value π. So we can write

$$\left.\frac{1-e^{i(\nu_0-\nu)t}}{\nu_0-\nu}\right|_{t\to\infty} \equiv \zeta(\nu_0-\nu) = \frac{\mathscr{P}}{\nu_0-\nu} - i\pi\delta(\nu_0-\nu). \tag{10'}$$

This ζ-function has been discussed in § 8. We substitute (10′) in (9) or (8). We find that γ has an imaginary contribution

$$\frac{\hbar}{2}\mathscr{I}(\gamma) = \mathscr{P}\int \frac{|H|^2\rho_k\,d\Omega dk}{k_0-k}. \tag{11}$$

This imaginary part has a simple physical significance. As is clear from (6), for instance, $\mathscr{I}(\gamma)$ means a correction to the frequency of the emitted line ν_0. This line shift is usually very small for allowed transitions. In the present context we are interested in the line breadth given by the real part of γ and we shall neglect the line shift here. It is discussed in detail in § 16.3 and § 34. The real part of γ (simply denoted by γ) is due to the second term of (10′). The integration of $\delta(\nu_0-\nu)$ over ν gives 1, and $\nu = \nu_0$ is to be inserted in $|H|^2$. Thus

$$\gamma = \frac{2\pi}{\hbar}\int \rho_k |H|^2\, d\Omega = w_{ab}. \tag{12}$$

According to § 17 eq. (5) γ *is just equal* to the *total spontaneous transition probability per unit time for emission* $b \to a$. This was to be expected since, according to (4), γ was defined as the reciprocal of the lifetime of the excited state.

The intensity distribution of the emitted line is given by the probability function for the final state $b_{a1\lambda}$. After a time $t \gg 1/\gamma$, when the atom has certainly jumped down, the probability that a quantum $\hbar\nu_\lambda$ has been emitted is given by

$$|b_{a1\lambda}(\infty)|^2 = \frac{|H|^2}{\hbar^2}\frac{1}{(\nu_\lambda-\nu_0)^2+\gamma^2/4}, \tag{13}$$

or, integrating over all directions of propagation, etc., according to (12),

$$I(\nu)\,d\nu = \hbar\nu\rho_k\,dk \int |b_{a1\lambda}(\infty)|^2\, d\Omega = \frac{\gamma}{2\pi}\frac{\hbar\nu\,d\nu}{(\nu-\nu_0)^2+\gamma^2/4}. \tag{13′}$$

The total intensity is $\hbar\nu = I_0$. Formula (13′) is therefore identical with the classical formula § 4 eq. (28), the only difference being that γ now represents the transition probability per unit time as given by (12) instead of $2e^2\nu_0^2/3mc^3$.

In the quantum theory a spectral line has therefore the same intensity distribution as in the classical theory (shown in Fig. 2, p. 33). *The breadth at half maximum is equal to the total transition probability per unit time.* The maximum intensity lies at the frequency ν_0 given by the energy difference of the two states of the atom (7) (corrected by a small line shift).

The relation between the half-value breadth and the transition probability can be understood from the *uncertainty relation* for energy and time:

$$\Delta E \Delta t = \hbar,$$

which states that the energy of a system is only known with an accuracy ΔE if, for the measurement of the energy, a time Δt is available. In our case the excited state of the atom has a lifetime $1/\gamma$ due to the radiative transition probability. Therefore the energy of the excited state is only defined with an uncertainty $\Delta E = \hbar\gamma$, or the energy level E_b has a *breadth* $\Delta E_b \simeq \hbar\gamma$. The frequency of the emitted line will then have the same breadth $\Delta\nu \sim \gamma$, which is just our relation.

The above results could have been derived directly from the general theory of § 16. For the sake of clarity we have derived them once more from the start for this simple example. In fact here the general equation § 16 eq. (10) for the probability amplitude reduces simply to

$$U_{a1_\lambda|b0} = H_{a1_\lambda|b0}$$

and the probability for $t = \infty$ is given by § 16 eq. (20) which is identical with (13), if we replace $\Gamma(E_n)$ by a constant γ. For the more complicated problems concerning line breadth (§ 20) we shall use the theory of § 16.

2. *Several atomic states.* The case where the atom has several states $a, b, c,...$ is more complicated. It has also been treated by Weisskopf and Wigner (loc. cit.). The result cannot, however, be determined unambiguously from the classical analogy in this case. One would perhaps expect the intensity distribution of a line emitted in a transition $b \to a$ to be determined as before by equation (11), with a half-value breadth γ_{ab} equal to the transition probability $b \to a$. But this conclusion is not in agreement with the above considerations about the uncertainty relation. The quantum theory leads to a different result.

If we denote the atomic levels in the order of their energies by $a_1, a_2,...$, we can, according to (13), attribute to each *level*, a_i say, a certain *breadth* given by the sum of all *transition probabilities* from a_i to all lower levels:

$$\Delta E_{a_i}/\hbar \equiv \gamma_i = \sum_{j<i} w_{a_j|a_i}, \tag{14}$$

where $w_{a_j|a_i}$ represents the transition probability for the transition $a_i \to a_j$. The breadth of a certain line, $a_i \to a_k$ say, is then given by the *sum of the breadths of the two levels a_i and a_k:*

$$\gamma_{ki} = \gamma_i + \gamma_k. \tag{15}$$

The intensity distribution is again the classical one, equation (12), with $\gamma = \gamma_{ki}$. These results are easily derived by the method of § 20.

This result is quite different from that suggested by the correspondence principle, according to which one might conclude that the breadth of a line would be proportional to its intensity. This is, however, not true in the quantum theory. Here it may happen that even a weak line is rather broad. Consider, for instance, a case as shown in Fig. 5 with three levels a, b, c. From the highest level c the transition probabilities are all small, and therefore c is a narrow level. From b a strong line leads to the ground-level a (which is always sharp). Therefore b will be broad. According to equation (15) the line $c \rightarrow b$ must also be broad although the transition probability is very small. The line $c \rightarrow a$, on the other hand, is narrow since it connects two narrow levels.

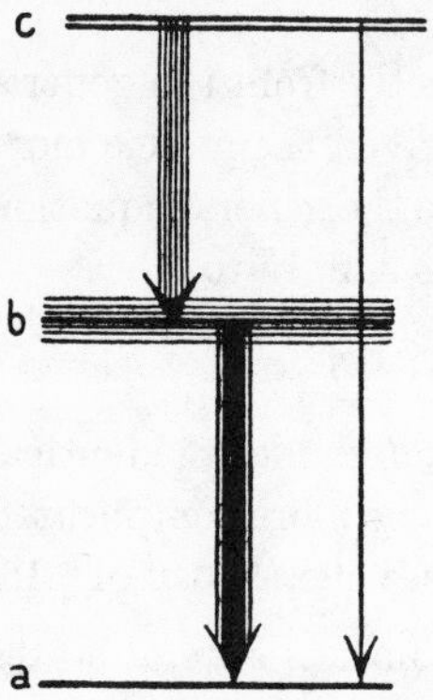

FIG. 5. 'Weak but broad' lines in the quantum theory of line breadth.

3. *Absorption*. The shape of the *absorption line* must be the same as the emission line if the incident light beam has a constant intensity in the region of the line breadth. This follows from general equilibrium considerations (Kirchhoff's law). If $I_0(\nu)\,d\nu = I_0(\nu_0)\,d\nu$ represents the intensity of the primary beam, the energy in the frequency range ν to $\nu+d\nu$ absorbed per unit time due to a transition $a \rightarrow b$ will be

$$S(\nu)\,d\nu = w_{ab}\frac{\pi^2c^2}{\nu_0^2}\frac{\gamma}{2\pi}\frac{I_0(\nu_0)\,d\nu}{(\nu-\nu_0)^2+\gamma^2/4} \quad (\gamma = \gamma_a+\gamma_b), \tag{16}$$

where w_{ab} is the transition probability for the spontaneous emission $b \rightarrow a$. The factor in (16) has been determined so that the total energy absorbed per unit time is identical with formula (19) § 17.

If we consider a layer of thickness Δx containing N atoms per cm.^{3} in the absorbing state a, we can define an *absorption coefficient per cm.* $\tau(\nu)$ for the primary light beam

$$\tau(\nu) = \frac{S(\nu)}{I_0(\nu_0)}N = Nw_{ab}\frac{\pi^2c^2}{\nu_0^2}\frac{\gamma}{2\pi}\frac{1}{(\nu-\nu_0)^2+\gamma^2/4}. \tag{17}$$

For frequencies at large distances from the maximum $(\nu-\nu_0)^2 \gg \gamma^2$ the absorption coefficient decreases with the square of the distance $\Delta\nu = \nu-\nu_0$, or in terms of the wave-length $2\pi\Delta\lambda = 2\pi c\Delta\nu/\nu^2$.

The ratio of the absorbed intensity to the primary intensity is then given by†

$$\tau(\lambda)\,\Delta x = N\,\Delta x\,w_{ab}\,\gamma\,\frac{\pi\lambda^6}{2c^2\,\Delta\lambda^2}. \tag{18}$$

(In (18) the wave-length is $2\pi\lambda$.)

† (18) is valid only if $\tau(\lambda)\,\Delta x$ is small. For large values of $\tau(\lambda)\,\Delta x$ the left-hand side of (18) has to be replaced by $1-\exp\{-\tau(\lambda)\,\Delta x\}$.

4. *Other causes for the line breadth.* Besides the damping due to the emission of the radiation itself there are several other causes which actually broaden a line:

(*a*) In a gas of temperature T the atoms (mass M) move with velocities distributed according to Maxwell's law: $\exp(-v_x^2 M/2kT)$. If we observe the light emitted in the x-direction, the line will be shifted because of the *Doppler effect* by an amount (§ 7 eq. (21), for $v \ll c$)

$$\Delta\nu = \nu_0 v_x/c. \tag{19}$$

Averaging, we obtain, therefore, a broad line with an intensity distribution

$$I(\nu)\,d\nu = \text{const.}\,d\nu\,e^{-Mc^2\Delta\nu^2/2\nu_0^2kT} \tag{20 a}$$

and a *breadth at half maximum*:

$$\delta = \nu_0\sqrt{\left(\frac{2kT}{Mc^2}\log 2\right)}. \tag{20 b}$$

In general the Doppler breadth δ is much greater than the natural breadth γ. The intensity distribution is, however, exponential and therefore decreases rapidly with the distance from the maximum $\Delta\nu$ in contrast to the natural breadth which has a very large tail decreasing only with $1/\Delta\nu^2$. The intensity observed at large distances $\Delta\nu$ from the maximum (i.e. if $\Delta\nu \gg \delta$) is therefore due to the natural breadth (and causes (*b*)–(*d*)).

(*b*) In a gas of finite density an excited atom undergoes *collisions* with neighbouring atoms which may induce a transition to the ground-level. The effect of these collisions on the line breadth can be described as follows: if the number of effective collisions per sec. is Γ the *lifetime* of the excited state b will be *shortened*. The total number of transitions per sec. (radiative+collisions) is now equal to $\gamma+\Gamma$. The breadth of the level b will therefore be

$$\Delta E_b/\hbar = \gamma+\Gamma. \tag{21}$$

The line emitted has the same intensity distribution as the natural line (16), the only difference being that γ has to be replaced by $\gamma+\Gamma$. For very low densities the effect of broadening by collisions becomes small.

(*c*) The excited atom will have all kinds of interaction with the neighbouring atoms, which causes a shifting and splitting of the excited state (Stark effect, resonance coupling, etc.), leading to an effective broadening. For very low densities these effects are likewise small.

(*d*) Auger effect: if an atom with several electrons is excited in that a K-electron is raised to a high level, an X-ray line is usually emitted while, for example, an L-electron jumps into the K-vacancy. Instead the interaction with the electrons in the higher shells may lead to the ejection of one of the latter electrons into the continuous spectrum, in place of the emission of a photon. This process competes with the emission of the photon and contributes to the width of the L-level. In contrast to the causes (a)–(c) the Auger width is independent of external circumstances (temperature, density) and cannot be separated from the radiation width. Radiation and Auger widths should be combined to give the total 'natural width'. The Auger width exists, of course, only for X-rays.

5. *Experimental check.* Measurements of the natural width of spectral lines are very few because the broadening influences (a)–(c) usually supersede the natural width in the optical region. A few measurements have, however, been made.† For X-rays one can eliminate the causes (a)–(c) more easily. A single example must suffice here:

The natural widths of the L-series of Au have been measured.‡ From the differences of the line widths it could be verified that the width of a line is the sum of widths of the two levels concerned, as predicted by the theory. From a further measurement of the shape of the L-absorption edge the widths of several X-ray levels could be determined individually. The transition probabilities from these levels into all other states were calculated for both radiative and Auger transitions. These calculations can be none too accurate owing to the lack of accurate wave functions for complicated atoms. The theoretical figures in the following table have been calculated with wave functions of the Thomas–Fermi model,§ those in brackets with hydrogen-like wave functions.|| Table I shows

TABLE I

Observed and calculated natural level widths (radiation and Auger effect) in eV for various X-ray levels in Au

Level (vacant place)	*Rad.*	*Calculated Auger*	*Total*	*Observed*
K	66	> 0·8	67	54
L I	1·8 (1)	> 11·9 (5·5)	> 13·7 (6·5)	8·7
L II	(0·9)	(2·2)	(3·1)	3·7
L III	(1·6)	(2·6)	(4·2)	4·4
M I	0·1	> 10·2	> 10·3	15·5

† Reports on the breadth of spectral lines, including measurements: V. Weisskopf, *Phys. Zs.* **34** (1933), 1; H. Margenau and W. W. Watson, *Rev. Mod. Phys.* **8** (1936), 22; J. H. v. Vleck and V. Weisskopf, ibid. **17** (1945), 227.

‡ F. K. Richtmyer, S. W. Barnes, and E. Ramberg, *Phys. Rev.* **46** (1934), 843.

§ E. G. Ramberg and F. K. Richtmyer, ibid. **51** (1937), 913.

|| L. Pincherle, *Nuovo Cimento*, **12** (1935), 162; *Physica*, **2** (1935), 596.

the observed and calculated widths for several levels. The > sign in some of the figures for the Auger width means that the figure is to be increased slightly because some of the transitions with probably very small probability have not been calculated. In view of the difficulty of the calculation the agreement must be considered as satisfactory.

Accurate measurements of the line shape and width for hydrogen are highly desirable. Such measurements seem to be possible with the help of the radio-frequency technique used for the measurement of the line displacement (§ 34) but at the time of writing no accurate comparison between theory and experiment has been made.

19. Dispersion and Raman effect

In this section we consider the scattering of light by an atom. The scattering process consists of the absorption of a primary light quantum $\mathbf{k}_0$ and the simultaneous emission of a secondary quantum $\mathbf{k}$. The scattering atom may be left in either its initial state (coherent scattering), or, as in the Raman effect, in some other state.

The general character of the scattering processes depends on whether the energy of the primary quantum k_0 is of the same order of magnitude as the binding energy of the electron in the atom, or large compared with the binding energy. In the latter case the electron can be considered as free. The scattering by free electrons will be discussed in detail in § 22. Let k_0 be of the same order of magnitude as the energy of the electron. This is roughly the region of visible light up to frequencies of soft X-rays. We can then neglect all relativistic corrections. Furthermore we can assume that the wave-length of the primary quantum λ_0 and of the scattered quantum λ are both large compared with the dimensions of the atom.

We denote the states of the atom by n_i, in particular the initial state by n_0, the final state by n (energies E_i, E_0, E). The conservation of energy states that the frequency of the scattered quantum k differs from k_0 by the energy difference of the atom

$$k-k_0 = E_0-E. \tag{1}$$

We assume that k_0 is not near a resonance frequency of the atom E_i-E_0 (see § 20).

In the case of *coherent* scattering (compare subsection 2) the atom is left in the same state $n_0 = n$. The frequency of the scattered quantum k is then the same as that of the primary quantum k_0. The case $E_0 \neq E$ represents the Raman effect.

On the other hand, the momentum will not in general be conserved in the interaction of light with a bound electron.

1. *The dispersion formula.* According to § 14 the non-relativistic interaction between an electron and the radiation is given by

$$H_{\text{int}} = -\frac{e}{\mu}(\mathbf{pA})+\frac{e^2}{2\mu}A^2 = H^{(1)}+H^{(2)}. \tag{2}$$

In contrast to the theory of emission, we must not neglect the second term, which is $\sim A^2$. As has been shown in § 14 this is just a second-order term having matrix elements for direct transitions in which the total number of light quanta changes by two.

The matrix elements of $H^{(2)}$ are given in § 14 eq. (24). In our case, i.e. for the transition $E_0, k_0 \rightarrow E, k$, we have

$$H^{(2)} = \frac{e^2}{\mu}\frac{2\pi\hbar^2c^2}{\sqrt{(k_0 k)}}\int \psi_n^* e^{i(\boldsymbol{\kappa}_0-\boldsymbol{\kappa},\mathbf{r})}\psi_{n_0}(\mathbf{e}_0\,\mathbf{e}), \tag{3}$$

where $\mathbf{e}_0$, $\mathbf{e}$ represent unit vectors in the direction of polarization of the two quanta $\mathbf{k}_0$, $\mathbf{k}$. ($\boldsymbol{\kappa} = \mathbf{k}/\hbar c$.)

The exponential function can be regarded as constant in the integral (3). The matrix element is then obviously only different from zero if $n_0 = n$, i.e. in the case of coherent scattering

$$H^{(2)} = \frac{e^2}{\mu}\frac{2\pi\hbar^2c^2}{\sqrt{(k_0 k)}} e^{i(\boldsymbol{\kappa}_0-\boldsymbol{\kappa},\mathbf{X})}\delta_{n_0 n}(\mathbf{e}_0\,\mathbf{e}), \tag{4}$$

where $\mathbf{X}$ represents a vector indicating the position of the atom.

The first term $H^{(1)}$ of equation (2) is a first-order term. It can only cause transitions involving two light quanta through the agency of *intermediate states* which differ from the initial and final states by having only one light quantum emitted or absorbed. In our case there are two kinds of such intermediate states differing in the order in which the emission of $\mathbf{k}$ and the absorption of $\mathbf{k}_0$, take place.

I. $\mathbf{k}_0$ is absorbed first, therefore no light quantum is present. In the transition to the final state $\mathbf{k}$ is emitted.

II. $\mathbf{k}$ is emitted first. Therefore both light quanta $\mathbf{k}_0$ and $\mathbf{k}$ are present. In the transition to the final state $\mathbf{k}_0$ is absorbed.

In both possible intermediate states the atom may be excited in any state n_i.

Denoting the initial state by O, the final state by F, and the two possible intermediate states by I and II, the matrix elements of the first term of $H^{(1)}$ for the transitions from O to I and II and then to F are given (according to § 14 eq. (23)) by:

$$\begin{aligned} H^{(1)}_{\mathrm{I}O} &= \\ H^{(1)}_{\mathrm{II}O} &= \\ H^{(1)}_{F\mathrm{I}} &= \\ H^{(1)}_{F\mathrm{II}} &= \end{aligned} \quad -\frac{e}{\mu}\sqrt{(2\pi\hbar^2c^2)} \begin{cases} \dfrac{1}{\sqrt{k_0}}\displaystyle\int \psi^*_{n_i} p_0\, e^{i(\boldsymbol{\kappa}_0 \mathbf{r})}\psi_{n_0} \\ \dfrac{1}{\sqrt{k}}\displaystyle\int \psi^*_{n_i} p\, e^{-i(\boldsymbol{\kappa}\mathbf{r})}\psi_{n_0} \\ \dfrac{1}{\sqrt{k}}\displaystyle\int \psi^*_{n} p\, e^{-i(\boldsymbol{\kappa}\mathbf{r})}\psi_{n_i} \\ \dfrac{1}{\sqrt{k_0}}\displaystyle\int \psi^*_{n} p_0\, e^{i(\boldsymbol{\kappa}_0 \mathbf{r})}\psi_{n_i}, \end{cases} \tag{5}$$

where p_0 and p denote the components of $\mathbf{p}$ in the direction of polarization of the two light quanta $\mathbf{k}_0$ and $\mathbf{k}$, respectively.

The energy differences of the initial state and the intermediate states are

$$E_O - E_{\mathrm{I}} = E_0 + k_0 - E_i, \qquad E_O - E_{\mathrm{II}} = E_0 - E_i - k. \tag{6}$$

The compound matrix elements $K_{F|O}$ for the transition $O \rightarrow F$ are given by the general formula § 14 eq. (18). The summation has to be carried out over all intermediate states.

Assuming again that the wave-lengths of $\mathbf{k}_0$ and $\mathbf{k}$ are large compared with the dimensions of the atom, we obtain in our case†

$$K_{F|O} = \frac{e^2}{\mu}\frac{2\pi\hbar^2c^2}{\sqrt{(k_0 k)}} \times$$
$$\times\, e^{i(\boldsymbol{\kappa}_0 - \boldsymbol{\kappa}, \mathbf{X})}\left[\frac{1}{\mu}\sum_i \left(\frac{p_{nn_i} p_{0n_i n_0}}{E_0 - E_i + k_0} + \frac{p_{0nn_i} p_{n_i n_0}}{E_0 - E_i - k}\right) + \delta_{n_0 n}\cos\Theta\right], \tag{7}$$

where $p_{0n_i n_0}$ is the matrix element of p_0 for the transition $n_0 \rightarrow n_i$ and Θ the angle between the directions of polarization of $\mathbf{k}_0$ and $\mathbf{k}$.

The transition probability per unit time is given by

$$w = \frac{2\pi}{\hbar}|K|^2 \rho_E, \tag{8}$$

where ρ_E is the number of final states per unit volume and per energy interval dE. In our case ρ_E is the number of radiation oscillators per unit volume in which the scattered quantum $\mathbf{k}$ can be placed, or

$$\rho_E = \rho_k = \frac{k^2\, d\Omega}{(2\pi\hbar c)^3}. \tag{9}$$

† The summation $\sum_i$ has also to be carried out over the states of the continuous spectrum of the atom. The contribution of the latter, however, is not very large for the optical region.

Dividing (8) by the intensity of the primary beam, i.e. for one light quantum by the velocity of light (see § 14 eq. (16), $L^3 = 1$), we obtain the differential cross-section for the scattering of a light quantum **k** into an element of solid angle $d\Omega$ and with a given polarization

$$d\phi = r_0^2 \frac{k}{k_0}\, d\Omega \left[\frac{1}{\mu} \sum_i \left(\frac{p_{nn_i} p_{0n_i n_0}}{E_0 - E_i + k_0} + \frac{p_{0nn_i} p_{n_i n_0}}{E_0 - E_i - k}\right) + \delta_{nn_0} \cos\Theta\right]^2. \quad (10)$$

This formula is not valid in the case of resonance, i.e. if $k_0 \simeq E_i - E_0$.

In the case of coherent scattering we obtain the well-known dispersion formula

$$d\phi = r_0^2\, d\Omega \left[\frac{1}{\mu} \sum_i \left(\frac{p_{n_0 n_i} p_{0n_i n_0}}{E_0 - E_i + k_0} + \frac{p_{0n_0 n_i} p_{n_i n_0}}{E_0 - E_i - k_0}\right) + \cos\Theta\right]^2. \quad (11)$$

Equation (11) was first obtained by Kramers and Heisenberg† by an application of the correspondence principle to the classical formula § 5 eq. (11) (for $\gamma = 0$). The existence of the second term $\cos\Theta$ in the dispersion formula was first shown by Waller.‡ It is identical with the formula describing the scattering by a free electron (§ 5 (4)).§ If

$$k_0 \gg E_i - E_0$$

(but λ_0 is still large compared with the dimensions of the atom), the first term of (11) is small and the dispersion formula goes over into the classical formula for the scattering by a free electron.

For $n_0 \neq n$ we obtain from (10) the well-known formula for the Raman scattering as was predicted by Smekal and Heisenberg. The existence of a scattered radiation with a frequency shifted by an amount corresponding to the energy difference between two quantum states was discovered experimentally by Landsberg and Mandelstamm|| (in solids) and Raman and Krishnan†† (in liquid solutions).

2. *Coherence.* In the classical theory the radiation scattered by an atom is *coherent* with the primary radiation. This has its origin in the fact that, except in the resonance case, the *phase* of the scattered wave is the same as the phase of the primary beam (the phase difference δ is given by § 5 eq. (9); δ is zero except in the case of resonance).

The same is true in the quantum theory also, though of course only if the scattered frequency is the same as the primary one. The applica-

† H. A. Kramers and W. Heisenberg, *Zs. f. Phys.* **31** (1925), 681.
‡ I. Waller, ibid. **51** (1928), 213.
§ In § 5 Θ denotes the angle between **k** and the direction of polarization of $\mathbf{k}_0$. In § 5 eq. (4) the summation is taken over the directions of polarization of **k**.
|| G. Landsberg and L. Mandelstamm, *Naturw.* **16** (1928), 557 and 772.
†† C. V. Raman and K. S. Krishnan, *Nature*, **121** (1928), 501.

tion of the idea of coherence, however, requires some care. As we have seen in § 7 the phase of a quantized light wave ϕ is only determined if the number of light quanta is undetermined corresponding to the uncertainty relation

$$\Delta N \Delta \phi \geqslant 1. \tag{12}$$

In the case of scattering of a single light quantum, as considered in subsection 1, the phases of the two waves are entirely undetermined.

We could check the phase relations if we considered a primary wave with a large enough number of quanta to allow the determination of both the phases and the number of quanta with a comparatively high accuracy. This is the ordinary transition to the classical theory.

But for the scattering of a single quantum also, we can give the idea of coherence a simple physical meaning, if we consider the scattering by *two atoms* A, B, situated, say, at a distance $\mathbf{R}$ apart. In the classical theory the scattered waves of the two atoms interfere with each other, giving a maximum or minimum intensity according to the difference of light path of the two scattered waves. For this classical result, only the phase *difference* of the two waves scattered by the two atoms is essential. In quantum theory the latter can have a definite value, even if the total number of light quanta is determined. In this case, however, we do not know from *which atom* the light quantum is scattered.

The same (classical) intensity distribution also follows from the quantum theory. We consider two atoms at the positions $\mathbf{X}_A$, $\mathbf{X}_B$ and with quantum states n_i, m_i. The distance between A and B may be denoted by

$$\mathbf{X}_A - \mathbf{X}_B = \mathbf{R} \tag{13}$$

(see Fig. 6). The transition probability for the scattering of a primary quantum $\mathbf{k}_0$ giving a secondary quantum $\mathbf{k}$ can be calculated in the same way as the scattering by a single atom. The radiation interacts with either of the two atoms, the interaction function is therefore given by

$$K = K_A + K_B. \tag{14}$$

The matrix element K (14) can then be written down immediately. Instead of (7) we obtain two similar terms

$$K = \frac{e^2}{\mu}\frac{2\pi\hbar^2c^2}{\sqrt{(k_0 k)}}\left\{e^{i(\mathbf{\kappa}_0-\mathbf{\kappa},\mathbf{X}_A)}\left[\frac{1}{\mu}\sum_i\left(\frac{p_{n_0 n_i}p_{0n_i n_0}}{E_0-E_i+k_0}+...\right)+\cos\Theta\right]+\right.$$
$$\left.+e^{i(\mathbf{\kappa}_0-\mathbf{\kappa},\mathbf{X}_B)}\left[\frac{1}{\mu}\sum_i\left(\frac{p_{m_0 m_i}p_{0m_i m_0}}{E_0-E_i+k_0}+...\right)+\cos\Theta\right]\right\}. \tag{15}$$

The two brackets [ ] are equal if the two atoms are assumed to be alike. (15) is therefore identical with the matrix element (7), the only difference being that the factor $\exp i(\boldsymbol{\varkappa}_0-\boldsymbol{\varkappa}, \mathbf{X})$ is replaced by the factor

$$e^{i(\boldsymbol{\kappa}_0-\boldsymbol{\kappa},\mathbf{X}_A)}+e^{i(\boldsymbol{\kappa}_0-\boldsymbol{\kappa},\mathbf{X}_B)} = e^{i(\boldsymbol{\kappa}_0-\boldsymbol{\kappa},\mathbf{X}_B)}(1+e^{i(\boldsymbol{\kappa}_0-\boldsymbol{\kappa},\mathbf{R})}). \tag{16}$$

For the transition probability (11) we obtain, therefore, a factor

$$|1+e^{i(\boldsymbol{\kappa}_0-\boldsymbol{\kappa},\mathbf{R})}|^2 = 2[1+\cos(\boldsymbol{\varkappa}_0-\boldsymbol{\varkappa}, \mathbf{R})]. \tag{17}$$

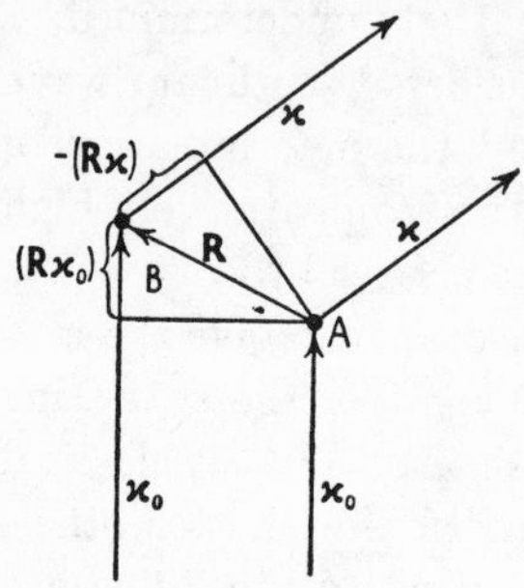

FIG. 6. Phase differences in the scattering of two atoms A, B.

This is exactly the result which is to be expected from the classical theory. The scalar product $(\boldsymbol{\varkappa}_0-\boldsymbol{\varkappa}, \mathbf{R})$ represents the difference of light path for the two scattered waves in units $1/\lambda$, i.e. the classical difference of the phases (Fig. 6). Thus, the two waves scattered by the atoms A and B can be considered as *coherent* in the same sense as in the classical theory.

3. *Scattering of X-rays.* Finally, we shall discuss qualitatively the behaviour of the scattering if the energy of the primary quantum $\mathbf{k}_0$ increases so that the wave-length becomes comparable with or smaller than the dimensions of the atom. In this case the exponential function $\exp i(\boldsymbol{\varkappa}_0 \mathbf{r})$ representing the light wave cannot be considered as constant in the integrals of the matrix elements (3) and (5).

The numerical value of the matrix elements *decreases* when these exponential functions vary appreciably inside the atom. Thus the *scattered intensity decreases* in the same way. This is true for the coherent scattering as well as for the Raman scattering (at least if the atom is left in a discrete quantum state). Finally, if the wave-length λ_0 is small compared with the dimensions of the atom, the matrix elements and therefore the intensity of the scattered wave *vanish*. This will roughly be the case if

$$k_0 \gg \frac{\hbar c}{a} \sim 2\times 137\frac{I}{Z}, \tag{18}$$

where I is the ionization energy and a the radius of the atom. For light elements (18) is satisfied in the region of hard X-rays.

On the other hand, as k_0 increases to values larger than I, another process becomes progressively more important. For $k_0 \gg I$ the electron can be left after the scattering in a state of the *continuous spectrum* with, say, a momentum $\mathbf{p}$ and an energy E. This is a certain kind of Raman effect. The frequency of the scattered radiation is then displaced relative to k_0,

$$k = k_0-(E-E_0). \tag{19}$$

Since the electron has a continuous energy spectrum, we obtain besides the ordinary undisplaced line (coherent scattering) another—displaced—line with a broad intensity distribution. The total intensity will, however, at this stage be small.

If k_0 now increases further, becoming large compared with I, the displaced line becomes increasingly sharper and more intense. This can be seen in the following way: We choose for the final state of the electron a state for which the momentum is determined by

$$\mathbf{p} \simeq \mathbf{k}_0 - \mathbf{k}. \tag{20}$$

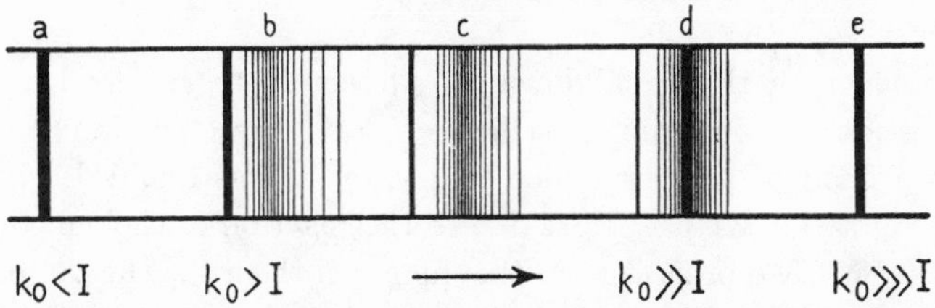

FIG. 7. Qualitative scheme of the coherent scattering and Compton scattering for increasing primary frequency k_0 (I = ionization energy of the atom) for a given angle. Case (a), coherent scattering only; case (e), Compton scattering of a free electron.

Since then $\psi_n = \exp(i\mathbf{p}/\hbar c)$ in the integral (3), the factor $\exp i(\boldsymbol{\varkappa}_0 - \boldsymbol{\varkappa}, \mathbf{r})$ is just compensated and (3) becomes *large*, however small the primary wave-length may be. For a given angle of scattering, k is determined entirely by (19) and (20). Thus we obtain an *intense and sharp displaced line*.

For very short wave-lengths, the process considered here becomes identical with the *scattering by a free electron*. (20) expresses then the law of conservation of momentum (which is always satisfied in the interaction with free electrons), since for $k_0 \gg I$ the momentum of the electron in the bound state is relatively very small. The breadth of the displaced (Compton) line is determined by the fluctuation of the momentum in the bound state.

The continuous transition from the coherent scattering of a bound electron to the Compton scattering of a free electron is shown qualitatively in Fig. 7. The intensity of the undisplaced (coherent) line decreases as the sharpness and intensity of the displaced line increases.

The intermediate case (c) or (d) of Fig. 7 is realized approximately when X-rays of 50,000 electron volts energy are scattered in light elements (carbon, beryllium). According to measurements of Du Mond†

† J. Du Mond, *Rev. of Modern Physics*, **5** (1933), 1. P. A. Ross and P. Kirkpatrick, *Phys. Rev.* **46** (1934), 668.

the displaced (Compton) line has a broad intensity distribution with a breadth of the order of magnitude of the displacement itself. The breadth has been shown to agree with that which one would expect from the momentum distribution of the electrons in the atom.

20. Resonance fluorescence

The theory of dispersion of § 19 breaks down if the frequency of the primary radiation k_0 approaches a resonance frequency of the atom $E_i - E_0$. In this case one of the denominators in the dispersion formula § 19 (11) vanishes and the intensity of the scattered radiation becomes infinite.

The reason for this breakdown can be seen from the classical theory of dispersion by a harmonic oscillator, § 5.2. There the infinity in the neighbourhood of the resonance frequency ν_0 was avoided by taking into account the *damping force* due to the reaction of the emitted light on the atom. The procedure in the quantum theory is the same. Since this damping force is very small the intensity of the scattered radiation will in any case be very large compared with the ordinary scattering.

The radiative damping can be introduced in the dispersion formula in much the same way as in the theory of line breadth, § 18. The general theory of damping phenomena, § 16, gives us the results ready made and we shall use this theory here. The solution of the relevant equations will be quite simple.

1. *General solution of the equations.*† Since the resonance fluorescence will depend decidedly upon the intensity distribution of the primary radiation in the region of the natural line breadth, we shall for the present assume a general form for the primary intensity distribution $I_0(k)\,dk$ (energy per cm.^{2} and sec.) which will be specified later.

As we see from the dispersion formula, § 19 (11), in the case of resonance the only important intermediate state is that which has a vanishing denominator, and the scattering due to the quadratic term A^2 can be neglected. Denoting the ground state of the atom by n_0 (energy E_0) and the excited state in question by n (energy E_n) (we assume that neither state is degenerate), we can confine ourselves to those intermediate states where the atom is excited and one light

† V. Weisskopf, *Ann. d. Phys.* **9** (1931), 23. See also E. Segré, *Rend. Ac. Linc.* **9** (1929), 887. We consider here the scattering of light by an atom in the ground state. The scattering by an excited atom has also been investigated by Weisskopf, *Zs. f. Phys.* **85** (1933), 451. For the method used here see W. Heitler and S. T. Ma, *Proc. Roy. Ir. Ac.* **52** (1949), 109.

quantum k_λ is absorbed. k_λ will *nearly* coincide with the resonance frequency of the atom, which we shall denote by

$$k_\lambda \sim (E_n - E_0) \equiv k_0. \tag{1}$$

In the final state the atom will again be in the state E_0 and another light quantum k_σ will be emitted. For the present we do not know whether the frequency k_σ is *exactly* identical with the absorbed frequency k_λ, but, of course, k_σ can differ from k_λ only by an amount of the order of magnitude of the natural line breadth.

Thus, three types of states have to be considered: (i) the initial state O (energy E_O), consisting of the primary intensity distribution I_0 and the atom in E_0; (ii) the intermediate state λ (energy E_λ): atom in E_n, one quantum k_λ absorbed out of I_0; (iii) the final state $\lambda\sigma$ (energy $E_{\lambda\sigma}$): atom in E_0, quantum k_σ emitted, k_λ remains absorbed. Evidently,

$$E_{\lambda\sigma} - E_\lambda = k_\sigma - k_0, \qquad E_{\lambda\sigma} - E_O = k_\sigma - k_\lambda. \tag{1'}$$

According to § 16, the probabilities for these states depend on the amplitudes $U_{\lambda|O}$, $U_{\lambda\sigma|O}$. They depend on a variable energy E, although they will only be used for $E = E_\lambda$ and $E = E_{\lambda\sigma}$ respectively. U satisfies the general equations (10) § 16, which in our case reduce to†

$$U_{\lambda|O}(E) = H_{\lambda|O} + \sum_\sigma H_{\lambda|\lambda\sigma} U_{\lambda\sigma|O}(E)\zeta(E - E_{\lambda\sigma}), \tag{2a}$$

$$U_{\lambda\sigma|O}(E) = H_{\lambda\sigma|\lambda} U_{\lambda|O}(E)\zeta(E - E_\lambda), \tag{2b}$$

where

$$\zeta(E - E_\lambda) = \frac{\mathscr{P}}{E - E_\lambda} - i\pi\delta(E - E_\lambda). \tag{2c}$$

Furthermore, a damping constant Γ occurs given by § 16 eq. (12)

$$\tfrac{1}{2}\hbar\Gamma(E) = i\sum_\lambda H_{O|\lambda} U_{\lambda|O}(E)\zeta(E - E_\lambda). \tag{3}$$

According to § 16 eq. (27), (28), the real part of Γ taken at $E = E_O$ is the total transition probability per sec. from the initial state. In (2) and (3) we have taken into account that H only has matrix elements where one quantum is involved, therefore $H_{\lambda\sigma|O} = 0$.

To solve (2) we insert (2b) into (2a):

$$U_{\lambda|O}(E) = H_{\lambda|O} - \frac{i\hbar\gamma}{2} U_{\lambda|O}(E)\zeta(E - E_\lambda) \tag{4}$$

with the abbreviation

$$\frac{\hbar}{2}\gamma(E) = i\sum_\sigma |H_{\lambda\sigma|\lambda}|^2 \zeta(E - E_{\lambda\sigma}). \tag{5}$$

† In (2b) no summation over λ occurs because k_λ in $U_{\lambda|O}$ is the same as in the first index $\lambda\sigma$.

Multiplying (4) by $E-E_\lambda$ and observing that $x\zeta(x) = 1$ we obtain

$$(E-E_\lambda)U_{\lambda|O}(E) = (E-E_\lambda)H_{\lambda|O} - \frac{i\hbar\gamma(E)}{2}U_{\lambda|O}(E)$$

or

$$U_{\lambda|O}(E) = \frac{(E-E_\lambda)H_{\lambda|O}}{E-E_\lambda+\frac{1}{2}i\hbar\gamma(E)} \tag{6}$$

and from (2 b)

$$U_{\lambda\sigma|O}(E) = \frac{H_{\lambda\sigma|\lambda}H_{\lambda|O}}{E-E_\lambda+\frac{1}{2}i\hbar\gamma(E)}. \tag{7}$$

U has no singularity for real E. Finally Γ becomes from (3)

$$\frac{\hbar}{2}\Gamma(E) = i\sum_\lambda \frac{|H_{\lambda|O}|^2}{E-E_\lambda+\frac{1}{2}i\hbar\gamma(E)}. \tag{8}$$

(5)–(8) is the complete solution. γ has a simple significance: $H_{\lambda\sigma|\lambda}$ is the matrix element for the emission of k_σ and is really independent of λ. For $E = E_\lambda$, γ becomes identical with the damping constant met with in the theory of the emission line breadth, § 18. In fact, (5) is identical with formula (8) § 18 (the time-dependent factor occurring there has been shown to be just the ζ-function). The real and imaginary parts of γ are thus the total transition probability for emission from the excited state and the level shift of the excited state, respectively. By splitting the ζ-function according to (2 c), one also obtains the formulae (11) and (12) of § 18 directly. This is true for $E = E_\lambda$, but we shall see that γ is practically independent of E.

Similarly, if we neglect the small γ in (8) we see that the imaginary part of Γ for $E = E_O$ is a contribution to the self-energy of the ground state, due to the absorption of photons out of the primary spectrum. In this section we are not interested in the level shifts (see § 34) and we shall replace γ and Γ by their real parts. Denoting these again by γ, Γ we have

$$\gamma(E) = \frac{2\pi}{\hbar}\sum_\sigma |H_{\lambda\sigma\,\lambda}|^2\delta(E-E_{\lambda\sigma}) = \frac{2\pi}{\hbar}\int d\Omega_\sigma |H_{\lambda\sigma|\lambda}|^2\rho_{k\sigma}, \tag{9 a}$$

$$\Gamma(E) = \gamma\sum_\lambda \frac{|H_{\lambda|O}|^2}{(E-E_\lambda)^2+\hbar^2\gamma^2/4}. \tag{9 b}$$

In (9 a) the sum over σ has been replaced by the usual integral, Ω_σ is the solid angle of the emitted k_σ. It is seen that γ is practically independent of E.† The same will be true for Γ after the summation over λ.

† Strictly speaking, the value of k_σ to be inserted on the right of (9 a) is determined from $E-E_{\lambda\sigma} = 0$ and depends on E. However, (9 a) varies very slowly with k_σ ($\rho \sim k_\sigma^2$), and γ will be needed only for $E = E_\lambda$ or $E = E_{\lambda\sigma}$. Both energies differ from E_0 by at most an order γ. Thus one can insert, for example, $k_\sigma = k_0$.

From (6) and (7) we could calculate the probabilities for all states at any time $t > 0$, according to equations (13), (14) of § 16. We are only interested in these probabilities for large times when the process has been completed. For $t \to \infty$ the probabilities are given by the general formula (20) of § 16, in our case:

$$b_\lambda(\infty) = \frac{U_{\lambda|O}(E_\lambda)}{E_\lambda - E_O + \frac{1}{2}i\hbar\Gamma(E_\lambda)}, \qquad b_{\lambda\sigma}(\infty) = \frac{U_{\lambda\sigma|O}(E_{\lambda\sigma})}{E_{\lambda\sigma} - E_O + \frac{1}{2}i\hbar\Gamma(E_{\lambda\sigma})}. \tag{10}$$

Note that the value of E to be inserted, as the argument of Γ and elsewhere in (10), is the energy of the state whose probability is to be calculated (E_n in § 16), i.e. E_λ in b_λ and $E_{\lambda\sigma}$ in $b_{\lambda\sigma}$. Inserting (6) and (7) we see that

$$b_\lambda(\infty) = 0. \tag{11}$$

This is quite clear because after a long time the atom cannot be in an excited state. Using (1′), the probability distribution of the re-emitted quanta becomes

$$|b_{\lambda\sigma}(\infty)|^2 = \frac{|H_{\lambda|O}|^2 |H_{\lambda\sigma|\lambda}|^2}{[(k_\lambda - k_\sigma)^2 + \hbar^2\Gamma^2/4][(k_\sigma - k_0)^2 + \hbar^2\gamma^2/4]}. \tag{12}$$

Actually, γ and Γ should be taken at $E = E_{\lambda\sigma}$, but both are practically constants.

The further discussion depends on the form of the primary intensity distribution $I_0(k)\,dk$. We shall confine ourselves to two important cases: (a) the primary intensity I_0 is constant in the region of the natural line breadth, i.e. we irradiate with a continuous spectrum; (b) the primary radiation consists of a monochromatic line which is sharp compared with the natural line width γ of the atom.

2. *Case* (a). *Continuous absorption.* If $I_0(k)$ is constant the sum $\sum_\lambda$ in (9b) can be replaced by an integral. $H_{\lambda|O}$ is the matrix element for absorption of k_λ. According to § 14 $|H_{\lambda|O}|^2$ is proportional to the number of quanta n_λ. We denote the average of $|H_{\lambda|O}|^2$ over all radiation oscillators from which a quantum k can be absorbed by $d\Omega|H(k)|^2\bar{n}_k$. We integrate (9b) over k_λ. Since only values E near E_O are needed, only quanta k near k_0 (within $\sim\gamma$) can be absorbed. H and ρ_k vary slowly with k and we can insert $k = k_0$ in H and ρ. Γ is then independent of E and we obtain

$$\Gamma = \frac{2\pi}{\hbar}\rho_{k_0}\,d\Omega|H(k_0)|^2\bar{n}_{k_0}. \tag{13}$$

$\bar{n}_{k_0}$ can be expressed by the primary intensity (§ 17, eq. (16))

$$\rho_{k_0}\, d\Omega\, \bar{n}_{k_0} = \frac{I_0(k_0)}{k_0 c}, \tag{14}$$

and hence
$$\Gamma = \frac{2\pi}{\hbar k_0 c} |H(k_0)|^2 I_0(k_0) = w_{nn_0}. \tag{15}$$

(15) is identical with the formula (17) § 17 for the *total probability for absorption* per unit time. Since Γ was the total transition probability of the initial state we see that the total probability for resonance fluorescence is equal to the total probability of absorption. In general, i.e. if the primary intensity is not extremely high, Γ will be very small compared with the transition probability γ for the spontaneous emission. Γ is the natural breadth of the ground state of the atom due to the probability of absorption. In fact, if I_0 contains only a limited number of photons, $\bar{n}_k$ and Γ vanish in the limit $L^3 \to \infty$ (see § 14).

We shall now discuss the formula (12) for the probability distribution of the emitted quantum k_σ.

Since Γ is very small the first factor of the denominator has practically the properties of a $\delta(k_\lambda - k_\sigma)$-function and shows that k_λ may differ hardly at all from k_σ, the difference being at most of the order $\hbar\Gamma$ which is the breadth of the ground state. The *energy* is therefore *conserved* in so far as this is allowed by the uncertainty relation.

If we integrate equation (12) over all quanta k_σ which can be emitted, we obtain the shape of the absorption line, and if we integrate over all k_λ, we get the shape of the emission line.

Using (13), the probability for the emission of a quantum k_σ becomes

$$\sum_\lambda |b_{\lambda\sigma}(\infty)|^2 = w(k_\sigma) = \frac{|H_{\lambda|\lambda\sigma}|^2}{(k_0 - k_\sigma)^2 + \hbar^2\gamma^2/4}. \tag{16}$$

This is the same formula as that deduced in § 18 for the *shape* of a line emitted *spontaneously*. Thus, if we irradiate an atom with a continuous radiation, we obtain the same emission line as if we excite the atom in any other way, say by collisions.

The probability that after a time $t = \infty$ a quantum k_λ has been *absorbed* is given by the summation of (12) over all k_σ. Since $\Gamma \ll \gamma$ we can consider the second factor in the denominator as nearly constant in the region in which the first factor has its maximum, $k_\sigma = k_\lambda$. We obtain, using (9 a),

$$\sum_\sigma |b_{\lambda\sigma}(\infty)|^2 = w(k_\lambda) = \frac{\gamma}{\Gamma} \frac{|H_{\lambda|O}|^2}{(k_0 - k_\lambda)^2 + \hbar^2\gamma^2/4}. \tag{17}$$

The total probability of absorption and emission for $t = \infty$, $\sum_{\sigma,\lambda} |b_{\lambda\sigma}(\infty)|^2$, is of course equal to one, by (13).

(17) is again identical with the formula deduced in § 18 for the *shape* of the *absorption line* (and also identical with the shape of the emission line).

We can therefore conclude the following: In the case where we irradiate with a continuous spectrum, the resonance fluorescence behaves with regard to the shape of the line, which is absorbed and

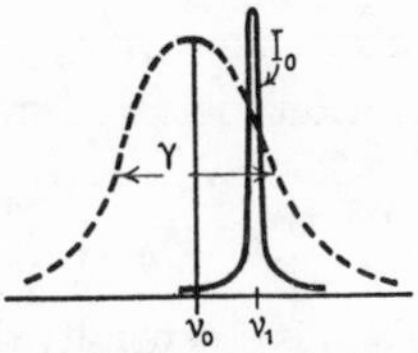

FIG. 8. Resonance fluorescence through excitation with a line $I_0(\nu)$ which is sharp compared with the natural breadth γ. The re-emitted line has the same shape $I_0(\nu)$.

emitted exactly as if two *independent processes, an absorption and a subsequent emission*, took place. For a single absorption-emission act, we have, however, to keep in mind that *energy is always conserved* within Γ and that therefore the atom 'remembers' before the emission which quantum it has absorbed. This is expressed by the fact that the formula (12) for $|b_{\lambda\sigma}(\infty)|^2$ is not the product of the probabilities of emission and absorption, since the first factor of the denominator connects, in a nearly δ-function-like manner, the frequencies of the two quanta k_λ and k_σ and not k_λ with k_0.

The dependence of the emitted on the absorbed quantum becomes even more significant in the case of *monochromatic absorption*.

3. *Case* (*b*). *Excitation by sharp line.* We assume now that the primary line is *sharp* compared with the natural emission line. Then $I_0(\nu)$ is different from zero only, say, at a frequency ν_1 (see Fig. 8). We denote the total primary intensity $\int I_0(k)\,dk$ by $\bar{I}_0$.

The total transition probability from the ground state is again $\Gamma(E_O)$. The $\sum_\lambda$ in (9 b) is to be extended only over the primary line which is practically sharp. We then get

$$\Gamma(E_O) = \frac{\gamma}{k_1 c}\frac{|H(k_1)|^2\bar{I}_0}{(k_1-k_0)^2+\hbar^2\gamma^2/4}. \tag{18}$$

Equation (18) gives the *total probability for the resonance fluorescence* per unit time. It decreases with the distance of the primary frequency

k_1 from the resonance frequency k_0 according to a line shape formula with the half-value breadth γ (= breadth of emission line).

The intensity of the re-emitted line is again given by integrating (12) over all quanta which can be absorbed. Since $\Gamma \ll \gamma$ the first factor in the denominator shows again that k_σ is practically equal to k_λ, or, since only quanta of frequency k_1 can be absorbed, $k_\sigma = k_1$. Hence the emitted line must have the *same breadth as the primary line* and must therefore be much *sharper than the natural line.* Since Γ vanishes for small primary intensities we can assume that it is even small compared with the extension of I_0 (although, of course, I_0 is sharper than γ). The integration yields then, with (14),†

$$w(k_\sigma) = \sum_\lambda |b_{\lambda\sigma}(\infty)|^2 = \frac{2\pi}{\hbar\Gamma k_0 c}\,\frac{|H(k_0)|^2|H_{\lambda|\lambda\sigma}|^2 I_0(k_\sigma)}{(k_0-k_\sigma)^2+\hbar^2\gamma^2/4}. \tag{19}$$

The intensity distribution $w(k_\sigma)$ is essentially determined by two factors: Firstly, $w(k_\sigma)$ is proportional to $I_0(k_\sigma)$. This shows that *the emitted line has the same shape* as the *primary line* and is therefore much sharper than the natural line (see Fig. 8). Secondly, the denominator of (19) is practically constant in the region where I_0 is different from zero. This factor therefore determines the total intensity. The total intensity decreases with increasing distance from the maximum k_0 in the same proportion as the intensity of the natural line.

In the case (*b*) of monochromatic excitation the emitted line has quite a different shape from that which the atom emits spontaneously. Emission and absorption cannot therefore be regarded as two subsequent *independent* processes, because the atom would not then 'remember' which light quantum it had absorbed before and would emit the natural line. In this case, the resonance fluorescence has to be considered as a *single quantum process.*

It can also be shown that the *re-emitted radiation is coherent with the primary radiation.*‡ This again would not be the case if absorption and emission were independent.

On the other hand, it may be asked, in which *state*—ground state or excited state—*the atom* is during the process of resonance fluorescence. This question can only be decided by a *measurement* of the

† $\Gamma(E)$ would be given by (18) with k_1-k_0 replaced by $E-E_0+k_1-k_0$ and appears to vary with E more strongly. In (12) and (19) the value $E = E_{\lambda\sigma} = E_0+k_\sigma-k_\lambda$ is to be inserted in $\Gamma(E)$. Since, however, k_σ and k_λ both practically coincide with the frequency of the primary beam k_1, $\Gamma(E_0)$ and $\Gamma(E_{\lambda\sigma})$ are again practically identical and can be replaced by the same constant Γ.

‡ The phase of the scattered radiation is, however, shifted against the phase of the primary radiation, as is the case in the classical theory (§ 5).

energy of the atom. We shall see that all phase relations are destroyed by such a measurement. The energy of the atom can be measured by inelastic collisions with electrons. To decide unambiguously whether the atom is excited or not the measurement must be carried out in a time which is shorter than the lifetime $1/\gamma$ of the excited state (otherwise, the atom would jump down spontaneously during the measurement). Therefore we must have at least one collision in the time $1/\gamma$. At the instant of time at which the collision takes place the *coherence* of the wave is *interrupted*. But if we interrupt a light wave $1/\gamma$ times per sec. the line is no longer monochromatic but has a breadth $\hbar\gamma$ (collision broadening, § 18). This is just the breadth of the natural line. As a result of the measurement of the energy we see, therefore, that the emitted line becomes broader and has at least the breadth of the natural line. The process of resonance fluorescence now behaves as if the atom emitted the light quantum spontaneously after having been excited by absorption of a primary quantum.

Thus we have found the following:

Resonance fluorescence represents a single coherent quantum process if the atom is undisturbed. For excitation by a sharp line the emitted line then has the same shape as the primary one. The energy of the atom is undetermined. As soon as the quantum state of the atom is determined the process behaves as an independent absorption and emission of a light quantum. The emitted line then has the natural shape.

The following problems are solved by the same method, and in an almost identical way:

(i) An atom is excited in the highest of 3 states (c) and can jump down either into level (b) or level (a) (ground state), and also from (b) to (a) (see Fig. 5, p. 186). The most important result is that described in § 18.2.

(ii) Two like atoms are placed at a distance R from each other. At $t = 0$, the first atom is excited, the second in the ground state. The first atom can jump down and emit a photon which can be absorbed by the second atom. We ask what is the probability w of finding the second atom in an excited state at any later time t. The result is the following: w is *zero exactly* for any $t < R/c$, and begins to rise gradually when $t > R/c$. This must be so because the photon cannot travel faster than light.†

† S. Kikuchi, *Zs. f. Phys.* **66** (1930), 558. See also reference on p. 196 and J. Hamilton, *Proc. Phys. Soc.* **62** (1949), 12. Also the scattering of a free particle by several centres of force requires the use of the theory of § 16. The problem was raised by G. Wentzel, *Helv. Phys. Acta*, **21** (1948), 49.

(iii) Excitation by electrons.† Here also a certain amount of coherence exists. If the electron beam is monochromatic (energy ϵ) the emitted light has frequencies $k < \epsilon$ and the shape of the line is roughly that of the natural line but cut off at $k = \epsilon$. Conservation of energy, of course, forbids the emission of photons $k > \epsilon$.

21. Photoelectric effect

If the quantum energy $\hbar\nu$ of light falling on an atom is greater than the ionization energy I of the atom, the electron is raised into a state of the continuous spectrum. In this case light of all frequencies can be absorbed, and the absorption spectrum is *continuous*. The kinetic energy T of the electron after leaving the atom is determined by Einstein's equation

$$T = \hbar\nu - I. \tag{1}$$

The photoelectric effect plays an important part in the absorption of X-rays and γ-rays in matter (cf. Chapter VII). It gives an appreciable absorption even if the primary energy is very much greater than the ionization energy of the atom. Since we are especially interested in this book in the *high* energy region, we shall keep in view the absorption of high-frequency radiation rather than of radiation in the optical region.‡

We shall carry out the calculations for a very simple case only and shall quote the results obtained for other cases. Since a free electron cannot absorb light, we should expect the probability for the photoelectric absorption to be greater the more strongly the electron is bound. We may therefore confine ourselves to the absorption by a *K-electron*. Furthermore we shall assume:

(*a*) The quantum energy of the incident light is large compared with the ionization energy I of the K-electron. For an atom with nuclear charge Z this condition can be written as follows:

$$T = \frac{p^2}{2\mu} \gg I = \frac{Z^2\mu}{2\times 137^2}, \quad \text{or} \quad \xi \equiv \frac{Ze^2}{\hbar v} \ll 1. \tag{2}$$

According to well-known facts of the theory of collisions, (2) is identical with the condition for the validity of Born's approximation.§ Therefore in the matrix elements we can replace the wave function of the electron in the continuous spectrum by a plane wave. Our results will

† W. Heitler, *Z. Phys.* **82** (1933), 146.

‡ For a detailed treatment and discussion of the optical region including the angular dependence, etc., the reader is referred to the literature, in particular to A. Sommerfeld, *Atombau und Spektrallinien II*, Braunschweig 1939.

§ Cf., for instance, N. F. Mott and H. S. W. Massey, *Theory of Atomic Collisions*, Oxford 1949, chap. vii.

not, of course, be valid in the neighbourhood of the absorption edge ($\hbar\nu \sim I$).

(*b*) The energy of the electron in the continuous spectrum, on the other hand, will be assumed to be small compared with mc^2, so that relativistic corrections are not important, i.e.

$$\hbar\nu \ll mc^2. \tag{3}$$

Actually the error will not be very large for energies up to about $0{\cdot}5mc^2$.

1. *Non-relativistic case, great distances from absorption edge.* The transition probability per unit time for photoelectric absorption is given by the general formula § 14 eq. (9). Since the final state of the electron belongs to the continuous spectrum, ρ_E is the number of quantum states per unit volume of the electron

$$\rho_E\, dE = \frac{pE\, dEd\Omega}{(2\pi\hbar c)^3} = \frac{\mu p\, dEd\Omega}{(2\pi\hbar c)^3}. \tag{4}$$

We may then assume that the primary radiation consists of a single light quantum $\hbar\nu$ only.†

For the matrix elements H we have to insert eq. (23a), § 14 for the absorption of a single quantum

$$H = -\frac{e}{\mu}\sqrt{\left(\frac{2\pi\hbar^2c^2}{k}\right)} \int \psi_b^*\, p_e\, e^{i(\boldsymbol{\kappa}\mathbf{r})}\psi_a, \tag{5}$$

where p_e is the component of the momentum in the direction of polarization of the primary quantum, ψ_a the wave function of the electron in the K-shell, and ψ_b the wave function of the electron in the continuous spectrum with the momentum $\mathbf{p}$:

$$\psi_a = \frac{1}{\sqrt{(\pi a^3)}} e^{-r/a}, \qquad \psi_b = e^{i(\mathbf{pr})/\hbar c}, \qquad a = \frac{a_0}{Z}, \tag{6}$$

where $a_0 = \hbar^2/me^2$ is the Bohr radius.

Introducing instead of a a quantity $\alpha = Z\hbar c/a_0$, which has the dimensions of an energy,

$$\alpha = \frac{\hbar c Z}{a_0} = 2\frac{137}{Z} I = \sqrt{(2\mu I)} = \frac{Z}{137}\mu, \tag{7}$$

and a vector $\mathbf{q}$ representing the momentum transferred to the atom,

$$\mathbf{q} = \mathbf{k} - \mathbf{p}, \tag{8}$$

† In the case of discrete absorption by an atom a transition probability per unit time exists only if the primary radiation consists of many quanta with a continuous intensity distribution.

the integration of (5) yields

$$H = -\frac{e}{\mu}p_e\sqrt{\left(\frac{\alpha^3}{\pi\hbar^3c^3}\right)}\sqrt{\left(\frac{2\pi\hbar^2c^2}{k}\right)}\frac{8\pi\alpha\hbar^3c^3}{(\alpha^2+q^2)^2}. \qquad (9)$$

If we divide the transition probability per unit time by the velocity of the primary beam, i.e. by c, we obtain the *differential cross-section* (differential, because it refers to an ejection of the electron into an element of the solid angle $d\Omega$), according to § 14, eq. (16). Thus

$$d\phi = \frac{2\pi}{\hbar c}|H|^2\rho_E\,d\Omega = \frac{32.137.r_0^2\,\mu pp_e^2\,\alpha^5\,d\Omega}{(\alpha^2+q^2)^4k} \qquad (r_0 = e^2/mc^2). \qquad (10)$$

(10) gives the angular distribution of the ejected photoelectrons. Denoting by θ the angle between the direction of the light quantum $\mathbf{k}$ and of the electron $\mathbf{p}$, and by ϕ the angle between the ($\mathbf{pk}$)-plane and the plane formed by $\mathbf{k}$ and the direction of polarization $\mathbf{e}$, i.e.

$$\theta = \angle(\mathbf{pk}),$$

$$\phi = \angle \text{ between } (\mathbf{pk})\text{-plane and } (\mathbf{ek})\text{-plane},$$

p_e and q can be expressed as follows

$$q^2 = p^2+k^2-2pk\cos\theta, \qquad (11\,\text{a})$$

$$p_e = p\sin\theta\cos\phi. \qquad (11\,\text{b})$$

The denominator α^2+q^2 of (10) can be simplified. Since our formula (10) is in any case only correct for non-relativistic velocities we can make use of equation (3). According to (1) and (7) we have

$$k = \frac{\alpha^2+p^2}{2\mu} \ll \mu,$$

and hence†

$$\alpha^2+q^2 = \alpha^2+p^2+k^2-2pk\cos\theta = k(2\mu+k-2p\cos\theta)$$
$$\simeq 2\mu k(1-\beta\cos\theta), \qquad \beta = v/c = p/\mu. \qquad (12)$$

Finally, α according to (7) and (2) is assumed to be small compared with p. We can therefore put $p^2 = 2k\mu$. Inserting again for α its value $Z\mu/137$, we obtain for the differential cross-section

$$d\phi = r_0^2\frac{Z^5}{137^4}\left(\frac{\mu}{k}\right)^{7/2}\frac{4\sqrt{2}\sin^2\theta\cos^2\phi}{(1-\beta\cos\theta)^4}\,d\Omega. \qquad \text{N.R.} \quad (13)$$

Most of the photoelectrons are therefore emitted in the direction of polarization of the primary light quantum ($\theta = \frac{1}{2}\pi$, $\phi = 0$). In the

† The relativistic correction only gives terms in $(v/c)^2$. It is therefore justifiable to keep terms in v/c in (12) and (13).

direction of **k** itself the intensity of photoelectric emission is zero. The denominator of (13) gives, however, a slight preponderance in the forward direction which becomes more important for increasing velocities of the electron. In the relativistic case the maximum is strongly displaced in the forward direction.

The total cross-section for the ejection of photoelectrons in any direction may be obtained by integrating (13) over all angles. Here we may neglect the term $\beta\cos\theta$ in the denominator. Multiplying by a factor 2 to allow for the fact that the K-shell contains two electrons, the cross-section ϕ_K for the photo effect of the K-shell becomes (expressed in terms of the ratio k/I and of k/μ)

$$\phi_K = \phi_0 \frac{Z^5}{137^4} 4\sqrt{2}\left(\frac{\mu}{k}\right)^{7/2} = \phi_0\, 64 \frac{137^3}{Z^2}\left(\frac{I}{k}\right)^{7/2}, \qquad \text{N.R.} \quad (14)$$

where $\phi_0 = 8\pi r_0^2/3$ is the cross-section for the Thomson scattering (§ 5 eq. (5)), which we may use as a convenient unit.

From (14) we obtain for the K-shell the *absorption coefficient* τ_K per cm., for radiation of frequency ν by multiplying by the number N of atoms per cm.[3]

$$\tau_K = N\phi_0 \frac{Z^5}{137^4} 4\sqrt{2}\left(\frac{\mu}{k}\right)^{7/2}. \qquad \text{N.R.} \quad (14')$$

The absorption coefficient decreases rapidly, with the 7/2th power of the frequency. This, however, is only true as long as our assumptions (2) and (3) are justified.

In Fig. 9 we have plotted $\log_{10}\phi_K$ (in units of ϕ_0) for C, Al, Cu, Sn, Pb on a logarithmic scale in order to cover a large frequency range. Formula (14) gives a straight line with a gradient $-3{\cdot}5$ (dotted curves for $\hbar\nu < 0{\cdot}5mc^2$). The deviations from the straight lines are due to the two corrections considered in subsections 2 and 3.

2. *Neighbourhood of absorption edge.* By (2), Born's approximation is no longer valid for heavy elements or if $\hbar\nu$ is so small that the energy of the ejected electron is of the same order of magnitude as the ionization energy I. In this case the exact wave functions of the continuous spectrum must be used instead of plane waves. Except for the very heaviest elements the non-relativistic approximation is sufficient.

The matrix elements (5) with the correct wave functions of the continuous spectrum have been computed by Stobbe.† The total

† M. Stobbe, *Ann. d. Phys.* **7** (1930), 661.

cross-section ϕ_K, eq. (14), has to be multiplied by a factor

$$f(\xi) = 2\pi\sqrt{\left(\frac{I}{k}\right)}\frac{e^{-4\xi\operatorname{arccot}\xi}}{1-e^{-2\pi\xi}}, \qquad \xi = \sqrt{\left(\frac{I}{k-I}\right)} = \frac{Ze^2}{\hbar v}, \tag{15}$$

thus

$$\frac{\phi_K}{\phi_0} = 128\pi\frac{137^3}{Z^2}\left(\frac{I}{k}\right)^4\frac{e^{-4\xi\operatorname{arccot}\xi}}{1-e^{-2\pi\xi}}. \qquad \text{N.R.} \tag{16}$$

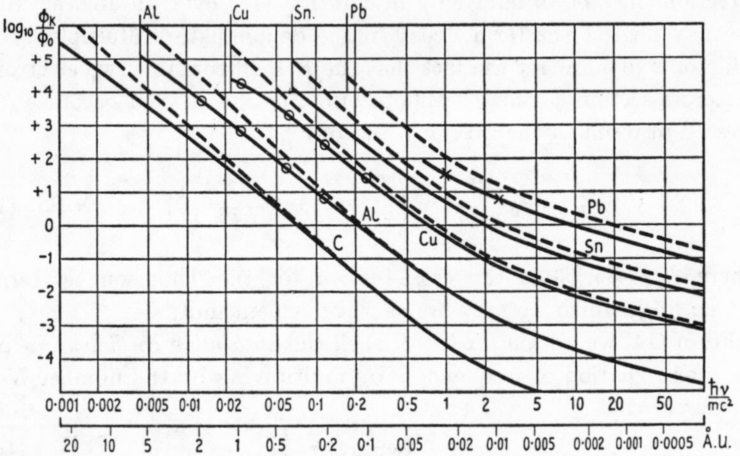

FIG. 9. $\log_{10}$ of the cross-section for photoelectric absorption in the K-shell for C, Al, Cu, Sn, Pb on a logarithmic scale. Above are shown the K-absorption edges. (For C the absorption edge lies just outside the diagram.) The dotted curves (straight lines for $h\nu < 0{\cdot}5mc^2$) are calculated with Born's approximation (formulae (14) and (17)). The deviation from the straight lines for $h\nu > 0{\cdot}5mc^2$ is due to relativistic effects. The full-drawn curves are exact. They are interpolated from formula (16) and the exact numerical calculations given in Table III. The circles refer to measurements of Allen, the crosses to those of Gray.

ξ^2 is the ratio of the ionization energy to the kinetic energy of the ejected electron. The factor $f(\xi)$ diminishes the cross-section in the immediate neighbourhood of the K-absorption edge ($\xi \to \infty$) by a factor $2\pi\exp(-4) = 0{\cdot}12$. Even for a distance of fifty times the ionization energy from the absorption edge $f(\xi)$ is still only $0{\cdot}66$.

The character of these deviations due to the factor (15) can be seen from Fig. 9. The correct curves approach rather slowly the straight lines calculated with Born's approximation. For Cu and Al some results of measurements by Allen† have been plotted. The agreement is actually much better than can be seen from the diagram in which only $\log_{10}\phi_K$ is plotted.

† S. J. M. Allen, *Phys. Rev.* **27** (1926), 266; **28** (1926), 907. From the measured absorption coefficient we have subtracted the part which is due to scattering (cf. § 22).

For the soft X-ray region, however, one has to keep in mind that Fig. 9 gives the absorption coefficient of the K-shell only. The outer shells, of course, also make some contributions which have to be considered, especially if $\hbar\nu$ is smaller than the K-absorption edge (see subsection 3).

3. *Relativistic region.* If, on the other hand, the energy of the primary quantum is of the order mc^2 or larger, the *relativistic* wave functions for ψ_K and $\psi_\mathbf{p}$ must be used. Using Born's approximation which can be applied for light elements the result is:†

$$\frac{\phi_K}{\phi_0} = \frac{3}{2}\frac{Z^5}{137^4}\left(\frac{\mu}{k}\right)^5(\gamma^2-1)^{\frac{3}{2}}\left[\frac{4}{3}+\frac{\gamma(\gamma-2)}{\gamma+1}\left(1-\frac{1}{2\gamma\sqrt{(\gamma^2-1)}}\log\frac{\gamma+\sqrt{(\gamma^2-1)}}{\gamma-\sqrt{(\gamma^2-1)}}\right)\right]; \tag{17}$$

$$\gamma = \frac{1}{\sqrt{(1-v^2/c^2)}} = \frac{k+\mu}{\mu}, \quad k \gg I. \tag{17'}$$

γ is the ratio of the total energy (kinetic energy$+mc^2$) of the electron to the rest energy. For $\gamma \to 1$ formula (17) goes over into our non-relativistic formula (14).

For extremely high energies, $k \gg \mu$, (17) becomes

$$\frac{\phi_K}{\phi_0} = \frac{3}{2}\frac{Z^5}{137^4}\frac{\mu}{k}. \qquad \text{E.R.} \tag{18}$$

ϕ_K decreases more slowly in the relativistic region than in the non-relativistic one, for large values of k/μ only as μ/k (instead of as $(\mu/k)^{7/2}$ for $k \ll \mu$).

The curves in Fig. 9 therefore, begin to turn up for $k \sim \mu$ and finally tend to straight lines inclined to the axis with a gradient -1. The slower falling off of ϕ_K in the relativistic region has the effect that even for $k \sim 10mc^2$ the photo effect of heavy elements contributes an appreciable part to the total absorption (cf. § 36).

Finally, we give in Table II the values of ϕ_K in units $\phi_0 Z^5/137^4$ for the region where, at least for light elements, the factor (15) is not appreciable. In the units chosen ϕ_K depends only on the ratio k/μ but not on Z.

For heavy elements Sauter's formula breaks down. Hulme, Fowler, and others‡ have made exact numerical calculations of ϕ_K for a few

† F. Sauter, *Ann. d. Phys.* **9** (1931), 217; ibid. **11** (1931), 454.

‡ H. R. Hulme, J. McDougall, R. A. Buckingham, and R. H. Fowler, *Proc. Roy. Soc.* **149** (1935), 131.

TABLE II

Theoretical values of $\phi_K\, 137^4/\phi_0\, Z^5$ (eq. (17)). (Born's approximation)

k/μ	0·1	0·25	0·5	1	2	3
$\dfrac{\phi_K\, 137^4}{\phi_0\, Z^5}$	$1{\cdot}84\times10^4$	793	81	10·4	2·04	0·96

k/μ	5	10	20	50	100
$\dfrac{\phi_K\, 137^4}{\phi_0\, Z^5}$	0·45	0·183	$8{\cdot}35\times10^{-2}$	$3{\cdot}13\times10^{-2}$	$1{\cdot}54\times10^{-2}$

elements and for two values of k in the range where relativistic effects are important. They found the following values for ϕ_K instead of those given in Table II (values $k/\mu = 1, 5$ are partially interpolated).

TABLE III

Exact theoretical values for $\phi_K\, 137^4/\phi_0\, Z^5$

k/μ	0·69	1	2·2	5
Al	22·3	8·1	1·24	0·35
Fe	17·8	6·5	1·05	0·30
Sn	12·3	4·5	0·79	0·24
Pb	7·9	3·2	0·60	0·19

Finally, Hall has deduced a formula which is a good approximation for all Z and $k \gg \mu$.† He obtains

$$\frac{\phi_K}{\phi_0} = \frac{3}{2}\frac{Z^5}{137^4}\frac{\mu}{k}e^{-\pi\alpha+2\alpha^2(1-\log\alpha)} \qquad \left(\alpha = \frac{Z}{137}\right). \quad \text{E.R.} \quad (19)$$

At very high energies (19) differs from the formula (18), obtained by using Born's approximation, by a factor 2·2 for Pb and 1·5 for Cu.

In Fig. 9 the full curves are interpolated from the values of Table III and those given by equations (16) and (19). ϕ_K has been determined experimentally for lead by Gray.‡ His results fit the theoretical curve excellently.

To obtain the *absorption coefficient* per cm. one has to multiply the values given in Tables II and III by $N\phi_0\, Z^5/137^4$. The values of the latter quantity are given in appendix 8 for several elements.

The above calculations represent the absorption coefficient for the K-shell only. To obtain a rough estimate of the contribution of the

† H. Hall, *Rev. Mod. Phys.* **8** (1936), 358. Even for $k \to \infty$ (19) is not quite exact. The difference amounts to about 4 per cent. for Pb. An approximate formula, valid when $k \gg I$, is obtained by multiplying (17) by the exponential factor of (19).

‡ L. H. Gray, *Proc. Camb. Phil. Soc.* **27** (1931), 103.

higher shells, we may use the experimental result that about 80 per cent. of the total photo-electric absorption at high energies ($\sim mc^2$) is due to the K-shell. This is also borne out by theoretical calculations of the photo-electric effect of the L-shell, at least approximately.† In Chapter VII we use, therefore, the theoretical values obtained in this section multiplied by 5/4 to give the photo-electric absorption of the atom.

22. Scattering by free electrons

1. *The Compton formula.* Scattering by free electrons is of fundamental importance in all phenomena connected with the absorption of γ-rays, cosmic radiation, etc., and we shall therefore investigate it in some detail.

The process discussed here is the following: A primary light quantum $\mathbf{k}_0$ collides with a free electron which we can assume to be initially at rest:

$$\mathbf{p}_0 = 0, \qquad E_0 = \mu \quad (\mu = mc^2). \tag{1}$$

The general case $\mathbf{p}_0 \neq 0$ can be obtained from the special case (1) by a Lorentz transformation. In the final state the light quantum has been scattered, so that we have a quantum $\mathbf{k}$ instead of $\mathbf{k}_0$. Since according to § 14.3 the *momentum* is *conserved* in the interaction of light with free electrons, in the final state the electron has a momentum $\mathbf{p}$ (energy E)

$$\mathbf{p} = \mathbf{k}_0 - \mathbf{k}. \tag{2}$$

The conservation of energy states that

$$E + k = k_0 + \mu. \tag{3}$$

According to (2) and (3) the frequency of the scattered quantum cannot be the same as that of the primary quantum. Using the relativistic relation between momentum and energy $p^2 = E^2 - \mu^2$ and denoting the *angle* between $\mathbf{k}_0$ and $\mathbf{k}$ by θ, we obtain from (2) and (3)

$$k = \frac{k_0 \mu}{\mu + k_0(1 - \cos\theta)}, \tag{4}$$

which is the well-known formula for the frequency shift of the scattered radiation. It shows that in the non-relativistic case, $k_0 \ll \mu$, the scattered and primary frequencies are the same. In the relativistic case the frequency shift increases with the angle of scattering θ. In the 'extreme relativistic' case, i.e. if the primary quantum k_0 is large compared with the rest energy of the electron ($k_0 \gg \mu$) we can distinguish

† For details of the theory and comparison with experiments we refer to the summary by Hall, loc. cit., also *Phys. Rev.* **84** (1951), 167.

between two regions of θ. For very small angles k is again nearly equal to k_0:

$$k \sim k_0 \quad \text{if} \quad k_0(1-\cos\theta) \ll \mu. \qquad \text{E.R.} \quad (5)$$

For large angles, i.e. for $(1-\cos\theta)k_0 \gg \mu$, we have

$$k = \frac{\mu}{1-\cos\theta}. \qquad \text{E.R.} \quad (6)$$

In this case the scattered quantum is always only of the order μ whatever the primary frequency. The wave-length is of the order

$$\lambda = \frac{\hbar c}{k} \sim \frac{\hbar}{mc} \equiv \lambda_0. \tag{7}$$

λ_0 is the universal Compton wave-length. The angular region (5), in which k is appreciably larger than μ, becomes smaller as the primary frequency k_0 increases.

2. *Intermediate states, transition probability.* To compute the transition probability from the initial state O ($\mathbf{k}_0, \mathbf{p}_0 = 0$) to the final state F ($\mathbf{k}, \mathbf{p}$), we must remember that our process can happen only by passing through an intermediate state which can differ by one quantum only from the initial and the final states. Since for these intermediate states the momentum is conserved (but not the energy), the following two intermediate states are the only ones possible:

I. $\mathbf{k}_0$ is first absorbed. No light quantum is present. The electron has a momentum

$$\mathbf{p}' = \mathbf{k}_0. \tag{8a}$$

$\mathbf{k}$ is emitted in the transition to the final state.

II. $\mathbf{k}$ is first emitted. Both quanta $\mathbf{k}_0$ and $\mathbf{k}$ are present. The electron has a momentum

$$\mathbf{p}'' = -\mathbf{k}. \tag{8b}$$

$\mathbf{k}_0$ is absorbed in the transition to the final state.

An electron moving with relativistic velocity with a given momentum $\mathbf{p}$ can exist in *four states*, corresponding to the fact that the electron may have either of two *spin directions* and also a *positive* or *negative energy* (§ 11):

$$E = \pm\sqrt{(p^2+\mu^2)}. \tag{9}$$

All these four states must be taken into account as *intermediate states.* Each of the two intermediate states I and II is therefore actually *fourfold*, because by (8a) and (8b) only the momentum of the electron is determined. On the other hand, in the initial and final states the electron has of course a positive energy, and we shall assume for the present that it also has a given spin direction.

The compound matrix element which determines the transition probability is then given by

$$K_{FO} = \sum \left(\frac{H_{F\mathrm{I}}\,H_{\mathrm{I}O}}{E_O - E_{\mathrm{I}}} + \frac{H_{F\mathrm{II}}\,H_{\mathrm{II}O}}{E_O - E_{\mathrm{II}}}\right), \tag{10}$$

where $\sum$ denotes the summation over all four intermediate states, i.e. over both spin directions and both signs of the energy. E_O, E_{I},... represent the total energies in the initial and the intermediate states. The energy differences occurring in the denominator of (10) are, according to (2), (3), (8), (9), given by

$$\left.\begin{aligned} E_O - E_{\mathrm{I}} &= \mu + k_0 - E' \\ E_O - E_{\mathrm{II}} &= \mu + k_0 - (E'' + k_0 + k) = \mu - E'' - k \end{aligned}\right\}, \tag{11}$$

where E', E'' represent the energy of the electron in the states I, II,

$$E' = \pm\sqrt{(p'^2 + \mu^2)}, \qquad E'' = \pm\sqrt{(p''^2 + \mu^2)}.$$

If we denote the Dirac amplitudes of the electron with the momenta $\mathbf{p}_0$, $\mathbf{p}$, $\mathbf{p}'$, $\mathbf{p}''$ by u_0, u, u', u'' and the components of the matrix vector $\boldsymbol{\alpha}$ in the direction of the polarization of the two light quanta $\mathbf{k}_0$ and $\mathbf{k}$ simply by α_0 and α, respectively, the matrix elements for the transitions $O \rightarrow \mathrm{I}$, etc., are given by § 14 eq. (27):

$$\begin{aligned} H_{F\mathrm{I}} &= -e\sqrt{\left(\frac{2\pi\hbar^2c^2}{k}\right)}(u^*\alpha u'), & H_{\mathrm{I}O} &= -e\sqrt{\left(\frac{2\pi\hbar^2c^2}{k_0}\right)}(u'^*\alpha_0 u_0), \\ H_{F\mathrm{II}} &= -e\sqrt{\left(\frac{2\pi\hbar^2c^2}{k_0}\right)}(u^*\alpha_0 u''), & H_{\mathrm{II}O} &= -e\sqrt{\left(\frac{2\pi\hbar^2c^2}{k}\right)}(u''^*\alpha u_0). \end{aligned} \tag{12}$$

Actually, the transitions to negative energy states are forbidden according to the hole theory, § 11. Then, instead of the intermediate states with negative energy, intermediate states containing positive electrons appear:

I. The final photon $\mathbf{k}$ is first emitted by an electron in the negative energy state with momentum $\mathbf{p}'$ which goes over into the state $\mathbf{p}$ with positive energy. In other words: $\mathbf{k}$ is emitted creating a pair with momenta $\mathbf{p}^+ = -\mathbf{p}'$, $\mathbf{p}^- = \mathbf{p}$. In the transition to the final state the primary electron $\mathbf{p}_0$ jumps into the hole (or annihilates with $\mathbf{p}^+$) absorbing the photon $\mathbf{k}$. The matrix elements for the two stages of the process are identical with $H_{F\mathrm{I}}$ (first stage) and $H_{\mathrm{I}O}$ (second stage), eq. (12), although their order is reversed. u' belongs to a negative energy state. The energy in the intermediate state is $E_{\mathrm{I}} = \mu + k_0 + k + |E'| + E$, and hence, by the conservation of energy, $E_O - E_{\mathrm{I}} = -\mu - k_0 + E'$, $E' = -|E'|$. This is minus the difference (11). This minus sign is now compensated by the fact that the primary electron is *exchanged* against the electron in the negative energy state which becomes the final electron. Since the two electrons have antisymmetrical wave functions their exchange produces a minus sign in the matrix element. The contribution to K of this intermediate state is therefore just the part due to negative energies of the first term of (10) (with (11) and (12)).

II. $\mathbf{k}_0$ is absorbed by the negative energy electron $\mathbf{p}''$ which goes over into $\mathbf{p}$, or a pair $\mathbf{p}^+ = -\mathbf{p}''$, $\mathbf{p}^- = \mathbf{p}$ is created. In the transition to the final state the primary electron jumps into the hole, emitting a quantum $\mathbf{k}$. Here also the matrix elements and energy denominators agree with (11) and (12) for the intermediate state II with E'' negative.

It is seen that we obtain the same results whether we use hole theory or allow negative energy states to occur as intermediate states† and we can proceed applying the latter procedure.

The transition probability per unit time for our scattering process is, according to § 14 eq. (14),

$$w = \frac{2\pi}{\hbar}|K_{FO}|^2\rho_F, \tag{13}$$

where ρ_F denotes the number of final states per energy interval dE_F. The evaluation of ρ_F requires some care. By the conservation of momentum the final state is determined completely by the frequency of the scattered quantum $\mathbf{k}$ and the angle of scattering. Therefore we have

$$\rho_F\,dE_F = \rho_k\,dk, \tag{14}$$

where ρ_k denotes the number of states for the scattered quantum per energy interval dk. It would be incorrect, however, to equate the energy intervals dk and dE_F. Since the final energy is given as a function of k and θ by

$$E_F = k+\sqrt{(p^2+\mu^2)} = k+(k_0^2+k^2-2k_0 k\cos\theta+\mu^2)^{\frac{1}{2}}, \tag{15}$$

we obtain

$$\left(\frac{\partial k}{\partial E_F}\right)_\theta = \frac{Ek}{\mu k_0}, \tag{16}$$

and hence

$$\rho_F = \rho_k\left(\frac{\partial k}{\partial E_F}\right)_\theta = \frac{d\Omega\,k^2}{(2\pi\hbar c)^3}\frac{Ek}{\mu k_0}. \tag{17}$$

$d\Omega$ is the element of solid angle for the scattered quantum. Collecting our formulae (10), (11), (12), (13), (17) and dividing by the velocity of

† This is also the case for most radiative processes in first approximation but should be checked in each case. An example where this is not the case is the self-energy of the electron (§ 29) where the hole theory gives a very different result. The hole theory is, of course, always *the* correct theory to be used (or the theory of positrons of § 12).

light, we obtain the *differential cross-section for the scattering process*,

$$d\phi = e^4 \frac{Ek^2}{\mu k_0^2} d\Omega \left[\sum \left(\frac{(u^*\alpha u')(u'^*\alpha_0 u_0)}{\mu+k_0-E'} + \frac{(u^*\alpha_0 u'')(u''^*\alpha u_0)}{\mu-k-E''} \right) \right]^2. \quad (18)$$

(18) is valid for a given polarization of both light quanta and a given spin direction of the electron in the initial and final states. The summation $\sum$ is over all spin directions and both signs of the energy for the intermediate states.

3. *Deduction of the Klein–Nishina formula.* Our next task is the evaluation of the matrix elements occurring in (18). The summation $\sum$ can easily be carried out if we make use of the general formula § 11 eq. (14 a). This formula cannot, however, be applied directly to (18) since the denominators of (18) depend upon the sign of the energy E'. We therefore multiply the numerator and denominator of the first term of (18) by $\mu+k_0+E'$. The denominator does not then depend on the sign of E'. For the numerator, we make use of the wave equation

$$E'u' = [(\boldsymbol{\alpha}\mathbf{p}')+\beta\mu]u' \equiv H'u', \quad (19)$$

$\mathbf{p}'$ is of course constant for the summation $\sum$. Thus we obtain†

$$\sum (\mu+k_0+E')(u^*\alpha u')(u'\alpha_0 u_0) = (\mu+k_0) \sum (u^*\alpha u')(u'^*\alpha_0 u_0) + \\ + \sum (u^*\alpha H'u')(u'^*\alpha_0 u_0). \quad (20)$$

Now applying our general formula § 11 eq. (14 a)

$$\sum\nolimits^{p'} (u_0^* Q_1 u')(u'^* Q_2 u) = (u_0^* Q_1 Q_2 u),$$

and taking into account the fact that $E'^2 = p'^2+\mu^2 = k_0^2+\mu^2$, we obtain for the first term of (18)

$$\sum\nolimits^{p'} \frac{(u^*\alpha u')(u'^*\alpha_0 u_0)}{\mu+k_0-E'} = \frac{(\mu+k_0)(u^*\alpha\alpha_0 u_0)+(u^*\alpha H'\alpha_0 u_0)}{2\mu k_0}. \quad (21)$$

The second term of (18) can be evaluated in the same way. If we introduce the abbreviations

$$K' = \mu+k_0+H' = \mu(1+\beta)+k_0+(\boldsymbol{\alpha}\mathbf{k}_0), \quad (22\text{ a})$$

$$K'' = \mu-k+H'' = \mu(1+\beta)-k-(\boldsymbol{\alpha}\mathbf{k}), \quad (22\text{ b})$$

where we have replaced $\mathbf{p}'$ and $\mathbf{p}''$ by (8 a) and (8 b), we obtain for the summation $\sum$ in (18)

$$\sum = \frac{1}{2\mu}\left[\frac{(u^*\alpha K'\alpha_0 u_0)}{k_0} - \frac{(u^*\alpha_0 K''\alpha u_0)}{k}\right]. \quad (23)$$

† Compare for this method, H. Casimir, *Helv. Phys. Acta*, 6 (1933), 287.

This can be further simplified by using the wave equation for u_0, viz. $[(\boldsymbol{\alpha}\mathbf{p}_0)+\beta\mu]u_0 = E_0 u_0$. Since $\mathbf{p}_0 = 0$, $E_0 = \mu$,

$$(1-\beta)u_0 = 0.$$

This equation can, of course, be used only when β acts directly on u_0. Now β anticommutes with α and α_0 and hence the terms $1+\beta$ of (22) vanish and we obtain

$$\sum = \frac{1}{2\mu}\left(u^*\left[2(\mathbf{e}_0\,\mathbf{e})+\frac{1}{k_0}\alpha(\boldsymbol{\alpha}\mathbf{k}_0)\alpha_0+\frac{1}{k}\alpha_0(\boldsymbol{\alpha}\mathbf{k})\alpha\right]u_0\right), \tag{24}$$

where use has also been made of the fact that

$$\alpha\alpha_0+\alpha_0\alpha \equiv (\boldsymbol{\alpha}\mathbf{e})(\boldsymbol{\alpha}\mathbf{e}_0)+(\boldsymbol{\alpha}\mathbf{e}_0)(\boldsymbol{\alpha}\mathbf{e}) = 2(\mathbf{e}_0\,\mathbf{e}). \tag{25}$$

The differential cross-section (18) is proportional to the square of (24). (24) depends on the spin directions of the electron in the initial and final states. We are not, however, interested in the probability of finding the electron with a certain spin after the scattering process. We shall therefore sum $d\phi$ over all spin directions of the electron after the scattering process and shall average over the spin directions in the initial state.

We denote the summation over the spin directions only by S (or S_0 for the initial state), in contrast to the summation $\sum$ which is extended over both signs of energy too. Thus we must form

$$\tfrac{1}{2}\mathsf{S}_0\,\mathsf{S}\,|\text{expression (24)}|^2. \tag{26}$$

The summation S over the spin directions can be reduced to a summation $\sum^p$ over all four states having the same momentum p, according to § 11 eqs. (16), (17). After the substitution

$$u = \frac{H+E}{2E}u, \qquad u_0 = \frac{H_0+E_0}{2E_0}u_0 \tag{27}$$

the summations S, S_0 can be replaced by the summations $\sum^p$, $\sum^{p_0}$ over all four states:

$$\tfrac{1}{2}\mathsf{S}_0\,\mathsf{S}|(u^*Qu_0)|^2 \equiv \tfrac{1}{2}\mathsf{S}_0\,\mathsf{S}(u_0^*\,Q^\dagger u)(u^*Qu_0)$$
$$= \frac{1}{8E_0\,E}\sum^p\sum^{p_0}(u_0^*\,Q^\dagger(H+E)u)(u^*Q(H_0+E_0)u_0)$$
$$= \frac{1}{8E_0\,E}\,\mathrm{Sp}\,Q^\dagger(H+E)Q(H_0+E_0). \tag{28}$$

For Q the operator in (24) is to be inserted. $Q^\dagger$ is the hermitian conjugate operator arising from Q (since the α's are hermitian) by reversing the order of the factors. Thus explicitly (with $\mathbf{p}_0 = 0$, $E_0 = \mu$, $\mathbf{p} = \mathbf{k}_0 - \mathbf{k}$, $E = \mu + k_0 - k$):

$$\tfrac{1}{2}S_0\, S|(24)|^2 = \frac{1}{8E\mu^2}\,\tfrac{1}{4}\mathrm{Sp}\Big[2(\mathbf{e}_0\,\mathbf{e}) + \frac{1}{k_0}\,\alpha_0(\boldsymbol{\alpha}\mathbf{k}_0)\alpha + \frac{1}{k}\,\alpha(\boldsymbol{\alpha}\mathbf{k})\alpha_0\Big] \times$$
$$\times [k_0 - k + \mu + \beta\mu + (\boldsymbol{\alpha}\mathbf{k}_0) - (\boldsymbol{\alpha}\mathbf{k})] \times$$
$$\times \Big[2(\mathbf{e}_0\,\mathbf{e}) + \frac{1}{k_0}\,\alpha(\boldsymbol{\alpha}\mathbf{k}_0)\alpha_0 + \frac{1}{k}\,\alpha_0(\boldsymbol{\alpha}\mathbf{k})\alpha\Big][1+\beta]. \quad (29)$$

The spur can easily be evaluated by the method of § 11. Contributions arise only when the number of factors β and α are both even. In a spur the order of factors can be reversed or permuted cyclically. Furthermore, $\alpha^2 = \alpha_0^2 = 1$, $(\boldsymbol{\alpha}\mathbf{k})^2 = k^2$, etc., $\mathbf{k} \perp \mathbf{e}$, $\mathbf{k}_0 \perp \mathbf{e}_0$, and therefore

$$\alpha_0(\boldsymbol{\alpha}\mathbf{k}_0) = -(\boldsymbol{\alpha}\mathbf{k}_0)\alpha_0, \quad \text{etc.}$$

(29) then reduces to

$$\tfrac{1}{2}S_0\, S|(24)|^2 = \frac{1}{4E\mu^2}\,\tfrac{1}{4}\mathrm{Sp}\Big\{2(\mathbf{e}_0\,\mathbf{e})^2(2\mu + k_0 - k) + k_0 - k -$$
$$-2(\mathbf{e}_0\,\mathbf{e})(k_0 - k)\alpha_0\,\alpha + \frac{k_0 - k}{k_0\,k}[(\boldsymbol{\alpha}\mathbf{k}_0)\alpha_0\,\alpha(\boldsymbol{\alpha}\mathbf{k})\alpha_0\,\alpha - 2(\mathbf{e}_0\,\mathbf{e})(\boldsymbol{\alpha}\mathbf{k}_0)\alpha_0\,\alpha(\boldsymbol{\alpha}\mathbf{k})]\Big\}. \quad (30)$$

The last two terms (square bracket) can be combined with the help of (25) and yield

$$-\tfrac{1}{4}\mathrm{Sp}(\boldsymbol{\alpha}\mathbf{k}_0)\alpha_0\,\alpha(\boldsymbol{\alpha}\mathbf{k})\alpha\alpha_0 = -\tfrac{1}{4}\mathrm{Sp}(\boldsymbol{\alpha}\mathbf{k}_0)(\boldsymbol{\alpha}\mathbf{k}) = -(\mathbf{k}_0\,\mathbf{k}). \quad (31)$$

Also

$$\tfrac{1}{4}\mathrm{Sp}\,\alpha_0\,\alpha = (\mathbf{e}_0\,\mathbf{e}). \quad (32)$$

Hence (30) becomes

$$\tfrac{1}{2}S_0\, S|(24)|^2 = \frac{1}{4E\mu^2}\Big[4\mu(\mathbf{e}_0\,\mathbf{e})^2 + \frac{k_0 - k}{k_0\,k}\,(k_0\,k - (\mathbf{k}_0\,\mathbf{k}))\Big]. \quad (33)$$

We denote the angle between the directions of polarization of the primary and secondary quanta by Θ, $(\mathbf{e}_0\,\mathbf{e}) = \cos\Theta$. Furthermore (from (2) and (3))

$$k_0\,k - (\mathbf{k}_0\,\mathbf{k}) = \mu(k_0 - k). \quad (34)$$

The differential cross-section now becomes according to (18), (24), and (33):

$$d\phi = \tfrac{1}{4}r_0^2\,d\Omega\,\frac{k^2}{k_0^2}\Big[\frac{k_0}{k} + \frac{k}{k_0} - 2 + 4\cos^2\Theta\Big]. \quad (35)$$

(35) represents the well-known Klein–Nishina formula.† It gives, for all primary light quanta of a given frequency and polarization, the

† O. Klein and Y. Nishina, *Zs. f. Phys.* **52** (1929), 853; Y. Nishina, ibid. **52** (1929), 869. The same formula has also been deduced by I. Tamm, ibid. **62** (1930), 545.

intensity of the scattered radiation at a given angle θ and with a given direction of polarization. (35) can be expressed by (4) as a function of k_0, θ, and Θ.

We have discussed in detail the computation of the matrix elements because the method used here will serve as a model for similar calculations in many other quantum processes.

4. *Polarization, angular distribution.* We discuss first the angular distribution and polarization of the scattered radiation as given by formula (35).

It will be convenient to consider the scattered radiation as composed of two linearly polarized components $\perp$ and $||$. Denoting the directions of polarization of $\mathbf{k}_0$ and $\mathbf{k}$ by $\mathbf{e}_0$ and $\mathbf{e}$ respectively, we can choose the following two directions for $\mathbf{e}$:

($\perp$) $\mathbf{e}$ perpendicular to $\mathbf{e}_0$, $\cos\Theta \equiv (\mathbf{e}_0\mathbf{e}) = 0$,

($||$) $\mathbf{e}$ and $\mathbf{e}_0$ in the same plane (i.e. in the ($\mathbf{k}$, $\mathbf{e}_0$) plane),

$$\cos^2\Theta = 1-\sin^2\theta\cos^2\phi,$$

where ϕ is the angle between the ($\mathbf{k}_0,\mathbf{k}$) plane and the ($\mathbf{k}_0, \mathbf{e}_0$) plane and θ the angle of scattering ($\mathbf{k}_0,\mathbf{k}$). According to (35) the $||$ component is always more intense than the $\perp$ component.

In the non-relativistic case (N.R.) we have $k_0 = k$ and

$$d\phi_\perp = 0, \quad d\phi_{||} = r_0^2\, d\Omega\, (1-\sin^2\theta\cos^2\phi). \qquad \text{N.R.} \quad (36)$$

(36) is identical with the classical (Thomson) formula § 5 eq. (4) when the damping constant κ is neglected. This is to be expected because the condition $k_0 = \hbar\nu_0 \ll mc^2$ can also be interpreted as $\hbar \to 0$, which represents the transition to the classical theory. For polarized primary radiation the scattered radiation is completely polarized. If the primary radiation is unpolarized, we have to take the average over ϕ. The intensity of the scattered radiation is then given by

$$d\phi = \tfrac{1}{2}r_0^2\, d\Omega\, (1+\cos^2\theta). \qquad \text{N.R.} \quad (37)$$

In the other extreme case, where the energy of the primary quantum is large compared with mc^2 ($k_0 \gg \mu$, *extreme relativistic* case), we have according to subsection 1 to distinguish between small angles θ (5) and large angles (6). In these two cases (i.e. for $k_0 \sim k$ and $k_0 \gg k$) the differential cross-section becomes

$$d\phi_\perp = 0, \quad d\phi_{||} \simeq r_0^2\, d\Omega\, (1-\sin^2\theta\cos^2\phi) \quad \text{(small angles)}, \quad \text{E.R.} \quad (38\text{a})$$

$$d\phi_\perp = d\phi_{||} = \frac{r_0^2\, d\Omega}{4}\frac{k}{k_0} = \frac{r_0^2\, d\Omega\, \mu}{4k_0(1-\cos\theta)} \quad \text{(large angles)}. \quad \text{E.R.} \quad (38\text{b})$$

From (38) we see that for very small angles the intensity distribution is the same as in classical theory. For large angles, the scattered radiation is *unpolarized*, even if the primary radiation is polarized, and has roughly a uniform intensity distribution. The intensity decreases, however, with increasing energy of the primary quantum.

From (38 b) we can see already that, for $k_0 \gg \mu$, the total probability of scattering decreases $\sim \mu/k_0$. Both regions (θ small and large) give about the same contribution to the total cross-section, since the region in which (38 a) is valid is of the order $\theta^2 \sim \mu/k_0$ and $d\Omega = \theta\, d\theta$.

To obtain finally the total intensity scattered into an angle θ we have to take the sum $d\phi = d\phi_\perp + d\phi_\parallel$. If the primary radiation is unpolarized (average over ϕ), we obtain

$$d\phi = \frac{r_0^2\, d\Omega}{2}\, \frac{k^2}{k_0^2}\left(\frac{k_0}{k} + \frac{k}{k_0} - \sin^2\theta\right). \tag{39}$$

Here k is a function of θ, given by (4). If we insert (4) in (39) the differential cross-section becomes:

$$d\phi = r_0^2\, d\Omega\, \frac{1+\cos^2\theta}{2}\, \frac{1}{[1+\gamma(1-\cos\theta)]^2}\left\{1 + \frac{\gamma^2(1-\cos\theta)^2}{(1+\cos^2\theta)[1+\gamma(1-\cos\theta)]}\right\}, \tag{40}$$

$$\gamma = k_0/\mu.$$

The angular distribution (40) is plotted in Fig. 10 as a function of the angle of scattering θ for various ratios $\gamma = k_0/\mu$. For small angles the scattered intensity has nearly the classical value, whereas for large angles the higher the primary frequency the smaller is the intensity. In the relativistic region the forward direction becomes more and more preponderant. Even for hard X-rays ($\gamma \sim 0{\cdot}2$) the relativistic deviation from the Thomson formula is rather large for large angles.

In Fig. 10 we have plotted the result of measurements by Friedrich and Goldhaber† of the angular distribution for X-rays with wave-length 0·14 Å or $\gamma = 0{\cdot}173$ in carbon. The agreement with the theoretical curve is exact (within the experimental error).

The differential cross-section can also be expressed in terms of the energy k of the scattered quantum instead of in terms of θ. Using (4) we obtain (after integration over the azimuth angle)

$$\phi_k\, dk = \pi r_0^2 \frac{\mu\, dk}{k_0^2}\left[\frac{k_0}{k} + \frac{k}{k_0} + \left(\frac{\mu}{k} - \frac{\mu}{k_0}\right)^2 - 2\mu\left(\frac{1}{k} - \frac{1}{k_0}\right)\right]. \tag{41}$$

k ranges from $k_0\mu/(\mu+2k_0)$ to k_0.

† W. Friedrich and G. Goldhaber, *Zs. f. Phys.* **44** (1927), 700. See also G. E. M. Jauncy and G. G. Harvey, *Phys. Rev.* **37** (1931), 698.

5. *Recoil electrons.* The scattering is accompanied by a recoil of the electron originally at rest. This has an energy $E-\mu = k_0-k$, and the direction of the recoil forms an angle β with that of the primary photon, both being functions of the scattering angle. From the conservation laws (2), (3) one finds

$$E-\mu = \frac{k_0^2(1-\cos\theta)}{\mu+k_0(1-\cos\theta)}, \qquad \cos\beta = (1+\gamma)\sqrt{\left(\frac{1-\cos\theta}{2+\gamma(\gamma+2)(1-\cos\theta)}\right)}. \tag{42}$$

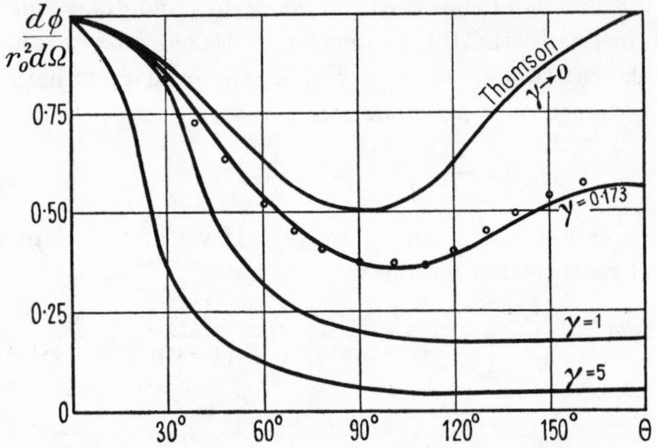

FIG. 10. Angular distribution of the Compton scattering as a function of the angle of scattering for various primary frequencies $\gamma = k_0/\mu$. Measurements of Friedrich and Goldhaber for $\gamma = 0{\cdot}173$.

When θ ranges from 0 to π, $E-\mu$ ranges from 0 to the maximum value $E_{\max}-\mu = 2k_0^2/(\mu+2k_0)$, and β ranges from $\frac{1}{2}\pi$ to 0. The recoil electrons are never scattered backwards. The angular distribution of the recoil electrons is found by inserting (42) into (40):

$$d\phi = 4r_0^2\frac{(1+\gamma)^2\cos\beta\,d\Omega_\beta}{[1+2\gamma+\gamma^2\sin^2\beta]^2}\times$$
$$\times\left\{1+\frac{2\gamma^2\cos^4\beta}{[1+2\gamma+\gamma^2\sin^2\beta][1+\gamma(\gamma+2)\sin^2\beta]}-\frac{2(1+\gamma)^2\sin^2\beta\cos^2\beta}{[1+\gamma(\gamma+2)\sin^2\beta]^2}\right\}, \tag{43}$$

where $d\Omega_\beta$ is now the element of the solid angle of the recoil electron. In the extreme relativistic case $\gamma \gg 1$, one can again distinguish between small and large angles β:

$$d\phi = \frac{\gamma r_0^2\,d\Omega_\beta}{\gamma^2\beta^2+1} \qquad (\gamma\beta^2 \ll 1), \qquad \text{E.R.} \tag{44 a}$$

$$d\phi = \frac{4r_0^2}{\gamma^2}\frac{\cos\beta\,d\Omega_\beta}{\sin^4\beta} \qquad (\gamma\sin^2\beta \gg 1). \qquad \text{E.R.} \tag{44 b}$$

Also here both regions give contributions to the total scattering of an equal order of magnitude. In the region of small angles the distribution decreases rapidly within a region of the order $\beta \sim 1/\gamma$.

6. *Total scattering.* To obtain the total scattering we have to integrate over all angles. An elementary integration yields

$$\frac{\phi}{\phi_0} = \frac{3}{4}\left\{\frac{1+\gamma}{\gamma^3}\left[\frac{2\gamma(1+\gamma)}{1+2\gamma}-\log(1+2\gamma)\right]+\frac{1}{2\gamma}\log(1+2\gamma)-\frac{1+3\gamma}{(1+2\gamma)^2}\right\}, \tag{45}$$

$$\phi_0 = 8\pi r_0^2/3, \qquad \gamma = k_0/\mu.$$

The unit ϕ_0 used for ϕ is the classical cross-section for the Thomson scattering (§ 5 eq. (5)).

In the non-relativistic case $\gamma \ll 1$ we again obtain $\phi = \phi_0$. The first terms of an expansion of the right-hand side of (45) in powers of γ are

$$\phi = \phi_0\left(1-2\gamma+\frac{26}{5}\gamma^2+...\right). \qquad \text{N.R.} \quad (46)$$

In the extreme relativistic case, on the other hand, we obtain from (45)

$$\phi = \phi_0\frac{3}{8}\frac{\mu}{k_0}\left(\log\frac{2k_0}{\mu}+\frac{1}{2}\right). \qquad \text{E.R.} \quad (47)$$

(47) agrees (apart from the logarithm) with the estimate of subsection 4, based on (38). *Thus for very high energies the number of scattered quanta decreases with the frequency of the primary radiation.* This is the reason why the penetrating power of γ-rays increases with increasing frequency as long as no other absorption processes, such as the production of pairs, are important.

The cross-section (45) is plotted in Fig. 11 as a function of the primary energy on a logarithmic scale in order to cover a large energy region. The values of ϕ/ϕ_0 are given in the following table.

TABLE IV

Cross-section for Compton scattering in units ϕ_0 for various primary energies

γ	0·05	0·1	0·2	0·33	0·5	1	2	3	
ϕ/ϕ_0	0·913	0·84	0·737	0·637	0·563	0·431	0·314	0·254	
γ	5	10	20	50	100	200	500	1,000	
ϕ/ϕ_0	19·1	12·3	7·54	3·76	2·15	1·22	0·556	0·304	$\times 10^{-2}$

An *experimental test* of the theory is provided by measurements of the total absorption coefficient of X-rays or γ-rays in various materials.

This absorption coefficient τ per cm. is given by

$$\tau = NZ\phi, \tag{48}$$

where N is the number of atoms per cm.^{3} and Z the number of electrons per atom. *The absorption coefficient is proportional to the total number of electrons NZ per cm.^{3}* In (48) we have of course assumed that the primary radiation is so hard that all electrons can be considered as free. To obtain τ one has to multiply the values given in Table IV or in Fig. 11 by $NZ\phi_0$. This quantity is given in appendix 8 for various substances.

For a comparison of the theory with experiment one has, however, to remember two points:

(i) For X-rays the total absorption is not only due to scattering. We have seen in § 21 that the *photo-electric effect* gives also a strong absorption which, however, decreases rapidly with the primary energy. To compare the photo-electric absorption with the scattering we have also plotted in Fig. 11 (dotted curves) the photo-electric cross-section ϕ_K of the K-shell (multiplied by 5/4, see p. 211) as computed in § 21 in the same units. (Since ϕ_K refers to a whole atom we have plotted $\phi_K/Z\phi_0$.) We see, for instance, that for carbon the photo-electric absorption is much greater than the scattering for a wave-length $\lambda > 500$ X.U., whereas for $\lambda < 300$ X.U. the scattering only is appreciable.

(ii) On the other hand, for γ-rays, the absorption is largely due to *pair production*, as we shall see in § 26. The pair production increases with the primary energy and also with the atomic number Z. For carbon the pair production is negligible for $k_0 < 10\mu$. For a comparison with the theory we have used only measurements for which the photo-electric absorption and the pair production is small compared with the absorption due to scattering. In Fig. 11 the experimental results have been plotted for three regions of wave-length:

(1) for X-rays of the wave-length 100–300 X.U.† (scattered by carbon);

(2) for hard X-rays $\lambda \simeq 20$–50 X.U.‡ (carbon and aluminium);

(3) for the ThC″ γ-radiation of wave-length $\lambda = 4{\cdot}7$ X.U.§ (carbon).

† C. W. Hewlett, *Phys. Rev.* **17** (1921), 284; S. J. M. Allen, ibid. **27** (1926), 266; **28** (1926), 907.

‡ J. Read and C. C. Lauritsen, ibid. **45** (1934), 433.

§ L. Meitner and H. Hupfeld, *Zs. f. Phys.* **67** (1930), 147; C. Y. Chao, *Phys. Rev.* **36** (1930), 1519; *Proc. Nat. Ac.* **16** (1930), 431; G. T. B. Tarrant, *Proc. Roy. Soc.* **128** (1930), 345.

For the first region the photo-electric absorption, which is about 10 per cent. of the total absorption, has been subtracted. For the second and third regions no corrections were necessary.

The experiments fit the theoretical curve excellently. We may therefore consider the *Klein–Nishina formula as proved*, at least for energies up to $10mc^2$.

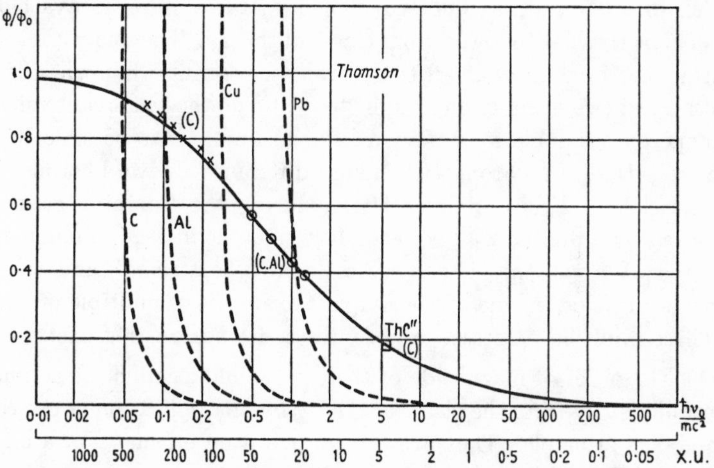

FIG. 11. Total cross-section for Compton scattering (Klein–Nishina formula) as a function of the primary frequency ν_0 (lower scale = primary wave-length). For comparison the cross-section for the photo-electric absorption per electron (dotted curves $\phi_K/Z\phi_0$). Measurements: × Hewlett, Allen (corrected for photo-absorption) in carbon; ○ Read and Lauritsen (carbon and aluminium); □ Meitner and Hupfeld, Chao (carbon).

We shall return to the question of the absorption of γ-rays in § 36, where also measurements for higher energies will be quoted.

For all these three regions the electrons are practically free and the coherent scattering is entirely negligible. This would, of course, no longer be true for softer radiation, for which the binding of the electrons has to be taken into account.† As we have seen in § 19 the binding changes the intensity of the scattered radiation and gives rise to a coherent scattering (unshifted line). Furthermore, the shifted (Compton) line becomes broader, for a given angle of scattering. The maximum of the shifted line is also displaced by a very small amount as compared with the value given by the Compton formula (4). (Bloch, loc. cit.)

† The effect of binding on the Compton scattering has been computed by H. Casimir, *Helv. Phys. Acta*, **6** (1933), 287; W. Franz, *Zs. f. Phys.* **90** (1934), 623; G. Wentzel, ibid. **43** (1927), 1 and 779; F. Bloch, *Phys. Rev.* **46** (1934), 674; A. Sommerfeld, *Ann. Phys.* **29** (1937), 715; W. Franz, ibid. 721.

The theoretical results of this section are based on first-order perturbation theory. Corrections (radiative and damping corrections) exist and will be considered in § 33. They are small for all energies.

23. Multiple processes

In the preceding sections we have considered various processes in which one photon was emitted. Their features were always closely analogous to what followed from classical theory. The quantum theory of the field predicts that all these processes can also happen with the emission of two or more quanta, instead of one, the two quanta sharing the energy available for emission. For example, it is possible that in the transition of an atom with energy difference E_0 two photons k_1, k_2 with $k_1+k_2 = E_0$ are emitted.† Or, during the Compton scattering initiated by a photon $\mathbf{k}_0$, two secondary photons instead of one may be emitted. We call the latter process 'double scattering', and we shall consider it below as one of the simplest examples of multiple processes.

These multiple processes are of interest for various reasons:

(1) The order of magnitude of their probability cannot be estimated unambiguously from the classical theory by an application of the correspondence principle. They are typical quantum effects. For a critical understanding of the present theory it is important to see whether the experiments confirm the predictions of the theory. Also, when one of the two photons has very small energy a peculiar situation arises, the so-called 'infra-red problem' (subsection 3), which is of great theoretical importance.

(2) At very high energies in cosmic radiation the striking phenomenon of the '*showers*' occurs. In these showers, together with photons, a large number of electron pairs are emitted by a single fast particle or light quantum in passing through matter. We know now that most of these showers are of a cascade nature and that the theory of Bremsstrahlung and pair production fully accounts for them (§ 38). From the experimental evidence it cannot, however, wholly be excluded that true multiple processes might also occur.

1. *Double Compton effect.* As a simple but typical example we consider here the 'double Compton scattering'.‡ A primary light quantum $\mathbf{k}_0$ is 'scattered' by a free electron initially at rest in such a way that

† In the optical region the simultaneous emission of two quanta has been studied by M. Mayer-Göppert, *Naturw.* **17** (1929), 932; *Ann. Phys.* **9** (1931), 273. The transition probability is very small. Here we are interested more in the region of high energies.

‡ W. Heitler and L. Nordheim, *Physica*, **1** (1934), 1059.

instead of one scattered quantum $\mathbf{k}$ two quanta $\mathbf{k}_1$ and $\mathbf{k}_2$ are emitted. The transition probability for this process can be computed in the same way as for the ordinary scattering, § 22.

The conservation laws in this case are expressed by the equations

$$\mathbf{k}_0 = \mathbf{k}_1+\mathbf{k}_2+\mathbf{p}, \tag{1a}$$

$$k_0+\mu = k_1+k_2+E. \tag{1b}$$

From (1) it follows that, if $k_0 < \mu$, the sum of both quanta k_1 and k_2 is of the order k_0:

$$k_1+k_2 \sim k_0+O\left(\frac{k_0^2}{\mu}\right), \tag{2}$$

where $O(x)$ denotes a term of the order x at most. If, on the other hand, $k_0 \gg \mu$ and if both quanta k_1, k_2 are scattered through large angles, then

$$k_1+k_2 \sim \mu; \qquad p,\, E \sim k_0. \tag{3}$$

In this subsection we consider the case where both quanta k_1 and k_2 are of the same order of magnitude, or, if $k_0 \geqslant \mu$, at least of order μ. This is the only case of interest in connexion with true multiple processes. The case where one of the two quanta has very small energy (so that the process is practically identical with the single scattering) will be treated in subsection 3.

The transition from the initial state O $(\mathbf{k}_0, \mathbf{p}_0 = 0)$ to the final state F $(\mathbf{k}_1, \mathbf{k}_2, \mathbf{p})$ can occur only by passing through *two subsequent intermediate states*, for instance:

I. $\mathbf{k}_0$ absorbed. The electron has a momentum

$$\mathbf{p}' = \mathbf{k}_0. \tag{4}$$

II. $\mathbf{k}_1$ emitted. The electron has a momentum

$$\mathbf{p}'' = \mathbf{k}_0-\mathbf{k}_1. \tag{5}$$

Other pairs of intermediate states are obtained by permuting the order in which the three quanta are emitted or absorbed. Altogether there exist six pairs of intermediate states.

The compound matrix element which determines the transition probability $O \rightarrow F$ is now, according to § 14, given by

$$K_{FO} = \sum \frac{H_{F\,\mathrm{II}}\, H_{\mathrm{II}\,\mathrm{I}}\, H_{\mathrm{I}\,O}}{(E_O-E_{\mathrm{I}})(E_O-E_{\mathrm{II}})}, \tag{6}$$

where the summation $\sum$ is over all six intermediate states, the spin directions, and the sign of the energy of the electron in the intermediate states. The energy differences in the denominator of (6) are, according to (4) and (5),

$$E_O - E_{\mathrm{I}} = k_0 + \mu - E'; \qquad E'^2 = p'^2 + \mu^2;$$
$$E_O - E_{\mathrm{II}} = k_0 + \mu - k_1 - E''; \qquad E''^2 = p''^2 + \mu^2. \tag{7}$$

The summation $\sum$ in (6) can be carried out in the same way as in § 22. The transition probability is proportional to $|K_{FO}|^2$. Taking the summation over the spin directions of the electron in the final state and the average value in the initial state, we obtain (dropping all numerical factors),

$$\tfrac{1}{2}\mathsf{SS}_0|K_{FO}|^2$$
$$= \frac{(e\hbar c)^6}{k_0 k_1 k_2} \frac{1}{E\mu^2}\left[\mathrm{Sp}\frac{\alpha_0 K'\alpha_1 K''\alpha_2(H+E)\alpha_2 K''\alpha_1 K'\alpha_0(1+\beta)}{k_0^2[\mu(k_0-k_1)-k_0 k_1(1-\cos\theta_1)]^2}+\ldots\right], \tag{8}$$

where θ_1 represents the angle between $\mathbf{k}_0$ and $\mathbf{k}_1$ and

$$K' = \mu(1+\beta)+k_0+(\boldsymbol{\alpha}\mathbf{k}_0),$$
$$K'' = \mu(1+\beta)+k_0-k_1+(\boldsymbol{\alpha}, \mathbf{k}_0-\mathbf{k}_1). \tag{9}$$

For an estimate of the order of magnitude it is sufficient to consider the first term in (8) arising from the first pair of intermediate states. All other terms are of the same order of magnitude.†

Thus the differential cross-section is given by

$$d\phi \sim \frac{1}{\hbar c}|K_{FO}|^2 \frac{k_1^2 k_2^2\, dk_2}{(\hbar c)^6} = \frac{r_0^2}{137}\frac{k_1 k_2\, dk_2}{k_0 E}[\mathrm{Sp}\,(8)]. \tag{10}$$

In (10) we have also dropped the factor $\partial E_F/\partial k$ (see § 22 eq. (16)) because it is of the order of magnitude 1 for all energies.

For the evaluation of the spur in (8) we consider the two cases $k_0 \ll \mu$ and $k_0 \gg \mu$:

(i) $k_0 \ll \mu$, $k_1, k_2 \sim k_0$. For $H+E$ we can write, according to (1),

$$H+E = E+\beta\mu+(\boldsymbol{\alpha}\mathbf{p}) = \mu(1+\beta)+(\boldsymbol{\alpha}\mathbf{p})+\frac{p^2}{2\mu}$$
$$= \mu(1+\beta)+(\boldsymbol{\alpha}\mathbf{p})+O\left(\frac{k_0^2}{\mu}\right). \tag{11}$$

The largest contribution to the spur (8) will be from those terms which are proportional to μ. Since, however,

$$(1+\beta)\alpha(1+\beta) = (1-\beta^2)\alpha = 0,$$
$$(1+\beta)\alpha(\boldsymbol{\alpha}\mathbf{p})\alpha(1+\beta) = 0, \tag{12}$$

† It can be shown that the different intermediate states do not 'interfere' so as to reduce the order of magnitude of the result. The order of magnitude of the whole matrix element is given by the first term of (8).

and since by (9) and (11) μ only occurs in the form $\mu(1+\beta)$, only one term $\mu(1+\beta)$ of $H+E$ or K' or K'' gives a contribution to the spur. The numerator of the spur is therefore of the order μk_0^4. The denominator is of the order $\mu^2 k_0^4$ and therefore

$$\text{Sp}\,(8) \sim 1/\mu. \tag{13}$$

Inserting (13) in (10) we obtain for the cross-section

$$\phi \sim \frac{r_0^2}{137}\left(\frac{k_0}{\mu}\right)^2 \qquad (k_0 \ll \mu). \qquad \text{N.R.} \tag{14}$$

(ii) $k_0 \gg \mu$. For large angle scattering k_1, $k_2 \sim \mu$ and $H+E$, K', K'' according to (1) are all of the order k_0. The numerator of Sp (8) is therefore of the order k_0^5 whereas the denominator is $\sim \mu^2 k_0^4$ and therefore

$$\text{Sp}\,(8) \sim k_0/\mu^2.$$

For large angles of scattering we have from (1) $E \sim k_0$; $k_1, k_2 \sim \mu$. Thus the cross-section amounts to

$$\phi \sim \frac{r_0^2}{137}\frac{\mu}{k_0} \qquad (k_0 \gg \mu). \qquad \text{E.R.} \tag{15}$$

We should also consider the case where one or both quanta are scattered into small angles so that their energy is comparable with k_0. However, as in the case of the single scattering, it can be seen that this region of small angles does not contribute a higher order of magnitude to the total cross-section than the large angles, except for the possible occurrence of a factor $\log k_0/\mu$ which we do not consider as of a higher order of magnitude here.

Comparing our results (14), (15) with the corresponding formulae (46) and (47) of § 22 for single scattering, we obtain the following order of magnitude of the cross-section (units r_0^2)

	$k_0 \ll \mu$	$k_0 \gg \mu$
single scattering	1	μ/k_0
double scattering	$k_0^2/137\mu^2$	$\mu/137k_0$

(15′)

For small energies, $k_0 \ll \mu$, the double process is thus extremely rare. Its probability differs from that for the single process by a factor $k_0^2/137\mu^2$. It tends to zero in the transition to the classical theory because $k_0^2/137 = e^2\nu_0^2\hbar/c$ is proportional to $\hbar$.

At high energies, $k_0 \gg \mu$, the double scattering is comparatively more likely. Its probability, however, is still *smaller by a factor* 1/137 than that of the single process.

The results obtained can easily be generalized. For instance, one can show that the probability that a photon passing through the field of an atom will create two electron pairs instead of one is 137 times smaller than the probability for the creation of a single pair. (Pair creation will be treated in detail in § 26.) The cross-section for the simultaneous creation of 'shower' of n pairs by a multiple process is of the order (when $k_0 \gg \mu$)†

$$\phi_{\text{pair}}^{(n)} \simeq \frac{Z^2 r_0^2}{137^n}. \qquad \text{E.R.} \quad (16)$$

Multiple processes are therefore *rare events* and the cosmic ray showers are not due to them (see § 38).

2. *Experimental evidence.* Experimental evidence for the existence of multiple processes cannot be obtained easily. Not only are these processes rare but it is also difficult to ascertain that a particular phenomenon has really happened in one single elementary act and not in a succession of steps. One piece of evidence, however, has been observed in cosmic radiation.‡ High energy γ-rays were allowed to pass through the emulsion of an electron sensitive photographic plate. In passing through the field of an atom, a photon normally produces a pair of electrons visible as a narrow angle fork track. About 1,400 such forks were observed. Two cases have now been found where four tracks originate from one point (also forming a very narrow angle), or at any rate within a distance of a few 10^{-4} cm. The majority of the tracks could be identified as electrons whilst the rest had energies too high to make an identification possible but there is no reason why they should not be electrons. It is extremely improbable, that these double forks could be due to two successive events. It might be thought that a single pair is created first, and then that one of the electrons emits a hard photon which eventually creates a second pair. It will be shown in §§ 25, 26 that the average length of path of an electron before it creates a hard photon is of the order of 1 cm. and the same is true for a photon before it creates a pair. The chance is therefore very remote that the two events could have happened within say 10^{-3} cm. It is therefore highly probable that the two events are really the *formation of a double pair* in one elementary act. The relative frequency of occurrence of the double process compared with that of the single process appears to be of the order of 1:700, which is quite in accord

† Closer inspection shows that the factor in the denominator of (14)–(16) is $\pi 137$ rather than 137, which makes the result still smaller.

‡ J. E. Hooper and D. T. King, *Phil. Mag.* **41** (1950), 1194.

with the estimates of subsection 2 considering that numerical factors like π, etc., have been left out. A further example for a multiple process has been found in connexion with the annihilation of positrons (§ 27).

Multiple processes are a very valuable test for quantum electrodynamics. They are a characteristic result of the combination of quantum theory and relativity applied to the electromagnetic field, in realms where the correspondence principle can no longer be applied. The test is all the more valuable because quantum electrodynamics is as yet far from wholly satisfactory (compare Chapter VI) and it is not *a priori* obvious that all its predictions agree with the facts. However, so far no contradiction between theory and experiment has been found.

3. *Emission of an infra-red quantum.* When one of the photons into which the primary photon splits up is very soft the situation is quite different from that found in subsection 2. Let the energy of this photon be k_r. We now consider the case $k_r \ll k_0$ and therefore also $k_r \ll k$ (the second scattered quantum, called the *main quantum*). In particular the case $k_r \to 0$ will be of interest. From the theoretical point of view this process may still be called a multiple process, but from the practical point of view it is hardly distinguishable from the ordinary scattering, as the scattered main photon $\mathbf{k}$ has practically the same energy and momentum as for the single process. In particular we should expect that in the limit $k_r \to 0$, the process gives a correction, of order e^6, to the Compton scattering. This, however, is not quite so simple. It will be seen below that this correction, as far as it is due to the double Compton effect diverges like dk_r/k_r for $k_r \to 0$. This apparent difficulty occurs for many radiation processes (compare § 25) and is characteristic for a general situation. It is called the *infra-red problem.* The difficulty will be removed when we consider further contributions to the e^6-corrections to the single scattering, arising from the higher terms of the matrix elements. This will be shown in § 33. In order to be able to treat the corrections to the Compton scattering, we shall require the cross-section for the double scattering for very low (infra-red) frequencies k_r.

We calculate the cross-section for the double scattering in the limit of very small k_r and retain only terms of the highest order in $1/k_r$.† The compound matrix element again consists of six terms corresponding to the six orders in which $\mathbf{k}_0$ can be absorbed and $\mathbf{k}$, $\mathbf{k}_r$ emitted. It is now seen that the energy denominators $E_O - E_{\mathrm{I}}$, $E_O - E_{\mathrm{II}}$ are of different orders of magnitude in k_r, according to whether the emission

† Cf. C. J. Eliezer, *Proc. Roy. Soc.* **187** (1946), 210; R. Jost, *Phys. Rev.* **72** (1947), 815.

of k_r stands in the middle of the sequence, or at the beginning or end. If k_r is emitted first and the electron has positive energy, we have from the conservation laws

$$E_O - E_{\mathrm{I}} = \mu - k_r - \sqrt{(\mu^2 + k_r^2)} \sim -k_r, \tag{17}$$

whereas $E_O - E_{\mathrm{II}}$ is, apart from negligible modifications, identical with the denominator occurring for the single scattering, and is of the order k_0 or k. Similarly, if k_r is emitted last,

$$E_O - E_{\mathrm{II}} = k_0 + \mu - k - \sqrt{\{\mu^2 + (\mathbf{k}_0 - \mathbf{k})^2\}} \sim k_r - \frac{(\mathbf{pk}_r)}{E}$$

$$(\mathbf{k}_0 - \mathbf{k} = \mathbf{p} + \mathbf{k}_r, \qquad E = \sqrt{(\mu^2 + \mathbf{p}^2)} = k_0 + \mu - k - k_r), \tag{18}$$

also of order k_r. $E_O - E_{\mathrm{I}}$ is the energy denominator occurring for the single scattering. If k_r is emitted in the middle of the sequence both denominators are large, of order k_0 or k. We can therefore confine ourselves to the two types of intermediate states where k_r is emitted first or last, and only if the electron has positive energy before and after this emission.

The matrix element for the emission of k_r has the factor $(u_0'^*(\boldsymbol{\alpha}\mathbf{e}_r)u_0)$ when k_r is emitted first. u_0' is the wave function after the emission of k_r. For very small k_r, u_0' is practically identical with u_0, so, to our approximation this factor is $(u_0^*(\boldsymbol{\alpha}\mathbf{e}_r)u_0)$ which is the expectation value of $(\boldsymbol{\alpha}\mathbf{e}_r)$ in the state where the electron has momentum $\mathbf{p}_0$ and positive energy. According to § 11 this is $(\mathbf{p}_0\,\mathbf{e}_r)/E_0 = 0$ because $\mathbf{p}_0 = 0$ (electron initially at rest). Similarly, if k_r is emitted last,

$$(u^*(\boldsymbol{\alpha}\mathbf{e}_r)u) = \frac{(\mathbf{pe}_r)}{E}. \tag{19}$$

So the only contribution of importance arises from the term where k_r is emitted last. The remaining factors are just the compound matrix element for the ordinary Compton effect, which we need not write down explicitly. Denoting this by K_C we find for the matrix element of the double Compton effect

$$K_{d.C.} = -e\sqrt{\left(\frac{2\pi\hbar^2c^2}{k_r}\right)}\frac{(\mathbf{pe}_r)}{Ek_r - (\mathbf{pk}_r)}K_C. \tag{20}$$

To form the cross-section, the square of (20) is to be taken and multiplied by the density function. The latter is the product of the density function for the ordinary Compton effect (including the factor $\partial E_F/\partial k$) and that for the infra-red photon, viz., $k_r^2\,dk_r/(2\pi\hbar c)^3\,d\Omega_{k_r}$. Summing over the two directions of $\mathbf{e}_r$ and integrating over the directions of k_r, we obtain

$$d\phi_{d.C.} = \frac{e^2}{\pi\hbar c}\left(\frac{E}{p}\log\frac{E+p}{E-p}-2\right)\frac{dk_r}{k_r}d\phi_C; \tag{21}$$

E, p are the energy and momentum of the electron after the scattering. (21) gives the differential probability for a Compton scattering accompanied by the emission in any direction of an infra-red photon in the energy range dk_r. We see that $d\phi_{d.C.}$ is proportional to the Klein–Nishina cross-section $d\phi_C$. The most characteristic features are (i) the factor $e^2/\hbar c$ which makes the double Compton effect improbable but, on the other hand, (ii) the dependence on k_r, dk_r/k_r. This diverges when $k_r \to 0$ or when integrated over an interval from 0 to a finite value. It appears that $\phi_{d.C.}$. increases above the estimates (15′), when $k_1 \ll k_2$. However, unless extremely small values of k_r are included, the integral of (21) over k_r is still of the order (15′).

When $k_0 \ll \mu$ ($p \ll \mu$, $E \sim \mu$), (21) reduces to

$$d\phi_{d.C.} = \frac{4}{3\pi}\frac{e^2}{\hbar c}r_0^2\frac{k_0^2}{\mu^2}(1-\cos\theta)\frac{1+\cos^2\theta}{2}\frac{dk_r}{k_r}d\Omega, \tag{22}$$

with $p^2 \simeq 2k_0^2(1-\cos\theta)$. θ is the scattering angle of the main quantum.

24. The scattering of two electrons

1. *Retarded interaction.* The scattering of two charged particles with Coulomb interaction can be treated *exactly* in wave mechanics provided that these particles move so slowly that their interaction is, at each instant of time, given by the *static* interaction and all retardation effects can be neglected.

The situation is different when retardation effects become important. In this case, as we shall see, the problem becomes one of quantum electrodynamics, and, up to the present, a solution has only been found by expansion according to powers of the interaction constant, i.e. $Z_1 Z_2 e^2$, where $Z_1 e$, $Z_2 e$ are the charges of the two particles under consideration. The reason for this necessity can be seen as follows:

In § 6 we have shown that the interaction of a set of charged particles between themselves and the electromagnetic field can be decomposed into two parts, namely: (i) a static, instantaneous, Coulomb interaction between the particles, and (ii) the interaction of each particle with the *transverse light field*, using Coulomb gauge. If we assume that light is not actually emitted during the scattering of two charged particles (which is, of course, an approximation), any departure from the static law of interaction, such as is represented by the retarded interaction, must be due to the mutual exchange of light waves between the two charged particles. Since we could not treat even the emission of light by a single particle otherwise than by expansion according to e^2, we cannot hope to treat the retarded interaction in a better manner. To be consistent, we must then also treat the static part of the interaction by means of expansions, i.e. by the well-known Born approximation method. Fortunately, from the quantitative point of view, this is a good way of expanding because retardation effects are important for high velocities only, where Born's method can also be applied.

Consider two charged particles both obeying the Dirac equation and let their wave functions be ψ_1, ψ_2. The static interaction is then $V = Z_1 Z_2 e^2/r_{12}$. In the first Born approximation V has a matrix element for scattering, from an initial state O where the two particles have momenta p_{01}, p_{02}, say, to a final state with momenta p_1, p_2. We use the plane wave solutions

$$\psi_{01} = u_{01} \exp i(\mathbf{p}_{01}\, \mathbf{r}_1)/\hbar c, \qquad \psi_1 = u_1 \exp i(\mathbf{p}_1\, \mathbf{r}_1)/\hbar c, \quad \text{etc.}$$

Then the matrix element of V is†

$$V_{FO} = Z_1 Z_2 e^2 \int \frac{d\tau_1\, d\tau_2}{|\mathbf{r}_1-\mathbf{r}_2|} e^{i(\mathbf{p}_{01}-\mathbf{p}_1,\mathbf{r}_1)/\hbar c} e^{i(\mathbf{p}_{02}-\mathbf{p}_2,\mathbf{r}_2)/\hbar c} (u_1^* u_{01})(u_2^* u_{02})$$

$$= Z_1 Z_2 e^2 \int \frac{d\tau_{12}\, d\tau_2}{|\mathbf{r}_1-\mathbf{r}_2|} e^{i(\mathbf{p}_{01}-\mathbf{p}_1,\mathbf{r}_1-\mathbf{r}_2)/\hbar c} e^{i(\mathbf{p}_{01}+\mathbf{p}_{02}-\mathbf{p}_1-\mathbf{p}_2,\mathbf{r}_2)/\hbar c} (u_1^* u_{01})(u_2^* u_{02})$$

$$= \frac{4\pi\hbar^2 c^2 Z_1 Z_2 e^2}{|\mathbf{p}_{01}-\mathbf{p}_1|^2} (u_1^* u_{01})(u_2^* u_{02}) \tag{1}$$

($d\tau_{12}$ = volume element of $\mathbf{r}_1-\mathbf{r}_2$), provided that the total momentum is preserved, i.e. that

$$\mathbf{p}_{01}+\mathbf{p}_{02} = \mathbf{p}_1+\mathbf{p}_2.$$

Otherwise V_{FO} vanishes.

Our task is now to generalize (1) for the case of a retarded interaction.

As an illustration of the methods of quantum electrodynamics and their relation with classical theory, it will be useful to evaluate in

† The integral is not properly convergent. Insert first a factor

$$\exp(-\alpha r_{12}) \qquad (r_{12} \equiv |\mathbf{r}_1-\mathbf{r}_2|)$$

and afterwards let α go to zero. Also take into account what was said in § 14.3 about the conservation of momentum and the normalization of the wave functions.

In the non-relativistic limit (1) leads, when the cross-section is formed, to the Rutherford scattering formula. It is remarkable that the latter is also the result of an *exact* treatment. The fact that the first Born approximation coincides with the exact solution (for the cross-section) is a peculiarity of the Coulomb law and does not, of course, hold for other laws of force.

various ways the matrix element for the retarded interaction: (i) we shall first follow Møller† and determine the matrix element for the transition $\mathbf{p}_{01}, \mathbf{p}_{02} \rightarrow \mathbf{p}_1, \mathbf{p}_2$ of the retarded interaction between the two particles, in a semi-classical way. Next we show that the same result follows by using the methods of this book. We can either (ii) use Coulomb gauge and thus add to (1) the contribution which arises from the mutual exchange of light quanta. Or, simpler still, we may (iii) use Lorentz gauge. Then the *whole* interaction can be represented as due to the exchange of photons, but there are four types of them, including longitudinal and scalar photons.

(i) Let the particles 1, 2 be represented by their charge and current densities $\rho(\mathbf{r}_1,t)$, $\rho(\mathbf{r}_2,t)$, $\mathbf{i}(\mathbf{r}_1,t)$, $\mathbf{i}(\mathbf{r}_2,t)$. Then the retarded potentials produced by the particle 1 at the position $\mathbf{r}_2$ are given in classical theory by § 1 eqs. (14)

$$\begin{aligned} \phi(\mathbf{r}_2,t) &= \int \frac{1}{r_{12}}\rho(\mathbf{r}_1,t-r_{12}/c)\,d\tau_1, \\ \mathbf{A}(\mathbf{r}_2,t) &= \frac{1}{c}\int \frac{1}{r_{12}}\mathbf{i}(\mathbf{r}_1,t-r_{12}/c)\,d\tau_1, \end{aligned} \tag{2}$$

and the interaction of particle 2 with this field is

$$-\frac{1}{c}\int i_\mu A_\mu\,d\tau = \int \rho(\mathbf{r}_2,t)\phi(\mathbf{r}_2,t)\,d\tau_2 - \frac{1}{c}\int (\mathbf{i}(\mathbf{r}_2,t)\mathbf{A}(\mathbf{r}_2,t))\,d\tau_2.$$

Thus the retarded interaction between the two particles is

$$K_{\text{ret}} = \iint \frac{\rho(\mathbf{r}_1,t-r_{12}/c)\rho(\mathbf{r}_2,t)}{r_{12}}\,d\tau_1\,d\tau_2 - \iint \frac{\mathbf{i}(\mathbf{r}_1,t-r_{12}/c)\mathbf{i}(\mathbf{r}_2,t)}{c^2 r_{12}}\,d\tau_1\,d\tau_2 \tag{3}$$

where the potentials do not occur explicitly. We can now go over to the wave mechanical description of the particles and substitute for ρ, $\mathbf{i}$ the expressions of the Dirac theory, viz. $(\psi^*\psi)$ and $c(\psi^*\boldsymbol{\alpha}\psi)$. Moreover, we can define a *transition element* for $\rho(\mathbf{r},t)$ by substituting for the factor ψ the plane wave $u_{01}\,e^{i(\mathbf{p}_{01}\mathbf{r}_1)/\hbar c}e^{-iE_{01}t/\hbar}$ and for ψ^* the plane wave $u_1^*\,e^{-i(\mathbf{p}_1\mathbf{r}_1)/\hbar c}e^{+iE_1t/\hbar}$, etc. In this way we obtain transition elements for $\rho(\mathbf{r}_1)$, $\rho(\mathbf{r}_2)$, $\mathbf{i}(\mathbf{r}_1)$, $\mathbf{i}(\mathbf{r}_2)$ corresponding to the transition considered in (1).‡ It is essential, of course, to use the time-dependent wave

† C. Møller, *Ann. Phys.* **14** (1932), 531.

‡ This procedure can be justified rigorously by using the method of second quantization for the electron field (§ 12). K_{ret} is given by (3) also in quantum theory, but $\rho = \psi^*\psi$, etc., are operators, arising from the second quantization of ψ. Thus K_{ret} is an operator. One then forms the matrix element of K with the state vector Ψ of the quantized field for the transition of (1), namely, two electrons with momenta $\mathbf{p}_{01}$ and $\mathbf{p}_{02}$ are absorbed and two with momenta $\mathbf{p}_1$, $\mathbf{p}_2$ are emitted. The result is eq. (5).

functions

$$\begin{aligned}
\rho(\mathbf{r}_1, t-r_{12}/c)_{p_1|p_{01}} &= Z_1\, e(u_1^* u_{01}) e^{i(\mathbf{p}_{01}-\mathbf{p}_1,\mathbf{r}_1)/\hbar c} e^{-i(E_{01}-E_1)t/\hbar} e^{i(E_{01}-E_1)r_{12}/\hbar c},\\
\mathbf{i}(\mathbf{r}_1, t-r_{12}/c)_{p_1|p_{01}} &= Z_1\, ec(u_1^* \boldsymbol{\alpha}_1 u_{01}) e^{i(\mathbf{p}_{01}-\mathbf{p}_1,\mathbf{r}_1)/\hbar c} e^{-i(E_{01}-E_1)t/\hbar} e^{i(E_{01}-E_1)r_{12}/\hbar c},\\
\rho(\mathbf{r}_2, t)_{p_2|p_{02}} &= Z_2\, e(u_2^* u_{02}) e^{i(\mathbf{p}_{02}-\mathbf{p}_2,\mathbf{r}_2)/\hbar c} e^{-i(E_{02}-E_2)t/\hbar},\\
\mathbf{i}(\mathbf{r}_2, t)_{p_2|p_{02}} &= Z_2\, ec(u_2^* \boldsymbol{\alpha}_2 u_{02}) e^{i(\mathbf{p}_{02}-\mathbf{p}_2,\mathbf{r}_2)/\hbar c} e^{-i(E_{02}-E_2)t/\hbar}.
\end{aligned} \tag{4}$$

Since we are dealing with two particles in configuration space the α-operators carry the index of the particle concerned. Substituting in (3), the transition matrix element of the retarded interaction becomes

$$K_{FO} = Z_1 Z_2 e^2[(u_2^* u_{02})(u_1^* u_{01}) - (u_2^* \boldsymbol{\alpha}_2 u_{02})(u_1^* \boldsymbol{\alpha}_1 u_{01})] e^{-i(E_{01}+E_{02}-E_1-E_2)t/\hbar} \times$$
$$\times \iint \frac{d\tau_1\, d\tau_2}{r_{12}} e^{i[(\mathbf{p}_{01}-\mathbf{p}_1,\mathbf{r}_1)+(\mathbf{p}_{02}-\mathbf{p}_2,\mathbf{r}_2)]/\hbar c} e^{+i(E_{01}-E_1)r_{12}/\hbar c}. \tag{5}$$

The second term in the square bracket is to be understood as the scalar product of the two matrix vectors $\boldsymbol{\alpha}_2$, $\boldsymbol{\alpha}_1$. (5) differs from (1) in two respects: the interaction of the currents contributes the term $\boldsymbol{\alpha}_1$, $\boldsymbol{\alpha}_2$. Secondly, the retardation effect is contained in the additional factor $\exp i(E_{01}-E_1)r_{12}/\hbar c$. Both are relativistic effects.

The integral (5) vanishes, like (1), unless the momenta are conserved. Also in the actual scattering process energy will be conserved and we need only consider transitions for which

$$E_{01}-E_1 = -E_{02}+E_2. \tag{6}$$

The integration of (5) yields finally

$$K_{FO} = \frac{Z_1 Z_2 e^2 4\pi\hbar^2 c^2}{k^2-\epsilon^2} [(u_2^* u_{02})(u_1^* u_{01}) - (u_2^* \boldsymbol{\alpha}_2 u_{02})(u_1^* \boldsymbol{\alpha}_1 u_{01})], \tag{7}$$

where we have put

$$\mathbf{p}_{01}-\mathbf{p}_1 = \mathbf{k}, \qquad E_{01}-E_1 = \epsilon. \tag{8}$$

The relativistic invariance of (7) is evident: $k^2-\epsilon^2$ is the square of the 4-vector, $p_{0\mu}-p_\mu$. The u-factor can also be combined relativistically. The transition element of the 4-current of particle 1 is

$$i_\mu(1)_{p_1|p_{01}} \equiv ec(u_1^\dagger \gamma_\mu u_{01}),$$

or
$$i_{4p_1|p_{01}} = iec(u_1^* u_{01}), \qquad i_{xp_1|p_{01}} = ec(u_1^* \alpha_x u_{01}).$$

Then evidently

$$(u_2^* u_{02})(u_1^* u_{01}) - (u_2^* \boldsymbol{\alpha}_2 u_{02})(u_1^* \boldsymbol{\alpha}_1 u_{01}) = -\frac{1}{e^2 c^2} (i_\mu(1) i_\mu(2))_{p_1 p_2|p_{01} p_{02}}. \tag{9}$$

The present derivation of (7) is semi-classical in the sense that no use is made of the quantization of the field. It is then not evident that (7) only gives the first term of an expansion $\sim e^2$. We show in the following subsection that the same result is obtained from the quantum theory of the field.

2. *Derivation from quantized field.* (ii) *Coulomb gauge.* If this gauge is used we have to add to (1) a contribution arising from the exchange of transverse photons between the two particles. Evidently the transition $p_{01}, p_{02} \to p_1, p_2$ can only be due to the exchange of a photon and is of second order. There are two intermediate states:

I. The particle 1 emits a photon, $\mathbf{k} = \mathbf{p}_{01} - \mathbf{p}_1$. This photon is then absorbed by particle 2 so that its momentum becomes $\mathbf{p}_{02} + \mathbf{k} = \mathbf{p}_2$.

II. The particle 2 emits first a photon $-\mathbf{k}$, $\mathbf{p}_{02} = -\mathbf{k} + \mathbf{p}_2$, and this photon is then absorbed by particle 1. Since

$$E_O = E_F = E_{01} + E_{02} = E_1 + E_2,$$

we have
$$E_{\mathrm{I}} = E_1 + E_{02} + k, \qquad E_{\mathrm{II}} = E_{01} + E_2 + k.$$

The compound matrix element of second order becomes, using the formulae § 14 eq. (13),

$$K_{FO}^{\mathrm{tr}} = \frac{H_{F\mathrm{I}} H_{\mathrm{I}O}}{E_O - E_{\mathrm{I}}} + \frac{H_{F\mathrm{II}} H_{\mathrm{II}O}}{E_O - E_{\mathrm{II}}}$$
$$= e^2 Z_1 Z_2 \frac{2\pi\hbar^2 c^2}{k} \sum_e (u_2^*(\boldsymbol{\alpha}_2 \mathbf{e}) u_{02})(u_1^*(\boldsymbol{\alpha}_1 \mathbf{e}) u_{01}) \left(\frac{1}{E_{01} - E_1 - k} + \frac{1}{E_{02} - E_2 - k} \right). \tag{10}$$

Here $\mathbf{e}$ is the direction of polarization of the photon, which is taken to be the same for both intermediate states. $\sum_e$ indicates a summation over both directions of polarization:

$$\sum_e (\boldsymbol{\alpha}_1 \mathbf{e})(\boldsymbol{\alpha}_2 \mathbf{e}) = (\boldsymbol{\alpha}_1 \boldsymbol{\alpha}_2) - \frac{(\boldsymbol{\alpha}_1 \mathbf{k})(\boldsymbol{\alpha}_2 \mathbf{k})}{k^2}. \tag{11}$$

The second term can be rewritten by making use of the wave equation:

$$\begin{aligned} (u_1^*(\boldsymbol{\alpha}_1 \mathbf{k}) u_{01}) &= (u_1^*(\boldsymbol{\alpha}_1, \mathbf{p}_{01} - \mathbf{p}_1) u_{01}) = (E_{01} - E_1)(u_1^* u_{01}), \\ (u_2^*(\boldsymbol{\alpha}_2 \mathbf{k}) u_{02}) &= -(E_{02} - E_2)(u_2^* u_{02}). \end{aligned} \tag{12}$$

Thus, with (6) and (8),

$$K_{FO}^{\mathrm{tr}} = e^2 Z_1 Z_2 \frac{4\pi\hbar^2 c^2}{k^2 - \epsilon^2} \left[\frac{\epsilon^2}{k^2} (u_2^* u_{02})(u_1^* u_{01}) - (u_2^* \boldsymbol{\alpha}_2 u_{02})(u_1^* \boldsymbol{\alpha}_1 u_{01}) \right], \tag{13}$$

K_{FO}^{tr} has now to be added to V_{FO}. The total interaction is

$$K_{FO} = K_{FO}^{\mathrm{tr}} + V_{FO}.$$

The term proportional to ϵ^2/k^2 of (13) combines with (1), and (7) is obtained immediately.

(iii) *Lorentz gauge.* Finally, as a simple application of the quantum treatment of the longitudinal and scalar fields, we treat the same problem in Lorentz gauge. Here four types of 'photons' exist, with 'polarizations' 1,..., 4. These photons all have the same energy k and momentum $\mathbf{k}$. No static interaction exists. The *whole* interaction is given by a compound matrix element of second order of the type (10). There are again the two types of intermediate states as in case (ii), and the energy denominators are the same. It has now been shown in § 10 that the matrix elements for emission or absorption of a photon of polarization α (= 1,..., 4) are formally the same as for transverse photons, the difference being only that the polarization vector $\mathbf{e}$ is replaced by unit vectors with the direction of one of the four space-time directions, e.g. (§ 14 eq. (29)):

$$H_{1\mu|0} = -eZ\sqrt{\left(\frac{2\pi\hbar^2c^2}{k}\right)}(u^\dagger\gamma_\mu u_0)$$

when the electron changes from state $\mathbf{p}_0$ to $\mathbf{p}$. Hence the compound matrix element is, in complete analogy to (10), given by (combining the two energy denominators)

$$K_{FO} = -e^2Z_1Z_2\frac{4\pi\hbar^2c^2}{k^2-\epsilon^2}\sum_{\mu=1}^{4}(u_1^\dagger\gamma_{\mu1}u_{01})(u_2^\dagger\gamma_{\mu2}u_{02}). \tag{14}$$

The relativistic invariance has been maintained from the start. To see that (14) is identical with (7) we have only to write out the $\sum_\mu$ remembering that $u^\dagger = iu^*\beta$, $\gamma_4 = \beta$, $\gamma_i = -i\beta\alpha_i$,

$$\sum_\alpha(u_1^\dagger\gamma_{\mu1}u_{01})(u_2^\dagger\gamma_{\mu2}u_{02}) = -(u_1^*u_{01})(u_2^*u_{02})+(u_1^*\boldsymbol{\alpha}_1u_{01})(u_2^*\boldsymbol{\alpha}_2u_{02}). \tag{15}$$

Thus (14) is identical with (7).

In the present derivation no notice has been taken of the Lorentz condition. This is not necessary for collisions between free particles because, as was shown in § 13.2 and appendix 3, it is possible to choose before and after the collision, when the particles are infinitely far apart, a state of complete absence of all types of photons.

It is evident that the last derivation (iii) is the most concise and is the most satisfying from a relativistic point of view.

3. *Exchange effects.* In the above derivations of K_{FO} we have not taken any notice of the fact that if the two colliding particles are alike the wave function in configuration space must be antisymmetrical. In fact (7) and (14) are valid only if the two particles are different. The

modification due to the antisymmetry of the wave function is very easily found.

The Dirac amplitude u depends on a discrete variable ρ which plays the role of a coordinate. Each particle (affixed with a number, cf. § 11) possesses such a variable. In addition u carries an index denoting the state of the particle, characterized by the momentum $\mathbf{p}$, spin, and sign of energy. Above, we have denoted the wave function of particle 1 with momentum $\mathbf{p}_1$ after the scattering by u_1. Written out fully, this should be called $u_{p_1}(1)$. Evidently, the wave function of the final state should now be replaced by the antisymmetrical combination

$$\frac{1}{\sqrt{2}}\{u_{p_1}(1)u_{p_2}(2)e^{i[(\mathbf{p}_1\mathbf{r}_1)+(\mathbf{p}_2\mathbf{r}_2)]/\hbar c}-u_{p_1}(2)u_{p_2}(1)e^{i[(\mathbf{p}_1\mathbf{r}_2)+(\mathbf{p}_2\mathbf{r}_1)]/\hbar c}\}. \tag{16}$$

The second term of (16) can also be regarded as arising through an interchange $\mathbf{p}_1 \rightleftarrows \mathbf{p}_2$. The same has to be done for the initial state, i.e. a second term has to be added with a minus sign interchanging $\mathbf{p}_{01} \rightleftarrows \mathbf{p}_{02}$. The modified matrix element K_{FO} then arises from (14) by adding terms interchanging $\mathbf{p}_1 \rightleftarrows \mathbf{p}_2$ and $\mathbf{p}_{01} \rightleftarrows \mathbf{p}_{02}$, both with a minus sign and a term $\mathbf{p}_1 \rightleftarrows \mathbf{p}_2$, $\mathbf{p}_{01} \rightleftarrows \mathbf{p}_{02}$ with a plus sign, and dividing by 2. Terms in which both interchanges are made are, of course, equal. Thus (we now put $Z_1 = Z_2 = 1$)

$$K_{FO} = -e^2 4\pi\hbar^2 c^2 \sum_\alpha \left[\frac{(u_{p_1}^\dagger \gamma_\mu u_{p_{01}})(u_{p_2}^\dagger \gamma_\mu u_{p_{02}})}{k^2-\epsilon^2} - \frac{(u_{p_1}^\dagger \gamma_\mu u_{p_{02}})(u_{p_2}^\dagger \gamma_\mu u_{p_{01}})}{k'^2-\epsilon'^2}\right]$$

$$(k' = \mathbf{p}_{01}-\mathbf{p}_2,\ \epsilon' = E_{01}-E_2). \tag{17}$$

The second term of (17) describes the exchange effects. Strictly speaking each bracket should read, when written out fully, $(u_{p_1}^\dagger(1)\gamma_{\mu 1} u_{p_{01}}(1))$, etc., but since such a bracket is a pure number it is independent of the number of the particle concerned.

4. *Cross-section.* We may take advantage of the fact that the theory is Lorentz invariant (and the same will be true of the cross-section) and work out the cross-section in a particular frame of reference. It will be most convenient to choose this so that the centre of mass of the two particles is at rest.

This means that

$$\mathbf{p}_{02} = -\mathbf{p}_{01} = -\mathbf{p}_0,\quad \mathbf{p}_2 = -\mathbf{p}_1 = -\mathbf{p},\quad |\mathbf{p}_0| = |\mathbf{p}|. \tag{18}$$

Also, of course, $E_{01} = E_{02} = E_1 = E_2 = E$. The cross-section in any other frame of reference, for instance when one of the particles is originally at rest, can be obtained by a simple Lorentz transformation.

The transition probability per unit time is given by the general formula (§ 14) $2\pi/\hbar\,|K_{FO}|^2\rho_F$. Here $\rho_F\,dE_F$ is the number of states for which the final energy lies in the energy interval dE_F. Since the two electrons have, by (18), exactly opposite momenta, $\rho_F\,dE_F$ is equal to the number of states where *one* electron has energy E (in the interval dE):

$$\rho_F\,dE_F = \rho_E\,dE.$$

E_F is the total final energy, $E_F = 2E$, therefore

$$\rho_F = \tfrac{1}{2}\rho_E = \frac{1}{2}\frac{pE\,d\Omega}{(2\pi\hbar c)^3}. \tag{19}$$

The differential cross-section is the transition probability per sec. divided by the *relative* velocity of the two particles, i.e. $2v = 2(pc/E)$. Thus

$$d\phi = \frac{\pi}{2\hbar c}\frac{E^2\,d\Omega}{(2\pi\hbar c)^3}|K_{FO}|^2. \tag{20}$$

Our only task now is to evaluate the square of the bracket of (17). This we do by means of the spur method applied already in § 22. We sum over the spin directions of p_1 and p_2 and average over those of p_{01}, p_{02}. Using § 11 eq. (18) and noting that all u's belong to positive energies, we obtain

$$\tfrac{1}{4}S^{p_1}S^{p_2}S^{p_{01}}S^{p_{02}}\left|\sum_{\mu}\left[\frac{(u^\dagger_{p_1}\gamma_\mu u_{p_{01}})(u^\dagger_{p_2}\gamma_\mu u_{p_{02}})}{k^2-\epsilon^2}-\frac{(u^\dagger_{p_2}\gamma_\mu u_{p_{01}})(u^\dagger_{p_1}\gamma_\mu u_{p_{02}})}{k'^2-\epsilon'^2}\right]\right|^2$$

$$= \frac{1}{64E^4(k^2-\epsilon^2)^2}\mathrm{Sp}\,\gamma_\mu(\gamma\cdot p_{01}+i\mu)\gamma_\nu(\gamma\cdot p_1+i\mu)\times$$
$$\times\mathrm{Sp}\,\gamma_\mu(\gamma\cdot p_{02}+i\mu)\gamma_\nu(\gamma\cdot p_2+i\mu)-$$
$$-\frac{1}{64E^4(k^2-\epsilon^2)(k'^2-\epsilon'^2)}\mathrm{Sp}\,\gamma_\mu(\gamma\cdot p_{01}+i\mu)\gamma_\nu(\gamma\cdot p_2+i\mu)\times$$
$$\times\gamma_\mu(\gamma\cdot p_{02}+i\mu)\gamma_\nu(\gamma\cdot p_1+i\mu)+\text{terms } p_1\rightleftarrows p_2,\ E_1\rightleftarrows E_2. \tag{21}$$

Here we have denoted the 4-product $\gamma_\mu p_\mu$ by $\gamma\cdot p$ for the sake of brevity. The sum is to be taken over all relativistic indices that occur twice.

The evaluation of the spurs follows the same pattern as in § 22, using the rules given in § 11. It is also useful to note that

$$\sum_\alpha \gamma_\alpha\gamma_\mu\gamma_\alpha = -2\gamma_\mu.$$

In the centre of mass system, which we are using,

$$p_{14} = p_{24} = p_{014} = p_{024} = iE.$$

Thus $\epsilon = \epsilon' = 0$ and also

$$\left.\begin{aligned}(p_1 \cdot p_{01}) &= (p_2 \cdot p_{02}) = (\mathbf{p}_0 \mathbf{p}) - E^2 = p^2 \cos^2\theta - E^2 \\ (p_1 \cdot p_{02}) &= (p_2 \cdot p_{01}) = -p^2 \cos^2\theta - E^2 \\ (p_1 \cdot p_2) &= (p_{01} \cdot p_{02}) = -(p^2 + E^2) = -(2E^2 - \mu^2)\end{aligned}\right\} \tag{22}$$

$$k^2 = 2p^2(1-\cos\theta), \qquad k'^2 = 2p^2(1+\cos\theta). \tag{23}$$

θ is the scattering angle of one of the particles. We use (22) and (23) in the evaluation of the spurs (21). The cross-section (20) with (17) becomes then, after some straightforward calculation:

$$d\phi = \frac{1}{4}\frac{e^4\,d\Omega}{p^4 E^2 \sin^4\theta}[4(E^2+p^2)^2 - 3(E^2+p^2)^2 \sin^2\theta + p^4(\sin^4\theta + 4\sin^2\theta)]. \tag{24}$$

(24) is not of immediate practical use because it refers to a Lorentz-frame in which the scattering is not usually observed. We now transform (24) to a frame of reference where one of the two electrons, say that which had momentum $\mathbf{p}_{02}$, is initially at rest. The cross-section is invariant under such a transformation. The velocity of the Lorentz transformation is evidently $cp_{02}/E = -cp/E$. Thus, if we denote all quantities referring to the new Lorentz system by an asterisk, the incident particle now has a momentum and energy given by

$$p_{01}^* \equiv p_0^* = \frac{2pE}{\mu}, \qquad E_0^* = \frac{E^2+p^2}{\mu}, \qquad p^2 = \tfrac{1}{2}\mu(E_0^* - \mu). \tag{25}$$

θ was the angle between $\mathbf{p}_{01}$ and $\mathbf{p}_1$. The scattered particles now have momenta $\mathbf{p}_1^*$, $\mathbf{p}_2^*$ and energies E_1^*, E_2^*. The angle between $\mathbf{p}_1^*$ and $\mathbf{p}_0^*$ will be called θ^*. We find

$$p_1^* \sin\theta^* = p\sin\theta, \quad p_1^* \cos\theta^* = \frac{E}{\mu}p(1+\cos\theta), \quad E_1^* = \frac{1}{\mu}(E^2 + p^2\cos\theta). \tag{26}$$

It will be useful to express the cross-section by the energy transferred to the electron originally at rest, namely, $E_2^* = E_0^* + \mu - E_1^*$. θ is related to E_2^* directly by (26), and (25),

$$\cos\theta = \frac{E_0^* + \mu - 2E_2^*}{E_0^* - \mu}. \tag{27}$$

If we denote the recoil energy transferred to the electron originally at rest (energy lost by the incident particle) in units of μ by q, we obtain from (28) and (24) the cross-section (Møller, loc. cit.)

$$d\phi = 2\pi r_0^2 \frac{\gamma^2}{\gamma^2-1} \frac{dq}{q^2(\gamma-1-q)^2} \times$$
$$\times \left\{(\gamma-1)^2 - \frac{q(\gamma-1-q)}{\gamma^2}[2\gamma^2+2\gamma-1-q(\gamma-1-q)]\right\}$$
$$(q = E_2^*/\mu-1,\ \gamma = E_0^*/\mu). \tag{28}$$

It follows from (25), (26) that q is related to θ^* by

$$q = \frac{E_2^*}{\mu}-1 = \frac{(\gamma^2-1)\sin^2\theta^*}{2+(\gamma-1)\sin^2\theta^*}. \tag{29}$$

$\gamma-1-q$ is the kinetic energy left to the colliding electron. There is no distinction between colliding and recoil electron and, indeed, (28) is unchanged if we interchange q and $\gamma-1-q$. One would, however, always call the particle with the smaller energy the recoil particle. Then q varies from 0 to $(\gamma-1)/2$. The corresponding angles of scattering of the colliding (the faster) particle are $\theta^* = 0$ and a maximum angle θ_m^* which follows from (29), namely,

$$\cos 2\theta_m^* = \frac{\gamma-1}{\gamma+3}.$$

The recoil electron always forms an angle θ' (with the incident electron) which is larger than θ_m^*. Since there is complete symmetry between the two particles we obtain the relation between q and θ' from (29) by replacing q by $\gamma-1-q$ and θ^* by θ'. Thus

$$q = \frac{2(\gamma-1)\cos^2\theta'}{2+(\gamma-1)\sin^2\theta'}. \tag{29'}$$

The maximum angle at which the recoil particle is emitted is given by (29') when $q = 0$, i.e. $\theta' = \frac{1}{2}\pi$. So $\theta_m^* \leqslant \theta' \leqslant \frac{1}{2}\pi$, $0 \leqslant \theta^* \leqslant \theta_m^*$.

In the non-relativistic approximation $(\gamma \to 1)$ $d\phi$ reduces to a simple form. In the first place, we find from (26) that then $\theta = 2\theta^*$. θ^* then ranges from 0 to $\frac{1}{2}\pi$. $d\phi$ becomes (directly from (24))

$$d\phi = 2\pi r_0^2\left(\frac{c}{v_0}\right)^4 d\cos 2\theta^* \left[\frac{1}{\sin^4\theta^*}+\frac{1}{\cos^4\theta^*}-\frac{1}{\sin^2\theta^*\cos^2\theta^*}\right], \quad \text{N.R.} \tag{30}$$

v_0 is the velocity of the incident electron. The last two terms are due to the exchange and were first obtained by Mott.† The first term which is alone present when the particles are different, represents the well-known scattering formula of Rutherford and need not be discussed here.

† N. F. Mott, *Proc. Roy. Soc.* **126** (1930), 259.

When the energy loss is small (28) reduces to

$$d\phi = 2\pi r_0^2 \frac{\gamma^2}{\gamma^2-1}\frac{dq}{q^2} \quad (q \ll \gamma-1). \tag{31}$$

It is seen that $d\phi$ decreases rapidly with increasing energy loss so that the smallest energy losses are far the more frequent. (31) is a well-known formula of Bohr. It is also relativistically exact. Even when q assumes its maximum value $\frac{1}{2}(\gamma-1)$, the departure of the exact formula (28) from (31) is not very large. (28) and (31) are used to calculate the stopping power of matter for fast electrons (compare § 37).

The scattering formula (24) or (28) has been checked experimentally for energies of a few mc^2 and found to be in good agreement with the facts.†

For the sake of completeness we also give the cross-section for the scattering of an electron in a fixed Coulomb field, i.e. when one of the particles is regarded as infinitely heavy. No retardation effects arise then and the matrix element is just

$$V_{FO} = \frac{4\pi\hbar^2c^2e^2Z}{|\mathbf{p}-\mathbf{p}_0|^2}(u^*u_0), \qquad |\mathbf{p}_0| = |\mathbf{p}| = p. \tag{32}$$

The factor $(u_2^*u_{02})$ referring to the nucleus is now, of course, omitted. The relativistically exact cross-section is, in Born approximation,‡

$$d\phi = \frac{\pi r_0^2 Z^2}{(1-\cos\theta)^2}\frac{\mu^2}{p^2}\left(1+\cos\theta+\frac{2\mu^2}{p^2}\right)d\cos\theta. \tag{33}$$

This well-known formula need hardly be discussed here. Transforming to the 'opposite Lorentz system' where the electron is originally at rest one obtains the cross-section for a collision of a fast heavy particle with an electron at rest. Expressing this, as in (28), by the energy transferred to the electron, $q = (E^*/\mu)-1$, we find

$$d\phi = \frac{\pi r_0^2 Z^2c^2\,dq}{q^2v^2}[2-q(1-v^2/c^2)] \tag{34}$$

(v = velocity of heavy particle). The difference between (28) and (34) arises from spin and exchange effects and also from the fact that in the latter case the change of momentum of the heavy particle is negligible.

† For a detailed discussion see N. F. Mott and H. S. W. Massey, *Theory of Atomic Collisions*, 2nd ed., Oxford 1949, p. 369.

‡ N. F. Mott, *Proc. Roy. Soc.* A, **135** (1932), 429.

25. Bremsstrahlung

If an electron with primary energy E_0 (momentum $\mathbf{p}_0$) passes through the field of a nucleus (or atom) it is in general deflected. Since this deflexion always means a certain acceleration, the electron must emit radiation. There will be a certain probability that a light quantum $\mathbf{k}$ is emitted, the electron making a transition to another state with energy E (momentum $\mathbf{p}$), where

$$E+k = E_0. \tag{1}$$

Since the nucleus is heavy compared with the electron, the momentum of electron plus light quantum is not in general conserved; the nucleus can take any amount of momentum. We therefore obtain a finite transition probability to any final state E, $\mathbf{p}$ which satisfies (1).

1. *Differential cross-section.*† The interaction causing the transition from the initial state O ($\mathbf{p}_0$) to the final state F ($\mathbf{p},\mathbf{k}$) consists of two parts: (i) the interaction H_{int} of the electron with the radiation field giving rise to the emission of $\mathbf{k}$, and (ii) the interaction V of the electron with the atomic field. Thus the total interaction is given by

$$H = H_{\text{int}}+V. \tag{2}$$

We shall treat V as a perturbation in the same way as we treat the interaction H_{int} with the field. This means an expansion of the transition probability in powers of e^2 (or Ze^2). The first approximation (Born's approximation) of this expansion only gives correct results if

$$2\pi\xi_0 \equiv 2\pi\frac{Ze^2}{\hbar v_0} \ll 1 \quad \text{and} \quad 2\pi\xi \equiv 2\pi\frac{Ze^2}{\hbar v} \ll 1, \tag{3}$$

where v_0, v represent the velocities before and after the collision. For light elements (3) is always satisfied if the primary energy is of relativistic order of magnitude except in a small frequency range where the electron has given nearly all its kinetic energy $E_0-\mu$ to the light quantum and, after the process, has therefore a small velocity v. For small primary velocities Sommerfeld has given an exact theory,‡ the results of which are quoted in subsection 2. Also for high energies $v \sim c$ but *heavy elements* we must expect that some corrections will have to be made. (For lead $Ze^2/\hbar c = 0{\cdot}6$.) These will be given in subsection 5.

We consider first the case of a pure Coulomb field $V = Ze^2/r$. The interaction H_{int} of the electron with the radiation only has non-vanishing matrix elements for transitions in which the momentum is conserved.

† H. Bethe and W. Heitler, *Proc. Roy. Soc.* A, **146** (1934), 83.
‡ A. Sommerfeld, *Atombau und Spektrallinien II*, Braunschweig 1939.

On the other hand, the Coulomb interaction V has matrix elements for which the state of the radiation field remains unchanged while the momentum of the electron may change by any amount. The transition $O \to F$ occurs by passing through an intermediate state. There are obviously two possible intermediate states:

I. $\mathbf{k}$ is emitted. The electron has a momentum $\mathbf{p}'$:

$$\mathbf{p}' = \mathbf{p}_0 - \mathbf{k}. \tag{4}$$

The transition $O \to \mathrm{I}$ is caused by the term H_{int}. In the transition to the final state F (due to V) the momentum of the electron changes from $\mathbf{p}'$ to $\mathbf{p}$.

II. The electron has momentum

$$\mathbf{p}'' = \mathbf{p} + \mathbf{k}, \tag{5}$$

chosen so that in the transition $\mathrm{II} \to F$ $\mathbf{k}$ is emitted with conservation of momentum. The transition $O \to \mathrm{II}$ is caused by V.

The matrix elements of H_{int} and V for the transitions in question can be taken immediately from § 14 eq. (27) and § 24 eq. (32):

$$\begin{aligned} H_{\mathrm{I}O} &= -e\sqrt{\left(\frac{2\pi\hbar^2c^2}{k}\right)}(u'^*\alpha u_0), \quad V_{F\mathrm{I}} = \frac{Ze^2 4\pi\hbar^2c^2}{|\mathbf{p}'-\mathbf{p}|^2}(u^*u'), \\ V_{\mathrm{II}O} &= \frac{Ze^2 4\pi\hbar^2c^2}{|\mathbf{p}_0-\mathbf{p}''|^2}(u''^*u_0), \quad H_{F\mathrm{II}} = -e\sqrt{\left(\frac{2\pi\hbar^2c^2}{k}\right)}(u^*\alpha u''); \end{aligned} \tag{6}$$

α represents the component of $\boldsymbol{\alpha}$ in the direction of polarization of $\mathbf{k}$.

The denominators in the matrix elements of V are equal by (4) and (5); we denote them by q^2, so that

$$\mathbf{q} = \mathbf{p}_0 - \mathbf{p}'' = \mathbf{p}' - \mathbf{p} = \mathbf{p}_0 - \mathbf{p} - \mathbf{k}, \tag{7}$$

$\mathbf{q}$ represents the total momentum transferred to the nucleus.

Denoting by $\sum$ the summation over the spin directions and signs of energy of the intermediate states† the matrix element responsible for the transition $O \to F$ is given by

$$K_{FO} = \sum\left(\frac{V_{F\mathrm{I}}H_{\mathrm{I}O}}{E_O - E_{\mathrm{I}}} + \frac{H_{F\mathrm{II}}V_{\mathrm{II}O}}{E_O - E_{\mathrm{II}}}\right), \tag{8}$$

where the energy differences $E_O - E_1$ and $E_O - E_{\mathrm{II}}$ are, by (4) and (5):

$$\begin{aligned} E_O - E_{\mathrm{I}} &= E_0 - k - E', \qquad E'^2 = p'^2 + \mu^2, \\ E_O - E_{\mathrm{II}} &= E_0 - E'', \qquad E''^2 = p''^2 + \mu^2. \end{aligned} \tag{9}$$

The transition probability per unit time is given by

$$w = \frac{2\pi}{\hbar}|K_{FO}|^2 \rho_F, \tag{10}$$

† Here also intermediate states with negative energies can be used in place of those with positive electrons (compare p. 214).

where ρ_F is the number of final states per energy interval dE_F. In the final state we have an electron with momentum $\mathbf{p}$ (energy E) and a light quantum $\mathbf{k}$, $E_F = E+k$. Since $\mathbf{k}$ and $\mathbf{p}$ are independent of each other, the number of final states ρ_F is equal to the product of two density functions ρ_E and ρ_k for the electron and the light quantum, and for a given k, we can put $dE_F = dE$. Thus we have

$$\rho_F = \rho_E \rho_k \, dk = \frac{pE\, d\Omega\, k^2\, d\Omega_k}{(2\pi\hbar c)^6}\, dk. \tag{11}$$

Dividing by the velocity of the incident electron cp_0/E_0, we obtain from (6), (8), (9), (10), (11) the differential cross-section for the process in question:

$$d\phi = \frac{Z^2e^4}{137\pi^2}\frac{pEE_0}{p_0 q^4}\, d\Omega d\Omega_k\, k\, dk \left| \sum \left(\frac{(u^*u')(u'^*\alpha u_0)}{E-E'} + \frac{(u^*\alpha u'')(u''^*u_0)}{E_0-E''} \right) \right|^2. \tag{12}$$

(12) refers to given spin directions before and after the process. Since we are not interested in the direction of the spin, we carry out a summation S over the spin directions of the final state and average ($\frac{1}{2}\mathsf{S}_0$) over the spin directions in the initial state.

We do not derive the polarization of the radiation emitted (see below); we shall therefore also sum (12) over both directions of polarization of $\mathbf{k}$. All these summations can be carried out by the same method as that used in the theory of the Compton effect, § 22. The differential cross-section becomes

$$d\phi = \frac{Z^2e^4}{2\pi 137}\frac{dk}{k}\frac{p}{p_0}\frac{\sin\theta\, d\theta \sin\theta_0\, d\theta_0\, d\phi}{q^4} \times$$
$$\times \left\{ \frac{p^2\sin^2\theta}{(E-p\cos\theta)^2}(4E_0^2-q^2) + \frac{p_0^2\sin^2\theta_0}{(E_0-p_0\cos\theta_0)^2}(4E^2-q^2) - \right.$$
$$-2\frac{pp_0\sin\theta\sin\theta_0\cos\phi}{(E-p\cos\theta)(E_0-p_0\cos\theta_0)}(4E_0E-q^2+2k^2)+$$
$$\left. +2k^2\frac{p^2\sin^2\theta+p_0^2\sin^2\theta_0}{(E-p\cos\theta)(E_0-p_0\cos\theta_0)} \right\}, \tag{13}$$

where θ, θ_0 are the angles between $\mathbf{k}$ and $\mathbf{p}$, $\mathbf{p}_0$ respectively and ϕ is the angle between the planes $(\mathbf{pk})$ and $(\mathbf{p}_0\mathbf{k})$. q is a function of the angles, and according to (7) it is given by

$$q^2 = p_0^2+p^2+k^2-2p_0k\cos\theta_0+2pk\cos\theta-$$
$$-2p_0p(\cos\theta\cos\theta_0+\sin\theta\sin\theta_0\cos\phi). \tag{14}$$

(13) gives the probability that a quantum $\mathbf{k}$ is emitted in a direction

forming an angle θ_0 with the direction of the primary electron and that the electron is scattered in a direction with polar angles θ, ϕ referred to **k**. Before discussing the angular distribution given by (13), we integrate the differential cross-section over all angles and obtain the total cross-section for the emission of a quantum k with an energy between k and $k+dk$. It will be convenient to represent the cross-section as a function of the ratio of k to the initial *kinetic* energy $E_0-\mu$ (i.e. of $k/(E_0-\mu)$, which then ranges from 0 to 1). We therefore define the cross-section ϕ_k for the emission of a quantum k in the range $dk/(E_0-\mu)$ by

$$\phi_k\, d\frac{k}{E_0-\mu} = \int d\phi d\Omega d\Omega_k. \tag{15}$$

The integration of (15) over the angles is elementary but rather tedious. It yields

$$\phi_k\, d\frac{k}{E_0-\mu} = \bar{\phi}\frac{dk}{k}\frac{p}{p_0}\Bigl\{\frac{4}{3}-2E_0 E\frac{p^2+p_0^2}{p^2p_0^2}+\mu^2\Bigl(\frac{\epsilon_0 E}{p_0^3}+\frac{\epsilon E_0}{p^3}-\frac{\epsilon\epsilon_0}{p_0 p}\Bigr)+$$
$$+L\Bigl[\frac{8}{3}\frac{E_0 E}{p_0 p}+\frac{k^2}{p_0^3 p^3}(E_0^2 E^2+p_0^2 p^2)+$$
$$+\frac{\mu^2 k}{2p_0 p}\Bigl(\frac{E_0 E+p_0^2}{p_0^3}\epsilon_0-\frac{E_0 E+p^2}{p^3}\epsilon+\frac{2kE_0 E}{p^2p_0^2}\Bigr)\Bigr]\Bigr\}, \tag{16}$$

where the following abbreviations have been used

$$L = \log\frac{p_0^2+p_0 p-E_0 k}{p_0^2-p_0 p-E_0 k} = 2\log\frac{E_0 E+p_0 p-\mu^2}{\mu k},$$
$$\epsilon_0 = \log\frac{E_0+p_0}{E_0-p_0} = 2\log\frac{E_0+p_0}{\mu},\quad \epsilon = 2\log\frac{E+p}{\mu},\quad \bar{\phi} = \frac{Z^2r_0^2}{137}. \tag{16'}$$

$\bar{\phi}$ is a suitable unit in which to express the cross-section for Bremsstrahlung and similar processes. It is proportional to the *square of the nuclear charge*.

2. *Continuous X-ray spectrum*. The first application that we shall make is to the continuous X-ray spectrum. Here the primary energies are small compared with the rest energy of the electron μ, i.e. we have a non-relativistic problem.

Equating E_0 and E to μ and neglecting all the p's and k compared with μ, we obtain for the differential cross-section (13) simply

$$d\phi = \frac{2Z^2e^4}{\pi 137}\frac{dk}{k}\frac{p}{p_0}\frac{\sin\theta\, d\theta\sin\theta_0\, d\theta_0\, d\phi}{q^4}\times$$
$$\times\{p^2\sin^2\theta+p_0^2\sin^2\theta_0-2pp_0\sin\theta\sin\theta_0\cos\phi\}. \quad \text{N.R.} \tag{17}$$

Since $k = (p_0^2 - p^2)/2\mu$ is small compared with p_0, we may write for q^2, according to (14),

$$q^2 = p^2 + p_0^2 - 2pp_0(\cos\theta\cos\theta_0 + \sin\theta\sin\theta_0\cos\phi)$$
$$= (\mathbf{p}_0 - \mathbf{p})^2. \qquad \text{N.R.} \quad (17')$$

$\mathbf{q}$ does not then depend upon the direction of $\mathbf{k}$.

For a given direction of deflexion (angle between $\mathbf{p}_0$ and $\mathbf{p}$) the maximum intensity is emitted perpendicular to the plane of motion of the electron (($\mathbf{p}_0\mathbf{p}$)-plane). This corresponds to the classical theory, where the maximum intensity is emitted perpendicular to the acceleration. The intensity emitted in a certain direction θ_0 is given by integrating (17) over all directions of $\mathbf{p}$.

The total cross-section for the emission of k is given by (15). In the non-relativistic case we obtain

$$\phi_k d\left(\frac{k}{T_0}\right) = \bar{\phi}\frac{16}{3}\frac{dk}{k}\frac{\mu^2}{p_0^2}\log\frac{p_0+p}{p_0-p} = \bar{\phi}\frac{8}{3}d\left(\frac{k}{T_0}\right)\frac{\mu}{k}\log\frac{\{\sqrt{T_0}+\sqrt{(T_0-k)}\}^2}{k} \qquad \text{N.R.} \quad (18)$$

where $T_0 = E_0 - \mu = p_0/2\mu$ is the kinetic energy of the primary electron. (18) shows that the probability for the emission of a quantum k decreases roughly as $1/k$. At the short wave-length limit $k = p_0^2/2\mu$, ϕ_k vanishes (see below), whereas for very long waves the intensity $k\phi_k$ diverges logarithmically. This, however, is only true in a pure Coulomb field and in Born's approximation. As we shall see in subsection 3, for a screened field, $k\phi_k$ tends to a finite value for $k \to 0$. On the other hand, we shall see (eq. (19)) that ϕ_k remains finite for $k = p_0^2/2\mu$ in an exact theory.

The formulae (17), (18) are only valid as long as the condition (3) for the applicability of Born's approximation is satisfied. For very small energies it is no longer legitimate to compute the matrix elements of V by inserting plane waves for the wave functions of the electron. One has then to use the exact wave functions of the continuous spectrum. This has been done by Sommerfeld (loc. cit.). From his results one can derive the following approximate formula:†

The angular distribution remains the same; the total intensity given by (17) and (18) has to be multiplied by the factor

$$f(\xi,\xi_0) = \frac{\xi}{\xi_0}\frac{1-e^{-2\pi\xi_0}}{1-e^{-2\pi\xi}}, \qquad \xi = \frac{Ze^2}{\hbar v}, \qquad \xi_0 = \frac{Ze^2}{\hbar v_0}. \qquad (19)$$

† G. Elwert, *Ann. Phys.* **34** (1939), 178. For criticism and later work see P. Kirkpatrick and L. Wiedmann, *Phys. Rev.* **67** (1945), 321. The factor of (19) depending on ξ is the square of the exact wave function $|\psi|^2$ for the electron in the final state at the position of the nucleus.

For high v_0 (i.e. for small ξ_0) the factor (19) deviates from the value 1 chiefly at the short wave-length limit. For $p \to 0$ (19) becomes infinite and, as (18) gave zero, the cross-section ϕ_k tends to a finite value. The factor (19) is always larger than unity, because $\xi > \xi_0$, and $x/\{1-\exp(-x)\}$ is monotonic.

In Fig. 12 we have plotted the *intensity* $k\phi_k$ of the emitted radiation in units $E_0\bar{\phi}$ (in the non-relativistic case $E_0 = \mu$) as a function of the ratio $k/(E_0-\mu)$. In the non-relativistic case (formula (18), dotted part of the curve) gives an intensity distribution which is independent of the primary energy T_0. Approaching the region of hard quanta, the Sommerfeld factor (19) leads to a deviation from the curve calculated with Born's approximation, depending upon the primary energy and upon the nuclear charge Z (we disregard, of course, the factor Z^2 contained in the unit $\bar{\phi}$). In the graph the full curve refers to aluminium ($Z = 13$) $T_0/\mu = 0{\cdot}125$ or $2\pi\xi_0 = 1{\cdot}2$. For heavy elements the deviation due to (19) is considerably larger. The intensity is then very much smaller than that given by the dotted curve (except for $k/T_0 \sim 1$).

In the region of soft quanta also the 'exact' curve (19) is extended into a dotted part, because in this region the intensity distribution will be altered by the screening.

For further discussion of the continuous X-ray spectrum, especially of the polarization and angular distribution and of the comparison with the experiment, the reader is referred to Sommerfeld's book.

3. *High energies, effect of screening.* At high energies the maximum in the angular distribution in the forward direction becomes more pronounced. This can clearly be seen if we consider the extreme relativistic case E, $E_0 \gg \mu$. Then p_0 is very nearly equal to E_0 and the denominators $E_0-p_0\cos\theta_0$, etc., become very small for small angles θ_0. Also q has its minimum value for small angles θ_0 and θ. The electron and the light quantum are then both projected in the forward direction within an average angle of the order $\theta \sim \mu/E_0$, as a closer inspection of (13) shows.†

Roughly, the angular distribution is at high energies of the form

$$\phi(\theta_0)\,d\theta_0 = A\frac{\theta_0\,d\theta_0}{[\theta_0^2+(\mu/E_0)^2]^2}\left[\log\left(1+\frac{\theta_0^2\mu^2}{E_0^2}\right)+B\right]. \qquad \text{E.R.} \quad (20)$$

† See also: P. V. C. Hough, *Phys. Rev.* **74** (1948), 80; M. Stearns, ibid. **76** (1949), 836. For the Compton effect the average angle is of the order $\theta \sim (\mu/E_0)^{\frac{1}{2}}$ when the secondary photon k is of order k_0. The reason for the difference in order of magnitude is the fact that the recoil electron takes up energy. This is not the case for the recoil nucleus in Bremsstrahlung (assumed to be infinitely heavy).

At the angle μ/E_0 the emitted radiation is also partially polarized.† The most favoured direction of polarization is perpendicular to the $\mathbf{p}_0, \mathbf{k}$-plane. The ratio of intensities perpendicular and parallel to this plane turns out to be

$$\frac{d\phi_\perp}{d\phi_\parallel} = \frac{E_0^2+E^2}{k^2} \quad \left(\theta_0 \sim \frac{\mu}{E_0}\right). \qquad \text{E.R.} \quad (20')$$

For all other angles θ_0 the ratio is closer to 1. The effect is particularly large when $k \ll E_0$ (but $k \gg \mu$).

The frequency distribution is given by equation (16). In the extreme relativistic case E_0, $E \gg \mu$ this formula becomes

$$\phi_k\, d\left(\frac{k}{E_0}\right) = 2\bar{\phi}\frac{dk}{k}\frac{E}{E_0}\left[\frac{E_0^2+E^2}{E_0 E} - \frac{2}{3}\right]\left[2\log\frac{2E_0 E}{\mu k} - 1\right]. \qquad \text{E.R.} \quad (21)$$

For a given ratio k/E_0 the probability of emission increases roughly with the logarithm of E_0/μ. For small quanta $k \sim 0$ the intensity $k\phi_k$ diverges logarithmically.

The above formulae for the cross-section have been derived on the assumption that the field of the nucleus is a pure Coulomb field. The question arises as to whether the screening of the Coulomb field due to the charge distribution of the outer electrons necessitates any important alterations. To decide this question one would ask, in a classical treatment, whether the field is screened appreciably for those impact parameters r which give the main contribution to the effect. In quantum theory the idea of impact parameter has no exact meaning because the electron is represented by a plane wave. Actually the averaging over all impact parameters is already contained in the integral representing the matrix element of V:

$$V_{F\mathrm{I}} \simeq \int \frac{e^{i(\mathbf{p}'-\mathbf{p},\mathbf{r})/\hbar c}}{r}\, d\tau = \int \frac{e^{i(\mathbf{q}\mathbf{r})/\hbar c}}{r}\, d\tau.$$

We can give a rough meaning to the idea of impact parameter by asking from which distances r the main contributions to the integral arise. They obviously come from a distance

$$r \sim \hbar c/q. \qquad (22)$$

For distances larger than (22) the contribution is small because the exponential function oscillates rapidly in a region where r is practically constant. For smaller distances $d\tau \sim r^2\, dr$ is small. We may therefore consider the quantity (22) as the most important *impact parameter*.

† M. May and G. C. Wick, *Phys. Rev.* **81** (1951), 628.

Now we have just seen that the differential cross-section (13) becomes very large if q is very small. For high primary energies, $E_0 \gg \mu$, the minimum value of q is given by

$$q_{\min} = p_0 - p - k \sim \frac{\mu^2 k}{2E_0 E}. \tag{23}$$

According to (22) therefore, we still obtain a large contribution to the cross-section from distances of the order

$$r_{\max} = \hbar c/q_{\min} = \frac{\hbar}{mc}\frac{2E_0 E}{\mu k} \sim \lambda_0 \frac{E_0 E}{\mu k}. \tag{24}$$

For a given ratio k/E, $r_{\max}$ is larger the *higher* the primary energy. If k is of the order of magnitude of E, $r_{\max}$ becomes of the order of magnitude of the radius of the K-shell for $E_0 \sim 137\mu/Z$. We should expect, therefore, that just for *high energies* the *screening* of the Coulomb field by the outer electrons will lead to a *decrease* in the cross-section. For soft quanta k this will even be the case, according to (24), for somewhat smaller energies.

We can get a rough idea of the effect of screening by considering the case where $r_{\max}$ is large compared with the atomic radius. We shall then call the *screening 'complete'*. For the atomic radius we may assume the value given by the Thomas–Fermi model:

$$a \sim a_0 Z^{-\frac{1}{3}} \sim 137\lambda_0 Z^{-\frac{1}{3}}. \tag{25}$$

(a_0 = Bohr's radius of the hydrogen atom.) If now $r_{\max}$ is large compared with a, we shall certainly obtain the right order of magnitude for the cross-section if we *replace the maximum impact parameter* $r_{\max}$ by the *atomic radius* a (25). In the formula (21) for the frequency distribution $r_{\max}$ occurs under the logarithm. This logarithm has now to be replaced by $\log(137Z^{-\frac{1}{3}})$. For a given value of k/E_0, ϕ_k will then tend to a finite value as $E_0/\mu \to \infty$. Also, for low-energy quanta ($k \sim 0$), the intensity $k\phi_k$ tends to a finite value instead of diverging logarithmically.

Assuming a Thomas–Fermi model for the atom, the exact calculations† lead, for the case of complete screening, to the following formula for the frequency distribution:

$$\phi_k d\left(\frac{k}{E_0}\right) = 2\bar{\phi}\frac{dk}{k}\frac{E}{E_0}\left[\left(\frac{E_0^2+E^2}{E_0 E} - \frac{2}{3}\right)2\log(183Z^{-\frac{1}{3}}) + \frac{2}{9}\right] \quad \text{E.R.} \tag{26}$$

$$\left(\text{for} \frac{E_0 E}{\mu k} \gg a/\lambda_0 = 137Z^{-\frac{1}{3}}\right).$$

† H. Bethe, *Proc. Camb. Phil. Soc.* **30** (1934), 524; H. Bethe and W. Heitler, loc. cit.

If the screening is not complete, the theory gives a continuous transition from formula (21) to formula (26).

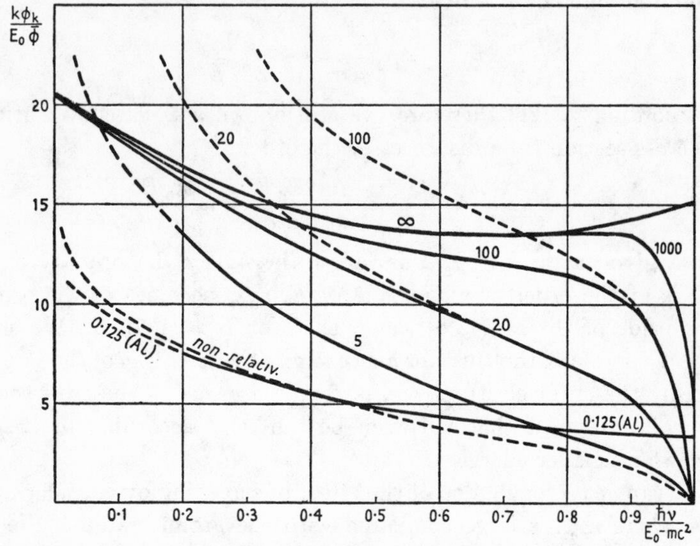

FIG. 12. Intensity distribution of the Bremsstrahlung as a function of $\hbar\nu/(E_0-mc^2)$. The numbers affixed to the curves refer to the primary kinetic energy $E_0-\mu$ in units of $\mu = mc^2$. The dotted portions are calculated using Born's approximation and neglecting the screening (formula (16)), and are valid for all elements. The full curves, where deviating from the dotted curves, are calculated for lead (except that for the non-relativistic case which is calculated for aluminium). The deviations represent (i) at high energies and for soft quanta the effect of screening, (ii) for small energies (non-relativistic curve) the effect of the Sommerfeld factor (19). Units: $\bar{\phi} = Z^2 r_0^2/137$.

Since ϕ_k is roughly proportional to E_0/k we have plotted in Fig. 12 the intensity $k\phi_k$ in units $E_0\bar{\phi}$ for various primary kinetic energies $E_0-\mu$. The dotted curves (screening neglected) are valid for all elements: Z is contained in the quantity $\bar{\phi}$ only. The full curves are calculated with screening for lead ($Z = 82$), except that for non-relativistic energies which is valid for aluminium; they approach the curve for complete screening $E_0 \sim \infty$ (formula (26)) in the region of low-energy quanta. Here, also, the non-relativistic curve would tend to a finite value if the screening were taken into account. In the region of high-energy quanta the screening has little influence. On the other hand, at the limit for high-energy quanta ($k = E_0-\mu$) the curves would probably tend to a finite value if the correct wave functions of the continuous spectrum were used, as is the case in Sommerfeld's exact theory for non-relativistic energies.

For light elements the screening becomes less appreciable. For heavy elements the use of Born's approximation leads to some error even for medium and soft quanta (compare subsection 5).

As we see from the graph the intensity distribution is roughly uniform over the whole frequency range.

A particular difficulty seems to arise when we consider the cross-section rather than the energy distribution. This behaves like dk/k for small k. One would think that the integral over k gives rise to a correction to the scattering of electrons in a Coulomb field (§ 24) due to emission of very soft photons. This correction would diverge logarithmically. A similar difficulty was encountered for the double Compton (§ 23) effect and the solution of the problem is also the same. There are further corrections to the radiation-less scattering arising from the higher approximations of radiation theory and these cancel the diverging part exactly. This infra-red problem will be treated in detail in connexion with the Compton and double Compton effect in § 33.

4. *Energy loss.* An appreciable fraction of the energy lost by an electron in passing through matter is due to Bremsstrahlung with k of the order of magnitude of the kinetic energy E_0. The average energy lost in one collision may be obtained by integrating the intensity $k\phi_k$ over all frequencies from 0 to $E_0-\mu$. The average energy lost per cm. path is given by

$$-\frac{dE_0}{dx} = N\int\limits_0^1 k\phi_k\, d\Big(\frac{k}{E_0-\mu}\Big), \tag{27}$$

where N is the number of atoms per cm.^{3} Fig. 12 shows that the area under the curves is of the same order of magnitude for all energies. Since in Fig. 12 we have plotted the cross-section divided by E_0, the energy lost per cm. will be roughly proportional to the primary energy E_0 for high energies $E_0 \gg \mu$, and constant for small kinetic energies $E_0-\mu \ll \mu$. It is therefore convenient to define a cross-section, ϕ_{rad}, for the energy lost by radiation:

$$-\frac{dE_0}{dx} = NE_0\phi_{\text{rad}}, \qquad \phi_{\text{rad}} = \frac{1}{E_0}\int\limits_0^1 k\phi_k\, d\Big(\frac{k}{E_0-\mu}\Big). \tag{28}$$

From (16) we obtain,† after an elementary but rather tedious calculation,

$$\phi_{\text{rad}} = \bar{\phi}\Big\{\frac{12E_0^2+4\mu^2}{3E_0 p_0}\log\frac{E_0+p_0}{\mu} - \frac{(8E_0+6p_0)\mu^2}{3E_0 p_0^2}\Big(\log\frac{E_0+p_0}{\mu}\Big)^2 - \\ -\frac{4}{3}+\frac{2\mu^2}{E_0 p_0}F\Big(\frac{2p_0(E_0+p_0)}{\mu^2}\Big)\Big\}, \tag{29}$$

† G. Racah, *Nuovo Cimento*, **11** (1934), N. 7.

where the function F is defined by the integral†

$$F(x) = \int\limits_0^x \frac{\log(1+y)}{y}\,dy. \tag{29'}$$

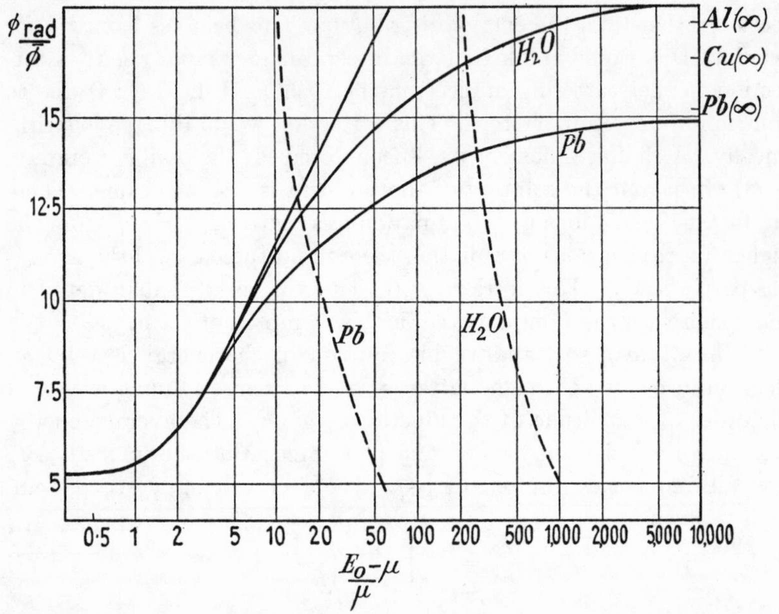

FIG. 13. Cross-section ϕ_{rad} for the energy loss of an electron per cm. path by radiation (ϕ_{rad} is defined in equation (28)). Born approximation. Units $\bar{\phi} = r_0^2 Z^2/137$. The straight line at high energies is calculated neglecting the effect of screening and is valid for all elements. The dotted curves show in the same units the energy loss by inelastic collisions for various materials. On the right: asymptotic values for Al, Cu, Pb.

For small x the function F can be expanded in a power series:

$$F(x) = x - \frac{x^2}{4} + \frac{x^3}{9} - \frac{x^4}{16} + \dots. \tag{30}$$

For large x one can use the (exact) formula:

$$F(x) = \tfrac{1}{6}\pi^2 + \tfrac{1}{2}(\log x)^2 - F(1/x). \tag{31}$$

From (29), (30), and (31) we obtain for the two limiting cases

$$\phi_{rad} = \tfrac{16}{3}\bar{\phi}, \qquad \text{N.R.} \quad (32)$$

$$\phi_{rad} = 4\left(\log\frac{2E_0}{\mu} - \frac{1}{3}\right)\bar{\phi}. \qquad \text{E.R.} \quad (33)$$

† This function is tabulated: cf. E. O. Powell, *Phil. Mag.* **34** (1943), 600; K. Mitchell, ibid. **40** (1949), 351.

For small energies the average energy radiated is independent of the primary energy. For high energies the ratio of the radiated energy to the initial energy increases logarithmically with E_0. This, however, is only true if the screening is neglected. In the case of complete screening we obtain from equation (26) by integration

$$\phi_{\mathrm{rad}} = \bar{\phi}\{4\log(183Z^{-\frac{1}{3}})+\tfrac{2}{9}\} \qquad \text{E.R.} \quad (34)$$

$$(\text{for } E_0 \gg 137\mu Z^{-\frac{1}{3}}).$$

ϕ_{rad} is then constant.

The cross-section ϕ_{rad} for energy loss by emission of radiation is plotted in Fig. 13 in a logarithmic scale. Formula (33) for high energies with neglect of screening gives a straight line. The curves with screening are obtained from Fig. 12 by numerical integration. They tend to the asymptotic value given by equation (34).

The curves are quantitatively correct for light elements. For heavy elements we must bear in mind that the use of Born's approximation leads to some error in the numerical values, but the error is not large, even for lead (see subsection 5).

The values of $\phi_{\mathrm{rad}}/\bar{\phi}$ are given in the following table:

TABLE V

Cross-section for the energy loss by radiation. (Born approximation)

<table>
<tr><th colspan="2">$\frac{E_0-mc^2}{mc^2}$</th><th>0</th><th>1</th><th>2</th><th>5</th><th>10</th><th>20</th><th>50</th><th>100</th><th>200</th><th>1000</th><th>∞</th></tr>
<tr><td rowspan="2">$\frac{\phi_{\mathrm{rad}}}{\bar{\phi}}$</td><td>$H_2O$</td><td rowspan="2">5·33</td><td rowspan="2">5·5</td><td rowspan="2">6·5</td><td>9·1</td><td>11·2</td><td>12·9</td><td>14·6</td><td>15·6</td><td>16·4</td><td>17·5</td><td>18·3</td></tr>
<tr><td>Pb</td><td>8·75</td><td>10·3</td><td>11·4</td><td>12·6</td><td>13·3</td><td>13·8</td><td>14·5</td><td>15·2</td></tr>
</table>

To give a rough idea of the practical significance of our results, in Fig. 13 we have plotted for comparison in the same units the cross-section for the ordinary energy loss of an electron by inelastic collisions (ionization of atoms). The general question of the energy loss of particles in passing through matter will be considered in § 37 in context. Here we only mention that the energy loss per cm. by collisions is proportional to Z, whereas our effect is proportional to Z^2 and roughly independent of the energy. The ratio of the energy loss to the primary energy (which is plotted in Fig. 13) *decreases* therefore as $1/E_0$ whereas the same quantity for the energy loss by radiation *increases* logarithmically. We obtain, therefore, the striking result, that for *energies higher than a certain limit, the energy loss is almost entirely due to emission of radiation* and reaches a *value which is much higher than that due to ionization.* For lead this limit lies at about $20mc^2$, for water at $250mc^2$.

5. *Corrections, experimental checks.* All the above results are approximate in that (i) the Born approximation has been used. Minor corrections are due to the fact that (ii) the nucleus is not infinitely heavy but can take up some recoil energy and (iii) is not a point charge but has a finite extension. Finally, it ought to be borne in mind that (iv) all higher order corrections of radiation theory (including damping effects) have been neglected. It will be shown in Chapter VI that the corrections (iv) are likely to be very small, at most a fraction of a per cent. The main correction is (i). It will also be seen below that the effect of recoil energy is likewise very small. The finite size of the nucleus makes itself felt only when q (cf. (22)) is large, i.e. at large angles θ, θ_0. Here this effect is large but for the integrated cross-section the finite size of the nucleus makes very little difference.

According to subsection 2 the departure from the Born approximation is large for heavy elements and low energies. One may expect from (19) that for large Z this continues to be the case at high energies also. However, a proper treatment shows that this is true only when the angle of emission is large, whereas the corrections are small for the main angle of order μ/E_0. In the E.R. region the results are the following†: The correction is essential only for small impact parameters (large q) where the screening is negligible. The screening effect and the departure from the Born approximation are therefore additive. The correction to (21) or (26) is an additional term

$$\phi'_k d\left(\frac{k}{E_0}\right) = -2\bar{\phi}\frac{dk}{k}\frac{E}{E_0}\left(\frac{E_0^2+E^2}{E_0 E}-\frac{2}{3}\right)Q(Z), \tag{35}$$

$$Q = 2{\cdot}414(Z/137)^2 \quad (Z/137 \ll 1); \qquad Q = 0{\cdot}67 \text{ (Pb)}.$$

The correction is $\sim Z^4$ and has the *opposite sign* of that in the N.R. region where (19) is larger than unity. The correction to ϕ_{rad} ((33) or (34)) is $\phi'_{\text{rad}} = -2\bar{\phi}Q(Z)$, which amounts to about 9 per cent. for Pb.

When considering the passage of an electron through an atom one must finally take into account the fact that the electron may also collide with the atomic electrons and produce Bremsstrahlung. We shall consider the emission of Bremsstrahlung in electron-electron collisions in appendix 6 by a simplified method. The chief theoretical

† L. C. Maximon and H. A. Bethe; H. Davies and H. A. Bethe, *Phys. Rev.* **87** (1952), 156.

point is the fact that the recoil electron can take up a large fraction of energy and momentum (a fact which makes an exact calculation exceedingly laborious). It is shown, however, that the final formula for electron-electron collisions for $k \gg \mu$ is hardly different from that derived in subsections 1–4 (except, of course, that the factor Z^2 is missing), although the cross-section is probably slightly smaller. This must be understood in the sense that large momentum transfers to the

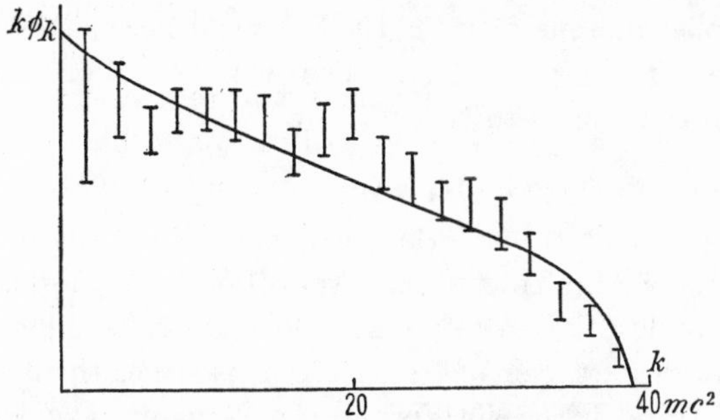

FIG. 14. Relative energy distribution of Bremsstrahlung following the passage of 19·5 MeV electrons through a Pt target. Measurements and theory.

electron originally at rest are, though possible, rare and contribute little to the total cross-section. This also shows that the neglect of the recoil in the above treatment does not cause any noticeable error. The contribution of the atomic electrons to Bremsstrahlung can then be taken into account with sufficient accuracy, by simply replacing the factor Z^2 by $Z(Z+A)$ where A is close to 1 (probably slightly smaller; compare also § 26.2). Thus merely the unit $\bar{\phi}$ is changed.

Measurements of Bremsstrahlung at high energies are not very plentiful. The chief verification of our results lies in the phenomena connected with the penetration of very fast electrons through matter and the ensuing cascade showers, discussed in Chapter VII. The experiments, though qualitatively in full agreement with the predictions, are not very accurate yet. We mention, as an attempt at a quantitative check of the theory a determination of the energy spectrum of the γ-rays produced in Pt by electrons of energy 19·5 MeV $= 39\,\mu$.† The γ-rays were observed in the forward direction (within a small solid

† H. W. Koch and R. E. Carter, *Phys. Rev.* **77** (1950), 165. At 330 MeV it has been shown that the theory agrees at least roughly with the experiments: W. Blocker, R. W. Kenney, and W. K. H. Panofsky, ibid. **79** (1950), 419.

angle). The results cannot be compared directly with the curves of Fig. 12 because the latter refer to the integral over the whole solid angle. The intensity distribution to be expected under the experimental conditions was calculated and is plotted in Fig. 14. It is similar to the curves of Fig. 12. In the figure the results of measurements together with their statistical standard errors are shown. It is seen that the agreement is, on the whole, good, and within the rather large errors of the experiment.

At about the same energy ($34\,\mu$) the Z-dependence has been measured† and found to be in very good agreement with the theory if the electron contribution is assumed to be $A = 0{\cdot}75$.

26. Creation of positive electrons

According to § 11 the creation of a pair of positive and negative electrons must be interpreted as a transition of an ordinary electron from a state of negative energy to a state of positive energy. The energy necessary to create a pair of *free* electrons is larger than $2mc^2$. It can be supplied through the absorption of a γ-quantum or by impact of a particle with kinetic energy greater than $2mc^2$. Energy and momentum conservation, however, are only possible if another particle is present (for instance, a nucleus). Thus pairs will be created by γ-rays or fast particles in passing through matter. We consider first the most important case; that is:

1. *Creation of pairs by γ-rays in the presence of a nucleus with charge* Z. Denoting the energy and momentum of the two electrons by E_+, $\mathbf{p}_+$, E_-, $\mathbf{p}_-$, the process in question is the following: a γ-quantum passing through the Coulomb field of the nucleus is absorbed by an electron in the negative energy state $E = -E_+$, $\mathbf{p} = -\mathbf{p}_+$, the electron going over into a state of positive energy E_-, $\mathbf{p}_-$.

This process is closely allied to Bremsstrahlung (§ 25). The reverse process to the creation of a pair is obviously the transition of an ordinary electron in the presence of a nucleus from a state with energy $E_0 = E_-$ to a state $E = -E_+$ emitting a light quantum

$$k = E_0 - E = E_+ + E_-. \tag{1}$$

This process differs from ordinary Bremsstrahlung only in that the energy in the final state is negative. Now, the matrix elements for the reverse process are the conjugate complex expressions of those for the direct process. We can therefore take the cross-section for the creation

† L. H. Lanzl and A. O. Hanson, *Phys. Rev.* **83** (1951), 959.

of a pair directly from the calculations of § 25. In calculating this, however, we must insert another density function ρ_F for the final state. Since in the case of pair production we have in the final state a positive and a negative electron instead of an electron and a light quantum, our density function is given by

$$\rho_F = \rho_{E_+}\rho_{E_-}\,dE_+, \tag{2}$$

instead of $\rho_E\,\rho_k\,dk$ as in § 25. Furthermore, we have now to divide by the velocity of the incident light quantum (i.e. by c), instead of by the velocity, v_0, of the incident electron. Thus formula (13) § 25 for the differential cross-section has to be multiplied by

$$\frac{\rho_{E_+}\rho_{E_-}\,dE_+}{\rho_E\,\rho_k\,dk}\frac{p_0}{E_0} = \frac{p_-^2\,dE_+}{k^2\,dk}. \tag{3}$$

Since $\mathbf{p}_0 = \mathbf{p}_-$, $\mathbf{p} = -\mathbf{p}_+$, the angles θ, θ_0, ϕ, denoting the direction of the electron in the initial and final state, are connected with the angles θ_+, θ_-, ϕ_+, denoting the direction of the positive and negative electron, by:

$$\theta_+ = \pi-\theta, \qquad \theta_- = \theta_0, \qquad \phi_+ = \pi+\phi, \tag{4}$$

$\theta_\pm = \angle\,(\mathbf{kp}_\pm)$, $\phi_+ = \angle$ between $(\mathbf{kp}_+)$-plane and $(\mathbf{kp}_-)$-plane).

Then putting

$$E_0 = E_-, \qquad E = -E_+, \qquad p_0 = p_-, \qquad p = p_+, \tag{5}$$

and inserting (3)–(5) in the formula (13) § 25 we obtain the differential cross-section for the creation of a pair $\mathbf{p}_+$, $\mathbf{p}_-$:

$$d\phi = -\frac{Z^2}{137}\frac{e^4}{2\pi}\frac{p_+p_-\,dE_+}{k^3}\frac{\sin\theta_+\sin\theta_-\,d\theta_+d\theta_-d\phi_+}{q^4}\times$$

$$\times\left\{\frac{p_+^2\sin^2\theta_+}{(E_+-p_+\cos\theta_+)^2}(4E_-^2-q^2)+\frac{p_-^2\sin^2\theta_-}{(E_--p_-\cos\theta_-)^2}(4E_+^2-q^2)+\right.$$

$$+\frac{2p_+p_-\sin\theta_+\sin\theta_-\cos\phi_+}{(E_--p_-\cos\theta_-)(E_+-p_+\cos\theta_+)}(4E_+E_-+q^2-2k^2)-$$

$$\left.-2k^2\frac{p_+^2\sin^2\theta_++p_-^2\sin^2\theta_-}{(E_--p_-\cos\theta_-)(E_+-p_+\cos\theta_+)}\right\}, \tag{6}$$

$$q^2 = (\mathbf{k}-\mathbf{p}_+-\mathbf{p}_-)^2. \tag{7}$$

The integration over the angles is also just the same as for Bremsstrahlung. The cross-section for the creation of a positive electron with energy E_+ and a negative one with energy E_- then becomes:†

$$\phi_{E_+}\,dE_+ = \bar{\phi}\frac{p_+p_-}{k^3}\,dE_+\Big\{-\frac{4}{3}-2E_+E_-\frac{p_+^2+p_-^2}{p_+^2p_-^2}+$$
$$+\mu^2\Big(\frac{E_+\epsilon_-}{p_-^3}+\frac{\epsilon_+E_-}{p_+^3}-\frac{\epsilon_+\epsilon_-}{p_+p_-}\Big)+L\Big[\frac{k^2}{p_+^3p_-^3}(E_+^2E_-^2+p_+^2p_-^2)-$$
$$-\frac{8}{3}\frac{E_+E_-}{p_+p_-}-\frac{\mu^2k}{2p_+p_-}\Big(\frac{E_+E_--p_-^2}{p_-^3}\epsilon_-+\frac{E_+E_--p_+^2}{p_+^3}\epsilon_++\frac{2kE_+E_-}{p_+^2p_-^2}\Big)\Big]\Big\}, \tag{8}$$

$$\epsilon_+ = 2\log\frac{E_++p_+}{\mu},\qquad L = 2\log\frac{E_+E_-+p_+p_-+\mu^2}{\mu k}, \tag{8'}$$
$$\bar{\phi} = Z^2r_0^2/137.$$

In the extreme relativistic case where all energies are large compared with the rest energy of the electron (8) becomes

$$\phi_{E_+}\,dE_+ = 4\bar{\phi}\,dE_+\frac{E_+^2+E_-^2+\frac{2}{3}E_+E_-}{k^3}\Big(\log\frac{2E_+E_-}{k\mu}-\tfrac{1}{2}\Big).\qquad \text{E.R.} \tag{9}$$

Formula (8) (and (9)) is symmetrical as between the positive and negative electrons. This is a consequence of the use of *Born's approximation*, on which the calculations of § 25 were based. In this approximation V occurs squared and the sign of the charge disappears.

The *limits of validity* of (8) and (9) are the same as for the equations (16), (21) in § 25, viz.:

(1) The velocities of both electrons v_+ and v_- must be so large and the nuclear charge Z so small that

$$2\pi\frac{Ze^2}{\hbar v_+},\qquad 2\pi\frac{Ze^2}{\hbar v_-} \ll 1 \tag{10}$$

(condition for the applicability of Born's approximation).

(2) On the other hand, the energies of both electrons must not be so high that the *screening* of the Coulomb field by the outer electrons becomes effective, viz.:

$$\frac{2E_+E}{k\mu} \ll 137Z^{-\frac{1}{3}}. \tag{11}$$

If (10) is not satisfied, the matrix elements have to be computed using the exact wave functions of the continuous spectrum. In

† H. Bethe and W. Heitler, *Proc. Roy. Soc.* A, **146** (1934), 83; G. Racah, *Nuov. Cim.* **11** (1934), No. 7; **13** (1936), 69.

the non-relativistic approximation (i.e. if v_+, $v_- \ll c$) the exact formula† differs from (8) approximately by the factor

$$f(\xi_+, \xi_-) = \frac{2\pi\xi_+ \, 2\pi\xi_-}{(e^{2\pi\xi_+}-1)(1-e^{-2\pi\xi_-})}, \qquad \text{N.R.} \quad (12)$$

$$\xi_+ = \frac{Ze^2}{\hbar v_+} = \frac{Z\mu}{137p_+}.$$

The factor (12) destroys the symmetry in E_+ and E_-; this is because the positive electron is repelled and the negative electron attracted by the nucleus. The probability for pair creation becomes small for small p_+ and large for small p_-. A similar correction must be made in the relativistic case for *heavy elements*. It is of the same type and order of magnitude as for Bremsstrahlung (§ 25.5). The correction is large when k is a few mc^2 but decreases (in per cent.) and then changes sign for increasing energies. In the E.R. case the correction to (9) or (13) is‡

$$\phi'_{E_+} dE_+ = -2\bar{\phi}\, dE_+ \frac{E_+^2+E_-^2+\frac{2}{3}E_+E_-}{k^3} Q(Z), \quad \text{E.R.} \quad (12')$$

where $Q(Z)$ is given by § 25 eq. (35).

The screening is only effective if both electrons have energies that are large compared with mc^2. In the case of complete screening, i.e. if $2E_+E_-/k\mu \gg 137Z^{-\frac{1}{3}}$, we obtain a formula§ corresponding to eq. (26) § 25

$$\phi_{E_+} dE_+ = 4\bar{\phi}\, dE_+ \left\{ \frac{E_+^2+E_-^2+\frac{2}{3}E_+E_-}{k^3} \log(183Z^{-\frac{1}{3}}) - \frac{1}{9}\frac{E_+E_-}{k^3} \right\}.$$

$$\text{E.R.} \quad (13)$$

2. *Discussion. Total number of pairs.* The angular distribution for both electrons is again given by a formula of the type (20) § 25, when E_+, $E_- \gg mc^2$. The average angle into which the electrons are emitted is of the order $\theta \sim mc^2/k$. For smaller energies the concentration in the forward direction is less marked.

An interesting point is the recoil transferred to the nucleus. Although we have neglected the recoil energy throughout it is clear that the *recoil momentum*, which is $\mathbf{q}$ eq. (7), is not necessarily small. One can express the cross-section in terms of q and the angle θ_q between $\mathbf{q}$ and $\mathbf{k}$, instead of by θ_+, θ_-, and thus obtain the distribution in magnitude

† Y. Nishina, S. Tomonaga, S. Sakata, *Scient. Pap. Inst. Phys. and Chem. Research, Japan*, **24** (1934), No. 17.

‡ L. C. Maximon and H. A. Bethe; H. Davies and H. A. Bethe, *Phys. Rev.* **87** (1952), 156.

§ H. Bethe, *Proc. Camb. Phil. Soc.* **30** (1934), 524.

and direction of the recoil momentum.† The result is shown in Fig. 15. The recoil momentum is in most cases less than μ and the direction of $\mathbf{q}$ is mainly sideways. For Bremsstrahlung the result would be quite similar. The smallness of the recoil is quite in accord with the considerations in appendix 6.

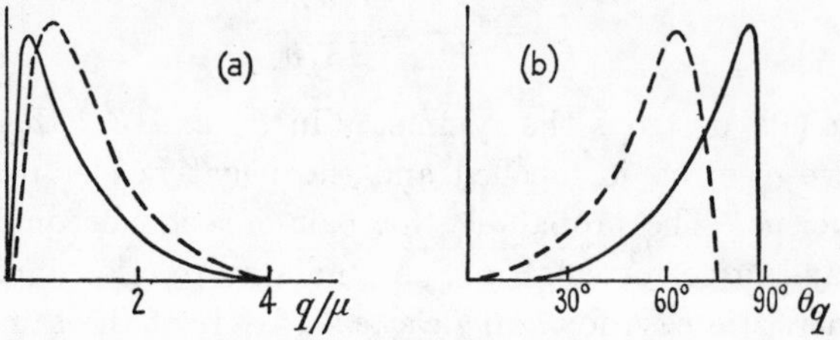

FIG. 15. Pair creation: distribution of (*a*) the recoil momentum of the nucleus q and (*b*) the direction of q against that of the incident photon $\mathbf{k}$. Full-drawn curves for $k = 33\mu$, dotted curves for $k = 8{\cdot}2\mu$.

The energy distribution as given by the formulae (8), (9), (13) is shown in Fig. 16. For convenience of representation we have plotted the cross-section in units $\bar{\phi} = Z^2r_0^2/137$ as a function of the *kinetic* energy of the positive electron divided by the total kinetic energy $k-2mc^2$.

For light quanta of small energy the distribution has a broad maximum if both electrons receive the same amount of energy. For higher energies the maximum becomes flatter. For very high energies the distribution has two maxima if one of the electrons obtains a very small and the other one a very large energy. Finally, the distribution tends to an asymptotic curve (∞) represented by formula (13).

The symmetry in E_+ and E_- is due to the application of Born's approximation. In an exact calculation the maximum of the distribution would be displaced to the right-hand side. This displacement is greatest for high nuclear charge Z and small k.

The total number of pairs created is found by integration of (8)–(13) over all possible energies of the positive electron. In the extreme relativistic case, if the screening is either negligible or complete, we obtain

$$\phi_{\text{pair}} = \bar{\phi}\left(\frac{28}{9}\log\frac{2k}{\mu}-\frac{218}{27}\right); \qquad \text{E.R.} \quad (14)$$

$$\phi_{\text{pair}} = \bar{\phi}\left(\frac{28}{9}\log(183Z^{-\frac{1}{3}})-\frac{2}{27}\right). \qquad \text{E.R.} \quad (15)$$

† R. Jost, J. M. Luttinger, and M. Slotnick, *Phys. Rev.* **80** (1950), 189. Attempts at measuring the recoil distribution have been made by G. E. Modesitt and H. W. Koch, ibid. **77** (1950), 175.

For small energies and for energies where the screening is not complete the integration was carried out numerically.

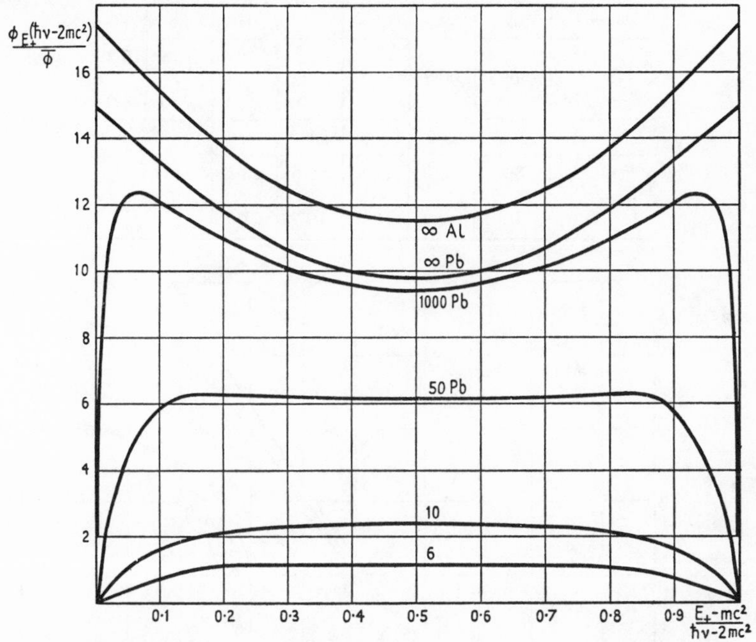

FIG. 16. Energy distribution of pairs (positive and negative electrons). $\phi_{E+}\, dE_+$ is the cross-section for the creation of a positive electron with an energy between E_+ and E_++dE_+. The numbers affixed to the curves refer to the energy of the primary quantum k in units of mc^2. The curves for $k = 6mc^2$ and $k = 10mc^2$ are valid for any element (screening neglected). The rest are calculated for lead (for $k = \infty$ also for aluminium). Units: $\bar{\phi} = Z^2r_0^2/137$. Born approximation.

The results are shown in Fig. 17 where we have plotted in a logarithmic scale the total cross-section ϕ_{pair}, in units $\bar{\phi}$, as a function of $\hbar\nu$. Fig. 17 shows that the probability of the pair formation increases rapidly with $\hbar\nu$ until it reaches a constant value at very high energies. It is very nearly proportional to Z^2 (because $\bar{\phi} \sim Z^2$). The values of $\phi_{\text{pair}}/\bar{\phi}$ are given in Table VI. These results refer to the Born approximation.

We can compare the probability for the creation of a pair by a γ-quantum k in the field of a nucleus with charge Z with the probability, for the same γ-quantum, of Compton scattering by the extra nuclear electrons. The latter is given by the Klein–Nishina formula § 22 eq. (45) multiplied by the number of electrons Z. We have therefore also plotted in Fig. 17 the cross-section for the Compton scattering in the same units $\bar{\phi}$. In these units we obtain, of course, different curves for different

TABLE VI

Cross-section for the creation of pairs by γ-rays (Born approximation)

<table>
<tr><td colspan="2">$\hbar\nu/mc^2$</td><td>3</td><td>4</td><td>5</td><td>6</td><td>10</td><td>20</td><td>50</td></tr>
<tr><td rowspan="2">$\dfrac{\phi_{\text{pair}}}{\bar{\phi}}$</td><td>Al</td><td rowspan="2">0·085</td><td rowspan="2">0·32</td><td rowspan="2">0·61</td><td rowspan="2">0·89</td><td rowspan="2">1·94</td><td>3·75</td><td>6·2</td></tr>
<tr><td>Pb</td><td>3·60</td><td>6·0</td></tr>
<tr><td colspan="2">$\hbar\nu/mc^2$</td><td>100</td><td>200</td><td>500</td><td>1000</td><td>∞</td><td></td><td></td></tr>
<tr><td rowspan="2">$\dfrac{\phi_{\text{pair}}}{\bar{\phi}}$</td><td>Al</td><td>8·2</td><td>10·0</td><td>11·8</td><td>12·6</td><td>13·4</td><td></td><td></td></tr>
<tr><td>Pb</td><td>7·7</td><td>9·0</td><td>10·3</td><td>10·7</td><td>11·5</td><td></td><td></td></tr>
</table>

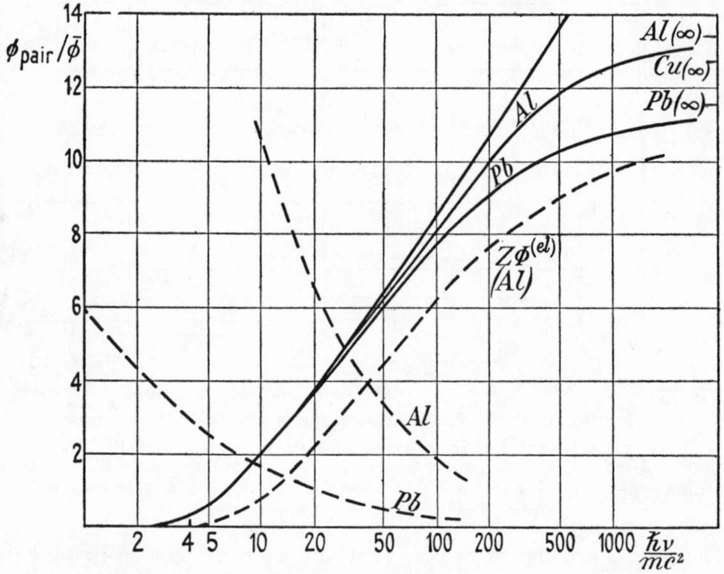

FIG. 17. Cross-section for the production of pairs in units $\bar{\phi} = Z^2 r_0^2/137$ as a function of the primary frequency. (Born approximation.) The straight line continuation for high energies corresponds to the neglect of screening. On the right asymptotic values for Al, Cu, Pb. The dotted curves show the total cross-section of the atom for Compton scattering, in the same units. The curve marked $Z\phi^{(el)}$ is a very rough estimate of the contribution of the atomic electrons (multiplied by Z) for Al.

elements. As may be seen from the graph, the behaviour of the cross-section for pair formation is entirely different from that for Compton scattering. For small energies the probability of pair formation is generally much smaller than that for Compton scattering, whereas for high energies the pair formation becomes much more frequent than the scattering. The energy at which both effects are equally probable depends on Z and is roughly equal to $10mc^2$ for Pb and $30mc^2$ for Al.

The corrections to be made when exact wave functions are used instead of the Born approximation can be seen from Table VII.

TABLE VII

Exact cross-sections and distribution shift for pair creation (Pb)
(Units $\bar{\phi}$)

k/mc^2	*Exact*	*Born app.*	$(\bar{E}_+-\mu)/(\bar{E}_--\mu)$
3	0·17	0·085	2·0
5·2	0·73	0·64	1·4

where the numerical calculations for two fairly low energies are quoted† (for Pb). The decrease of the correction with increasing energy is evident. To show the shift of the energy distribution in favour of larger positron energies the ratio of the average kinetic energies of the positive and negative electron is also given. This ratio would be unity in the Born approximation and it is seen that it approaches this value when k increases.

For very high energies we obtain from (12′) a relatively small correction to (14) or (15) for the total number of pairs:

$$\phi'_{\text{pair}} = -\tfrac{14}{9}\bar{\phi}Q(Z), \qquad \text{E.R.} \quad (15')$$

which amounts to a *decrease* of about 10 per cent. in Pb.

As in the case of Bremsstrahlung, the atomic electrons contribute to pair creation. It can easily be seen from the conservation laws that the threshold value of k for producing a pair in the field of a free electron is 4μ instead of 2μ. After the collision we have one positive and two negative electrons. The cross-section has been worked out by direct methods‡ for some energy regions and can also be derived in the manner shown in appendix 6. The cross-section per electron is

$$\phi^{(\text{el})}_{\text{pair}} = \frac{r_0^2}{137}\left[\frac{28}{9}\log\frac{2k}{\mu} - 11{\cdot}3\right]. \qquad (16)$$

The additive constant (−11·3) is somewhat uncertain. (16) is very similar to (14) for $Z = 1$. The difference lies in the larger numerical value of the negative constant (−8·1 in (14)). It appears that the electron contribution (per electron) is slightly smaller than that of the nuclear contribution (divided by Z^2).

† H. R. Hulme and J. C. Jaeger, *Proc. Roy. Soc.* **153** (1936), 443.
‡ V. Votruba, *Bull. int. Acad. tschèque des sciences*, **49** (1948), No. 4; *Phys. Rev.* **73** (1948), 1468. Cf. also J. A. Wheeler and W. E. Lamb, ibid. **55** (1939), 858.

Also the Coulomb field of an atomic electron is screened at a distance comparable with that of the screening distance of the nuclear field. We shall not be far wrong if we assume a formula for complete screening by replacing $\log 2k/\mu$ by the Z-dependent log of (15). Thus $\phi_{\text{pair}}^{(\text{el})}$ will be smaller than (15) (divided by Z^2) by an additive constant -3, or so. For complete screening the total electron contribution is then

$$Z\phi^{(\text{el})}/\phi_{\text{nucleus}} \sim A/Z, \qquad A \sim 0{\cdot}7\text{–}0{\cdot}8. \tag{17}$$

We can account for the electron contribution, as in the case of Bremsstrahlung, by replacing the factor Z^2 in the unit $\bar{\phi}$ by $Z(Z+A)$. The value of A, however, is not very reliable. For smaller energies A must decrease, and in particular become zero at $k = 4\mu$. Interpolating roughly, we may draw a curve for the energy dependence of $\phi_{\text{pair}}^{(\text{el})}$. It is also shown in Fig. 17, but this is more qualitative than quantitative.

3. *Pair creation by charged particles.* Electron pairs can also be created in a collision between any two charged particles with sufficient energy. In the following we quote the theoretical results for various cases of interest. With the exception of (*a*) the results have been derived by the method of appendix 6 or similar considerations.

(*a*) A *heavy particle* (mass $M_0 \gg \mu$, charge Z_0) collides with a *heavy particle* (M, Z) or atom at rest. The kinetic energy T_0 is small compared with its rest energy $T_0 \ll M_0 c^2$. The total cross-section for the creation of a pair of any energy is of the order†

$$\phi \sim \left(\frac{ZZ_0 r_0}{137}\right)^2 \frac{\mu^2}{M_0 c^2 T_0} \left(\frac{ZM_0 - Z_0 M}{M}\right)^2 \quad (T_0 \ll M_0 c^2). \tag{18}$$

This decreases with increasing T_0 and the numerical value is very small. However, when E_0 exceeds $M_0 c^2$, the situation changes, and the cross-section rises again. It assumes a fairly large value when $E_0 \gg Mc^2$.

(*b*) A heavy particle $(M_0 \gg \mu)$ collides with another heavy particle at rest, but

$$E_0 \gg M_0 c^2.$$

The total cross-section is given by‡

$$\phi = \frac{28}{27\pi} \left(\frac{ZZ_0 r_0}{137}\right)^2 \log^3 \frac{\beta E_0}{M_0 c^2}, \tag{19}$$

where β is a constant of order unity and terms of the type $C \log \beta' E_0/M_0 c^2$

† W. Heitler and L. Nordheim, *J. d. Phys.* **5** (1934), 449; E. Lifshitz, *Phys. Zs. Sov. Un.* **7** (1935), 385.

‡ H. J. Bhabha, *Proc. Roy. Soc.* A, **152** (1935), 559; *Proc. Camb. Phil. Soc.* **31** (1935), 394; L. Landau and E. Lifshitz, *Phys. Zs. Sov. Un.* **6** (1934), 244; Y. Nishina, S. Tomonaga, and M. Kobayasi, *Sci. Pap. Inst. Phys. Chem. Research, Japan,* **27** (1935), 137; E. J. Williams, *Kgl. Dansk. Vid. Selsk.* **13** (1935), No. 4.

have been neglected. If the particle at rest is an atom with a screened field this formula is only valid for not too high energies,

$$E_0/M_0c^2 < 137Z^{-\frac{1}{3}}.$$

For complete screening one finds approximately:

$$\phi = \frac{28}{27\pi}\left(\frac{ZZ_0 r_0}{137}\right)^2 \log\frac{137}{Z^{\frac{1}{3}}}\left[3\log\frac{\beta E_0}{M_0c^2}\log\frac{\beta E_0 Z^{\frac{1}{3}}}{M_0c^2.137}+\log^2\frac{137}{Z^{\frac{1}{3}}}\right]$$

$$(E_0/M_0c^2 > 137Z^{-\frac{1}{3}}). \quad (20)$$

It is seen that ϕ increases first as $\log^3 E_0$ and, when screening is effective, like $\log^2 E_0$. β cannot be determined by the methods used but the more elaborate calculations of Nishina *et al.* suggest a value of about $\beta \sim \frac{1}{4}$. Putting $\beta = \frac{1}{4}$, ϕ is plotted in Fig. 18 (p. 268) as a function of E_0/M_0c^2. We see that screening has only a minor influence.

(*c*) One of the particles is an *electron*. We may consider either the collision of a fast electron, $E_0 \gg \mu$, with an atom, or the collision of a fast proton, $E_0 \gg M_0c^2$, with the atomic electrons at rest. The total cross-section in both cases must be the same, when considered for the same ratio E_0/M_0c^2 (apart from the trivial changes in Z), because the two cases arise from each other by a Lorentz transformation. However, the energy of the produced pair will be quite different in the two cases. If the fast particle is a heavy particle, it follows from appendix 6 that nearly all the pairs produced together with the recoil electron have small energies, less than $\mu E_0/M_0c^2 \ll E_0$. If the impinging particle is an electron, we carry out a Lorentz transformation transforming the heavy particle to rest. The produced pair will then become fast, and the pair energy will extend up to E_0 (energy of the colliding electron).

The total cross-section is in both cases again of the order (19), although the constant β may have a different numerical value. The curves of Fig. 18 also apply roughly to pair production by a fast electron in the field of an atom. The same is true for pair creation in *electron-electron* collisions. In the latter case the threshold energy is 7μ. For high energies formula (19) with a different β is again valid.

(*d*) It is of purely theoretical interest that pairs are also created in a collision between two photons, $k_1+k_2 > 2\mu$. For radiation densities occurring in the black-body radiation at reasonable temperatures the probability is extremely small.

We have seen that the cross-section for pair creation by a fast charged particle increases at least as $\log^2 E_0$, even for a screened field. Theoretically, when $E_0 \to \infty$, it becomes infinitely large. This is clearly impossible. It is not feasible that the cross-section should be larger than the size of the atom, and the indefinite increase also contradicts the elementary fact that all probabilities are normalized to unity and therefore the probability of the final state cannot increase indefinitely. The reason for this failure must be sought in the application of perturbation theory. In particular it is to be expected that the damping effects will eventually limit the cross-section. However, the question is academic because the energy where this becomes effective is so high (owing to the logarithmic increase) that the correction need not be considered. (Compare also § 33.5.)

4. *Experiments.* The most important test of the theory lies in the fact that γ-rays are absorbed in matter. The absorption coefficient per cm. due to pair creation is

$$\tau_{\text{pair}} = N\phi_{\text{pair}}, \tag{21}$$

where N is the number of atoms per cm.[3] The absorption coefficient due to all causes will be treated in detail in § 36 including the relevant experiments. Here we consider experiments checking some of the more specific features of pair creation.

Absolute measurements of the total ϕ_{pair} are few. In one of the earliest experiments the ratio $\phi_{\text{pair}}/\phi_{\text{compton}}$ was measured† with γ-rays of ThC″ ($k = 5{\cdot}2\mu$). The number of positrons obtained is equal to the number of pairs, whilst the number of negative electrons is greater, due to recoils from Compton scattering. If we assume that the latter are rendered correctly by the Klein–Nishina formula, it is found that $\phi_{\text{pair}} = 2{\cdot}8\times10^{-24}$ cm.$^2 = 0{\cdot}73\bar{\phi}$ (for Pb). This is in complete agreement with the figure of Table VII.

For the same γ-rays ($5{\cdot}2\,\mu$) the energy distribution of the positrons as well as the angular distribution was measured.‡ The preponderance of higher positron energies, as shown in Table VII, arising from the use of exact Coulomb wave functions, could be verified. Also the angular distribution of the positrons and negative electrons was shown to follow the law § 25 eq. (20) approximately.

A crucial point is the dependence on Z. Theoretically, the Z^2 law is modified by (i) the contribution of the atomic electrons leading to a $Z(Z+A)$ law, (ii) by the screening which is effective for $k > 50\mu$, and (iii) by the departure from the Born approximation. This is of the order of 10 per cent. for $k \sim 30\mu$, but is greater at low energies. The

† J. Chadwick, P. M. S. Blackett, and G. P. S. Occhialini, *Proc. Roy. Soc.* A, **144** (1934), 235.

‡ L. Simons and K. Zuber, ibid. A, **159** (1937), 383; K. Zuber, *Helv. Phys. Acta*, **11** (1938), 207.

Z-dependence has been measured for $\hbar\nu = 3$ and $5mc^2$. It was found that for elements up to Fe the values of the Born approximation hold and that ϕ_{pair} increases for larger Z up to the values of Table VII for Pb.† Relative measurements of ϕ_{pair} for various Z have also been made for $k = 34{\cdot}4\mu$.‡ These are shown in Table VIII. Screening is, for this energy, very small, according to Fig. 17. Thus $\phi_{\text{obs}}/Z(Z+A)$ should be constant (if the Born approximation were valid), and A should be somewhat smaller than 1. We have chosen A so that the

TABLE VIII

Z-dependence of pair creation. (Measurements, $k = 34{\cdot}4\mu$)

Z	3	13	29	50	82
ϕ_{obs}	34	530	2400	6,800	16,600
$\phi_{\text{obs}}/Z(Z+0{\cdot}8)$	3·0	3·0	2·8	2·7	2·5

ratio is constant for light elements. One finds $A = 0{\cdot}8$. This is slightly larger than the curve $\phi^{(\text{el})}$ in Fig. 17 suggests, but this curve is very crude and not meant as a quantitative prediction. It is seen now that $\phi_{\text{obs}}/Z(Z+A)$ is constant for $Z < 20$, say, but decreases slightly for increasing Z. Whilst the general $Z(Z+A)$ law is well established, this discrepancy is likely to be due to the departure from the Born approximation. The correction (15′) has the right sign and order of magnitude. A really quantitative comparison cannot be made because (15′) is calculated for the E.R. case and is not quite valid here, nor are the experiments sufficiently accurate. For a more accurate test at higher energies see § 36.

Evidence for the creation of pairs by fast electrons in the nuclear field has been found in cosmic radiation. These events can be seen in the photographic plate. They appear as triplets (pair+scattered primary) with an additional incoming track. Rough measurements of the energy dependence of the cross-section could be made.§ The result is shown in Fig. 18. Although the accuracy is not high it can clearly be seen that the $\log^3 E_0$ or $\log^2 E_0$ law holds.

In the same experiment the creation of 'double pairs' was observed. They appear as quadruplets with no incoming track and are one of the few examples of multiple processes (see § 23).

† B. Hahn, E. Baldinger, and P. Huber, ibid. **25** (1952), 505.
‡ R. L. Walker, *Phys. Rev.* **76** (1949), 1440.
§ J. E. Hooper, D. T. King, and A. H. Morrish, *Phil. Mag.* **42** (1951), 304.

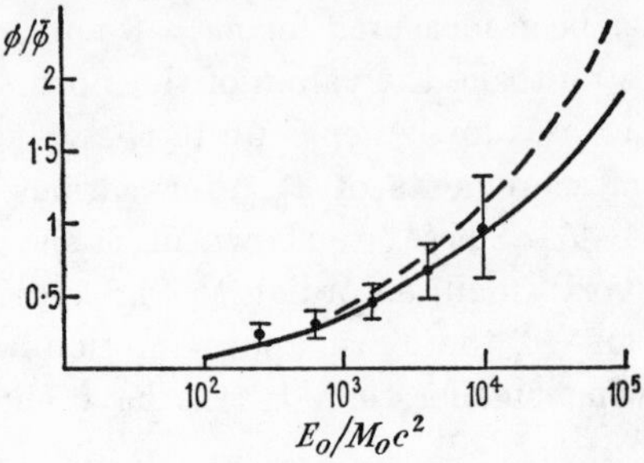

FIG. 18. Cross-section for the creation of pairs in the field of a nucleus by a charged particle with energy $E_0 \gg M_0 c^2$. Full curve: screened field; dotted curve: unscreened. Measurements for cosmic ray electrons.

27. The annihilation of positive electrons

The inverse process to the creation of pairs is the *annihilation* of a positive and a negative electron. According to the hole theory, this has to be understood as a transition of an ordinary electron from a positive energy state to a negative energy state. The energy that is given off in this process ($\geqslant 2mc^2$) may, for instance, be emitted in the form of light.

The most important of these processes is that in which a free positive electron is annihilated in a collision with a free negative electron. The conservation laws show that this can only happen if at least *two light quanta* are emitted.

1. *The two-quanta annihilation.* To deduce the probability of annihilation when a free positive and a free negative electron collide, it is convenient to carry out the calculations first in a Lorentz system, where the centre of mass of both electrons is at rest. The electrons then have momenta of equal magnitude and opposite direction:

$$\mathbf{p}_+ = -\mathbf{p}_-. \tag{1}$$

All other cases can be derived from this case by a Lorentz transformation. In this system of coordinates we have to consider the transition of an ordinary electron from a state with momentum $\mathbf{p}_0 = \mathbf{p}_-$ and energy $E_0 = E_-$ to a negative energy state with momentum $\mathbf{p} = -\mathbf{p}_+ = \mathbf{p}_0$ and energy $E = -E_+ = -E_-$. The conservation laws show that both light quanta have the same energy which is equal to the energy of one of the electrons and opposite directions of motion:

$$\mathbf{k}_1 = -\mathbf{k}_2, \qquad k_1 = k_2 = E_0. \tag{2}$$

The calculation of the transition probability is very similar to that for the Compton effect (§ 22) and can be carried out in the same way. To obtain the cross-section we remember that the final state is determined by the direction of $\mathbf{k}_1$ only, the direction of $\mathbf{k}_2$ being opposite to that of $\mathbf{k}_1$. The energy of the final state is $E_F = 2k_1$. Therefore the density function is $d\Omega\, k_1^2/2(2\pi\hbar c)^3$. Furthermore, we have to divide by the *relative* velocity of the two particles $2p_0 c/E_0$. Thus

$$d\phi = \frac{2\pi}{\hbar c}\frac{E_0}{2p_0}|K|^2\frac{d\Omega\, E_0^2}{2(2\pi\hbar c)^3}. \tag{3}$$

K is the compound matrix element for the process. If we average over the spin directions of the positive and negative electron, we obtain for the differential cross-section

$$d\phi = \frac{e^4\, d\Omega}{8p_0 E_0}\left[\frac{E_0^2-(E_0^2-p_0^2\cos^2\theta)(\mathbf{e}_1\,\mathbf{e}_2)^2+4(\mathbf{p}_0\,\mathbf{e}_1)(\mathbf{p}_0\,\mathbf{e}_2)(\mathbf{e}_1\,\mathbf{e}_2)}{E_0^2-p_0^2\cos^2\theta}-\right.$$
$$\left.-\frac{4(\mathbf{p}_0\,\mathbf{e}_1)^2(\mathbf{p}_0\,\mathbf{e}_2)^2}{(E_0^2-p_0^2\cos^2\theta)^2}\right]. \tag{4}$$

Here $\mathbf{e}_1$, $\mathbf{e}_2$ are the unit vectors of polarization of the two photons, and θ is the angle between the direction of the positron and one of the photons, $\mathbf{k}_1$, say.

To discuss the polarization it is convenient to choose $\mathbf{e}_1$, $\mathbf{e}_2$ either in the $(\mathbf{p}_+\mathbf{k}_1)$-plane or perpendicular to it. Accordingly the cross-section will be denoted by $d\phi_{\|\|}$ when both photons are polarized in the $(\mathbf{p}_+\mathbf{k}_1)$-plane, by $d\phi_{\perp\|}$ when one is polarized in this plane, the other perpendicular to it, etc. We shall see that the case of greatest interest is that where $p_+ \equiv p_0$ is vanishingly small. In this case (4) becomes

$$d\phi_{\|\|} = d\phi_{\perp\perp} = 0, \qquad d\phi_{\perp\|} = \frac{e^4\, d\Omega}{8p_0\mu}. \qquad \text{N.R.} \tag{5}$$

Thus the two annihilation photons are *polarized* at *right angles*. This result will be discussed further below. For high energies, $E_0 \gg \mu$, no marked polarization effect exists.

Summing over all directions of polarization the cross-section becomes

$$d\phi = \frac{e^4\, d\Omega}{4p_0 E_0}\left[\frac{E_0^2+p_0^2+p_0^2\sin^2\theta}{E_0^2-p_0^2\cos^2\theta}-\frac{2p_0^4\sin^4\theta}{(E_0^2-p_0^2\cos^2\theta)^2}\right]. \tag{6}$$

To obtain the total probability of annihilation we have to integrate over θ (from 0 to π) and over ϕ. The limits of the latter integration are 0 and π, because a permutation of the two light quanta does not lead to a new state. The integration yields

$$\phi = \frac{\pi e^4}{4p_0 E_0}\left[2(\beta^2-2)+\frac{3-\beta^4}{\beta}\log\frac{1+\beta}{1-\beta}\right], \qquad \beta = \frac{v_0}{c} = \frac{p_0}{E_0}. \tag{7}$$

(7) represents the cross-section for annihilation of a positive and negative electron with equal and opposite momenta $\mathbf{p}_0$. In the case of practical importance, however, the negative electron is practically at rest. For this case we obtain the probability of annihilation by a Lorentz transformation. Since the cross-section is perpendicular to the direction of motion, it is itself invariant under this Lorentz transformation. Thus, we have only to express the quantities E_0, p_0 occurring in (7) in terms of the energy E'_+ and velocity v'_+ of the positive electron in the system where the negative electron is at rest. Since the relative velocity of the two Lorentz systems is equal to $\beta = v_0/c = p_0/E_0$, we obtain

$$E'_+ = \frac{E_0+\beta p_0}{\sqrt{(1-\beta^2)}} = \frac{E_0^2+p_0^2}{\sqrt{(E_0^2-p_0^2)}} = \frac{2E_0^2-\mu^2}{\mu}$$

or

$$E_0^2 = \tfrac{1}{2}\mu(E'_+ + \mu). \tag{8}$$

β can also be expressed in terms of the energy E'_+:

$$\beta = \frac{p_0}{E_0} = \frac{\sqrt{(E_0^2-\mu^2)}}{E_0} = \sqrt{\left(\frac{E'_+-\mu}{E'_++\mu}\right)}. \tag{9}$$

Inserting (8) and (9) into (7) we obtain the cross-section for the annihilation of a positive electron of energy E'_+ with a negative electron at rest:

$$\phi = \pi r_0^2 \frac{1}{\gamma+1}\left[\frac{\gamma^2+4\gamma+1}{\gamma^2-1}\log\{\gamma+\sqrt{(\gamma^2-1)}\} - \frac{\gamma+3}{\sqrt{(\gamma^2-1)}}\right] \quad \left(\gamma = \frac{E'_+}{\mu}\right). \tag{10}$$

This formula was first deduced by Dirac.†

(10) *has its maximum for small energies* ($\gamma \sim 1$). For $E'_+ \to \mu$ the cross-section for annihilation diverges. This, however, does not mean that the probability of annihilation becomes infinite. The rate of annihilation per sec. in a substance with N atoms per unit volume for this case is given by

$$R = ZN\phi v_+ = NZ\pi r_0^2 c\,(\text{sec.}^{-1}). \qquad \text{N.R.} \tag{11}$$

For lead, for instance, we obtain $R = 2\times 10^{10}$ sec.$^{-1}$ The lifetime of a very slow positive electron in lead is therefore of the order 10^{-10} sec.

For high energies the cross-section decreases. For very high energies we obtain

$$\phi = \pi r_0^2 \frac{\mu}{E_+}\left(\log\frac{2E_+}{\mu} - 1\right). \qquad \text{E.R.} \tag{12}$$

In the Lorentz system in which the negative electron is at rest, the two annihilation quanta do not in general have the same frequency. For *high energies* of the positive electron, we see from the angular

† P. A. M. Dirac, *Proc. Camb. Phil. Soc.* **26** (1930), 361.

distribution function (6) that, in the initial Lorentz system, the two quanta are mainly emitted in the forward and backward directions. After the Lorentz transformation the *quantum* emitted in the *forward direction* takes up nearly all the *energy of the positive electron*, whereas the second quantum only has an energy of the order mc^2. If, however, the kinetic energy of the positron is small compared with mc^2 the two photons have an energy mc^2 each and are emitted in opposite directions with polarization at right angles.

From (10) we can deduce the probability of annihilation of a fast positron while passing through matter. This will be done in § 37 and it will be seen that this probability is quite small. So in most cases a fast positron will first lose all its energy and is *then* annihilated at the rate given by (11).†

2. *Experimental evidence.* The above theory can be verified in various ways. There are numerous experiments using coincidence methods which show that positrons stopped in matter give rise to two photons travelling in opposite directions. A precision measurement of the wave-length of the annihilation photons has been made by Dumond *et al.*‡ Since $k = \mu$, the wave-length should be the Compton wave-length $2\pi\lambda_0 = 0{\cdot}024265\,\text{Å}$. The wave-length of the annihilation photon was found to be $0{\cdot}02427 \pm 0{\cdot}00001$ Å. This shows that the positrons are, indeed, stopped first and then annihilated. The X-ray line also showed a certain width attributed to the velocity of the negative electrons in the metal (Cu) in which the positrons were stopped. This width corresponds to energies up to 16 eV in agreement with the average energy of the conduction electrons.

A very interesting point is the theoretical prediction that the two photons are polarized at right angles. This can be checked by making use of the fact that the scattering of a photon by a free electron depends on its polarization. We allow the two photons $\mathbf{k}_1$ and $\mathbf{k}_2 = -\mathbf{k}_1$ to be scattered by electrons, both by the same angle θ, and observe coincidences between the scattered photons $\mathbf{k}_1'$, $\mathbf{k}_2'$ ($\angle \mathbf{k}_1'\mathbf{k}_1 = \angle \mathbf{k}_2'\mathbf{k}_2 = \theta$), in the following two cases:

(i) $\mathbf{k}_2'$ lies in the same plane as $\mathbf{k}_1$, $\mathbf{k}_1'$ and (ii) $\mathbf{k}_2'$ is perpendicular to

† More precisely: a kind of thermal equilibrium is established within a time which is shorter than the life-time (11). Thus the positron retains a small energy. Also the atomic electrons have a small velocity. Consequently, the two annihilation quanta have energies slightly different from μ and are emitted at an angle slightly different from π. See S. de Benedetti, W. R. Konneker, and H. Primakoff, *Phys. Rev.* **77** (1950), 205. It is also possible that the positron, together with a negative electron, forms a bound 'hydrogen atom', called *positronium* (see below).

‡ J. W. M. Dumond, D. A. Lind, and B. B. Watson, ibid. **75** (1949), 1226.

the $(\mathbf{k}_1\mathbf{k}_1')$-plane. Now according to (5) the polarization of the annihilation photons is such that either (α) $\mathbf{e}_1$ is in the $(\mathbf{k}_1\mathbf{k}_1')$-plane and then $\mathbf{e}_2$ is perpendicular to this plane or else (β) $\mathbf{e}_1$ is perpendicular and $\mathbf{e}_2$ parallel to this plane. Using the scattering formula § 22 eq. (35) (which depends on the polarization) and summing over the two cases (α), (β) and over the (unobserved) polarization of the scattered photons, one finds after a simple consideration that the ratio of coincidences to be expected for the two experimental arrangements (i) and (ii) is

$$\frac{\phi(\text{ii})}{\phi(\text{i})} = \frac{b^2+(b-2\sin^2\theta)^2}{2b(b-2\sin^2\theta)}, \qquad b = \frac{1+(2-\cos\theta)^2}{2-\cos\theta}. \tag{13}$$

(Here $k_1 = k_2 = \mu$ is used.) This ratio is 2·6 when $\theta = \frac{1}{2}\pi$. The optimum angle at which (13) is a maximum is a little less than 90°, the optimum value being 2·85.

The experiment has been carried out.† Under the experimental conditions (with a solid angle comprising less favourable angles) the ratio (13) should be 2·00 and this was also found (2·04±0·08).

In addition to the annihilation process discussed above it is also to be expected that multiple processes occur in which three (or more) photons are emitted instead of two. According to § 23 the rate of occurrence of this process should be of the order of $1/137\pi$ compared with the rate of the 2-photon annihilation. In fact, the theoretical ratio for a free positron practically at rest turns out to be $\sim 1/370$ (see also subsection 4). The order of magnitude was verified in an experiment in which triple coincidences were observed.‡ The experiment provides a further interesting example of the occurrence of multiple processes.

3. *One-quantum annihilation.* Positive electrons can be annihilated with the emission of a *single* γ-quantum if the negative electron is bound to a nucleus. The probability for this *1-quantum annihilation* is in general much smaller than that for the 2-quanta process, and amounts in heavy elements (in which it is largest) to less than 20 per cent. of the 2-quanta annihilation.

We shall consider this process in the most simple case where the negative electron is bound in the K-shell of an atom, and where the kinetic energy of the positive electron is large compared with the ionization energy of the K-shell. The latter assumption makes the use of Born's

† C. S. Wu and I. Shaknov, *Phys. Rev.* **77** (1950), 136. This experiment was first suggested by J. A. Wheeler, *Ann. N.Y. Ac. Sc.* **48** (1946), 219. See also M. H. L. Pryce and J. C. Ward, *Nature*, **160** (1947), 435; H. S. Snyder, S. Pasternack, and J. Hornbostel, *Phys. Rev.* **73** (1948), 440.

‡ J. A. Rich, *Phys. Rev.* **81** (1951), 140.

approximation possible. The calculation is then very similar to that for the photo-electric absorption of a light quantum by the K-electron (see § 21.1, where the same assumptions are made).

The process in question consists of a transition of an electron from the K-state to a state of negative energy with momentum $\mathbf{p} = -\mathbf{p}_+$, emitting a light quantum

$$k = \sqrt{(\mu^2+p_+^2)}+\mu-\frac{\alpha^2}{2\mu}, \tag{14}$$

where $\alpha^2/2\mu$ represents the binding energy of the K-electron.

The matrix element differs from that for the photo-electric effect in that the electron goes over into a state of negative energy $E = -E_+$. The density function is $d\Omega\, k^2/(2\pi\hbar c)^3$ instead of $d\Omega\, pE/(2\pi\hbar c)^3$ and, furthermore, we have to divide by the velocity of the incoming positron instead of by c. This gives a factor $-k^2/p_+^2$. In the energy balance we can neglect the binding energy of the K-electron so that $k = E_+ + \mu$. We obtain then the cross-section of the K-shell (both K-electrons) from § 21 eq. (17) by replacing $\gamma \to -E_+/\mu$, $k \to E_+ + \mu$, and multiplying by the above factor:

$$\phi_K = 4\pi r_0^2 \frac{Z^5}{137^4} \frac{\mu^3}{p_+(E_+ + \mu)^2}\left[\frac{E_+^2}{\mu^2}+\frac{2}{3}\frac{E_+}{\mu}+\frac{4}{3}-\frac{E_+ + 2\mu}{p_+}\log\frac{E_+ + p_+}{\mu}\right]. \tag{15}$$

When the kinetic energy of the positron is either small or large compared with μ (15) becomes

$$\phi_K = \frac{4\pi}{3} r_0^2 \frac{Z^5}{137^4}\frac{p_+}{\mu}, \qquad \text{N.R.} \tag{16}$$

$$\phi_K = 4\pi r_0^2 \frac{Z^5}{137^4}\frac{\mu}{E_+}. \qquad \text{E.R.} \tag{17}$$

The cross-sections for 1- and 2-quanta annihilation are plotted in Fig. 19. For the same energy the cross-section for the 1-quantum annihilation is always considerably smaller than that for the 2-quanta annihilation (per atom). In contrast to the latter ϕ_K decreases for small energies.

The ratio of 1-quantum to 2-quanta annihilation is largest for about $E_+ \sim 10\mu$ where it amounts to $\sim$ 20 per cent. for lead. These values are obtained from Born's approximation. It is likely that the correct values are still smaller as is the case for the photo-electric effect.

(Compare Tables II and III, p. 210.) In particular the factor p_+/μ in (16) is to be replaced by

$$\frac{2\pi Z}{137}\Big/(e^{2\pi\xi}-1), \qquad \xi = \frac{Z}{137}\frac{\mu}{p_+},$$

for small energies.†

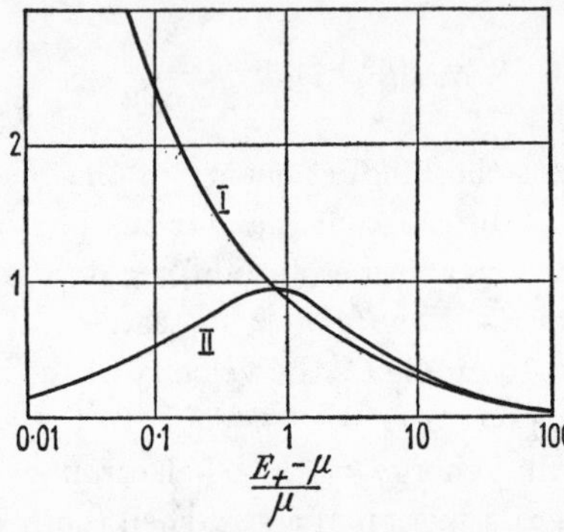

FIG. 19. Cross-sections for annihilation as function of the energy of the positron (negative electron at rest). I 2-quanta annihilation in units $Z\pi r_0^2$ per atom, II 1-quantum annihilation in units $\pi r_0^2 Z^5/137^4$.

4. *Positronium.* A positive and negative electron can form a bound state similar to the hydrogen atom. In first approximation the levels and wave functions are those of the hydrogen atom with the Bohr radius replaced by $a = 2a_0$, owing to the reduced mass. However, the fine structure is quite different owing to spin and exchange effects. The interesting level scheme cannot be discussed here.‡ We only mention that the ground state is a ^1S-state with the spins of positron and negative electron antiparallel, and this forms a fine structure with a ^3S-state which has an excitation energy of $8{\cdot}5.10^{-4}$ eV. The wave function of both states is practically that of the ground state of hydrogen.

This positronium atom is, of course, unstable and annihilates itself. The lifetime can be derived as follows: since the velocities in the atom are small we can use eq. (11) for the rate of annihilation. The density

† For a more detailed discussion of the 1-quantum annihilation see: E. Fermi and G. E. Uhlenbeck, *Phys. Rev.* **44** (1933), 510; H. R. Hulme and H. J. Bhabha, *Proc. Roy. Soc.* **146** (1934), 723; Y. Nishina, S. Tomonaga, and H. Tamaki, *Sc. Pap. Inst. Phys. Chem. Research*, Tokio, **24** (1934), No. 18; H. Bethe, *Proc. Roy. Soc.* A, **150** (1935), 129; J. C. Jaeger and H. R. Hulme, *Proc. Camb. Phil. Soc.* **32** (1936), 158.

‡ J. Pirenne, *Arch. d. Sc. Phys. et Nat.* **28** (1946), 233, **29** (1947), 121, 207; V. B. Berestetzky and L. D. Landau, *J. Exp. Theor. Phys. U.S.S.R.* **19** (1949), 673, 1130.

of electrons ZN must be replaced by the density of the negative electron at the position of the positron which is

$$|\psi(r=0)|^2 = \frac{1}{\pi a^3} = \frac{1}{8\pi a_0^3}, \tag{18}$$

where $\psi(r)$ is the hydrogen wave function. But there is a further point to be taken into account. (11) is the *average* over all four spin directions of the two particles. Conservation of angular momentum shows, however, that only the ^1S-state can decay into two photons with opposite momenta (and therefore angular momentum zero). The ^3S-state must decay into at least three photons, and does not contribute to (11) at all. To obtain the rate of decay for the ^1S-state we have therefore to multiply by 4. We obtain from (11) and (18)

$$R(^1S) = \frac{r_0^2 c}{2a_0^3} = \frac{1}{2.137^4}\frac{c}{a_0} = 8.10^9 \text{ sec.}^{-1} \tag{19 a}$$

The theoretical rate of decay into three photons has been found to be 370 times smaller than (11).† Therefore

$$R(^3S) = \frac{1}{3}\frac{R(^1S)}{370} = 7.10^6 \text{ sec.}^{-1} \tag{19 b}$$

The factor $\frac{1}{3}$ comes, of course, from the three spin states of ^3S.

Experimental evidence for the positronium has been found by Deutsch.‡

† A. Ore and J. L. Powell, *Phys. Rev.* **75** (1949), 1696; E. M. Lifshitz, *Dokl. Ak. Nauk U.S.S.R.* **60** (1948), 211; D. Ivanenko and A. Sokolov, ibid. **61** (1948), 51.

‡ M. Deutsch, *Phys. Rev.* **82** (1951), 455, **83** (1951), 866, and later papers.

VI

RADIATIVE CORRECTIONS, AMBIGUOUS FEATURES

In this chapter p, x, etc., stand for the 4-vectors p_μ, x_μ, etc. $p \cdot x$ denotes the 4-product $\sum_\mu p_\mu x_\mu$, $p^2 \equiv \sum_\mu p_\mu^2$.

THE applications of the theory made in Chapter V to various radiation processes were based on an expansion according to the electric charge e which occurs as coupling parameter between the charged particles and the radiation field. Only the first non-vanishing order in e was taken into account. Actually, it was not the probability amplitudes themselves which we expanded. This can, for example, be seen from our treatment of the line breadth, § 18 eq. (13), where the interaction occurs both in the numerator (total transition probability) and in the denominator (line breadth), but for both these quantities only the first order in e was considered. Generally, the type of expansion used appears from §§ 15 and 16: for collisions between free particles it was the *kernel* K of the wave equation § 15 eq. (3) that was expanded, and for bound state problems we expanded the amplitude U and the damping constant Γ. Broadly speaking, the probability amplitudes b differ from these quantities by the damping effects, which were seen to be closely analogous to classical theory. They are the result of a proper *solution of the wave equation* and are not due to the (neglected) higher terms of K, Γ, etc. Their treatment is quite unambiguous and straightforward.

In contrast to these damping effects we shall term the higher orders in the expansion according to e (or the dimensionless constant $e^2/\hbar c$) *radiative corrections*, or due to *radiative reaction*. These are typical quantum effects and are mixed up with the problem of the point charge (already faced with difficulties in classical theory). Their treatment is far from straightforward. The excellent agreement between theory and experiment found in Chapter V suggests, though, that the radiative corrections are small compared with the first order. Nevertheless, the question of their treatment and their order of magnitude is of prime importance for several reasons. Until about 1947 it was not possible to give these radiative corrections a proper theoretical treatment at all because it turned out that a straightforward evaluation led to an infinite result. The difficulty can be traced to the use of point charges. A great progress in the understanding of quantum electrodynamics has now been made by recognizing that all these infinities are caused by two quantities only, namely, an infinite *self-mass* of the electron, caused by its interaction with the radiation field and a similar infinite *self-charge*, due to the polarization of the vacuum (§ 11.4). In addition there are infinities, appropriate to a pure vacuum, like the self-energy of the vacuum and the fluctuations of the vacuum (see subsection 4).

All these quantities are, however, unobservable in principle. What is observable is the total mass and the total charge of the electron and these include the self-mass and self-charge. Although it is still a major unsolved difficulty of the theory that these quantities turn out to be infinite, they should, whatever their value, be combined with the 'original' mass and charge (i.e. the theoretical mass and charge when no interaction with the radiation field existed at all). For the original plus the self-mass and charge the observed finite values of mass and charge should then be substituted. This procedure will be called the *re-normalization*† of mass and charge. For a free electron the entire self-energy thus reduces to an unobservable self-mass. By separating off the infinite self-effects, finite results will be obtained for the radiative corrections of any particular problem and these are, as was to be expected, in general small compared with the first order effects of Chapter V.

In some cases the radiative corrections are also of *fundamental practical importance*. Whilst the self-energy of a free electron is unobservable, this is no longer so when the electron is placed in an external field. The self-energy is then modified and (if due account is taken of the self-mass and charge) a finite change of energy due to the interaction with the radiation field results. If the external field is, for example, the Coulomb field of a nucleus, the *energy levels* will be *displaced* as compared with the levels of the ordinary wave-mechanical problem. This displacement was previously encountered in §§ 16, 18, 20 but can only be treated in connexion with the developments of this chapter. Similarly, if the external field is a constant magnetic field, the electron will have a slightly different energy from that corresponding to its magnetic moment $e\hbar/2mc$. This can be regarded as a correction to the Bohr value of the magnetic moment and is also due to the interaction with the radiation field. The experimental discovery of the

† The idea of mass renormalization is old and was implied in all publications where the 'self-energy is omitted'. That *all* infinities of quantum electrodynamics reduce to the above unobservable quantities was first recognized by T. Tati and S. Tomonaga, *Prog. Theor. Phys.* (Japan), **3** (1948), 391 and subsequent papers; J. Schwinger, *Phys. Rev.* **73** (1948), 415, **74** (1948), 1439, **75** (1949), 651; R. P. Feynman (ref on p. 279).

radiative displacement of spectral lines and of the deviation of the magnetic moment from the usual Bohr value and the subsequent calculation of these effects, in perfect agreement with the experimental values, represents one of the major successes of quantum electrodynamics.

This chapter will be devoted to the treatment of the radiative corrections. We have to deal with a delicate mathematical situation, owing to the occurrence of infinite quantities. The mathematical analysis will be far from being wholly satisfactory and we shall take care to exhibit the difficulties and ambiguities rather than attempt to cover them up by a representation biased in favour of the theory. A very essential aid in the mathematical analysis to follow will be the relativistic covariance of the result. In order to use arguments of covariance it will be very convenient to develop the perturbation theory further in a covariant manner. This will be our first task (§ 28).

28. General evaluation of the matrix element

1. *The interaction diagrams.* In this section a general method will be developed for the evaluation of the matrix elements occurring in collisions between free electrons and photons. We have seen in § 15 eq. (13) that this is generally given by (denoting the various times by $t_1, t_2,..., t_n$)

$$\bar{K} = \bar{K}_2+\bar{K}_4+...,$$

where

$$\bar{K}_n = \frac{1}{2\pi\hbar}\int\limits_{-\infty}^{+\infty} K_n(t)\, dt$$

$$= \frac{1}{2\pi\hbar}\left(\frac{1}{i\hbar}\right)^{n-1}\int\limits_{-\infty}^{\infty} dt_n \int\limits_{-\infty}^{t_n} dt_{n-1}... \int\limits_{-\infty}^{t_2} dt_1\, H(t_n)H(t_{n-1}) ... H(t_1)+$$

$$+\text{ renormalization terms.} \quad (1)$$

K_n is the nth approximation of the kernel K occurring in the wave equation § 15 eq. (3). It was shown in § 15.3 that only the time integral $\bar{K}$ occurs in the solution.

The matrix product of factors H in (1) is restricted in so far as for any partial product of subsequent factors no elements on the energy shell are to be included. This was denoted in § 15 eq. (13) by subscripts n.d. (non-diagonal with respect to energy). This restriction only plays a role when the intermediate states belong to the continuous spectrum and a singularity occurs on the energy shell. This case will be considered below, and to start with we shall ignore this restriction. Furthermore, the 'renormalization terms' (of type $\frac{1}{2}S_1^2\bar{K}_2$, etc.) need not be evaluated separately. Part of them will eliminate certain vacuum effects (subsection 4). The rest is handled in a manner similar to that shown on p. 158, i.e. it can be included in the main term (1), if a properly defined limiting process is carried out. This will be shown in the subsequent applications (§ 30). The method for evaluating (1) used in § 15.2 and applied throughout in Chapter V is very simple if the number of intermediate states is small, but for the more complicated processes, to be considered now, it becomes very cumbersome. The following method due to Feynman and Dyson‡ permits us to combine certain groups of intermediate states and express the contribution of each group by a simple formula. The different groups are easily visualized and often have different physical significance. We can apply formula (1) also if an external field is present, replacing each factor H alternatively by $H^{(e)}$.

We can write (1) in a different form, making use of the ϵ-function, $\epsilon(t) = \pm 1$, if $t \gtrless 0$. Then

$$\bar{K}_n = \frac{1}{2\pi\hbar}\left(\frac{1}{i\hbar}\right)^{n-1} \times$$

$$\times \int_{-\infty}^{+\infty} dt_n \dots \int_{-\infty}^{+\infty} dt_1 \frac{1+\epsilon(t_n-t_{n-1})}{2} \dots \frac{1+\epsilon(t_2-t_1)}{2} H(t_n)\dots H(t_1). \quad (2)$$

The relativistic interaction of electrons and radiation $H(t)$ is given by

$$H(t) = -e\int d\tau\, \psi^\dagger(\mathbf{r},t)\gamma_\mu A_\mu(\mathbf{r},t)\psi(\mathbf{r},t). \quad (3)$$

Actually, we should use the expression $\frac{1}{2}(\psi^\dagger\gamma_\mu\psi - \psi\gamma_\mu\psi^\dagger)$ for the current, but the result would be precisely the same as for (3). We use Lorentz gauge throughout, and as we are dealing with free particles and photons, we need not take into account the auxiliary condition. For the electron field we use the method of second quantization (§ 12). $\bar{K}$ will then still be an operator. We expand $\psi^\dagger$, ψ, A_μ into a 4-dimensional Fourier

‡ R. P. Feynman, *Phys. Rev.* **74** (1948), 1430 and **76** (1949), 749 and 769; F. J. Dyson, ibid. **75** (1949), 486 and 1736. Much that is contained in this section has been anticipated by E. C. G. Stückelberg, *Helv. Phys. Acta,* **14** (1941), 51; **17** (1944), 3; **18** (1945), 195; **19** (1946), 242. See also: G. C. Wick, *Phys. Rev.* **80** (1950), 268.

series according to § 10 eq. (46) and § 12 eq. (21):

$$\psi(x) = \int \frac{d^4p}{(2\pi\hbar c)^{\frac{3}{2}}}\psi(p)e^{ip\cdot x/\hbar c}, \tag{4a}$$

$$\psi^\dagger(x) = \int \frac{d^4p}{(2\pi\hbar c)^{\frac{3}{2}}}\psi^\dagger(p')e^{-ip'\cdot x/\hbar c}, \tag{4b}$$

$$A_\mu(x) = \sqrt{(4\pi c^2)} \int \frac{d^4k}{(2\pi\hbar c)^{\frac{3}{2}}} A_\mu(k)e^{ik\cdot x/\hbar c}. \tag{4c}$$

(For reasons of symmetry we introduce $k_\mu = \kappa_\mu \hbar c$. Then $A(k) = (\hbar c)^{-\frac{5}{2}}A(\kappa)$ if $A(\kappa)$ is the quantity of § 10. $\kappa_4 = i\nu/c$.) If it were not for the ϵ-functions occurring in (2) the integration over the 4-dimensional space would yield immediately a factor $\delta^4(s)$ for each factor H, with

$$s = p+k-p'.$$

We denote the 4-dimensional integration variables in the factor $H(t_i)$ by x_i and the corresponding 4 vectors occurring in the expansion of ψ, $\psi^\dagger$, and A by p_i, p'_i, k_i, etc. Then (2) becomes ($x_0 = ct$)

$$\bar{K}_n = -(4\pi e^2)^{\frac{1}{2}n}\frac{1}{2\pi\hbar}\left(\frac{i}{\hbar}\right)^{n-1}\frac{1}{(2\pi\hbar c)^{9n/2}}\int d^4x_n...d^4x_1 \int d^4p_n\, d^4p'_n\, d^4k_n...\times$$

$$\times ...d^4p_1\, d^4p'_1\, d^4k_1 \frac{1+\epsilon(x_{0n}-x_{0n-1})}{2}...\frac{1+\epsilon(x_{02}-x_{01})}{2}e^{i(s_n\cdot x_n+...+s_1\cdot x_1)/\hbar c}\times$$

$$\times \psi^\dagger(p'_n)\gamma_\mu A_\mu(k_n)\psi(p_n)...\psi^\dagger(p'_1)\gamma_\mu A_\mu(k_1)\psi(p_1), \tag{5}$$

where
$$s_i = p_i+k_i-p'_i. \tag{6}$$

Let the product of the operators occurring in (5) be called L:

$$L = \psi^\dagger(p'_n)\gamma_\mu A_\mu(k_n)\psi(p_n)...\psi^\dagger(p'_1)\gamma_\mu A_\mu(k_1)\psi(p_1). \tag{7}$$

We now consider a certain specified process in which electrons (positive and negative) with momenta P, P',..., say, and photons with momenta K ... take part. These we call the *real* particles. P always denotes a negative electron to be absorbed or a positron to be emitted, P' a negative electron to be emitted or positron to be absorbed, K a photon to be emitted or absorbed. Then, if $\bar{K}_n$ is to have a matrix element for the process in question, a factor $\psi(P)$, $\psi^\dagger(P')$, $A(K)$ must occur once and once only somewhere in the product (7) for each real electron and photon P, P', K. In addition there will be further factors $\psi(p)$, $\psi^\dagger(p')$, $A(k)$, which describe the emission and absorption of *virtual particles and quanta*. These must necessarily occur in pairs:

each intermediate particle which is created must be absorbed again, before the process is completed. Therefore each $A(k)$, $k \neq K$, must occur together with $A(-k)$, and the $\psi(p)$, $\psi^\dagger(p')$, $p \neq P$, $p' \neq P'$, must occur in *pairs* $\psi^\dagger(p') \ldots \psi(p)$ with a particular $p' = p$. Since (5) is an integral over the whole 4-dimensional p, p', k-space, the various factors $\psi(P)$, $\psi(p)$, etc., will occur in *all possible places* in the product (7). There are always three factors (arising from one $H(t_i)$), $\psi^\dagger(p_i')A(k_i)\psi(p_i)$ joined together, giving rise to a factor $\exp\{i(p_i+k_i-p_i')\cdot x_i/\hbar c\}$. Some of these three 4-vectors p_i', p_i, k_i may or may not be momenta of the *real* particles, taking part in the process.

Let us assume for a moment that the values of the 4-momenta P, P', K of the real particles and p, p', k,... of the virtual particles be fixed. The integration in (5) first of all entails a summation of the product L (7) over all permutations Π, say, such that P, p,... occur alternatively in each factor ψ, K, k in each factor A, etc. K_n consists of a sum over these permutations. Later each term is integrated over the values of P, p, K,.... For the time being we denote the momenta of a pair $\psi^\dagger(p')$, $\psi(p)$ belonging to a virtual particle differently; $p = p'$ will follow automatically later.

The contributions to K_n arising from these permutations can be surveyed and classified graphically in the following manner:

(i) We represent each of the 4-momenta P, p, P',... of the electrons by a line $\rightarrow$ or $\leftarrow$, the momenta K, k,... for the photons by a line ----.

(ii) The three factors joined in one particular $H(t_i)$ will be represented by a point, called *joining point*. From this joining point three lines will start, namely, exactly one photon line and two electron lines, the one describing p_i by a line directed towards the joining point (factor $\psi(p_i)$) and that describing p_i' (factor $\psi^\dagger(p_i')$) by a line directed from the joining point outwards. An electron line p directed towards a joining point describes the absorption/emission of a negative/positive electron with 4-momentum $\pm p$; that directed outwards describes the emission/absorption of a negative/positive electron. The photon lines are not directed. The three factors ψ, $\psi^\dagger$, A giving rise to a joining point will be called a triplet. Some of the three lines converging on the joining point i may belong to real particles.

(iii) Evidently there are precisely n joining points in $\bar{K}_n$. Each line belonging to P, P', K joins only one joining point and has one free end, each line belonging to the virtual k, p has no free end but begins and ends in a joining point. We anticipate the fact that for a virtual line the 4-momenta in the two factors $\psi^\dagger(p')$ and $\psi(p)$ must be the same, $p = p'$, so one letter suffices to describe a virtual line. The same is true for a virtual k.

(iv) We can now construct diagrams for the process in question (i.e. P, P', K, and n fixed) connecting the free lines, virtual lines p, k, and joining points in all possible manners conforming to the above conditions.

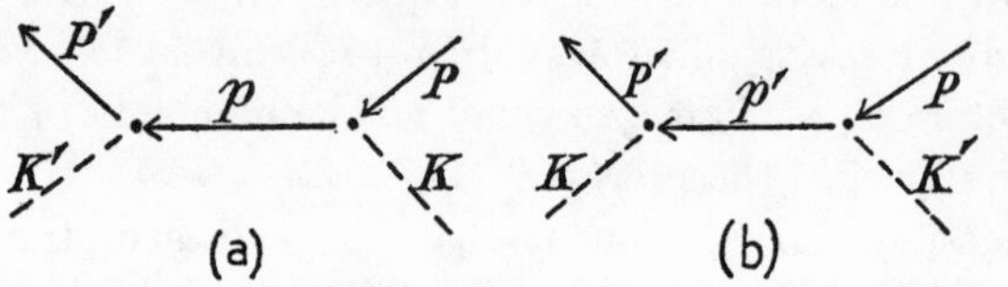

FIG. 20. Interaction diagrams: Compton effect in first approximation.

Examples are given in Fig. 20 for the Compton effect in the e^2-approximation, and in Fig. 21 for the scattering of an electron in an external field $A^{(e)}$ (here $A^{(e)}$ takes the place of an A with a free photon line, shown by -·-·-), in first and second approximation. Evidently the diagrams shown are exhaustive, and there are no others, except diagrams which consist of two disconnected parts. These will be discussed in subsection 4.

It is seen that each diagram represents a number of sequences of intermediate states. For the Compton effect, for example, the intermediate line in Fig. 20 (a) represents both a negative electron with momentum $\mathbf{p} = \mathbf{P}+\mathbf{K}$ and a positron with momentum

$$\mathbf{p}_+ = -\mathbf{p} = -\mathbf{P}-\mathbf{K} = -\mathbf{P}'-\mathbf{K}'.‡$$

These are the two intermediate states called I in § 22. The two intermediate states II are described by Fig. 20 (b).

Fig. 21 (o) describes the scattering of an electron in the lowest order ($\sim e$) and Figs. 21 (a)–($c2$) represent the first radiative corrections to the same process ($\sim e^3$). We leave it to the reader to enumerate the possible sequences of intermediate states represented, e.g. by Fig. 21 (a).§

Evidently (5) is a sum of contributions from each such diagram:

$$\bar{K}_n = \sum G_n. \tag{8}$$

‡ i.e. if we anticipate the fact that, at a joining point, momentum will be conserved (see below).

§ Each joining point represents a virtual process. To find the various sequences, go through the joining points in any possible order; this gives $3! = 6$ possible sequences for each diagram (a), (b), (c). Thus the number of intermediate states comprised in one diagram increases rapidly with the number of joining points.

We shall see that the various G often have a rather different physical significance. In each G_n some of the permutations Π are still included. We now consider one particular diagram and evaluate its contribution G.

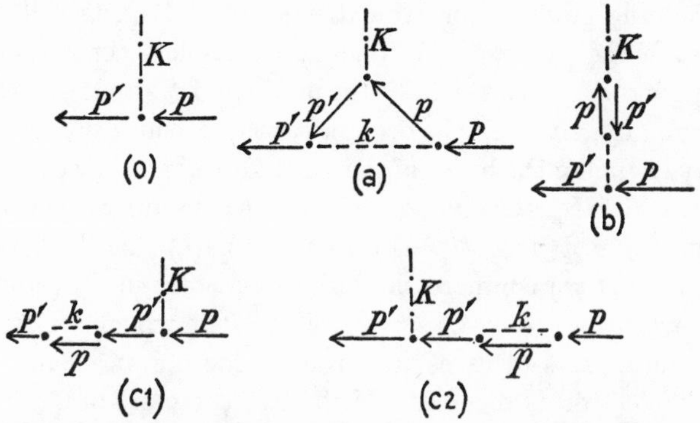

FIG. 21. Interaction diagrams: scattering of an electron in an external field: (*o*) in first approximation; (*a*), (*b*), (*c*1), (*c*2) first radiative correction.

2. *Evaluation of a particular interaction diagram.* For one particular diagram G, the momenta P, K, p, etc., of lines converging in each joining point are fixed. The various places in L in which each $\psi^\dagger(p')$ or $\psi^\dagger(P')$, etc., can occur is now very much restricted. Evidently, the only change of order that can still occur is a permutation of the triplets belonging to the joining points, keeping the three factors $\psi^\dagger A\psi$ of each joining point together.

Consider such a permutation, P, say. (There can be no confusion with the momentum of a real particle.) We may start from some conveniently chosen first arrangement I of the joining points 1, 2,..., n, where amongst the p_i, p'_i, k_i, some, well specified, momenta are real particle momenta P, P', K. A permutation P of the joining points changes the order of the ψ's and A's in (7), but keeps each triplet together, and also permutes the s_i in the exponential in (5). We can well assume that all the P, P', K are different, and are also different from all the intermediate p, p', k. It is true that an integration over all values of a virtual p may occur, but then the point where p is just equal to some P contributes a negligible amount to the integral. The exceptional cases where one P is identical with one P', or when the integrand becomes singular when a p becomes equal to a P (so that the integration over p cannot be carried out directly), will be considered below (subsections 3, 4). Hence all factors $\psi^\dagger(p')$, $\psi(p)$ will anticommute and all $A(k)$ commute *except* the pairs $\psi^\dagger(p'_i)$, $\psi(p_j)$ (later $p'_i = p_j$) belonging to the same virtual electron line and the pairs $A(k_i)$, $A(k_j)$ (later $k_i = -k_j$) belonging to the same virtual photon line. The two factors of a pair necessarily occur in different triplets belonging to different joining points. Consequently, any two triplets will commute (because each has a ψ *and a* $\psi^\dagger$) unless both triplets contain one factor each belonging to a pair. Let L_{I} be the product of ψ's and A's in the order chosen originally, and PL_{I} the order after the permutation P of joining points. By changing the order of the ψ's and A's in PL_{I} we can 'almost' restore the order L_{I}: the changes of order which are *not* permitted are the changes $\psi^\dagger(p')\psi(p) \rightleftarrows \psi(p)\psi^\dagger(p')$ or $A(k')A(k) \rightleftarrows A(k')A(k)$ belonging to a pair. If we compare the order of factors in PL_{I} and L_{I} and restore the order L_{I} as far as is possible, PL_{I} will differ from L_{I} in that in a number λ_P of cases the order of factors in a pair belonging to a virtual electron is reversed. If the two ψ-factors of a pair were to anticommute, the restoration of the order I would not involve a change of sign. Now since λ_P-times no interchange of two ψ-factors has taken place, this gives rise to a factor $(-1)^{\lambda_P}$. In addition, in a number of cases the order $A(k)A(k')$ of a pair belonging to a virtual photon will be interchanged, but this does not give rise to a change of sign (because the $A(k)$ not belonging to a pair commute).

We choose the order I in such a way that for each *virtual* electron the factor $\psi(p)$ stands directly to the left of $\psi^\dagger(p')$. This is always possible (by a suitable numbering of the joining points) if the virtual electron lines do not form a closed polygon.‡ We consider this case first. (Examples: Fig. 21 (*a*), (*c*) but not (*b*).) Then

$$L_{\mathrm{I}} = \ldots\psi(p_{j+1})\psi^\dagger(p'_j)\ldots\psi(p_{i+1})\psi^\dagger(p'_i)\ldots \tag{9}$$

where ... stands for factors belonging to *real* electrons only, and also, of course, for the factors $\gamma_\mu A_\mu(k)$, etc. The joining points are numbered $1\ldots i\ldots j\ldots n$ and each virtual electron line joins two joining points $j, j+1$ numbered by successive numbers. PL_{I} differs from L_{I} (i) by a factor $(-1)^{\lambda_P}$, (ii) in that in λ_P cases $\psi(p)\psi^\dagger(p')$ is replaced by $\psi^\dagger(p')\psi(p)$, and (iii) in that in some cases the order of $A(k)A(k')$ is interchanged. All other factors occur in the same order as in L_{I}.

Now consider the matrix element of a particular G for the transition in question. It is clear that, since each virtual particle is emitted and

‡ Follow along the open polygon of virtual electron lines in the direction →. The first virtual factor is then a $\psi^\dagger$. Read L from right to left.

absorbed, the *expectation value for the electron vacuum* can be taken for each pair $\psi(p)\psi^\dagger(p')$. According to § 12 eq. (26)‡

$$\mathsf{S}\langle\psi_\rho(p)\psi^\dagger_{\rho'}(p')\rangle_0 = +T_{\rho\rho'}\frac{1+\epsilon(p_0)}{2}\delta^4(p-p')\delta(p^2+\mu^2), \quad (10\,\mathrm{a})$$

$$-\mathsf{S}\langle\psi^\dagger_{\rho'}(p')\psi_\rho(p)\rangle_0 = +T_{\rho\rho'}\frac{1-\epsilon(p_0)}{2}\delta^4(p-p')\delta(p^2+\mu^2), \quad (10\,\mathrm{b})$$

$$T = \gamma\cdot p+i\mu \quad (p^2 \equiv p_\mu^2).$$

Thus, a pair contributes only when $p = p'$ as was to be expected.

We shall treat the corresponding photon contributions below. We absorb the factor $(-1)^{\lambda_P}$ in the minus sign of (10 b) for each $\psi^\dagger\psi$ that occurs. The permutation P has also to be applied to the s_i occurring in the exponential in (5). We may also change the notation of the integration variables $x_1,\ldots,x_n$ in the same way. Then the exponential

$$\exp i(s_n x_n+\ldots+s_1 x_1)/\hbar c$$

is unaltered and instead P acts on the factor

$$\frac{1+\epsilon(x_{0n}-x_{0n-1})}{2}\cdots\frac{1+\epsilon(x_{02}-x_{01})}{2}.$$

In the original order I the pair $\psi(p_{j+1})\psi^\dagger(p'_j)$ connects two joining points numbered j, $j+1$, say, and owing to the factor

$$\frac{1+\epsilon(x_{0j+1}-x_{0j})}{2}, \qquad x_{0j+1} > x_{0j}.$$

If the permutation P of joining points is such that it produces $\psi^\dagger(p'_j)\psi(p_{j+1})$ it will necessarily, when applied to the notation $x_1,\ldots,x_n$, also interchange x_{0j} and x_{0j+1} so that after the permutation $x_{0j} > x_{0j+1}$. Then the two formulae (10) can be combined by inserting a factor $\epsilon(x_{0j+1}-x_{0j})$ after $\epsilon(p_0)$. In the present connexion we have therefore:

$$\mathsf{S}\langle\psi_\rho(p_{j+1})\psi^\dagger_{\rho'}(p'_j)\rangle_0 = -\mathsf{S}\langle\psi^\dagger_{\rho'}(p'_j)\psi_\rho(p_{j+1})\rangle_0$$

$$= T_{\rho\rho'}\frac{1+\epsilon(p_{0j})\epsilon(x_{0j+1}-x_{0j})}{2}\delta^4(p_{j+1}-p'_j)\delta(p_j^2+\mu^2). \quad (11)$$

We now imagine the integration $d^4p_{j+1}\,d^4p'_j$ to be carried out. Since P only acts on $(1+\epsilon(x_{0j+1}-x_{0j}))/2$ which is independent of p and p', the only factor depending on p, p' besides (11) is $\exp i(s_{j+1}\cdot x_{j+1}+s_j\cdot x_j)/\hbar c$. This contains the factor $\exp ip_{j+1}\cdot(x_{j+1}-x_j)/\hbar c$, as the integration over p'_j merely makes $p'_j = p_{j+1}$, owing to the $\delta^4(p_{j+1}-p'_j)$ in (11). We need not carry out the integration explicitly. It can be seen from § 8.4 that

‡ In the integrals over the virtual p's a summation over the spins is, of course, included.

$$\int d^4p\ e^{ip\cdot x/\hbar c}\frac{1+\epsilon(p_0)\epsilon(x_0)}{2}\delta(p^2+\mu^2) = \frac{1}{2\pi i}\int d^4p\ e^{ip\cdot x/\hbar c}\zeta^*(p^2+\mu^2). \quad (12)$$

To see this we need only remark that the Fourier transform of $\epsilon(p_0)\delta(p^2+\mu^2)$ is $-iD$ and that, on the other hand (transforming back to p-space), the Fourier transform of $-i\epsilon(x_0)D$ is $(1/i\pi)\mathscr{P}/(p^2+\mu^2)$. Hence $\epsilon(p_0)\epsilon(x_0)\delta(p^2+\mu^2)$ has, *under the integral*, the same effect as $(1/i\pi)\mathscr{P}/(p^2+\mu^2)$ (cf. § 8 eqs. (29), (37)). (12) follows by adding $\frac{1}{2}\delta(p^2+\mu^2)$ on both sides. In the present connexion we can therefore put

$$\iint d^4p d^4p'\ \mathsf{S}\langle\psi_\rho(p)\psi^\dagger_{\rho'}(p')\rangle_0\ldots = -\iint d^4p d^4p'\ \mathsf{S}\langle\psi^\dagger_{\rho'}(p')\psi_\rho(p)\rangle_0\ldots$$
$$= \frac{1}{2\pi i}\iint d^4p d^4p'\ T_{\rho\rho'}\zeta^*(p^2+\mu^2)\delta^4(p-p')\ldots. \quad (13)$$

The dots stand for the exponentials under the integral.

The virtual photons can be treated in a similar way. Let there be a photon line connecting joining points j, l, $j<l$, say (where l, j are not, in general, consecutive numbers). Since k is an integration variable we can choose the signs of k, k' in the pair belonging to the two joining points so that $A(k)$, $A(-k')$, belong to l, j respectively. In the original arrangement we then have a factor $A_\mu(k)A_{\mu'}(-k')$, which only contributes its vacuum expectation value (§ 10 eq. (55))

$$\langle A_\mu(k)A_{\mu'}(-k')\rangle_0 = \hbar^2\delta_{\mu\mu'}\delta^4(k-k')\frac{1+\epsilon(k_0)}{2}\delta(k^2). \quad (14)$$

Now when the joining points are permuted, the factor $A(-k')$ will, in some cases, stand before $A(k)$. $\langle A(-k')A(k)\rangle_0$ differs from (14) in that $\epsilon(k_0)$ is replaced by $-\epsilon(k_0)$. (Reverse the signs of k, k', and $k \rightleftarrows k'$ but owing to the factor $\delta^4(k-k')$, $\epsilon(k_0') = \epsilon(k_0)$.) This will be the case if the permutation P is such that after the permutation $j>l$. Then since x_{0j} and x_{0l} are also permuted, $\epsilon(k_0)$ occurs with $\pm$ signs if $x_{0l} \gtrless x_{0j}$. Thus we can again write

$$\langle A_\mu(k)A_{\mu'}(-k')\rangle_0 = \langle A_{\mu'}(-k')A_\mu(k)\rangle_0$$
$$= \hbar^2\delta_{\mu\mu'}\delta^4(k-k')\delta(k^2)\frac{1+\epsilon(k_0)\epsilon(x_{0l}-x_{0j})}{2}. \quad (15)$$

k now occurs in the exponential just in the form $\exp ik\cdot(x_l-x_j)$ (taking into account $k=k'$). We can then argue in exactly the same way as for the virtual electron lines and state that, in the present connexion,

$$\iint d^4k d^4k'\langle A_\mu(k)A_{\mu'}(-k')\rangle_0\ldots = \iint d^4k d^4k'\langle A_{\mu'}(-k')A_\mu(k)\rangle_0\ldots$$
$$= \frac{1}{2\pi i}\hbar^2\delta_{\mu\mu'}\iint d^4k d^4k'\zeta^*(k^2)\delta^4(k-k')\ldots. \quad (16)$$

(13) and (16) are now independent of all the permutations P of the joining points. The only place where P still occurs is in

$$\sum_P P\frac{1+\epsilon(x_{0n}-x_{0n-1})}{2}\cdots\frac{1+\epsilon(x_{02}-x_{01})}{2}=1. \tag{17}$$

That this is true is easily seen from the fact that P can be built up by transpositions of neighbouring joining points alone, and for each such transposition ϵ changes its sign. The coordinates x (including x_0) have, now, through (13), (16), disappeared from the integrand, except in the exponentials, and the integration over $x_1,\dots,x_n$ in (5) can be performed immediately. The integration gives

$$\int d^4x_i\, e^{is_i\cdot x_i/\hbar c}=(2\pi\hbar c)^4\delta^4(s_i). \tag{18}$$

Before we summarize our results we must consider the case where a closed polygon of electron lines occurs. The simplest case is a configuration ---·⇄·---, as it occurs, for example, in Fig. 21 (*b*), but there are more complicated cases. Let us assume that only one closed polygon occurs. Passing along the electron lines, as before, it is clear that in the original order I there must be just one factor $\psi^\dagger(p')\psi(p)$ occurring instead of $\psi(p)\psi^\dagger(p')$. The factor will necessarily connect joining points $i\rightarrow j$, $i>j$, and therefore $x_{0i}>x_{0j}$. The expectation value is

$$\mathsf{S}\langle\psi^\dagger(p')\psi(p)\rangle_0=-\frac{1-\epsilon(p_0)}{2}\cdots,\qquad x_{0i}>x_{0j}. \tag{19 a}$$

The remaining factors on the right are the same as in (10). A permutation that changes the order of the factors will produce

$$-\mathsf{S}\langle\psi(p)\psi^\dagger(p')\rangle_0=-\frac{1+\epsilon(p_0)}{2}\cdots,\qquad x_{0j}>x_{0i}. \tag{19 b}$$

Both formulae can again be combined and the rest of the argument runs as before. The only difference is a *minus sign*. The case of several closed polygons is similar, but will not occur in our applications.

The contribution G to the matrix element, arising from one particular diagram, can now be expressed in momentum space in a rather simple way. Some care is only necessary as regards the order of factors. For the sake of clarity it will be best to write out the Dirac indices ρ of ψ, $\psi^\dagger$ and of the γ-operators explicitly. We number the joining points 1, 2,..., n in the order in which we pass along the virtual electron lines, if no closed polygons occur. When real electron lines join, the (open) polygon may, however, be interrupted. Each joining point i gives rise to a factor $\psi^\dagger_{\rho_i'}(p')\gamma_{\mu_i\rho_i'\rho_i}A_{\mu_i}(k)\psi_{\rho_i}(p)$, where we denote the Dirac indices occurring in this product by ρ_i' (for $\psi^\dagger$) and ρ_i (for ψ) and the 4-vector index by μ_i. Some of the $\psi^\dagger$, A, ψ may be absorbed in the expectation values. G_n then consists of factors belonging to (i) real photon lines $A(K)$, (ii) real electron lines (ingoing or outgoing) $\psi(P)$, $\psi^\dagger(P')$, (iii) virtual photon lines (factor (16)), (iv) virtual electron lines (factor (13)), and (v) a factor (18) and γ_μ for each joining point. We write down one example of each of these species in the product G_n, indicating the others by .... Let these lines join or connect the following joining points:

real photon line: h
incoming real electron line: g
outgoing real electron line: f
virtual electron line: $i \to i+1$
virtual photon line: j—l.

Let N_e, N_p be the number of real electron and photon lines respectively, n_e, n_p that of virtual electron and photon lines, then

$$n = N_p+2n_p = \tfrac{1}{2}(N_e+2n_e), \tag{20}$$

because exactly two electron lines and one photon line converge in each joining point, and each virtual line connects two joining points. If a closed polygon of virtual electron lines occurs a minus sign is to be added which we express by a factor $(-1)^{\text{c.p.}}$ The contribution G_n to the matrix element then becomes:

$$G_n = (-1)^{\text{c.p.}} \frac{i^{n+1}}{2\pi\hbar^{N_p}} \left(\frac{2e^2}{\hbar c}\right)^{\frac{1}{2}n} \int ... \int d^4P...d^4P'...d^4K...d^4p...d^4p'...d^4k... \times$$
$$\times \delta^4(s_1)...\delta^4(s_n)\psi^\dagger_{\rho_f'}(P')...\gamma_{\nu_h\rho_h'\rho_h} A_\nu(K)...\psi_{\rho_g}(P)... \times$$
$$\times (\gamma\cdot p+i\mu)_{\rho_{i+1}\rho_i'} \frac{\zeta^*(p^2+\mu^2)}{2\pi i} ...\gamma_{\mu_j\rho_j'\rho_j}\gamma_{\mu_j\rho_l'\rho_l} \frac{\zeta^*(k^2)}{2\pi i}. \tag{21}$$

If a closed polygon occurs one virtual electron contributes a factor $(\gamma\cdot p+i\mu)_{\rho_j\rho_i'}$ $(i > j)$.

A summation is, of course, to be carried out over all ρ_i', ρ_i, and also over all μ_i. In the factor arising from the virtual photon line we have used the fact that $\mu_l = \mu_j$, owing to the factor $\delta\mu_l\mu_j$. It is always possible to combine all factors depending on ρ, ρ' in the form of matrix products but the order depends on the diagram. It can usually be read off from the latter immediately. (Follow along the electron polygon and insert factors due to lines and joining points in the order in which they occur.)

As an example we write down the G_n for the three diagrams Fig.

21 (*o*) and Fig. 21 (*a*), (*b*) (the diagrams (*c*) are considered in subsection 3). A closed polygon occurs in (*b*). We get

$$\bar{K}_1^{(e)} \equiv G_1^{(e)} = -\frac{1}{2\pi\hbar}\left(\frac{2e^2}{\hbar c}\right)^{\frac{1}{2}} \int \dots \int d^4P\,d^4P'\,d^4K\,\delta^4(P'-P-K)\times$$
$$\times(\psi^\dagger(P')\gamma_\mu A_\mu^{(e)}(K)\psi(P)), \quad (22)$$

$$G_{3(a)}^{(e)} = \frac{1}{2\pi\hbar}\left(\frac{2e^2}{\hbar c}\right)^{\frac{3}{2}} \int \dots \int d^4P\,d^4P'\,d^4K\,d^4p\,d^4p'\,d^4k\times$$
$$\times\delta^4(P-p-k)\delta^4(p+K-p')\delta^4(p'+k-P')\frac{\zeta^*(p^2+\mu^2)\zeta^*(p'^2+\mu^2)\zeta^*(k^2)}{(2\pi i)^3}\times$$
$$\times(\psi^\dagger(P')\gamma_\mu(\gamma\cdot p'+i\mu)\gamma_\nu A_\nu^{(e)}(K)(\gamma\cdot p+i\mu)\gamma_\mu\psi(P)), \quad (23\,\text{a})$$

$$G_{3(b)}^{(e)} = -\frac{1}{2\pi\hbar}\left(\frac{2e^2}{\hbar c}\right)^{\frac{3}{2}} \int \dots \int d^4P\,d^4P'\,d^4K\,d^4p\,d^4p'\,d^4k\times$$
$$\times\delta^4(P-P'-k)\delta^4(p'-p+k)\delta^4(p-p'+K)\frac{\zeta^*(p^2+\mu^2)\zeta^*(p'^2+\mu^2)\zeta^*(k^2)}{(2\pi i)^3}\times$$
$$\times(\psi^\dagger(P')\gamma_\mu\psi(P))\mathrm{Sp}(\gamma\cdot p+i\mu)\gamma_\mu(\gamma\cdot p'+i\mu)\gamma_\nu A_\nu^{(e)}(K). \quad (23\,\text{b})$$

It is seen that, although the derivation of these formulae was somewhat complicated, we have achieved a simplification in so far as the contributions of a number of intermediate states are included in a single integral, one for each diagram. The simplification is substantial only when the number of intermediate states is large (which is not the case in any of the applications of Chapter V). The most important advantage, however, is the fact that G_n is expressed in a completely *covariant* form. Everywhere only 4-vectors occur.

3. *Singular cases.* We must now consider the important case where the 4-momentum of a virtual electron may become equal to that of a real electron. It is in this case that the 'n.d.-restriction' in (1) comes into play, i.e. where we have to take into account the fact that the matrix product of any subsequent factors of H in (1) must not lie on the energy shell of the initial or final state.‡ An example is the matrix element of the diagram Fig. 21 (*c*1) or (*c*2). If we were to apply formula (21) two of the δ^4-functions $\delta^4(p-k-P')\delta^4(p'+k-p)$ combine to a $\delta^4(P'-p')$ and thus $p' = P'$. There is also the factor $\zeta^*(p'^2+\mu^2)$. Since P' belongs to a real electron $P'^2+\mu^2 = 0$ and we would obtain the highly singular $\zeta^*(0)$. Actually $\delta^4(P'-p')\zeta^*(p'^2+\mu^2)$ is quite indeterminate. We confine ourselves to the case where in the initial state only one electron is present and the singularity is of the type just described.

‡ The actual non-anticommutability of the ψ's does not give rise to any changes, as will be shown in subsection 4.

In the energy representation of perturbation theory (§ 15) the n.d.-restriction is expressed in the fact that vanishing energy denominators do not actually occur, and instead certain new terms occur, the renormalization terms, for which the square of energy denominators is characteristic. We have also seen in § 15 that we may, formally, include these in the general formula with the vanishing denominator, but interpret the latter as the principal value and carry out a suitable limiting process for $(\mathscr{P}/x)\delta(x) = -\frac{1}{2}\delta'(x)$. It is with this situation that we are dealing here.‡ The singularities occur in the ζ^*-functions, which can be split up $\zeta^*(x) = (\mathscr{P}/x)+i\pi\delta(x)$.

The part of $\zeta^*(p'^2+\mu^2)$ which lies on the energy shell $(p'^2+\mu^2 = 0)$ is the δ-part, whereas the $\mathscr{P}$-part lies, by definition, outside the shell. The following modification has then to be made in the treatment of subsection 2. If the 4-momentum p' of a virtual particle becomes, by virtue of the δ^4-functions, equal to that of the real particle (so that $p'^2+\mu^2 = 0$), ζ^* has to be replaced by $\mathscr{P}/(p'^2+\mu^2)$, and the δ-part has to be omitted:

$$\zeta^*(p'^2+\mu^2) \to \frac{\mathscr{P}}{p'^2+\mu^2} \quad (\text{when } p' \to P). \tag{24}$$

All the other considerations of subsection 2 remain unaltered. We can then, for example, immediately write down the contribution from diagram Fig. 21 ($c1$). It is given by

$$G^{(e)}_{3(c1)} = \frac{1}{2\pi\hbar}\left(\frac{2e^2}{\hbar c}\right)^{\frac{3}{2}} \int \dots \int d^4P d^4P' d^4K d^4p d^4p' d^4k \times$$

$$\times \delta^4(P+K-p')\delta^4(p'+k-p)\delta^4(p-k-P')\frac{\zeta^*(p^2+\mu^2)\zeta^*(k^2)}{(2\pi i)^3}\frac{\mathscr{P}}{p'^2+\mu^2}\times$$

$$\times (\psi^\dagger(P')\gamma_\mu(\gamma\cdot p+i\mu)\gamma_\mu(\gamma\cdot p'+i\mu)\gamma_\nu A^{(e)}_\nu(K)\psi(P)). \tag{24 c1}$$

In this formula, when the δ^4-functions are combined, the apparently singular factor $\delta^4(P'-p')\mathscr{P}/(p'^2+\mu^2)$ still occurs. This, however, will be of the type $\delta(x)\mathscr{P}/x$ (when the self-energy corrections are made) and gives rise to no difficulties. It reduces to $-\frac{1}{2}\delta'(x)$ and just yields part of the renormalization terms. This will be shown in detail in § 30, where the formulae $G_{3(a)}$, $G_{3(b)}$, $G_{3(c1)}$ will be evaluated.

‡ It becomes clear here that the expansion of $\bar{K}$ is preferable to that of the $\mathscr{S}$-matrix, in spite of the more elegant appearance of the latter (§ 15, eqs. (35′), (36)). Also the advantage of the canonical transformation carried out in § 15 becomes evident. Without this transformation (i.e. if $K \to H$) the expansion of $\mathscr{S}$ would be given by (1) without any n.d.-restrictions and renormalization terms. We see now that this contains highly singular contributions, with no physical significance. They are merely the result of the expansion of an unsuitable quantity.

4. *Vacuum fluctuations.* In subsection 2 we assumed that the 4-momenta of the real electrons and photons were all different and were also different from the virtual electron momenta. Moreover, we had not yet considered the possibility that an interaction diagram may consist of two disconnected parts. If some of the electron momenta are equal, or become equal at certain points within the range of integration, the corresponding ψ's do not, as was assumed in subsection 2, anticommute. A simple example for the case where two real electron momenta are equal, is the self-energy of the electron which we shall consider in § 29. This is a diagonal element with real electron factors $\psi^\dagger(P')...\psi(P)$ and $P = P'$. Here $\psi^\dagger(P')$ and $\psi(P)$ do not anticommute any longer. One should then expect further contributions to the matrix element, due to the anticommutator $\{\psi^\dagger(P')\psi(P)\}$. However, this is not the case. The anticommutator $\{\psi^\dagger(P')\psi(P)\}$ is a c-number and no longer an operator influencing the state of the system. The additional contributions to the matrix element have the same value independent of whether the electron is there or not. In particular they also have the same value for the *electron vacuum*. They do not describe part of the self-energy of the electron but describe the *self-energy of the vacuum*. If we want to know the self-energy of the electron alone, we have to subtract that of the vacuum, i.e. to omit the contribution $\{\psi^\dagger\psi\}$ (see § 29.1). This is quite generally so. The non-commutability of the ψ's other than that occurring for the same virtual electron only gives rise to irrelevant vacuum effects which must be omitted in the calculation of the physical effects in question. The same applies to the case where two photons have equal momenta. The above formulae therefore continue to hold in these cases.

Similar considerations apply to the case where an interaction diagram consists of two disconnected parts. An example is shown in Fig. 22 referring to the scattering of an electron in an external field, in the approximation $\sim e^3$. The corresponding matrix element is clearly the product of two factors relating to the two parts of the diagram. The first factor is identical with $G_1^{(e)}$, in the lowest order of approximation (Fig. 21 (o)). The second factor, $-G_v$, say,‡ is independent of the external field and the electron and describes a process that can exist in a pure vacuum also. A virtual electron pair, together with a photon, is created and subsequently annihilated. This can be described as a *fluctuation of the vacuum*: an electron pair plus a photon appears some-

‡ The minus sign occurs because we have one closed polygon. The real part of G_v is then positive.

where in space (possibly very far from where our interest lies). These fluctuations are the necessary consequence of the fact that a pair creating interaction H exists and the complete absence of non-interacting electrons and photons is not an eigenstate of quantum electrodynamics. The true eigenstate of lowest energy properly called 'vacuum', if expanded into states of non-interacting particles, contains virtual electron pairs and photons. In this state there is therefore a certain probability for finding some electrons anywhere in space at

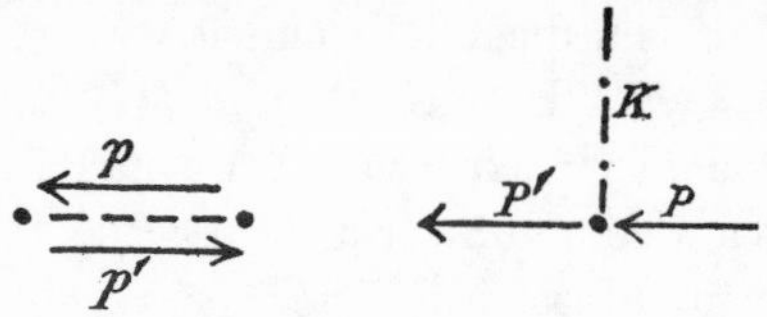

FIG. 22. Interaction diagram consisting of two disconnected parts.

at any time. Now the inclusion of such virtual pairs in the proper definition of a *real state* is just what we have achieved by the canonical transformation in § 15. The redefined states Ψ' include these virtual accompaniments. If we consider transitions between such redefined states (and this is what we always do) the above vacuum fluctuations should not occur in the final result and the contributions from diagrams like Fig. 22 should cancel.† This is in fact the case:

$-G_v G_1^{(e)}$ is a contribution from the main terms of (1), but the renormalization terms have not been considered yet. We show now that $-G_v G_1^{(e)}$ is compensated by part of the latter. This is easily seen in energy representation. For the transition $P \rightarrow P'$ we have

$$(-G_v G_1^{(e)})_{P'|P} = H_{P'|P'+\text{pair}} \frac{1}{E_{\text{pair}}} H^{(e)}_{P'+\text{pair}|P+\text{pair}} \frac{1}{E_{\text{pair}}} H_{P+\text{pair}|P}$$

$$= H^{(e)}_{P'|P} \frac{H_{V|\text{pair}} H_{\text{pair}|V}}{E^2_{\text{pair}}} \quad (E_P = E_{P'}). \tag{25}$$

Here V denotes the vacuum and 'pair' the state where a virtual pair +photon is present. The last equality sign holds because in each transition either P, P' or the virtual pair do not take part. The factor

† Observable charge fluctuations occur if we carry out a measurement, for example, of the charge contained in a certain volume element. This charge does not commute with the total energy and hence in the lowest energy state (vacuum) there is a certain probability for non-vanishing values of the charge in the volume element. The charge fluctuations were first computed by W. Heisenberg, *Ber. Sächs. Ak. Wissensch.* (1934), 317.

of $H^{(e)}_{P|P'}$ is just the square of the amplitude of a vacuum fluctuation $V \to$ pair, viz. $|H_{\text{pair}|V}/E_{\text{pair}}|^2$.

The corresponding contribution from the renormalization terms § 15 eq. $(13_2^{(e)})$ is

$$-\frac{1}{2}H^{(e)}_{P'|P}\frac{H_{P|P+\text{pair}}H_{P+\text{pair}|P}}{E^2_{\text{pair}}}-\frac{1}{2}\frac{H_{P'|P'+\text{pair}}H_{P'+\text{pair}|P'}}{E^2_{\text{pair}}}H^{(e)}_{P'|P}$$

$$= -H^{(e)}_{P'|P}\frac{H_{V|\text{pair}}H_{\text{pair}|V}}{E^2_{\text{pair}}}. \quad (26)$$

Thus (25) and (26) cancel. It is thanks to the canonical transformation that these vacuum effects disappear. The remaining contributions of the renormalization terms are essential parts of the process but can be included in the main terms with the help of a suitable limiting process (§ 30 and p. 158). For practical purposes we can state: *disconnected diagrams* of type Fig. 22 *describe irrelevant vacuum fluctuations* and *must be left out of consideration.*

The actual evaluation of G_v leads to a highly divergent result, but, as G_v does not occur in the evaluation of observable effects, this does not prevent us from calculating physically observable phenomena to any desired approximation.

In more complicated processes further 'topological' complications occur in the interaction diagrams but we need not consider these for our subsequent applications.

29. The self-energy of the electron

The occurrence of an infinite self-energy was, in classical theory (§ 4), the main obstacle to a consistent application of the idea of a point particle. In quantum theory the difficulty has so far not been removed but, in various respects, appears in a different way. We shall see below that, whilst the self-energy is still divergent, it is so to a lesser degree, so that the inertia of the electromagnetic field contained in the volume outside, say, the limits set by the classical electron radius, is negligible compared with the observed mass of the electron. Whereas in classical theory the idea could be entertained that the mass of the electron is entirely electromagnetic in nature, this appears now to be very unlikely. A further problem, already raised in § 4, which will have to be solved now, is the relativistic behaviour of the self-energy. If the electron with its field is to behave relativistically like a particle, the self-energy can merely have the effect of giving a contribution to the *mass* of the electron. It will be seen that, as in classical theory, this is not automatically the case, but can be achieved just by utilizing the divergent character of the self-energy, namely, by carrying out the integration to infinity in a certain prescribed way. At the same time the self-momentum $-(i/c)\int T_{4k}\,d\tau$ will acquire the correct transformation properties and all the other components of the energy momentum tensor of the electron's field will be made to disappear. (See appendix 7.)

The self-energy of the electron, being then merely a contribution to its mass, can be regarded as *unobservable in principle*, because it cannot be distinguished from other contributions to the electron mass.

True to our general programme we shall calculate the self-energy of the electron by expanding it according to powers of e^2 and we shall consider only the lowest order $\sim e^2$.

1. *The self-energy. Elementary calculation.*† For the sake of clarity we shall first apply an elementary method, similar to that used in Chapter V. It will be most convenient to use Lorentz gauge and to apply the method of quantization of the longitudinal and scalar photons explained in § 10. We shall then be able to treat the static self-energy simultaneously with the 'transverse self-energy', and consider it as due to emission and absorption of longitudinal and scalar photons.

According to § 14.2 the self-energy of a state A is given by

$$W = \sum_n \frac{H_{A|n}H_{n|A}}{E_A-E_n}, \tag{1}$$

where $H_{n|A}$ is the matrix element of the interaction for the transition $A \rightarrow n$. If A is the state of a free electron in the photon vacuum the states n must arise from A through the emission of a photon with a corresponding change of the state of the electron. For the intermediate states n the momentum is conserved but, in general, the energy is not. Now let $\mathbf{P}$ be the momentum of the electron in state A, and let the energy E be positive. The states n are then those with a photon $\mathbf{k}$ emitted, with *any of the four polarizations* $\alpha = 1,..., 4$. The contribution arising from transverse photons $\alpha = 1, 2$ gives rise to what is called the transverse self-energy, whereas $\alpha = 3, 4$ gives rise to the Coulomb self-energy, the analogue of the classical effect of § 4. We shall not, however, separate the two effects. The occurrence of a transverse self-energy (which is a quantum effect) can be understood in the following way. We have seen in § 7 that in a photon vacuum defined by the absence of photons, the field strengths do not vanish but carry out

† V. Weisskopf, *Phys. Rev.* **56** (1939), 72.

fluctuations about an average value zero, because the field strengths do not commute with the number of photons. An electron placed in such a photon vacuum interacts with these vacuum fields and the result of this interaction is the transverse self-energy.

In the state n the electron has momentum $\mathbf{p}' = \mathbf{P}-\mathbf{k}$ and could have positive or negative energy E'. Now the transitions to negative energy states $E' < 0$ are forbidden, according to the hole theory. On the other hand, there is just one contribution arising from these negative energy electrons.

The vacuum also has a self-energy, namely, the self-energy of all the electrons in the negative energy states. This is due to transitions of the negative energy electrons into states of positive energy. (No other transitions can occur.) The energy of a single free electron must clearly be defined as the energy difference of electron minus vacuum, so that the vacuum energy is normalized to zero. Now if an electron is present in the state A, the transitions of the negative energy electrons into that particular state will be forbidden. Accordingly, the presence of the electron in A changes the self-energy of the vacuum by the amount due to the transition into A. The change of the vacuum energy due to the presence of the electron in A appears with a minus sign, as a contribution to the self-energy of the electron in A. Written out explicitly (1) then reads:

$$W = \sum_{E'>0} \frac{H_{\mathbf{P}|\mathbf{p}'\mathbf{k}} H_{\mathbf{p}'\mathbf{k}|\mathbf{P}}}{E-E'-k} - \sum_{E''<0} \frac{H_{\mathbf{p}''|\mathbf{P}\mathbf{k}} H_{\mathbf{P}\mathbf{k}|\mathbf{p}''}}{-|E''|-E-k}. \tag{2}$$

It is convenient to denote the momentum in the negative energy states by $-\mathbf{p}''$, so that $\mathbf{p}''$ is the momentum of the positron, $|E''|$ its energy, and $\mathbf{p}'' = -\mathbf{P}-\mathbf{k}$.

The matrix elements for the emission and absorption of a photon with polarization $\alpha = 1,\ldots, 4$ have been given in § 14 eq. (29),

$$H_{\text{em}} = H_{\text{abs}} = -e\hbar c\sqrt{\left(\frac{2\pi}{k}\right)}(u_2^\dagger \gamma_\alpha u_1) \tag{3}$$

if the electron changes from state 1 to 2. Denoting the amplitudes u for the states $\mathbf{P}$, $\mathbf{p}'$, $\mathbf{p}''$ by u, u', v'' respectively (2) becomes:

$$W = 2\pi\hbar^2c^2e^2\left\{\sum_{E'>0} \frac{(u^\dagger\gamma_\alpha u')(u'^\dagger\gamma_\alpha u)}{(E-E'-k)k} + \sum_{E''<0} \frac{(u^\dagger\gamma_\alpha v'')(v''^\dagger\gamma_\alpha u)}{(|E''|+E+k)k}\right\} \tag{4}$$

changing the order of the factors in the second term. The summation is over all values of $\mathbf{k}$, and, of course, all four polarizations $\alpha = 1,\ldots, 4$. $\mathbf{p}'$, $\mathbf{p}''$ are determined by the conservation of momentum, whilst the signs of E' and E'' are fixed. In addition we have to sum over the spins of the electron in $\mathbf{p}'$, $\mathbf{p}''$. The latter summation can be carried out by the method of § 11 (eq. (18 b), the sign of p'' is reversed)

$$\mathrm{S}^{p'}(u^\dagger\gamma_\alpha u')(u'^\dagger\gamma_\alpha u) = \frac{1}{2E'}(u^\dagger\gamma_\alpha(\gamma\cdot p'+i\mu)\gamma_\alpha u), \tag{5a}$$

$$\mathrm{S}^{p''}(u^\dagger\gamma_\alpha v'')(v''^\dagger\gamma_\alpha u) = \frac{1}{2|E''|}(u^\dagger\gamma_\alpha(\gamma\cdot p''-i\mu)\gamma_\alpha u) \tag{5b}$$

$$(p_4 = iE',\, p_4'' = i|E''|).$$

The summation over α yields:

$$\gamma_\alpha(\gamma\cdot p')\gamma_\alpha = -2\gamma\cdot p', \qquad \gamma_\alpha\gamma_\alpha = 4. \tag{6}$$

We insert this in (5) and (4) and express the summation over $\mathbf{k}$ by the usual integral. Furthermore $\gamma\cdot p' = i\beta E'+(\boldsymbol{\gamma}\mathbf{p}')$. In the second term of (4) we can change the sign of $\mathbf{k}$. Then $\mathbf{p}'' = -\mathbf{p}+\mathbf{k} = -\mathbf{p}'$ and we obtain

$$W = \frac{e^2}{4\pi^2\hbar c}\int\frac{d^3k}{k}\left\{\frac{(u^\dagger(2i\mu-i\beta E'-(\boldsymbol{\gamma}\mathbf{p}'))u)}{E'(E-E'-k)}+\frac{(u^\dagger(-2i\mu-i\beta|E'|+(\boldsymbol{\gamma}\mathbf{p}'))u)}{|E'|(E+|E'|+k)}\right\}. \tag{7}$$

We could easily reduce these expressions further to spurs, as $u^\dagger$, u refer to the same state. However, we do not wish to do this, for the following reason. So far we have nowhere used the fact that $u^\dagger$ and u refer to the same states. We shall see below that in fact there is also a certain non-diagonal 'self-energy' described by the same formulae but with $u^\dagger$ and u referring to certain different states. The same applies to the generalization of W required in § 34. We therefore carry out the integration over $\mathbf{k}$ directly in (7). Henceforth we assume that u refers to the state $\mathbf{P}$ but leave it open whether $u^\dagger$ does also. Thus we evaluate essentially an operator, acting on u.

First we integrate over the azimuth angle ϕ of $\mathbf{k}$, using the direction of $\mathbf{P}$ as polar axis. The only term which depends on ϕ is $(\boldsymbol{\gamma}\mathbf{p}') = (\boldsymbol{\gamma}\mathbf{P})-(\boldsymbol{\gamma}\mathbf{k})$:

$$\int(\boldsymbol{\gamma}\mathbf{p}')\,d\phi = 2\pi(\boldsymbol{\gamma}\mathbf{P})(\mathbf{P}\mathbf{p}')/\mathbf{P}^2.$$

Since $(\boldsymbol{\gamma}\mathbf{P})$ acts directly on u we can use the wave equation

$$(\boldsymbol{\gamma}\mathbf{P})u = (-i\beta E+i\mu)u$$

and W becomes, after combining the two terms,

$$W = \frac{ie^2}{\pi\hbar c}\int\frac{k^2\,dk\,d\cos\theta}{kE'}\times$$

$$\times\frac{(u^\dagger\beta u)(-EE'+E(E'+k)(\mathbf{P}\mathbf{p}')/\mathbf{P}^2)+(u^\dagger u)(E'+k)\mu(2-(\mathbf{P}\mathbf{p}')/\mathbf{P}^2)}{E^2-(E'+k)^2}. \tag{8}$$

If we carried out the further integrations using the variables k, $\theta = \angle \mathbf{P}, \mathbf{k}$, the result could not possibly appear in relativistically covariant form, because k is not invariant. To achieve the desired covariant result we replace the variables k, $\cos\theta$ by different *invariant* variables. This can be done, by a slight generalization of a method due to Pauli and Rose.‡ We introduce the following variables:

$$v = E'-k, \qquad w = E'+k, \tag{9}$$

where $E' = [\mu^2+(\mathbf{P}-\mathbf{k})^2]^{\frac{1}{2}}$. The quantity

$$\epsilon^2 = \frac{w^2-\mathbf{P}^2}{\mu^2} = \frac{1}{\mu^2}[(E'+k)^2-(\mathbf{p}'+\mathbf{k})^2] \tag{9'}$$

is then an *invariant quantity*, because $\mathbf{p}'$, E' form a 4-vector, and the same is true for $\mathbf{p}'+\mathbf{k}$, $E'+k$. (9′) is the square of this 4-vector. We shall see that (8) reduces to an integral over ϵ only. In the variables (9), (9′) the volume element is (we write now $P \equiv |\mathbf{P}|$)

$$\frac{k^2\, dk\, d\cos\theta}{kE'} = \frac{1}{2P}\, dvdw, \tag{10}$$

and the limits of integration corresponding to $-1 \leqslant \cos\theta \leqslant +1$, $0 \leqslant k \leqslant \infty$, are:

$$\frac{\mu^2 w-P(w^2-E^2)}{w^2-P^2} \leqslant v \leqslant \frac{\mu^2 w+P(w^2-E^2)}{w^2-P^2}, \tag{10'}$$

$$E \leqslant w \leqslant \infty.$$

To express (8) in terms of v, w, we note that

$$E' = \frac{v+w}{2}, \qquad (\mathbf{P}\mathbf{p}') = \tfrac{1}{2}(vw+P^2-\mu^2). \tag{11}$$

The integration over v can be performed immediately:

$$\int dv = \frac{2P(w^2-E^2)}{w^2-P^2} = 2P\frac{\epsilon^2-1}{\epsilon^2},$$
$$\int v\, dv = 2\mu^2 wP\frac{w^2-E^2}{(w^2-P^2)^2} = 2Pw\frac{\epsilon^2-1}{\epsilon^4}. \tag{12}$$

Inserting (11), (12), into (8), the factor of $(u^\dagger \beta u)$ vanishes upon integration over v. The remaining term yields after a simple calculation

$$W = \frac{e^2\mu i}{2\pi\hbar c}(u^\dagger u)\int\limits_E^\infty dw\frac{w(\epsilon^2-1)(3\epsilon^2-1)}{(E^2-w^2)\epsilon^4} = \frac{\mu}{2\pi 137}(u^*\beta u)\int\limits_1^\infty \frac{3\epsilon^2-1}{\epsilon^3}\, d\epsilon. \tag{13}$$

‡ W. Pauli and M. E. Rose, *Phys. Rev.* **49** (1936), 462; S. N. Gupta, *Proc. Phys. Soc.* **63** (1950), 681; R. Kawabe and H. Umezawa, *Progr. Theor. Phys.* **4** (1949), 461.

We see, in the first place, that the result diverges like $\lim_{\epsilon\to\infty} \log\epsilon$. ϵ is a dimensionless quantity, an energy divided by μ. If we were to set a limit to the validity of the theory at some very high energy, we could do so in an *invariant way*, by choosing a very large finite value for the upper limit of ϵ. Even for an $\epsilon_{\max}$ exceeding the highest energies known $\log\epsilon_{\max}$ will still be of the order of magnitude unity (say 10 or 20). The factor $2\pi 137$ then makes W very small compared with μ ($(u^\dagger\beta u)$ is of the order unity). Although we must admit a serious failure in the theory (or in the expansion method?), it is very probable that 'in reality' the electromagnetic self-energy of the electron is small compared with its rest energy.

In the second place (13) has just the correct relativistic properties. It is a *correction to the mass of the electron*.‡ All the factors are invariant except $(u^*\beta u)$. Now the Hamiltonian of a single free electron is

$$H = \int (\psi^*[(\boldsymbol{\alpha}\mathbf{p})+\beta\mu]\psi)\, d\tau = (u^*[(\boldsymbol{\alpha}\mathbf{p})+\beta\mu]u).$$

If we make a correction to the *mass*, the change in H, $\delta_m H$ (which is what we have calculated) is

$$\delta_m H \equiv W = \delta\mu \int (\psi^*\beta\psi)\, d\tau = \delta\mu(u^*\beta u). \tag{14}$$

Thus (13) describes a change of mass, independent of P:

$$\delta\mu = \frac{\mu}{2\pi 137}\int\limits_1^\infty \frac{3\epsilon^2-1}{\epsilon^3}\, d\epsilon. \tag{14'}$$

For this result the vanishing of the term $(u^\dagger\beta u) = i(u^*u)$ in (8) is essential. It would not have the correct relativistic properties.

For a free particle with momentum $\mathbf{P}$ and positive energy $(u^*\beta u)$ can, of course, be evaluated immediately

$$(u^*\beta u) = \tfrac{1}{2}\,\mathrm{Sp}\frac{\beta(E+(\boldsymbol{\alpha}\mathbf{p})+\beta\mu)}{2E} = \frac{\mu}{E}. \tag{14''}$$

Again, if $\delta\mu$ is a correction to the *mass*, W should be of the form

$$W = \frac{\partial E}{\partial\mu}\,\delta\mu = \frac{\mu}{E}\,\delta\mu,$$

and (13) reduces indeed, by (14'), to this form.

The fact that W can be understood as a mass correction is a great success of the theory. It makes it possible to regard the whole effect as an unobservable one. Obviously, what we have hitherto called μ

‡ An invariant expression for the self-energy was first derived by J. Schwinger, *Phys. Rev.* **75** (1949), 651, in a more complicated way.

is not the mass as it is measured, but some unobservable theoretical quantity (the mass of an electron without electromagnetic field) and the observed mass is

$$\mu_{\text{exp}} = \mu + \delta\mu.$$

This will have to be borne in mind throughout the investigations of this chapter, and the fact that finite results can be obtained for the observable higher order corrections to physical phenomena rests on a proper analysis of the mass corrections $\delta\mu$ (and a similar charge correction δe, cf. § 32).

As the field energy of a free electron (which *is* the self-energy) reduces to a mere *self-mass*, it follows that the total energy of electron plus field behaves like the fourth component of a 4-vector. It would seem then that the other components of the energy momentum tensor must also have the correct particle properties. Again this is not the case automatically, but it will be shown in appendix 7 that the integrations can be carried out in such a way that the correct relativistic properties result. In this sense the problem of the relativistic behaviour of the field of a charged particle (§ 4) can be regarded as solved.

The logarithmic divergence found for the total self-energy contrasts sharply with the linear divergence of the static self-energy in classical theory ($1/r|_{r\to 0}$, § 4). One may ask how the static self-energy which should remain unaltered in quantum theory (using Coulomb gauge) can be thus reduced. This is a consequence of the hole theory and the Pauli principle. Consider an electron represented by a very small wave packet in coordinate space. In momentum space this would be represented by a distribution including negative energy states. The latter, however, are all filled with vacuum electrons. Consequently, the negative energy contributions to the wave function must be eliminated and the electron cannot be a wave packet of infinitely small size but must have a finite extension (of the order $\hbar/mc$, as one easily finds). Consequently the static self-energy will be diminished also.

2. *Non-diagonal elements.* We have seen that a change of mass $\delta\mu$ in the Hamiltonian of a free electron is described by an addition proportional to β to the Hamiltonian and that the self-energy of the electron has just this form. β does not commute with $H = (\boldsymbol{\alpha}\mathbf{p})+\beta\mu$, and therefore the 'self-energy' must also have non-diagonal elements (when H is diagonal). If ψ is subjected to second quantization, ψ contains the absorption operator of a negative electron as well as the emission operator of a positron. ψ^* contains the emission/absorption operator of a negative/positive electron. So (14) will, for example, have

matrix elements for the simultaneous creation or annihilation of a pair $0 \rightleftarrows e^+ + e^-$. Since momentum is conserved in (14) the momenta of the two electrons are $\mathbf{p}^+ + \mathbf{p}^- = 0$. In the single particle picture this means that (14) has a matrix element for the transition of an electron from a positive energy state with momentum $\mathbf{p}$ to a negative energy state with the same momentum $\mathbf{p} = -\mathbf{p}^+$. If we denote by v the 1-particle amplitude for negative energy states the non-diagonal element of (14) is

$$(\delta_m H)_{\text{n.d.}} = \delta\mu(v^* \beta u). \tag{15}$$

Naturally, for a free electron, the transition (15) cannot occur because of the conservation of energy. In intermediate transitions, however, this does occur (as we shall see in § 30), and it is important then to realize that such transitions involve an unobservable correction to the mass, which has to be handled in the same way as the diagonal term if finite results are to be achieved for the observable effects.

We can account for this mass correction in a general way, following the pattern of § 15 eqs. (15), (16). If μ is the 'theoretical mass' occurring in the unperturbed Hamiltonian H_0 and μ_{exp} the observed mass, the Hamiltonian describing free electrons with the observed mass and energy is evidently $H_0 + \delta_m H$. Consequently, if we wish to deal with the observed energies only and go over to interaction representation, we have to put

$$\Psi' = e^{i(H_0+\delta_m H)t/\hbar}\Psi, \qquad H'_{\text{int}} = e^{i(H_0+\delta_m H)t/\hbar} H_{\text{int}}\, e^{-i(H_0+\delta_m H)t/\hbar}, \tag{16}$$

and Ψ' satisfies
$$i\hbar\dot{\Psi}' = (H(t) + H^{(e)} - \delta_m H)\Psi'. \tag{16'}$$

We obtain an additional term in the interaction $-\delta_m H$. For a single free electron $-\delta_m H$ compensates the self-energy completely. In more complicated processes $-\delta_m H$ acts as an addition to the interaction, which will be treated in the following sections. It will have the effect of cancelling all 'mass-type infinities' in the theory.

3. *Alternative evaluation of the self-energy.* Since the addition $-\delta_m H$ will have to be embodied in the general perturbation theory, it will be useful to repeat the calculation of $\delta_m H$ employing the calculating technique of § 28. The simple application to $\delta_m H$ will serve as a first illustration of these methods.

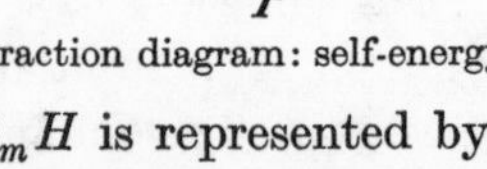

FIG. 23. Interaction diagram: self-energy of electron.

The diagonal part of $\delta_m H$ is represented by a single graph Fig. 23. P and P' are the 4-momenta of the electron in question. (It will turn out that $P' = P$.) The electron P, P' contributes factors $\psi^\dagger(P')...\psi(P)$ with some operator between, and clearly this is the same for both the diagonal and non-diagonal parts of $\delta_m H$.

The general formula § 28 eq. (21) gives at once ($n = 2$, $N_p = 0$):

$$\delta_m H = -\frac{ie^2}{\pi\hbar c}\int...\int d^4P d^4P' d^4p d^4k\, \delta^4(P'-p+k)\delta^4(p-k-P)\times$$
$$\times(\psi^\dagger(P')\gamma_\mu(\gamma\cdot p+i\mu)\gamma_\mu\psi(P))\frac{\zeta^*(k^2)\zeta^*(p^2+\mu^2)}{(2\pi i)^2}. \quad (17)$$

The summation over μ and the integration over k and P' can be performed immediately (by combining the two δ^4-functions):

$$\delta_m H = \frac{ie^2}{2\pi^3\hbar c}\int d^4P \int d^4p(\psi^\dagger(P)(2i\mu-\gamma\cdot p)\psi(P))\zeta^*((P-p)^2)\zeta^*(p^2+\mu^2). \quad (18)$$

The integral over p can be evaluated in various ways. We use now a method which retains the covariance throughout. We represent ζ^* by

$$\zeta^*(x) = i\int_0^\infty e^{-ix\tau}\,d\tau. \quad (19)$$

Since the arguments of ζ^* in (18) are invariants, τ is an invariant variable, of dimensions $1/\mu^2$. Then, using $P^2 = -\mu^2$,

$$\int d^4p(2i\mu-\gamma\cdot p)\xi^*((P-p)^2)\zeta^*(p^2+\mu^2)$$
$$= \int_0^\infty\int_0^\infty d\tau d\tau'\left[2i\mu Z(\tau,\tau')-\frac{1}{2i\tau}\gamma_\mu\frac{\partial Z}{\partial P_\mu}\right] \quad (20)$$

where the abbreviation is used

$$Z(\tau,\tau') = -\int d^4p\, e^{-ip^2(\tau+\tau')}e^{2iP\cdot p\tau}e^{+i\mu^2(\tau-\tau')}. \quad (21)$$

To evaluate Z we apply the general formula for Fresnel integrals:

$$\int_{-\infty}^{+\infty} e^{\pm(i\alpha x^2-2i\beta x)}\,dx = \sqrt{\frac{\pi}{\alpha}}\,e^{\mp i\beta^2/\alpha}e^{\pm i\pi/4} \quad \alpha > 0, \beta > 0. \quad (22)$$

In the 4-fold integral (21) for the four components of p_μ a negative sign in the exponential occurs three times (for p_i, $i = 1, 2, 3$) and a positive sign once (for p_0). Thus (21) yields

$$Z = +i\pi^2 e^{iP^2\tau^2/(\tau+\tau')}e^{i\mu^2(\tau-\tau')}\frac{1}{(\tau+\tau')^2}, \quad (23)$$

$$-\frac{1}{2i\tau}\gamma_\mu\frac{\partial Z}{\partial P_\mu} = -\gamma\cdot P\frac{\tau}{\tau+\tau'}Z. \quad (23')$$

It is seen that γ_μ only occurs in the combination $\gamma\cdot P$ and this acts on $\psi(P)$:

$$\gamma\cdot P\psi(P) = i\mu\psi(P).$$

After the differentiation with respect to P_μ, $P^2 = -\mu^2$ can again be used in the exponential (23). $\delta_m H$ now becomes, by (18), (20), (23), (23′),

$$\delta_m H = +\frac{e^2\mu}{2\pi\hbar c}\int d^4P(\psi^*(P)\beta\psi(P))\int\limits_0^\infty\int\limits_0^\infty d\tau d\tau'\,\frac{\tau+2\tau'}{(\tau+\tau')^3}e^{-i\mu^2\{\tau'^2/(\tau+\tau')\}}. \quad (24)$$

The factor $\psi^*\beta\psi$ again shows that $\delta_m H$ has the correct relativistic properties. Elsewhere, only invariant quantities occur. We may again put‡

$$\delta_m H = \delta\mu\int d^4P(\psi^*(P)\beta\psi(P)). \quad (25)$$

To see the degree of divergence of $\delta\mu$ we introduce the variables

$$\eta' = \mu^2\tau', \qquad \eta = \tau'/(\tau+\tau')$$

and obtain

$$\delta\mu = \frac{e^2\mu}{2\pi\hbar c}\int\limits_0^\infty d\eta'\int\limits_0^1 d\eta\,\frac{e^{-i\eta\eta'}(1+\eta)}{\eta'} = \frac{e^2\mu}{2\pi\hbar c}\left(\frac{5}{4}+\frac{3}{2}\int\limits_0^\infty\frac{e^{-i\eta'}\,d\eta'}{\eta'}\right). \quad (26)$$

The divergent part of the integral is $-\frac{3}{2}\log\eta'|_{\eta'\to 0}$. To compare this with the previous result (subsection 1) we remember that $\eta' = \mu^2\tau'$ and τ' is the square of a reciprocal energy. ϵ was an energy divided by μ. The proper comparison with subsection 1 will be obtained if we put $\epsilon^2 = 1/\eta'$. The divergent part of $\delta\mu$ then becomes

$$\frac{3e^2}{2\pi\hbar c}\log\epsilon|_{\epsilon\to\infty},$$

which is identical with the divergent part of (14′).§

‡ The occurrence of an integral over P is due to second quantization. For discrete waves we have (§ 12.3)

$$\int d^4P(\psi_P^*\beta\psi) \to \sum_P a_P^* a_P(u_P^*\beta u_P) + \text{terms with } b^*, b.$$

The diagonal matrix element of this for a state with one electron with momentum **P** is just $(u_P^*\beta u_P)$.

§ (26) appears to have a finite imaginary part also. This is solely due to the ambiguity in the evaluation of a divergent integral. Alternative evaluations give real results. For example, one can split up the ζ^*-functions into their $\mathscr{P}$- and δ-parts and it can easily be verified that only the imaginary cross-products $\mathscr{P}\cdot\delta$ give a contribution, leading to a real $\delta\mu$. In fact one can re-obtain the integral (7) of subsection 1.

4. *The role of the mass operator in perturbation theory.* The change in mass $\mu \to \mu+\delta\mu$, where the latter is the *observed mass*, has to be taken into account in all radiation processes of higher order. This is expressed by the fact that the additional term $-\delta_m H$ appears in the interaction (eq. (16′)). Henceforth we shall call $-\delta_m H$ the mass operator:

$$-\delta_m H = -\int \psi^* \beta\, \delta\mu\, \psi\, d\tau \equiv H^{(s)}. \tag{27}$$

For reasons which will become clear later we now write $\delta\mu$ as an operator between ψ^* and ψ. As far as we have calculated $\delta\mu$ above, $\delta\mu$ is proportional to e^2 but, naturally, (27) is to be supplemented by higher terms $\sim e^4,\ldots$. (27) includes the non-diagonal elements.

$-\delta_m H$ has to be treated as an additional perturbation in the same way as the first order interaction H in the general perturbation theory (§ 15). We confine ourselves here to collisions between free particles. $-\delta_m H$ is of the second order e^2, and has both diagonal and non-diagonal elements with respect to the energy: the latter are those discussed in subsection 2. The canonical transformation S applied in § 15, which removes all non-diagonal parts, will be supplemented by an additional term of second order

$$S_2^{(s)}(t)_{\text{n.d.}} = \frac{1}{i\hbar}\int H^{(s)}_{\text{n.d.}}(t')\,dt', \qquad (S_2^{(s)})^{-1} = -S_2^{(s)}.$$

Similarly, there is an additional term $S_3^{(s)}$ containing a product $H^{(s)}H$. Thus the formulae (13) for $\bar{K}$ in § 15 are to be supplemented by additions which are easily seen to be ($H_d = 0$):

$$2\pi\hbar \bar{K}_2^{(s)} = \int_{-\infty}^{+\infty} H^{(s)}(t)\,dt, \tag{28 a}$$

$$2\pi\hbar \bar{K}_3^{(s)} = \frac{1}{i\hbar}\int_{-\infty}^{\infty} dt \int_{-\infty}^{t} dt'\big(H^{(s)}_{\text{n.d.}}(t)H(t')+H(t)H^{(s)}_{\text{n.d.}}(t')\big), \tag{28 b}$$

$$2\pi\hbar \bar{K}_4^{(s)} = \left(\frac{1}{i\hbar}\right)^2 \int_{-\infty}^{+\infty} dt \int_{-\infty}^{t} dt' \int_{-\infty}^{t'} dt''\big(H^{(s)}_{\text{n.d.}}(t)H(t')H(t'')+H(t)H^{(s)}(t')H(t'')+$$
$$+H(t)H(t')H^{(s)}_{\text{n.d.}}(t'')\big)+\frac{1}{i\hbar}\int_{-\infty}^{+\infty} dt \int_{-\infty}^{t} dt' H^{(s)}(t)H^{(s)}_{\text{n.d.}}(t'), \tag{28 c}$$

etc., with further additions from the fourth order of $H^{(s)}$ which, however, will not be needed. $\bar{K}$ is automatically on the energy shell. $\bar{K}_2^{(s)}$ only has D-elements (completely diagonal) because the non-diagonal parts of $H^{(s)}$ are outside the energy shell. In $\bar{K}_4^{(s)}$ we have left out the renormalization terms containing factors $\frac{1}{2}(S_1^2)_D$, because these can, as in §§ 15 and 30, be included in the main terms with a certain limiting process. Also H may be replaced by $H^{(e)}$ in (28) if an external field is present. The total $\bar{K}$ consists of the contributions (13), ($13^{(e)}$) of § 15 and the above (28).

The contributions $\bar{K}^{(s)}$ can be treated in the same way as in § 28. $H^{(s)}$ is of the form $\psi^\dagger...\psi$. We can again construct the interaction diagrams consisting of electron and photon lines. $\psi^\dagger...\psi$ then gives rise to a joining point where one electron line starts and one ends but, owing to the absence of a factor $\gamma_\mu A_\mu$, *no photon line* joins this joining point. Since $H^{(s)}$ is already of order e^2 a matrix element $\bar{K}_n^{(s)}$ of order e^n in which $H^{(s)}$ occurs once will contain only $n-1$ joining points, and the numbers of real (N) and virtual (n) electron (N_e, n_e) and photon (N_p, n_p) lines are connected by

$$n-1 = \tfrac{1}{2}(N_e+2n_e), \qquad n-2 = N_p+2n_p. \tag{29}$$

$n-2$ joining points have a photon line joining. Instead of a factor e one factor $-\delta\mu/i$ appears. The general formula (21) of § 28 thus becomes, with the appropriate factors,

$$G_n^{(s)} = (-1)^{\text{c.p.}} \frac{i^{n-1}}{\hbar^{N_p}} \left(\frac{2e^2}{\hbar c}\right)^{\frac{1}{2}n-1} \int ... \int d^4P ... \times \delta^4(s_1)...\delta^4(s_{n-1}) \times$$
$$\times \psi^\dagger(P')...\delta\mu...\psi(P)... \times (\text{factors as in eq. (21), § 28}). \tag{30}$$

$G_n^{(s)}$ is to be added to the contributions G_n. Here also the considerations of § 28.3 apply, viz. if the 4-momentum p' of a virtual particle becomes, by virtue of the δ^4-functions, equal to that of a real particle, $\zeta^*(p'^2+\mu^2)$ must be replaced by $\mathscr{P}/(p'^2+\mu^2)$, in accordance with the general n.d.-prescriptions of the perturbation theory.

For the more complicated applications in which $H^{(s)}$ is involved, in particular in order to see clearly the cancellation of infinities effected by $H^{(s)}$, it is advisable not to consider $\delta\mu$ as evaluated explicitly, but to let it stand as the operator which stands between $-i\psi^\dagger(P')$ and $\delta^4(P-P')\psi(P)$ in (17) after combining the δ^4-functions:

$$\delta\mu \rightarrow \frac{e^2}{\pi\hbar c} \int d^4k \int d^4p\, \gamma_\mu(\gamma \cdot p+i\mu)\gamma_\mu\, \delta^4(P'-p+k) \times \frac{\zeta^*(k^2)\zeta^*(p^2+\mu^2)}{(2\pi i)^2}. \tag{31}$$

The position of this operator in the general formula (30) is clear from the interaction diagram. $\delta^4(P-P')$ is replaced by some $\delta^4(s)$.

As an example we consider the contribution of $H^{(s)}$ to the matrix element for an electron in an external field, in the e^3 approximation. There are two new diagrams, shown in Fig. 24. Diagrams (1), (2) are closely analogous to Fig. 21 ($c1$), ($c2$). We apply (30) ($n = 3$, $N_p = 1$)

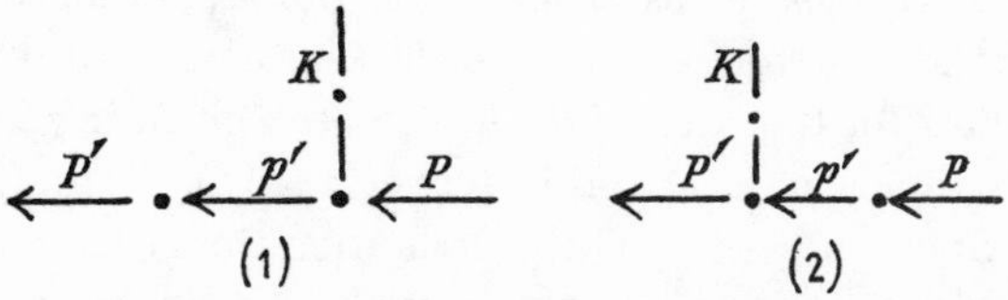

FIG. 24. Interaction diagrams for an electron in an external field: contribution of the mass operator.

using (31). P, P' now denote the momenta of the electron before and after the scattering; in ($c1$), $\delta\mu$ evidently stands after $\psi^\dagger(P')$. Then

$$G^{(s)}_{3(c1)} = -\frac{1}{2\pi\hbar}\left(\frac{2e^2}{\hbar c}\right)^{\frac{3}{2}} \int \dots \int d^4P d^4P' d^4K d^4k d^4p d^4p' \delta^4(P'-p+k) \times \\ \times \delta^4(P+K-p')\delta^4(p'-P')(\psi^\dagger(P')\gamma_\mu(\gamma\cdot p+i\mu)\gamma_\mu(\gamma\cdot p'+i\mu) \times \\ \times \gamma_\nu A_\nu^{(e)}(K)\psi(P)) \frac{\zeta^*(k^2)\zeta^*(p^2+\mu^2)}{(2\pi i)^3} \frac{\mathscr{P}}{p'^2+\mu^2}. \quad (32)$$

The n.d.-condition is effective, we have one factor $\mathscr{P}$ instead of ζ^*. The behaviour of an electron in an external field will be investigated in the following sections and it will be seen that (32) cancels against all 'mass-type-infinities' occurring in this problem.

A generalization of the mass operator will be required for bound particles. This will be considered in § 34.

30. Electron in an external field

1. *The matrix element of the interaction: diagrams* (a), (b). To study the radiative corrections we consider an electron in a given external field $A_\nu^{(e)}$ which may be either an electric or a magnetic field or both. We assume the electron to be in the continuous spectrum, so that the theory of 'collisions between free particles' can be applied. Various physical phenomena can then be studied. By letting $A_\nu^{(e)}(x)$ be a constant magnetic field we shall obtain the interaction energy of the electron with the field which, at least for a slow particle, can be described in terms of a magnetic moment of the electron. We shall see that the radiative corrections lead to a magnetic moment slightly in excess of its magneton value (§ 31). In § 32 we shall study the vacuum polarization with its most important feature, the self-charge of the electron. Finally, we could calculate the radiative corrections to the scattering of the electron, for example, in a Coulomb field, and our considerations

provide the theoretical basis for this process, but we shall not work this out explicitly. (Instead the corrections to the Compton effect will be derived in § 33.) This section will still be concerned with further formal considerations, in particular with the separation of physically significant effects from effects due to the self-mass of the electron.

The operator for the scattering $G^{(e)}$ has been given in § 28 eq. (22) to a first approximation ($\sim e$) and in § 28 eqs. (23 a), (23 b), (24 c1) to the next approximation ($\sim e^3$). The contribution from $c2$ is similar to that of $c1$. To this one has to add the contribution of the mass operator, given by § 29 eq. (32).

Consider now the matrix element of these operators for the transition of a negative electron from a specific state P to a state P' with amplitudes u_P, $u^\dagger_{P'}$. To obtain the matrix element of $\psi(P)$ we note that ψ was expanded in § 28 as a Fourier integral with coefficient $\psi(P)$, etc., and the integral $\int d^4P\psi(P)/(2\pi\hbar c)^{\frac{3}{2}}$ corresponds to a discrete sum $\sum_P a_P u_P$ (see § 12.3). The matrix element of a_P for the absorption of the electron in P is unity, and we can thus replace $\int d^4P\, \psi(P)/(2\pi\hbar c)^{\frac{3}{2}}$ by u_P. For simplicity we assume that $A_\nu^{(e)}(x)$ has only one Fourier component, and it is now simpler to write

$$A_\nu^{(e)}(x) = A_\nu^{(e)} e^{iK\cdot x/\hbar c}. \tag{1}$$

Since the external field, described classically, certainly satisfies the Lorentz condition $\partial A_\nu^{(e)}(x)/\partial x_\nu = 0$, we also have

$$K_\nu A_\nu^{(e)} = 0. \tag{2}$$

By comparison with § 28 eq. (4 c)

$$A_\nu^{(e)}(K') = A_\nu^{(e)} \frac{(2\pi\hbar c)^{\frac{3}{2}}}{(4\pi c^2)^{\frac{1}{2}}} \delta^4(K-K').$$

Furthermore, when going over to a sum over a discrete set of states a function $\delta^4(p-p')$ becomes $\delta_{pp'}/(2\pi\hbar c)^3$. Then the matrix element of $G_1^{(e)}$ becomes

$$G_1^{(e)} = -eA_\nu^{(e)}(u^\dagger_{P'}\cdot\gamma_\nu u_P), \quad P'-P = K. \tag{3}$$

Similarly, in all formulae for $G_3^{(e)}$ we replace $\int d^4P\,\psi(P)$ by u_P, etc., $A_\nu^{(e)}(K)$ by $A_\nu^{(e)}$, the numerical factor being $\pi\hbar\sqrt{(2\hbar c)}$.

We consider first the contributions to $G_{3(a)}^{(e)}$, $G_{3(b)}^{(e)}$ from the diagrams Fig. 21 (a) and (b). The expressions depending on the γ-operators are easily simplified. We use the relations implied by the δ^4-functions, namely:

$$K = P'-P = p'-p, \quad p' = P'-k, \quad p = P-k \quad (\text{for } G_{3(a)}^{(e)}), \tag{4 a}$$

$$k = P-P' = -K = p-p' \quad (\text{for } G_{3(b)}^{(e)}). \tag{4 b}$$

Further we can make use of $P^2 = P'^2 = -\mu^2$ and $\gamma\cdot P u_P = i\mu u_P$, $u^\dagger_{P'}\gamma\cdot P' = i\mu u^\dagger_{P'}$ and of (2). Then, carrying out the integration over p, p' in $G^{(e)}_{3(a)}$ and over p', k, in $G^{(e)}_{3(b)}$, we obtain, after a simple calculation,

$$G^{(e)}_{3(a)} = \frac{ie^3}{2\pi^3\hbar c} A^{(e)}_\nu \int d^4k\, u^\dagger_{P'}\{2(\gamma\cdot k)(P_\nu+P'_\nu-k_\nu)-2i\mu k_\nu+$$
$$+\gamma_\nu[k^2+2(P\cdot P')-2(P+P')\cdot k]\}u_P\times$$
$$\times\zeta^*(k^2)\zeta^*(k^2-2P\cdot k)\zeta^*(k^2-2P'\cdot k), \tag{5 a}$$

$$G^{(e)}_{3(b)} = -\frac{ie^3}{\pi^3\hbar c} A^{(e)}_\nu \int d^4p\, u^\dagger_{P'}\{2p_\nu(\gamma\cdot p)+p_\nu(\gamma\cdot K)+K_\nu(\gamma\cdot p)-$$
$$-\gamma_\nu(\mu^2+p^2+p\cdot K)\}u_P\,\zeta^*(K^2)\zeta^*(p^2+\mu^2)\zeta^*(\mu^2+(p+K)^2). \tag{5 b}$$

The term $K_\nu(\gamma\cdot p)$ in (5 b) actually vanishes, by (2), but is left standing for reasons of symmetry.

These expressions will be evaluated in the following sections. It will be seen that $G^{(e)}_{3(b)}$ describes the vacuum polarization and leads to unobservable effects only, when $K\to 0$. $G^{(e)}_{3(a)}$ will, for example, give a contribution to the magnetic moment of the electron.

2. *Cancellation of self-energy parts*. We consider next the contributions $G^{(e)}_{3(c1)}$ and $G^{(s)}_{3(c1)}$. Here care is required as we are dealing with a singular situation arising from a vanishing denominator. We shall show that, when the two contributions are combined, the divergences are cancelled, and that the singularity in the denominator $\mathscr{P}/(p'^2+\mu^2)$ disappears for $p' = P'$ giving a finite result.

It is convenient to introduce the 4-vector $q = p-k$ instead of p in $G^{(e)}_{3(c1)}$. $\delta^4(P+K-p')$ can be replaced by $\delta^4(P+K-P')$. Then

$$G^{(e)}_{3(c1)} = \frac{ie^3}{4\pi^3\hbar c} A^{(e)}_\nu \int d^4k d^4q d^4p' u^\dagger_{P'}\{\gamma_\mu[\gamma\cdot(q+k)+i\mu]\gamma_\mu(\gamma\cdot p'+i\mu)\gamma_\nu\}u_P\times$$
$$\times\zeta^*(k^2)\zeta^*((q+k)^2+\mu^2)\frac{\mathscr{P}}{p'^2+\mu^2}\delta^4(q-P')\delta^4(p'-q). \tag{6 a}$$

In $G^{(s)}_{3(c1)}$ the integration over p can be performed:

$$G^{(s)}_{3(c1)} = -\frac{ie^3}{4\pi^3\hbar c} A^{(e)}_\nu \int d^4k d^4p' u^\dagger_{P'}\{\gamma_\mu[\gamma\cdot(P'+k)+i\mu]\gamma_\mu(\gamma\cdot p'+i\mu)\gamma_\nu\}u_P\times$$
$$\times\zeta^*(k^2)\zeta^*((P'+k)^2+\mu^2)\frac{\mathscr{P}}{p'^2+\mu^2}\delta^4(P'-p'). \tag{6 b}$$

The two contributions almost cancel. If it were permitted to replace $\delta^4(q-P')\delta^4(p'-q)$ by $\delta^4(q-P')\delta^4(p'-P')$, the integration over q in (6 a) would just produce the expression (6 b) with the opposite sign. On account of the singular denominator $\mathscr{P}/(p'^2+\mu^2)$, however, this is not permitted and the sum (6 a)+(6 b) will be finite.

The limiting procedure necessary to deal with the singular denominator is very similar to that described on page 158 and will just produce the 'renormalization terms' of elementary perturbation theory. Actually there are various ways in which a limiting value can be ascribed to the difference in question, but there is one demand that makes the results unambiguous. This is that all the quantities, including the δ^4-functions in (6), are covariants and the result of the limiting procedure must also be covariant. The four components of p' must therefore be handled in a symmetrical way. The function $\delta^4(P'-p')$, when multiplied by a function $f(p')$ and integrated, has two effects: (i) it makes the vector p' parallel to P' so that $p'_\mu = P'_\mu(1+\epsilon)$, say ($\epsilon$ = invariant), and (ii) it makes the length $p'^2 = P'^2$, so that $\epsilon \to 0$. The dangerous singularity of $\mathscr{P}/(p'^2+\mu^2)$ is concerned with the length of p' only. The δ^4-functions in (6 a) and (6 b) will therefore have the (unambiguous) effect of making the 4-vectors q, p', P all parallel (exactly), whereas a certain limiting procedure will be required before the lengths of the vectors can be made equal. We can then put

$$p'_\mu = P'_\mu(1+\epsilon), \qquad q_\mu = P'_\mu(1+\epsilon') \tag{7}$$

and replace

$$\left.\begin{aligned} d^4p'\delta^4(P'-p') \quad &\text{by} \quad \delta(\epsilon)d\epsilon \\ d^4p'd^4q\,\delta^4(q-P')\delta^4(p'-q) \quad &\text{by} \quad \delta(\epsilon-\epsilon')\delta(\epsilon')d\epsilon d\epsilon' \end{aligned}\right\}. \tag{8}$$

Evidently, when multiplied by a regular function $f(p')$ or $f(p',q)$ the replacement (7) in the argument of f and (8) in that of the δ-functions produces the same effect as the δ^4-functions. Applying this procedure to (6) we obtain

$$G^{(e)}_{3(c1)} \equiv \int d^4p'd^4q\, f(p',q)\frac{\mathscr{P}}{p'^2+\mu^2}\delta^4(q-P')\delta^4(p'-q)$$

$$\to \int f(P'(1+\epsilon), P'(1+\epsilon'))\delta(\epsilon-\epsilon')\delta(\epsilon')\frac{\mathscr{P}}{-\mu^2\epsilon(2+\epsilon)}d\epsilon d\epsilon', \tag{9 a}$$

$$G^{(s)}_{3(c1)} \equiv \int d^4p' g(p')\frac{\mathscr{P}}{p'^2+\mu^2}\delta^4(P'-p') \to g(P'(1+\epsilon))\frac{\mathscr{P}}{-\mu^2\epsilon(2+\epsilon)}\delta(\epsilon)d\epsilon, \tag{9 b}$$

where $f(p',q)$, $g(p')$ are the remaining factors of (6 a), (6 b) not written down explicitly and $f(p', q = P') = -g(p')$. In (9 a) we can immediately integrate over ϵ (and afterwards write ϵ instead of ϵ')

$$G^{(e)}_{3(c1)} = -\int f(P'(1+\epsilon), P'(1+\epsilon))\frac{\mathscr{P}}{\mu^2\epsilon(2+\epsilon)}\delta(\epsilon)d\epsilon. \tag{10}$$

We can now split up

$$\frac{\mathscr{P}}{\epsilon(2+\epsilon)} = \frac{1}{2}\left(\frac{\mathscr{P}}{\epsilon} - \frac{1}{2+\epsilon}\right). \tag{11}$$

The second term is non-singular and can be integrated at once. The corresponding contributions from (9 b) and (10) are equal and opposite and therefore cancel. It was shown in § 15 that in the context of perturbation theory $(\mathscr{P}/\epsilon)\delta(\epsilon)$ has the value

$$\frac{\mathscr{P}}{\epsilon}\delta(\epsilon) = -\tfrac{1}{2}\delta'(\epsilon). \tag{12}$$

While in general the left-hand side of (12) is ambiguous (see § 8 eq. (18)), there is no ambiguity here because all singular functions arise from the same integrations in § 28 eqs. (1) and (4) and the same representation must be used for them.

Inserting (11), (12) into (9 b), (10), and (6) we obtain

$$G^{(e)}_{3(c1)} = \frac{ie^3}{16\pi^3\hbar c\mu^2} A^{(e)}_\nu \int d^4k \int d\epsilon \times$$
$$\times u^\dagger_{P'}\{\gamma_\mu[\gamma\cdot(P'+k)+i\mu+\gamma\cdot P'\epsilon]\gamma_\mu(\gamma\cdot P'+i\mu+\gamma\cdot P'\epsilon)\gamma_\nu\}u_P\times$$
$$\times\delta'(\epsilon)\zeta^*(k^2)\zeta^*[(P'+k)^2+\mu^2+2\epsilon P'\cdot(P'+k)-\mu^2\epsilon^2], \tag{13 a}$$

$$G^{(s)}_{3(c1)} = -\frac{ie^3}{16\pi^3\hbar c\mu^2} A^{(e)}_\nu \int d^4k \int d\epsilon \times$$
$$\times u^\dagger_{P'}\{\gamma_\mu[\gamma\cdot(P'+k)+i\mu]\gamma_\mu(\gamma\cdot P'+i\mu+\gamma\cdot P'\epsilon)\gamma_\nu\}u_P\times$$
$$\times\delta'(\epsilon)\zeta^*(k^2)\zeta^*[(P'+k)^2+\mu^2]. \tag{13 b}$$

The integration over ϵ is straightforward. We have to differentiate the integrands with respect to ϵ and put $\epsilon = 0$ afterwards (and add a minus sign). In (13 b) ϵ only occurs in one place and the differentiation cancels with that of the same factor in (13 a). ϵ^2 gives no contribution.

We are left with

$$G_{3(c1)} \equiv G^{(e)}_{3(c1)}+G^{(s)}_{3(c1)} = -\frac{ie^3}{16\pi^3\hbar c\mu^2} A^{(e)}_\nu \int d^4k\, \zeta^*(k^2)\times$$
$$\times\frac{\partial}{\partial\epsilon} u^\dagger_{P'}\{\gamma_\mu[\gamma\cdot(P'+k)+i\mu+\gamma\cdot P'\epsilon]\gamma_\mu(\gamma\cdot P'+i\mu)\gamma_\nu\}u_P\times$$
$$\times\zeta^*[(P'+k)^2+\mu^2+2\epsilon P'\cdot(P'+k)]\Big|_{\epsilon\to 0}. \tag{14}$$

For the differentiation of ζ^* we use the representation $\zeta^*(x) = 1/(x-i\sigma)$, and we bring the two terms arising from the differentiation of (14) on the same denominator $\{1/(x-i\sigma) \to x/(x-i\sigma)^2$ as $\sigma \to 0\}$. Then the two terms can be combined more elegantly to‡

$$G_{3(c1)} = \frac{ie^3}{16\pi^3\hbar c\mu^2} A_\nu^{(e)} \int d^4k \times$$
$$\times u_{P'}^\dagger \{\gamma_\mu[\gamma\cdot(P'+k)+i\mu](\gamma\cdot P')[\gamma\cdot(P'+k)+i\mu]\gamma_\mu(\gamma\cdot P'+i\mu)\gamma_\nu\}u_P \times$$
$$\times \zeta^*(k^2)\{\zeta^*[(P'+k)^2+\mu^2]\}^2. \quad (15)$$

This is the relativistic form of the 'renormalization terms' given in § 15.2 in energy representation. We notice the square of the energy denominators which occurs here in the form $\{\zeta^*\}^2$.

As for $G_{3(a)}^{(e)}$ and $G_{3(b)}^{(e)}$, the γ-expressions can easily be worked out. One finds

$$G_{3(c1)} = \frac{ie^3}{4\pi^3\hbar c} A_\nu^{(e)}(u_{P'}^\dagger \gamma_\nu u_P) \int d^4k \left\{2\mu^2-2(P'\cdot k)-k^2-\frac{2}{\mu^2}(P'\cdot k)^2\right\} \times$$
$$\times \zeta^*(k^2)\{\zeta^*[(P'+k)^2+\mu^2]\}^2. \quad (16)$$

Thus (16) is proportional to $G_1^{(e)}$, eq. (3).

It is of some interest to compare the present analysis with the perturbation theory of § 15.2. The perturbation due to the mass operator $\delta\mu(\psi^\dagger\psi)$ gives rise to a second-order compound matrix element with one energy denominator, $H^{(s)}H^{(e)}/(...)$. The diagonal elements of $H^{(s)}$ do not contribute, owing to the n.d.-prescription. The only contribution arises from the non-diagonal elements (§ 29.2). This is given here by the second term of (11) and the effect of the differentiation of the factor $\gamma\cdot P'+i\mu+\epsilon(\gamma\cdot P')$ in (13b). It has no singular denominator but diverges after the integration over k. This contribution cancels with a third-order matrix element of the form $H\cdot H\cdot H^{(e)}$. What is left are the renormalization terms $\frac{1}{2}(S_1^2)H^{(e)}$, given here by (16). We shall see that (16) is finite (except at $k \to 0$, see § 33.4).

In addition to (16) there is, of course, a contribution from the diagram $c2$. This is the same as $G_{3(c1)}$ with P replaced by P' under the integral:

$$G_{3(c2)} = G_{3(c1)}(P \rightleftarrows P'). \quad (17)$$

The radiative corrections to the scattering in an external potential are thus described by

$$G_{3(a)}^{(e)} + G_{3(b)}^{(e)} + G_{3(c1)} + G_{3(c2)},$$

given by (5a), (5b), (16), and (17) respectively.

‡ Compare R. P. Feynmann, *Phys. Rev.* **76** (1949), 769.

31. The anomalous magnetic moment of the electron

1. *Electron at rest in a magnetic field.* One of the most interesting consequences of the theory is the fact that the radiative corrections lead to a small magnetic moment of a free electron at rest, additional to the usual magneton $e\hbar/2mc$. To derive this excess moment we use the general theory of the scattering of an electron in an external field, § 30, and specialize the external field to be a magnetic field with vector potential $A_n^{(e)}(\mathbf{r}) = A_n^{(e)} e^{i(\mathbf{Kr})/\hbar c}$, where $\mathbf{n}$ is a spatial direction, $A_0^{(e)} = 0$, and with vanishing frequency $\mathbf{K} \to 0$, $K_0 = 0$. The magnetic field strength is $\mathbf{H}^{(e)} = i[\mathbf{KA}]/\hbar c$, so the linear terms in $\mathbf{K}$ will be important. Moreover, we shall treat the electron non-relativistically and with vanishing momentum $\mathbf{P} \sim \mathbf{P}' \to 0$. The matrix element of first order is given by § 30 eq. (3),

$$G_1^{(e)} = -eA_n^{(e)}(u_{P'}^\dagger \gamma_n u_P) = -eA_n^{(e)}(u_{P'}^* \alpha_n u_P). \tag{1}$$

In non-relativistic approximation this becomes, according to § 11 eq. (22),

$$G_1^{(e)} = -\frac{ei}{2\mu} A_n^{(e)}\, u_{P'}^*(\boldsymbol{\sigma}[\mathbf{P}'-\mathbf{P},\mathbf{n}])u_P - \frac{e}{2\mu} A_n^{(e)}\, u_{P'}^*(P_n+P'_n)u_P, \tag{2}$$

where u_P is now only a 2-row spinor in σ-space. The first term of (2) describes the interaction of the magnetic moment with the field. Since $\mathbf{P}'-\mathbf{P} = \mathbf{K}$, $(i/\hbar c)[\mathbf{P}'-\mathbf{P},\mathbf{n}]A_n^{(e)} = \mathbf{H}^{(e)}$, and (2) is the matrix element of $-(e\hbar/2mc)(\boldsymbol{\sigma}\mathbf{H}^{(e)})$, as should be the case. The second term of (2) describes scattering processes, etc.

The part of the radiative corrections, which gives rise to an additional magnetic moment will be proportional to the first term of (2), with some numerical factor. In particular it will be proportional to $\mathbf{P}'-\mathbf{P}$, whereas all parts proportional to $\mathbf{P}+\mathbf{P}'$ describe different physical effects. Also, terms of the form $(u_{P'}^\dagger u_P)$ do not describe a magnetic moment, since they are independent of $\boldsymbol{\sigma}$. We need therefore only consider terms containing γ, and linear in $\mathbf{P}'-\mathbf{P}$.

The e^2-correction to the matrix element of an electron in an external field ($\sim e^3$) consists of three parts, given in § 30, namely, $G_{3(a)}^{(e)}$, $G_{3(b)}^{(e)}$, $G_{3(c1)}+G_{3(c2)}$. We shall see in § 32 that $G_{3(b)}^{(e)}$ describes the vacuum polarization and when the field is almost static gives no contribution to the magnetic moment. $G_{3(a)}^{(e)}$ and $G_{3(c1)}+G_{3(c2)}$ give contributions to the effect in question. The magnetic moment arises only from terms $\sim\gamma$. γ is, in the non-relativistic approximation, already proportional to $\mathbf{P}'-\mathbf{P}$. For the remainder of the factors in $G_{3(a)}^{(e)}+G_{3(c1)}+G_{3(c2)}$ we can therefore put $\mathbf{P} = \mathbf{P}'$, $P_0 = P_0'$. We may also put $\mathbf{P} = 0$, $P_0 = \mu$, but it is convenient to do this later.

For $P_\mu = P'_\mu$, $G_{3(c1)} = G_{3(c2)}$ (which results in a factor 2), and the terms $G^{(e)}_{3(a)}$ and $2G_{3(c1)}$ can be combined, if the sign of k is reversed in $G_{3(c1)}$.

Using also $P\cdot P' = P^2 = -\mu^2$ we obtain

$$G_3 \equiv (G^{(e)}_{3(a)}+2G_{3(c1)})_{P=P'} = \frac{ie^3}{\pi^3\hbar c} A_n^{(e)} \int d^4k \times$$
$$\times u_P^\dagger\Big\{(\gamma\cdot k)(2P_n-k_n)-\gamma_n\Big(P\cdot k+\frac{1}{\mu^2}(P\cdot k)^2\Big)\Big\}u_P\zeta^*(k^2)\{\zeta^*(k^2-2P\cdot k)\}^2. \tag{3}$$

2. *Evaluation of the excess magnetic moment.* Before carrying out the integration a certain simplification is achieved by writing

$$\zeta^*(k^2)\{\zeta^*(k^2-2P\cdot k)\}^2 = 2\int_0^1 \frac{w\,dw}{[k^2-2(P\cdot k)w-i\sigma]^3}\bigg|_{\sigma\to 0} \tag{4}$$

which is easily verified. ζ^* is represented by $\zeta^*(x) = 1/(x-i\sigma)$. w is an invariant auxiliary variable.

Furthermore it is convenient to use a transformation of variables:

$$k_\mu = k'_\mu+P_\mu w. \tag{5}$$

Inserting (4) and (5) into (3) and dropping the dash on k' again we can use $\gamma\cdot P = i\mu$ and $P^2 = -\mu^2$ as γ is the only operator standing between $u^\dagger$, u. Then, retaining only terms $\sim\gamma$, we obtain

$$G_3 = \frac{2ie^3A_n^{(e)}}{\pi^3\hbar c}\int d^4k\int_0^1 dw\, u_P^\dagger\Big\{(\gamma\cdot k)[(2-w)P_n-k_n]-$$
$$-\gamma_n\Big[(P\cdot k)+\frac{1}{\mu^2}(P\cdot k)^2-2(P\cdot k)w+\mu^2(w^2-w)\Big]\Big\}u_P\frac{w}{(k^2+\mu^2w^2-i\sigma)^3}. \tag{6}$$

The advantage of the transformation (5) is that the denominator now only depends quadratically on each component of k and so after the integration every term linear in any one component of k (including k_0) vanishes. $\gamma\cdot k$ only contributes when multiplied by k_n and then only the component $\gamma_n k_n$ (no summation) contributes. Thus (6) is proportional to γ_n, and the factor $G_1^{(e)}$ eq. (1) can be taken out:

$$G_3 = \frac{2ie^2}{\pi^3\hbar c}G_1^{(e)}\int d^4k\int_0^1 dw\, w\frac{k_n^2+(P\cdot k)^2/\mu^2+\mu^2(w^2-w)}{(k^2+\mu^2w^2-i\sigma)^3}. \tag{7}$$

We confine ourselves now to an electron at rest $\mathbf{P} = 0$ $P_0 = \mu$ so that $(P\cdot k)^2/\mu^2 = k_0^2$. Comparing (7) with (2) we see that (7) describes an additional magnetic moment, Δg, in units of $e\hbar/2mc$, given by

$$\Delta g = \frac{2e^2 i}{\pi^3 \hbar c}(I_1+I_2), \tag{8}$$

$$I_1 = \int d^4k \int\limits_0^1 dw\, w^2 \frac{\mu^2(w-1)}{(k^2+\mu^2w^2-i\sigma)^3}, \tag{9_1}$$

$$I_2 = \int d^4k \int\limits_0^1 dw\, w \frac{k_n^2+k_0^2}{(k^2+\mu^2w^2-i\sigma)^3}. \tag{9_2}$$

We consider the integral I_1 first. It is properly convergent and easily worked out. We simplify (9_1) by integrating first over two components k_x, k_y using plane polar coordinates $k_x^2+k_y^2 = k_r^2$, $dk_x\,dk_y = \pi\, d(k_r^2)$,

$$I_1 = \frac{\pi}{2}\int\limits_{-\infty}^{+\infty} dk_z \int\limits_{-\infty}^{+\infty} dk_0 \int\limits_0^1 w^2\, dw \frac{\mu^2(w-1)}{(\epsilon^2-k_0^2-i\sigma)^2}, \qquad \epsilon^2 = k_z^2+\mu^2w^2.$$

Next we integrate over k_0 by contour integration through the upper half plane. The denominator can be written, with a slight redefinition of σ,

$$\frac{1}{(\epsilon^2-k_0^2-i\sigma)^2} = \frac{1}{(\epsilon-k_0-i\sigma)^2(\epsilon+k_0-i\sigma)^2},$$

which shows that there is a pole of second order in the upper half plane at $k_0 = -\epsilon+i\sigma$. Consequently, according to Cauchy's formula,

$$\int\limits_{-\infty}^{+\infty} dk_0 \frac{1}{(\epsilon^2-k_0^2-i\sigma)^2} = 2\pi i \frac{\partial}{\partial k_0}\frac{1}{(\epsilon-k_0)^2}\bigg|_{k_0=-\epsilon} = \frac{i\pi}{2}\frac{1}{\epsilon^3}.$$

The remaining integration is trivial and gives

$$I_1 = \frac{i\pi^2}{2}\int\limits_0^1 (w-1)\, dw = -\frac{i\pi^2}{4}. \tag{10}$$

The integral I_2 requires special attention. It is easy to see that the individual parts k_n^2 and k_0^2 diverge, but the sum converges, at least for not specially unsuitable methods of evaluation. Consequently, however, the value of I_2 is not unique but depends on the method of evaluation. The reader is advised to try, for instance, the following two easy methods of integration: (i) integrate (as for I_1) first over k_x, k_y, then over k_0, then over k_z, (ii) integrate first over k_0 in the complex plane (pole of third order), then integrate over $\mathbf{k}$ using spherical polar coordinates (and $k_x^2 = \frac{1}{3}\mathbf{k}^2$, averaged over the angles). In both cases integrate over w last. One finds two finite but *different* values. It is possible, therefore, by choosing different methods of evaluation, to assign any arbitrary numerical value to I_2, including also the value zero. How then can the value of I_2 be fixed? It is settled by the *requirement of relativistic covariance.* Consider an integral of the form

$$I_{\mu\nu} = \int k_\mu k_\nu \frac{d^4k}{(k^2+\mu^2w^2-i\sigma)^3}.$$

Since d^4k, μw, and σ are invariants, $I_{\mu\nu}$ is a tensor with indices μ, ν. Since, however, there is no vector or tensor on which $I_{\mu\nu}$ could depend when the integration is over the full k-space, the only possibility is

$$I_{\mu\nu} = I\delta_{\mu\nu},$$

where I is invariant.

Now I_2, eq. (9_2), is of the form ($k_0^2 = -k_4^2$)

$$I_2 = I_{xx} - I_{44} = I(\delta_{xx} - \delta_{44}) = 0 \tag{11}$$

(if $n = x$, for example, postponing the integration over w).

Thus the value $I_2 = 0$ is the only result that is in accord with the covariance properties of the integrand, and we must assign this value to the (otherwise ambiguous) integral I_2.

It must be admitted that this type of 'wishful mathematics', which is characteristic for many applications of the theory to higher order corrections, is far from satisfactory. One must expect from a satisfactory theory that it leads to unambiguous results which can never contradict the basic foundations of the theory (e.g. its Lorentz invariance). On the other hand, there can be no doubt that the argument employed leads to the correct value and that future improvements of the theory will change practically nothing in the result.

Inserting (10) and (11) into (8) we find

$$\Delta g = \frac{1}{2\pi}\frac{e^2}{\hbar c} = 0{\cdot}001161. \tag{12}$$

This result was first derived by Schwinger.†

† J. Schwinger, *Phys. Rev.* **76** (1949), 790. An interesting alternative derivation has been given by J. M. Luttinger, ibid. **74** (1948), 893. Luttinger starts from the *exact* wave functions of an electron in a homogeneous magnetic field. There is one state where (to a first approximation) the magnetic energy compensates the rest energy completely and the total energy vanishes. The radiative corrections to this state give the energy to a second approximation and this is entirely due to the excess moment. The advantage of this method is that no renormalization of the mass is required. See also J. Géhéniau and F. Villars, *Helv. Phys. Acta,* **23** (1950), 178.

Further contributions to the magnetic moment of the electron arise from corrections of order $\sim e^4, e^6,\ldots$. The e^4-term has also been worked out.† It is equal to $\Delta g' = -2{\cdot}97/\pi^2 137^2$ so that

$$\Delta g + \Delta g' = 0{\cdot}001145. \tag{13}$$

We shall see that this value is in excellent agreement with the experiments.

The calculation can also be extended in another direction. The excess magnetic moment due to the radiative corrections also includes parts depending on the field strength $\mathbf{H}^{(e)}$. The first term $\sim \mathbf{H}^{(e)}$ is interpreted as a *magnetic polarizability* of the electron. Actually, it turns out that the next term in an expansion according to $\mathbf{H}^{(e)}$ is $\sim H^{(e)}\log H^{(e)}$. The field-dependent part of the magnetic moment is given by‡

$$\Delta g_H = \frac{47}{45\pi 137}\frac{e\hbar H^{(e)}}{2mc\mu} + \frac{8}{3\pi 137}\frac{e\hbar H^{(e)}}{2mc\mu}\log\frac{2e\hbar H^{(e)}}{mc\mu} + O(H^{(e)3}). \tag{14}$$

The effect is too small to be measurable, but it is important to notice that the magnetic polarizability (and all higher terms in $H^{(e)}$) are also finite.

3. *Measurements.* The first indication of the fact that the electron has an anomalous magnetic moment came from measurements of the hyperfine structure of hydrogen and deuterium.§ Both turned out to be larger by about 0·3 per cent. than was to be expected from the known magnetic moments of the two nuclei. The fact that the discrepancy was the same percentage for both nuclei made it unlikely that it could be due to effects of nuclear structure or the like. Breit|| was the first to interpret the result in terms of an anomalous magnetic moment. The excess moment was definitely established by precision measurements†† of the Zeeman splittings of various atomic states for which Russell–Saunders coupling can be taken for granted. ($^2P_{\frac{3}{2}}$, $^2P_{\frac{1}{2}}$ of Ga and $^2S_{\frac{1}{2}}$ of Na were first used.) The splittings give the magnetic moments of these atomic states immediately, but owing to the difficulty of measuring $H^{(e)}$ very accurately, only the ratios for two different atomic states could be determined. The results showed slight disagreements with the values to be expected on the assumption that the electron moment was $e\hbar/2mc$ and equal to the orbital magnetic moment

† R. Karplus and N. M. Kroll, *Phys. Rev.* **77** (1950), 536.

‡ S. N. Gupta, *Nature*, **163** (1949), 686. For this result the method of Luttinger (footnote on p. 314) has been used.

§ J. E. Nafe, E. B. Nelson, and I. I. Rabi, *Phys. Rev.* **71** (1947), 914.

|| G. Breit, ibid. **72** (1947), 984; **73** (1948), 1410.

†† H. M. Foley and P. Kusch, ibid. **72** (1947), 1256; **73** (1948), 412; **74** (1948), 250.

for $l = 1$. Since the latter can hardly be different from the expected value, the only interpretation is a change in the magnetic moment of the electron. Later precision measurements‡ of the Zeeman levels of hydrogen gave the value

$$\Delta g = 0{\cdot}001146 \pm 0{\cdot}000012. \tag{15}$$

This is in perfect agreement with the theoretical value (13). We note that even the e^4-contribution is probably just outside the experimental error and seems to be confirmed by the measurements.

The theory of the excess magnetic moment, together with that of the line shift of spectral lines (§ 34), is one of the major successes of quantum electrodynamics, and is particularly important, because these two phenomena are so far the only ones where the radiative corrections are accessible to an experimental test.

32. Vacuum polarization§

1. *The induction tensor and its properties.* We have yet to discuss another contribution to the scattering of an electron in an external field, whose matrix element is given by $G^{(e)}_{3(b)}$ (§ 30 eq. (5 b)). Since P and P' are fixed, $K = P' - P$ is also fixed and different from zero. $\zeta^*(K^2)$ can then be replaced by $1/K^2$. We can write $G^{(e)}_{3(b)}$ in the form

$$G^{(e)}_{3(b)} = -\frac{i_\mu L_{\mu\nu} A^{(e)}_\nu}{cK^2}, \tag{1}$$

where i_μ is the matrix element of the electron current

$$i_\mu = ec(u^\dagger_{P'} \gamma_\mu u_P) \tag{2}$$

and the tensor $L_{\mu\nu}$ is defined by

$$L_{\mu\nu} = \frac{ie^2}{\pi^3 \hbar c} \int d^4p\, \{2p_\nu p_\mu + p_\nu K_\mu + p_\mu K_\nu - \delta_{\mu\nu}(\mu^2 + p^2 + p\cdot K)\} \times$$
$$\times \zeta^*(p^2+\mu^2)\zeta^*(\mu^2+(p+K)^2). \tag{3}$$

In (3) the term $\sim K_\nu$ actually vanishes because of the Lorentz condition for $A^{(e)}_\nu$ (§ 30 eq. (2)):

$$K_\nu A^{(e)}_\nu = 0. \tag{4}$$

However, for reasons of symmetry we leave it standing.

‡ S. Koenig, A. G. Prodell, and P. Kusch, *Phys. Rev.* **88** (1952), 191.

§ The most important publications on vacuum polarization due to electron pairs are the following: W. Heisenberg, *Z. Phys.* **90** (1934), 209; V. Weisskopf, *Det kgl. Dansk. Vid. Selskab*, **14** (1936), No. 6 (includes a summary of earlier work); W. Pauli and M. E. Rose, *Phys. Rev.* **49** (1936), 462; J. Schwinger, ibid. **75** (1949), 651; G. Wentzel, ibid. **74** (1948), 1070; Z. Koba and S. Tomonaga, *Prog. Theor. Phys.* **3** (1948), 290; Z. Koba and G. Takeda, ibid. **3** (1948), 407, **4** (1949), 60 and 130; R. P. Feynmann, *Phys. Rev.* **76** (1949), 769; S. T. Ma, *Phil. Mag.* **11** (1949), 1112; W. Pauli and F. Villars, *Rev. Mod. Phys.* **21** (1949), 434.

Before we proceed to evaluate $L_{\mu\nu}$ we consider its physical interpretation and state certain requirements which must be true if the results are to be physically sound. These requirements are connected with the Lorentz and gauge invariance of the result. They are to be understood in the following sense. If the theory were sound, and therefore the integral $L_{\mu\nu}$ properly convergent, the requirements would be fulfilled automatically because the theory has, from the start, all these invariance properties. Actually $L_{\mu\nu}$ has divergent and ambiguous parts, depending on the way it is evaluated. The situation is similar to that met with in the problem of the magnetic moment (§ 31) where the integral I_2 was ambiguous. Here the situation is aggravated further by divergencies. Thus the following 'requirements' will not necessarily be fulfilled when $L_{\mu\nu}$ is actually evaluated but they have to be enforced by special devices which will be discussed below.

$L_{\mu\nu}$ depends on K only (since p is an integration variable) and is *independent of* P. Moreover, it is seen that when the signs of K and p are both changed $L_{\mu\nu}$ remains unaltered. The change $p \to -p$ has no effect on the integral (integration variable) and it follows that $L_{\mu\nu}$ must be an even function of K. If we regard K as small, $L_{\mu\nu}$ can be expanded, and its tensor character shows that it must have the form

$$L_{\mu\nu} = L_0\,\delta_{\mu\nu} + L_1^{(1)} K_\mu K_\nu + L_1^{(2)} K^2 \delta_{\mu\nu} + L_2^{(1)} K^2 K_\mu K_\nu + L_2^{(2)} (K^2)^2 \delta_{\mu\nu} + \dots \quad (5)$$

Actually the term $L_1^{(1)}$ gives no contribution when inserted in (1), because of (4). Next we use the fact that $G^{(e)}_{3(b)}$ must not depend on the gauge used for $A_\nu^{(e)}$. It must remain unaltered if to $A_\nu^{(e)}(x)$ a term $\partial\chi(x)/\partial x_\nu$ is added, where χ is a scalar. Let $\chi(x)$ be expanded in a Fourier series (like $A_\nu^{(e)}(x)$) and let the 4-dimensional Fourier component be $\chi(K)$. The Fourier component of $\partial\chi/\partial x_\nu$ is then $\chi(K)K_\nu$. The addition of this term to $A_\nu^{(e)}$ in (1) must leave $G^{(e)}_{3(b)}$ unchanged. Therefore $L_{\mu\nu}$ must have the property:

$$L_{\mu\nu} K_\nu = 0. \quad (6)$$

Using this in (5) we find that

$$L_0 = 0, \quad L_1^{(1)} = -L_1^{(2)}, \quad L_2^{(1)} = -L_2^{(2)}, \dots$$

or

$$L_{\mu\nu} = -(K_\mu K_\nu - K^2 \delta_{\mu\nu}) L', \quad (7\,\text{a})$$

$$L' = L_1 + L_2 K^2 + \dots \equiv L'(K^2). \quad (7\,\text{b})$$

(3) shows that $L_{\mu\nu}$ is a relativistic tensor of second rank. It follows that L' is invariant and can therefore depend on K^2 only.

In any sound theory $L_{\mu\nu}$ must have the form (7). For the reasons explained below $L_{\mu\nu}$ is called the induction tensor.

2. *The induced currents.* We now insert (7) into (1) and obtain, using (4),

$$G^{(e)}_{3(b)} = -\frac{1}{c} L'(K^2) i_\nu A^{(e)}_\nu \equiv -\frac{1}{c} \Delta i_\nu A^{(e)}_\nu. \tag{8}$$

This equation is easily interpreted. It describes the *interaction of a new electron current*

$$\Delta i_\nu = i_\nu L' \tag{9}$$

with the external field. In Δi_ν or L' virtual photons and electron pairs are involved. Indeed, through the action of its electromagnetic field, the electron is capable of creating new electron pairs which constitute an 'induced current' Δi_ν. (The interaction of i_ν itself with $A^{(e)}_\nu$ is, of course, described by $G^{(e)}_1$.) It is as if the 'vacuum' behaves like a dielectric, and any charge or current placed into it produces a polarization current. Thus (8) is due to the vacuum polarization already described in § 11.4. Δi_ν will in general not depend on K in the same way as i_ν because L' depends on K^2. For the Fourier transform of Δi_ν, this means that Δi_ν is not the same function of the coordinates as the original current i_ν. The 'dielectric' is not homogeneous. The reason is obviously the fact that the field of the electron is inhomogeneous. Only the first term in the expansion (7 b), L_1, is independent of K^2, and the corresponding term $\Delta i_\nu(0)$ is exactly proportional to i_ν. This will be discussed further below.

We may in fact work out the polarization current produced by a free electron directly without considering the interaction with the external field, and we obtain exactly the result (9). That $G^{(e)}_{3(b)}$ describes the interaction of a polarization current with $A^{(e)}_\nu$ is also evident from the interaction diagram Fig. 21 (*b*). It is seen that the electron through the field produced by it (virtual photon line) creates a pair which interacts with $A^{(e)}_\nu$ (i.e. is annihilated by $A^{(e)}_\nu$).

There is also an alternative interpretation of (8). We can write (8) in the form

$$G^{(e)}_{3(b)} = -4\pi\hbar^2 \frac{i_\nu \Delta I^{(e)}_\nu}{K^2}, \qquad \Delta I^{(e)}_\nu = \frac{K^2 L' A^{(e)}_\nu}{4\pi\hbar^2 c}. \tag{10}$$

$K^2 A^{(e)}_\nu$ is the Fourier transform of $-\hbar^2 c^2 \Box A^{(e)}_\nu(x)$, and if we denote by $I^{(e)}_\nu$ the external current which produces the external field $A^{(e)}_\nu(x)$, namely,

$$\frac{4\pi}{c} I^{(e)}_\nu = -\Box A^{(e)}_\nu,$$

or

$$K^2 A^{(e)}_\nu = 4\pi\hbar^2 c I^{(e)}_\nu,$$

we obtain

$$\Delta I^{(e)}_\nu = I^{(e)}_\nu L'. \tag{11}$$

(10) is the retarded interaction between the two currents i_ν and $\Delta I_\nu^{(e)}$.† $\Delta I_\nu^{(e)}$ given by (11) can be interpreted as the 'polarization current' produced by the external field, and indeed, if we work out this polarization current directly (without considering the electron) we also find (11) exactly.

It is, at first sight, surprising that the two interpretations are *alternative* and not additive. One would think that if $A^{(e)}$ produces the current (11) and i_ν the current (9) (as the direct calculation in fact shows!), then in the interaction between $A_\nu^{(e)}$ and i_ν we should have to add the two effects, and thus obtain a factor 2. This factor 2 is, however, missing. The point is connected with the problem of measurement. If we wish to observe the polarization through its physical effects, we have to choose some known starting-point, for instance an external field $A^{(e)}$, which we must assume to be *known*. $A^{(e)}$ therefore already includes its polarization current. Then we could measure whatever physical effects are produced by Δi_ν. Or, *alternatively*, the electron may be chosen as a starting-point, and we could measure $\Delta I_\nu^{(e)}$, but we could not do both at the same time.‡

3. *Charge renormalization.* If we use the expansion (7 b) the first term of L' (L_1) is independent of K^2. The polarization current $\Delta i_\nu(0)$ is then exactly proportional to i_ν (or $\Delta I_\nu^{(e)}$ to $I_\nu^{(e)}$). $G_{3(b)}^{(e)}$ will then also be exactly proportional to $G_1^{(e)}$,

$$G_{3(b)}^{(e)}(K^2 \to 0) = L_1 G_1^{(e)}. \tag{12}$$

L_1 does not depend on anything and is therefore a universal constant and Lorentz invariant.§

The physical effects due to (12) are identical with the first-order effects $G_1^{(e)}$ and differ only by the numerical factor L_1. (12) can be completely accounted for by a *change* of one *universal constant*, namely, the *electronic charge*, or rather the interaction constant e^2 between the two currents i_ν and $I_\nu^{(e)}$ in question. If $G_1^{(e)}$ is written as the interaction

† See, for example, § 24 eqs. (7), (9), which give the matrix element of the retarded interaction in momentum space between two electron currents, for a transition where the energy–momentum change of one particle is ϵ, $\mathbf{k}$. In (10) the energy–momentum change is K_0, $\mathbf{K}$.

‡ The point is stated incorrectly in many papers on vacuum polarization in which merely the current $\Delta I_\nu^{(e)}$ is worked out, and from this the charge renormalization derived. This gives rise to an erroneous factor 2 in δe.

§ It is to be noted that the contributions $G_{3(a)}^{(e)}+G_{3(c1)}+G_{3(c2)}$ discussed in §§ 30, 31 do not give rise to a similar contribution when we merely put $K \to 0$. It is true, that there are contributions proportional to $G_1^{(e)}$ when we put $\mathbf{P}-\mathbf{P}' = 0$, but in general these depend on P in addition to K and therefore do not permit the same interpretation as will be given for (12).

of the two currents i_ν and $I_\nu^{(e)}$, it is proportional to e^2. (12) then describes a change of e^2, called *self-charge*, due to the second-order radiative corrections:

$$\delta(e^2) = e^2 L_1, \quad \text{or} \quad \delta|e| = \frac{|e|}{2} L_1. \tag{13}$$

Thus in the case $K^2 \to 0$ the vacuum behaves like a homogeneous constant dielectric which has the effect of changing all charges effectively by a constant amount. As in the case of an ordinary dielectric it is not difficult to see that the conservation of charge is maintained in that the balance, $-\delta|e|$ for each elementary charge placed in the dielectric, is removed to infinity.

Now any such change of charge is completely unobservable. We have no means of 'removing the dielectric', which is the vacuum, and of preventing the creation of virtual pairs by a charge. We can only measure 'e' by studying the behaviour of an electron in a given known external field (or the interaction with another particle). If 'e' is measured in this way, the measured value already includes the charge $\delta|e|$ due to the radiative corrections. This is much the same situation which we have already met in connexion with the radiative change of mass $\delta\mu$ which was also *unobservable in principle*.

If, in the discussion of physical phenomena, we always use the *observed values* of all charges (as we always do), it follows, therefore, that the contribution L_1 or $G_{3(b)}^{(e)}(K^2 \to 0)$ has to be omitted. The same applies, of course, to corresponding contributions of higher orders in e^2. We shall find that δe, like $\delta\mu$, is really *divergent* and the fact that it is an *unobservable quantity* is of fundamental importance for the applicability of quantum electrodynamics to physical phenomena.

For the consistency of this interpretation a further requirement is, however, necessary. Unlike the masses of the elementary particles, which have somewhat odd values, the charges of all elementary particles are exactly equal. Whilst we may well accept small and different changes of the masses due to electromagnetic interaction (assuming that eventually, in a correct theory $\delta\mu$ will be finite), a change δe can only be accepted if it turns out that the δe for all elementary particles is exactly the same. Although, at the time of writing, no such general proof has yet been given, it is almost certain that this is really the case. δe given by (13) is independent of the nature of the external current $I_\nu^{(e)}$. $I_\nu^{(e)}$ may be due to a proton, electron, meson, or any other particle. It is true, however, that L_1 depends on the mass and spin of the *virtual* pairs. In the general case, however, when two currents belonging to two different particles interact, pairs of *all* existing particles will be created and contribute to the polarization current, and the resulting δe, though it will have further contributions due to pairs of particles other than electrons, will be independent of the nature of the two original particles considered. It is also to be noted that no exchange effects between the original electron and the virtual electrons of the polarization current occur. This follows from the fact that the same polarization current can be interpreted as produced by either the electron or the external field. The problem is completely symmetrical in the two particles considered, and therefore, if all virtual pairs are taken into account, cannot depend on the particles considered. This probably also applies to the e^4...-corrections.

It is possible to eliminate the self-charge by embodying the change of e in the Hamiltonian in much the same way as we have done for $\delta\mu$ in § 29. Since for the further applications the charge correction results in a mere omission we refer the reader to the literature.†

4. *Evaluation of the induction tensor L.* We proceed now to evaluate the induction tensor $L_{\mu\nu}$. We shall meet with a situation even worse than in the case of the ambiguous integral I_2 of § 31. The ambiguous parts of $L_{\mu\nu}$ diverge, disagree with the general requirements (7), and thus violate both Lorentz and gauge invariance.

We first work out $L_{\mu\nu}$ explicitly by straightforward integration. In order to remove the linear terms in K in the numerator we rewrite (3) introducing $p_\mu+\frac{1}{2}K_\mu$ as integration variable (denoted again by p_μ):

$$L_{\mu\nu} = \frac{ie^2}{\pi^3\hbar c}\int d^4p\,\frac{2p_\nu p_\mu-\frac{1}{2}K_\mu K_\nu-\delta_{\mu\nu}(\mu^2+p^2-\frac{1}{4}K^2)}{[(p+\frac{1}{2}K)^2+\mu^2-i\sigma][(p-\frac{1}{2}K)^2+\mu^2-i\sigma]}. \tag{14}$$

Next we integrate over p_0 by contour integration. In the upper half plane, the denominator has two single poles at

$$\begin{gathered} p_0 = -\tfrac{1}{2}K_0-E_1 \qquad p_0 = +\tfrac{1}{2}K_0-E_2, \\ E_1 = (\mu^2+(\mathbf{p}+\tfrac{1}{2}\mathbf{K})^2)^{\frac{1}{2}}, \qquad E_2 = (\mu^2+(\mathbf{p}-\tfrac{1}{2}\mathbf{K})^2)^{\frac{1}{2}} \end{gathered} \tag{15}$$

both shifted above the real axis by an infinitesimal amount. The integration requires a distinction of cases (in $p_\mu p_\nu$) according as to whether μ and ν are space or time indices. After a short calculation

† S. N. Gupta, *Proc. Phys. Soc.* **44** (1951), 426.

(making use of (15) explicitly) we find:

$$L_{\mu\nu} = +\frac{e^2}{\pi^2\hbar c}\int d^3p\,\frac{1}{E_1E_2[(E_1+E_2)^2-K_0^2]}\Big\{(E_1+E_2)[-2p_\nu p_\mu\beta_{\nu4}\beta_{\mu4}+ +\tfrac{1}{2}K_\mu K_\nu-2\delta_{\mu4}\delta_{\nu4}(E_1E_2+\tfrac{1}{4}K_4^2)+\tfrac{1}{2}\delta_{\mu\nu}((E_1+E_2)^2-\mathbf{K}^2)]- -(E_1-E_2)K_4(p_\nu\beta_{\nu4}\delta_{\mu4}+p_\mu\beta_{\mu4}\delta_{\nu4})\Big\} \tag{16}$$

where

$$\beta_{\mu\nu} = 1-\delta_{\mu\nu}. \tag{16'}$$

On account of the factors $\beta_{\nu4}$, etc., p_4 (or p_0) no longer occurs. For the further integration we use invariant variables similar to those used in the calculation of the self-energy (§ 29.1). We put

$$\left.\begin{aligned} &E_1-E_2 = v, \qquad E_1+E_2 = w, \qquad z^2 = w^2-\mathbf{K}^2\\ &d^3p = \frac{E_1E_2}{2|\mathbf{K}|}\,dv\,dw\,d\phi\\ &-\frac{1}{z}|\mathbf{K}|\sqrt{(z^2-4\mu^2)} \leqslant v \leqslant \frac{1}{z}|\mathbf{K}|\sqrt{(z^2-4\mu^2)}\\ &\sqrt{(4\mu^2+\mathbf{K}^2)} \leqslant w \leqslant \infty, \quad \text{or} \quad 2\mu \leqslant z < \infty \end{aligned}\right\}. \tag{17}$$

ϕ is the azimuthal angle of $\mathbf{p}$ about the direction of $\mathbf{K}$. Since iE_1 and iE_2 can be regarded as fourth components of a vector with momenta $\frac{1}{2}\mathbf{K}+\mathbf{p}$ and $\frac{1}{2}\mathbf{K}-\mathbf{p}$ respectively, $i(E_1+E_2)$ and $\mathbf{K}$ form a 4-vector and therefore $z^2 = w^2-\mathbf{K}^2$ is *invariant*. Therefore also

$$(E_1+E_2)^2-K_0^2 = w^2-K_0^2 = z^2+K^2$$

is invariant.

The term $p_\nu p_\mu$ of (16) contains only spatial indices and, on account of the integration over ϕ, gives $-2p_\nu^2\delta_{\mu\nu}\beta_{\nu4}$ (no summation over ν). Likewise the last term (K_4p_ν, K_4p_μ) only gives a contribution when ν (or μ) is in the $\mathbf{K}$-direction. $\mathbf{p}^2$ and $(\mathbf{pK})$ can be expressed through (15) by E_1, E_2, and hence by v and z. The integration over ϕ and v is elementary. The result is readily expressed in the form:

$$L_{\mu\nu} = \bar{L}_{\mu\nu}+\bar{\bar{L}}_{\mu\nu},$$

$$\bar{L}_{\mu\nu} = \frac{2}{3}\frac{e^2}{\pi\hbar c}\int\limits_{2\mu}^{\infty}\frac{dz}{z^2+K^2}\,\frac{(z^2+2\mu^2)\sqrt{(z^2-4\mu^2)}}{z^2}\,(K_\mu K_\nu-K^2\delta_{\mu\nu}), \tag{18 a}$$

$$\bar{\bar{L}}_{\mu\nu} = \frac{2}{3}\frac{e^2}{\pi\hbar c}\int\limits_{2\mu}^{\infty}\frac{(z^2+2\mu^2)\,dz}{z^2}\sqrt{(z^2-4\mu^2)}\,\delta_{\mu\nu}\beta_{\mu4}. \tag{18 b}$$

The part $\bar{L}_{\mu\nu}$ has the form required by gauge and Lorentz invariance, whereas $\bar{\bar{L}}_{\mu\nu}$ violates both (factor $\beta_{\mu4}$). We discuss first $\bar{L}_{\mu\nu}$.

As was to be expected $\bar{L}_{\mu\nu}$ depends, apart from the necessary factor $K_\mu K_\nu - K^2\delta_{\mu\nu}$, on K^2 only, and the coefficient L' is invariant. If we expand it in powers of K^2 and compare with (7) we find

$$\bar{L}_1 = -\frac{2}{3\pi 137}\int\limits_{2\mu}^{\infty}\frac{dz}{z^4}(z^2+2\mu^2)\sqrt{(z^2-4\mu^2)}. \tag{19}$$

$\bar{L}_1$ diverges logarithmically for large z ($\sim \log z$). Its physical meaning is the charge renormalization given by (13)

$$\delta|e|/|e| = \tfrac{1}{2}\bar{L}_1. \tag{20}$$

It is seen that $\delta|e|$ is in the direction of a diminution of the absolute value of e. We can argue about this self-charge in much the same way as for the self-mass. If, in any future theory, the divergence of δe and $\delta\mu$ is removed, the actual value of both will presumably be very small on account of the factor 1/137. Since z is an invariant variable (of dimensions μ^2) some large invariant cutting-off value z_{max} may be introduced to make δe finite and then even if z_{max} is fairly large δe will still be very small.

$\bar{L}_2$ and all higher terms in the expansion according to K^2 converge. $\bar{L}_2$, $\bar{L}_3$,... lead to *observable physical effects*. The contribution to $G^{(e)}_{3(b)}$ is of order K^2 and higher, apart from any dependence on P and P' of i_ν. $\bar{L}_2$, $\bar{L}_3$,... do not therefore contribute to the magnetic moment of an electron at rest and we were justified in disregarding $G^{(e)}_{3(b)}$ for its computation (pending the proof of $\bar{\bar{L}} = 0$).

The evaluation of (18 a) gives†

$$\bar{L}'(K^2)-\bar{L}'(0) = \frac{1}{3\pi 137}\left[-\frac{5}{3}+\frac{1}{\gamma}+\frac{2\gamma-1}{\gamma}\sqrt{\frac{1+\gamma}{\gamma}}\log(\sqrt{\gamma}+\sqrt{(1+\gamma)})\right],$$

$$\gamma = \frac{K^2}{4\mu^2}. \tag{18 c}$$

For small K^2 the first term in the expansion is

$$\bar{L}_2 K^2 = \frac{1}{15\pi 137\mu^2}K^2. \tag{18 d}$$

Next we have to discuss $\bar{\bar{L}}$. The fact that from an obviously covariant expression $L_{\mu\nu}$ eq. (3), when evaluated, a non-covariant contribution (18 b) can be derived, means that the evaluation is ambiguous. In fact $\bar{\bar{L}}$ is even quadratically divergent. It is also independent of K. Even if gauge invariance were not required it would have to be of the form $L_0\delta_{\mu\nu}$, but (18 b) has only spatial elements L_{ii} and $L_{44} = 0$. That $\bar{\bar{L}}$

† Cf. E. A. Uehling, *Phys. Rev.* **48** (1935), 55; W. Pauli and M. E. Rose, loc. cit.

is ambiguous can be seen by another mode of evaluation. We form $L_{\mu\nu}K_\nu$ directly from (3). As $\bar{L}_{\mu\nu}K_\nu = 0$ automatically, we can 'prove' that $\bar{\bar{L}}_{\mu\nu}K_\nu = 0$ if we can show that $L_{\mu\nu}K_\nu = 0$. Since $\bar{\bar{L}}_{\mu\nu}$ is, by definition, not of a form for which this would be fulfilled automatically, it would follow that $\bar{\bar{L}}_{\mu\nu} = 0$. Apart from trivial factors we can write

$$L_{\mu\nu}K_\nu \simeq \int d^4p\, \frac{p_\mu((p+K)^2+\mu^2)-(K_\mu+p_\mu)(p^2+\mu^2)}{[(p+K)^2+\mu^2-i\sigma][p^2+\mu^2-i\sigma]}.$$

We can 'prove' immediately that this vanishes. In the limit $\sigma \to 0$ we can put $(p^2+\mu^2)/(p^2+\mu^2-i\sigma) = 1$. Likewise for the first term. Then

$$L_{\mu\nu}K_\nu \simeq \int d^4p \left(\frac{p_\mu}{p^2+\mu^2-i\sigma} - \frac{K_\mu+p_\mu}{(p+K)^2+\mu^2-i\sigma}\right).$$

This integral obviously vanishes, for if we change the variable in the second term $K+p \to p'$ the two terms cancel if the integration is first carried out over the same finite domain of p and p' and the limits of integration are afterwards allowed to go to infinity. Or, alternatively, each term vanishes, for reasons of symmetry, when integrated, first over p_μ (and p'_μ) up to finite limits $-X_\mu ... +X_\mu$, say, covering the point $p_\mu = 0$ symmetrically. However, the integral diverges strongly and the 'proof' is not cogent, but evidently ways of integration exist for which $\bar{\bar{L}} = 0$. It depends on the order in which the various infinite integration limits are reached whether we get zero or a divergent non-covariant and non-gauge invariant result.

Thus we find that in the present form of quantum electrodynamics ambiguities arise which, in actual computations, may destroy both Lorentz and gauge invariance. On the other hand, the ambiguities may be used to *enforce* the required invariance properties by not permitting ways of evaluation which destroy them. Clearly in our case $\bar{\bar{L}}$ must vanish.

In § 35 attempts at removing the above ambiguities will be discussed. It will be seen that so far they have met with only partial success.

5. *Photon self-energy.* A special application of our discussion is the self-energy of the photon. Through intermediate production and re-annihilation of electron pairs a free photon has, according to elementary calculations, a self-energy which proves to be quadratically divergent. The diagram representing it is – – –·⇄·– – –. We need not evaluate it separately because it is already implicit in our calculations (as diagram Fig. 21 (*b*) shows). Quite clearly, if the photon had a self-energy, i.e. a mass, gauge invariance would be destroyed, in contradiction to the starting-point of the theory. We shall see that the so-called photon

self-energy is just given by $\overline{\overline{L}}$ and vanishes as soon as ways have been found to make $\overline{L} = 0$, e.g. by postulating special ways of evaluation.

Nothing changes in the formalism, as we have seen in § 28, if the external field $A_\nu^{(e)}$ is replaced by the operator A_ν of a real photon. Only the Lorentz condition $K_\nu A_\nu = 0$ is not valid as an operator equation fulfilled identically. The real photon produces then a polarization current which, if (4) is not used, is given by (compare (1) and (10))

$$\Delta I_\mu = \frac{L_{\mu\nu} A_\nu}{4\pi\hbar^2 c}. \tag{21}$$

Now this current interacts again with the photon field and the interaction is $-(1/c)\Delta I_\mu A_\mu$. If we take the expectation value of this interaction for one photon with momentum **K** present, we obtain the self-energy of the photon

$$W = -\frac{L_{\mu\nu}}{4\pi\hbar^2 c^2} \langle A_\mu A_\nu \rangle_{1\,\mathrm{phot}}. \tag{22}$$

$\langle A_\mu A_\nu \rangle_{1\,\mathrm{phot}}$ is the product of the absorption and the emission operator for the photon considered. For this we can use the formulae of § 10.

$$\langle A_\mu A_\nu \rangle_{1\,\mathrm{phot}} = \frac{2\pi\hbar^2 c^2}{|\mathbf{K}|} \delta_{\mu e} \delta_{\nu e},$$

where **K** is now the momentum and **e** the direction of polarization of the photon. We then obtain

$$W = -\frac{1}{2|\mathbf{K}|} L_{ee}. \tag{23}$$

Since **K** now refers to a real photon, $K^2 = 0$. Furthermore, **K** is perpendicular to **e**, and so $K_e = 0$. We see at once that the part $\overline{L}$ of L, which is the physically sound part, gives no contribution; thus $W = -\overline{\overline{L}}_{ee}/2|\mathbf{K}|$, and this is what elementary calculations have also given before. Clearly, any device or postulate that makes $\overline{\overline{L}} = 0$ leads to a vanishing photon self-energy.

6. *Observable effects of vacuum polarization.* We have seen that any parts $\overline{\overline{L}}$, if they occur, have no physical meaning and must be discarded.

$\overline{L}'(K^2)$ is physically meaningful but the part $\overline{L}_1 = \overline{L}'(0)$ describes the change of charge and is unobservable. We are left with the contributions $\overline{L}_2 K^2$,..., etc., or $\overline{L}'(K^2) - \overline{L}'(0)$ which describe observable phenomena. These parts of L can be understood to describe a space-dependent induced current.

L_2, L_3,... contribute in general to the radiative corrections of collision processes, the displacement of atomic energy levels (§ 34), etc. The part played in these corrections by vacuum polarization is usually uninteresting. Phenomena which rest entirely on the observable parts of vacuum polarization are, however, of prime interest. The following is an example. A static external $A^{(e)}$ field gives rise to an inhomogeneous induced current distribution $\Delta I^{(e)}$, which is different from that producing the external field.

Now consider a photon passing through the field $A^{(e)}$. Whereas by Maxwell's theory the external field and the photon field are superposed without any interaction, the photon will now be scattered by the induced current. Thus, the photon is deflected by the external field. Or, in terms of virtual processes, the external field produces a virtual pair of electrons, the incoming photon is absorbed by one of the electrons (which changes its momentum), the pair is annihilated, emitting a secondary photon into another direction.

Similarly, replacing the external field by a photon, two photons will be scattered by each other. These qualitative results show that quantum electrodynamics, more precisely the interaction of the electromagnetic field with the virtual vacuum electrons, leads to a phenomenological *departure from the linearity of Maxwell's equations.* However, this non-linearity is not an inherent property of the electromagnetic field itself but is due to the interaction with positrons.†

The calculation of the scattering of light by light is quite straightforward (but lengthy) and as an illustration for these non-linear effects we give the result:‡

When two photons with the same frequency ν, the same polarization and opposite momenta collide, the cross-section for the deflexion (of both quanta) by an angle θ is given by (when $\hbar\nu \ll \mu$)

$$d\phi = \frac{r_0^2\, d\Omega}{2\pi^2 . 5^2 . 3^4} \frac{(3+\cos^2\theta)^2}{137^2}\left(\frac{\hbar\nu}{\mu}\right)^6 \quad \text{N.R. } (\hbar\nu \ll \mu). \tag{24}$$

Owing to the factor $1/137^2$ and the other numerical factors this is extremely small even for γ-rays $\hbar\nu \sim \mu$ ($\phi \sim r_0^2 . 10^{-6}$ for $\hbar\nu = \mu$) and completely negligible for visible light. It is interesting that ϕ appears to increase very rapidly with $\hbar\nu$ and so would increase beyond all limits

† A phenomenological non-linear theory of the Maxwell field has been established by M. Born, *Proc. Roy. Soc.* A, **143** (1934), 410 and M. Born and L. Infeld, ibid. **144** (1934), 425; **147** (1934), 522; **150** (1935), 141. This theory has many features in common with the non-linearities arising in quantum electrodynamics. See, for example, E. Schrödinger, *Proc. Roy. Ir. Ac.* **47** (1942), 77 (and subsequent papers); J. McConnell, ibid. **49** (1943), 149 (scattering of light by light).

‡ H. Euler, *Ann. Phys.* **26** (1936), 398.

if (24) continued to hold. However, this is not the case. In the extreme relativistic region it has been shown† that the total cross-section is

$$\phi = r_0^2 k\left(\frac{\mu}{\hbar\nu}\right)^2 \qquad \text{E.R. } (\hbar\nu \gg \mu), \quad (25)$$

where k is a small numerical constant. ϕ therefore never exceeds the order of magnitude $r_0^2.10^{-6}$. It is unlikely that the scattering of light by light and other non-linear effects will play a physical role, even in the interior of stars where the density of photons is very high.

33. Corrections to the Compton effect

1. *General.* As an example for the computation of higher-order corrections to collision processes we consider as the simplest case the Compton effect. This is typical for most collision processes, and what we shall find for the order of magnitude of the corrections will also apply, more or less, to the scattering of an electron in an external field, pair production by γ-rays, etc. There are two different types of corrections to be considered, which we call radiative and damping corrections respectively:

(i) *Radiative corrections.* The matrix element for the process consists of a series $K = K_2+K_4+...$ and in the calculations of § 22 only K_2 was taken into account. K_4 leads to a correction and since $K_2 \sim e^2$, $K_4 \sim e^4$, the lowest order in e, due to cross-terms $K_2 K_4^*+K_2^* K_4$ (when $|K|^2$ is formed), is of the order e^6. It will be seen that this correction is a pure quantum effect (it will be $\sim \hbar$ when otherwise expressed by ν, m, etc.), and has no analogue in classical theory. An important problem which will be solved through the effects due to K_4 is the *infra-red problem*, occurring whenever the emission of low-energy quanta occurs. We shall discuss it in this connexion in subsection 3.

(ii) *Damping corrections.* The cross-section is not given by $|K|^2$ but by $|U|^2$, where U satisfies (§ 15 eq. (33'))

$$U_{FO} = K_{FO} - i\pi K_{FB}\rho_B U_{BO}. \quad (1)$$

Even if K_4 is neglected, the second (damping) term of (1) gives a correction. It will be seen that in the non-relativistic approximation, this correction is the same as that caused by the classical damping force, and we shall in fact obtain the classical Thomson formula corrected by the damping. In the relativistic region the damping effects are, of course, modified by quantum effects and we shall see that they are very small for all energies.

† A. Achieser, *Phys. Z. Sowj. Un.* **11** (1937), 263. See also R. Karplus and M. Neuman *Phys. Rev.* **83** (1951), 776.

Whilst the damping effect rests on the lowest order K (with further corrections from the higher orders of K) and is from every point of view quite unambiguous, the computation of K_4 and higher orders becomes possible only when the technique of discarding the self-mass $\delta\mu$ and self-charge δe, treated in the preceding sections, is applied.

Although it is useful to make the distinction between the two corrections because they correspond to rather different physical effects, this distinction should not be understood to imply any difference in order of magnitude. There may be cases where the one or the other effect is actually larger, or both are of the same order of magnitude.

The technique by which K_4 can be computed is that developed in § 28. Nevertheless, the calculation is very long and will not be reproduced here. All interesting features can, however, be seen from the treatment of a simplified problem for which the calculation will reduce to a few lines. We shall confine ourselves to the 'non-relativistic case' $k_0 \ll \mu$ (k_0 = incident photon) and we shall only consider the lowest order in k_0/μ. It will be seen that this is† $\sim (k_0/\mu)^2 \log(\mu/k_0)$.

We use then the non-relativistic interaction in Coulomb gauge

$$H = \frac{e^2}{2\mu}\mathbf{A}^2 - \frac{e}{\mu}(\mathbf{pA}). \tag{2}$$

For the order of magnitude in question, this is really justifiable only when the spin of the particles is zero. For particles with spin $\frac{1}{2}$, (2) would in general not be sufficiently accurate, to the order considered, even when supplemented by the spin term $-(e\hbar c/2\mu)(\boldsymbol{\sigma}\mathbf{H})$.‡ The cross-section for particles with spin $\frac{1}{2}$ differs, if at all (see p. 332), from

† Strictly speaking, the term non-relativistic should be applied only to terms k_0/μ at most, and terms k_0^2/μ^2 include really relativistic effects. Compare footnote below.

‡ That (2) is sufficiently accurate to the order considered for spinless particles is by no means trivial. The fact can be shown from a study of the relativistic wave equation for such particles, which is outside the scope of this book. For the reader interested in this matter we only remark that the point in question can be shown most simply by using the relativistic wave equation in the form of Sakata–Taketani

$$H\psi = \left\{\beta\mu + \frac{1}{2\mu}\beta(1+\eta)(\mathbf{p}-e\mathbf{A})^2\right\}\psi,$$

where β, η are anticommuting 2-row, 2-column matrices, $\beta\eta+\eta\beta = 0$, $\beta^2 = \eta^2 = 1$. This wave equation, although relativistic, has an appearance that is rather similar to the non-relativistic form. ψ is normalized $\psi^*\beta\psi = \pm 1$ for positive/negative particles (cf. S. Sakata and M. Taketani, *Sc. Pap. Inst. Phys. Chem. Research*, Tokyo, **38** (1940), 1; *Proc. Mat. Phys. Soc. Japan*, **22** (1940), 757; W. Heitler, *Proc. Roy. Ir. Ac.* **49** (1943), 1). In particular it is necessary to show that intermediate states with production of pairs of particles contribute to a lesser order of magnitude. For spin $\frac{1}{2}$ in general no simple non-relativistic wave equation exists which is accurate to the order required. For the following derivation see; W. Heitler and S. T. Ma, *Phil. Mag.* **11** (1949), 651.

the case considered here only by different numerical factors in some terms, and the result that is obtained by an exact relativistic calculation, when afterwards we go to the limit $k_0/\mu \ll 1$, will be quoted below.

2. *Radiative corrections.* For our restricted problem it is by far the simplest to use the ready-made formulae § 15 eq. (22_4) for the fourth order contribution $\tilde{K}_4$ of the matrix element. In these formulae the self-energies are already accounted for (indicated by $\sim$) so that no mass renormalization will be required. The charge renormalization will be readily effected in the result.

We consider the particle to be initially at rest, $\mathbf{p}_0 = 0$ (notations as in § 22). Then the matrix element $H^{(1)}$ of $(\mathbf{pA})$ when it stands on the extreme right vanishes. Formula (22_4) of § 15 then reduces to three terms, viz.

$$\tilde{K}_{4FO} = \sum_n \frac{H^{(2)}_{Fn} H^{(2)}_{nO}}{E_O - E_n} + \sum_{l,n} \frac{H^{(1)}_{Fl} H^{(1)}_{ln} H^{(2)}_{nO}}{(E_O - E_l)(E_O - E_n)} - \frac{1}{2} \sum_j \frac{H^{(1)}_{Fj} H^{(1)}_{jF}}{(E_O - E_j)^2} H^{(2)}_{FO}. \quad (3)$$

The matrix elements of $H^{(1)}$ (term linear in $\mathbf{A}$ of (2)) and $H^{(2)}$ ($\mathbf{A}^2$-term) are given in § 14.3:

$$H^{(1)} = -\frac{e\hbar c}{\mu} \sqrt{\frac{2\pi}{k}} (\mathbf{pe}), \qquad H^{(2)} = \frac{2\pi e^2 \hbar^2 c^2}{\mu \sqrt{(kk')}} (\mathbf{ee}'). \quad (4)$$

To the lowest order we would simply have $\tilde{K}_{2FO} = H^{(2)}_{FO}$. $H^{(1)}$ applies to the emission or absorption of a photon $\mathbf{k}$ (polarization $\mathbf{e}$) and $\mathbf{p}$ is the momentum of the particle before or after the act. $H^{(2)}$ applies to any process in which two quanta $\mathbf{k}$, $\mathbf{k}'$ are involved (scattering $\mathbf{k} \to \mathbf{k}'$ or emission of k and k', etc.).

There are two types of intermediate states n, with either one photon with variable momentum $\mathbf{k}'$ or else three photons $\mathbf{k}$, $\mathbf{k}_0$, $\mathbf{k}'$ present:

$$n_1: \quad \mathbf{k}', \mathbf{p}' = \mathbf{k}_0 - \mathbf{k}',$$
$$n_2: \quad \mathbf{k}, \mathbf{k}_0, \mathbf{k}', \mathbf{p}' = -(\mathbf{k} + \mathbf{k}').$$

It is necessary for reasons of convergence to include the recoil energy of the electron in the energy denominators in non-relativistic approximation. This can also be shown to be correct to the order of magnitude considered.

So:

$$E_O - E_{n_1} = k_0 - k' - (\mathbf{k}_0 - \mathbf{k}')^2/2\mu, \qquad E_O - E_{n_2} = -k - k' - (\mathbf{k} + \mathbf{k}')^2/2\mu. \quad (5)$$

For each state n, there are just two isolated states l with no further variable momentum leading to the final state F. In the renormalization term (factor $-\frac{1}{2}$), however, j are states with a variable momentum $\mathbf{k}'$.

The first term of (3) becomes:

$$\sum_n \frac{H^{(2)}_{Fn} H^{(2)}_{nO}}{E_O - E_n} = \sum_{e'} \int \frac{k'^2 dk' d\Omega_{k'}}{(2\pi\hbar c)^3} \left(\frac{2\pi e^2 \hbar^2 c^2}{\mu}\right)^2 \frac{(\mathbf{e}_0 \mathbf{e}')(\mathbf{e}'\mathbf{e})}{k' \sqrt{(k_0 k)}} \times$$
$$\times \left\{\frac{1}{k_0 - k' - (\mathbf{k}_0 - \mathbf{k}')^2/2\mu} - \frac{1}{k + k' + (\mathbf{k}+\mathbf{k}')^2/2\mu}\right\}. \quad (6)$$

We evaluate (6) up to order $k_0/\mu^3 \log(\mu/k_0)$, neglecting even k_0/μ^3. It is then seen that we can neglect $(\mathbf{k}_0\mathbf{k}')/\mu$ and $(\mathbf{k}\mathbf{k}')/\mu$ in the denominators because, if we expand in these terms, the integration over the direction of $\mathbf{k}'$ gives zero and the quadratic terms are of order $1/\mu^4$. The denominators, wherever they can vanish, are principal values. The integration is very simple, and, taking into account the conservation laws

$$k_0 - k = \frac{(\mathbf{k}_0 - \mathbf{k})^2}{2\mu} = \frac{k_0^2}{\mu}(1 - \cos\theta),$$

where θ is the scattering angle ($\sphericalangle\, \mathbf{k}_0, \mathbf{k}$), we obtain

$$\sum_n \frac{H^{(2)}_{Fn} H^{(2)}_{nO}}{E_O - E_n} = -\frac{4}{3} \frac{e^4 \hbar c}{\mu \sqrt{(k_0 k)}} (\mathbf{e}_0 \mathbf{e}) \left\{4 \log \frac{k'}{2\mu}\bigg|_{k'\to\infty} - \frac{k_0^2}{\mu^2}(3 + \cos\theta) \log \frac{\mu}{k_0}\right\}. \quad (7)$$

The first, logarithmically divergent, term will be discussed below. In much the same way, we obtain for the second term of (3), which converges:

$$\sum_{l,n} \frac{H^{(1)}_{Fl} H^{(1)}_{ln} H^{(2)}_{nO}}{(E_O - E_l)(E_O - E_n)} = -\frac{8}{3} \frac{e^4 \hbar c}{\mu^3} \frac{(\mathbf{e}_0 \mathbf{k})(\mathbf{e}\mathbf{k}_0)}{\sqrt{(k k_0)}} \log \frac{\mu}{k_0}. \quad (8)$$

A new and important feature arises in the last (the renormalization) term. There are two intermediate states

$$j_1: \quad \mathbf{k}, \mathbf{k}', \mathbf{p}_1 = \mathbf{k}_0 - \mathbf{k} - \mathbf{k}', \qquad E_O - E_{j_1} = k_0 - k - k' - (\mathbf{k}_0 - \mathbf{k} - \mathbf{k}')^2/2\mu$$
$$= -k' - k'^2/2\mu + (\mathbf{k}', \mathbf{k}_0 - \mathbf{k})/\mu,$$
$$j_2: \quad \mathbf{p}_2 = \mathbf{k}_0, \qquad E_O - E_{j_2} = k_0 - k_0^2/2\mu.$$

j_2 is an isolated state, and it is easily seen that it gives a vanishing contribution when the volume L^3, in which the process is considered to take place, goes to infinity. Now $E_O - E_{j_1}$ is proportional to k' and vanishes when $k' \to 0$. The volume element of the integration in k'-space is $k'^2\, dk'$ and there is an additional factor $1/k'$ from the matrix element $H^{(1)}H^{(1)}$ eq. (4). Since the square of $(E_O - E_{j_1})$ occurs here it is seen that the integral behaves for low values of k' like dk'/k' and therefore *diverges at the lower limit*. This fact will give us the solution of the infra-red problem, and we call the terms arising the *infra-red terms*. For the time being we replace the lower limit by ω, say. The integration is again elementary. We give only the highest order, which this time is $\sim k_0^2/\mu^2\log(\mu/\omega)$ instead of $k_0^2/\mu^2\log(\mu/k_0)$. Then again neglecting $(\mathbf{k}',\mathbf{k}_0-\mathbf{k})/\mu$ in the denominator, we obtain

$$-\frac{1}{2}\frac{H_{Fj}^{(1)}H_{jO}^{(1)}}{(E_O-E_j)^2}H_{FO}^{(2)} = -\frac{4}{3}\frac{e^4\hbar c}{\mu\sqrt{(k_0 k)}}(\mathbf{e}_0\,\mathbf{e})(1-\cos\theta)\frac{k_0^2}{\mu^2}\log\frac{\mu}{\omega}. \tag{9}$$

Before we collect our results, we discuss the divergent term of (7). We notice that it is proportional to $H_{FO}^{(2)} = \tilde{K}_2$, i.e. to the second-order matrix element for the process, with a divergent but purely numerical factor which is independent of $\mathbf{k}_0$, $\mathbf{k}$. If we denote this term for a moment by $\tilde{K}_{4\text{s.ch.}}$ (because it will describe the self-charge) we have

$$\tilde{K}_{4\text{s.ch.}} = -\frac{8}{3\pi}\frac{e^2}{\hbar c}\log\frac{k'}{2\mu}\bigg|_{k'\to\infty}\tilde{K}_2. \tag{10}$$

(10) would merely lead to a change in the numerical values of the constants occurring in the Thomson formula for scattering, derived from $\tilde{K}_2$, but the energy and angular dependence (due to (10)) would be precisely the same. It was just this situation which we found, in a more general way, for the effects of the self-charge (§ 32).† If we attribute (10) to the self-charge, and observe that $H_{FO}^{(2)}$ is $\sim e^2$, we obtain in fact

$$\frac{\delta(e^2)}{e^2} = -\frac{8}{3\pi}\frac{e^2}{\hbar c}\log\frac{k'}{2\mu}\bigg|_{k'\to\infty}. \tag{11}$$

This is almost precisely the formula (19), (20) of § 32. There z was an invariant variable of dimensions k^2 and the diverging part would be

$$\int\frac{dz}{z} \sim 2\log\frac{k}{\mu}.$$

We are not using invariant variables. Otherwise (11) differs from the rigorous result of § 32 only by a factor 2, and more cannot be expected from such a non-relativistic calculation. The sign of $\delta(e^2)$ agrees with that of § 32. In accordance with what was said there, the term $\tilde{K}_{4\text{s.ch.}}$ must be *discarded*.

Now collecting the results we obtain from (7), (8), (9) we find:

$$\tilde{K}_4 = -\frac{4}{3}\frac{e^4\hbar c}{\mu\sqrt{(k_0 k)}}\left\{\left[-(\mathbf{e}_0\,\mathbf{e})\frac{k_0^2}{\mu^2}(3+\cos\theta)+\frac{2(\mathbf{e}_0\mathbf{k})(\mathbf{e}\mathbf{k}_0)}{\mu^2}\right]\log\frac{\mu}{k_0}+\right.$$
$$\left.+(\mathbf{e}_0\,\mathbf{e})\frac{k_0^2}{\mu^2}(1-\cos\theta)\log\frac{\mu}{\omega}\right\}. \tag{12}$$

† (10) cannot be due to a change of mass because the self-energies have already been discarded.

The cross-section is according to § 22 eq. (17) ($E/\mu = 1$ in our approximation):

$$d\phi = \frac{2\pi}{\hbar c}|\tilde{K}_2+\tilde{K}_4|^2\rho_k\frac{k}{k_0}\,d\Omega.$$

This gives rise to a contribution proportional to e^6, in addition to the Thomson formula given by $|\tilde{K}_2|^2 \sim e^4$:†

$$d\phi_4 = r_0^2\frac{1+\cos^2\theta}{2}\frac{k^2}{k_0^2}\,d\Omega.$$

After summing over **e**, averaging over $\mathbf{e}_0$, and also putting $k = k_0$, we obtain for $d\phi_6$:

$$d\phi_6 = -\tfrac{8}{3}r_0^2\frac{1}{137}\frac{d\Omega}{4\pi}\frac{k_0^2}{\mu^2}\Big\{[-3(1+\cos\theta+\cos^2\theta)+\cos^3\theta]\log\frac{\mu}{k_0}+$$
$$+(1+\cos^2\theta)(1-\cos\theta)\log\frac{\mu}{\omega}\Big\} \quad (\text{spin } 0,\ k_0 \ll \mu). \quad (13)$$

The formula (13) is identical with that derived from an exact relativistic treatment (which is much more complicated).‡

For spin $\frac{1}{2}$ two different formulae have been published: (i) § a formula where the factor [ ] of $\log(\mu/k_0)$ in (13) is replaced by

$$\tfrac{3}{4}-\tfrac{1}{2}\cos\theta+\tfrac{3}{4}\cos^2\theta+\tfrac{1}{2}\cos^3\theta,$$

(ii) a formula‖ identical with (13) (for unknown reasons). The infra-red term is in both cases the same. The calculations are extremely complicated and, at the time of writing, the discrepancy has not been cleared up.

The total cross-section is

$$\phi_6 = -\tfrac{8}{3}r_0^2\frac{1}{137}\frac{k_0^2}{\mu^2}\Big\{-4\log\frac{\mu}{k_0}+\frac{4}{3}\log\frac{\mu}{\omega}\Big\} \quad (\text{spin } 0,\ k_0 \ll \mu). \quad (14)$$

There is nothing objectionable in ϕ_6 being negative, as this is only a correction to ϕ_4.

Pending a discussion of ω, which will have to be chosen so that $\log(\mu/\omega)$ is not excessively large, we see that ϕ_6 is very small indeed. Compared with ϕ_4 it not only contains a factor $1/137$ but also a factor $k_0^2/\mu^2 \ll 1$.†† There is, with the present accuracy of measurements, very little chance

† It is consistent with a non-relativistic approximation to keep the factor k^2/k_0^2, which to a first approximation is $1-\frac{2k_0}{\mu}(1-\cos\theta)$. Terms neglected are of order k_0^2/μ^2 and higher.

‡ E. Corinaldesi and R. Jost, *Helv. Phys. Acta*, **21** (1948), 183.

§ M. R. Schafroth, ibid. **22** (1949), 501.

‖ L. M. Brown and R. P. Feynman, *Phys. Rev.* **85** (1952), 231.

†† It has been shown quite generally that all radiative corrections vanish in the limit of small energies: W. Thirring, *Phil. Mag.* **41** (1950), 1193.

that the departure from the Klein–Nishina formula might be discovered.

If expressed in terms of frequencies, $k_0 = \hbar\nu_0$, $1/137 = e^2/\hbar c$, ϕ_6 is proportional to $\hbar$, which shows that it is a pure quantum effect.

In the extreme relativistic region, $k_0 \gg \mu$, the correction is relatively larger.† For some angles of scattering it amounts to as much as a few per cent. of the Klein–Nishina cross-section.

Our results are important as a proof of the fact that the radiative corrections are finite and also of the fact that the first approximation used throughout Chapter V is an exceedingly good approximation.

3. *The infra-red problem.*‡ In Chapter V we found on several occasions that every process can be accompanied by the emission of a photon with very low, and in the limit, vanishing frequency. For example the Compton effect with a scattered quantum k can be accompanied by the emission of a second, infra-red, quantum k_r (§ 23). While in general the double Compton effect is very improbable, it behaves like $\log(\mu/k_r)$ for small k_r, and becomes infinite when $k_r \to 0$. Strictly speaking this would make it impossible to calculate even the probability for the ordinary scattering, because as a correction (of order e^6) we should have to include the probability for the double process with $k_r = 0$. The problem is purely academic, however, because, even if k_r corresponds to a wave-length of laboratory size L, $\log(\mu/k_r) \sim \log(L\mu/c\hbar) \sim \log 10^{13}$ is still smaller than 137 (and there are other small factors as well). Nevertheless, a theoretical difficulty existed, and we show now that this is solved by considering the correction ϕ_6 for the single Compton effect.

In § 23 we obtained the formula (22) for the double Compton effect. If we integrate this over values of k_r extending from a lower limit $\omega \to 0$ to a limit $k_m \ll k_0$, say, we obtain

$$\phi_{\text{d.c.}} = \tfrac{8}{3}r_0^2 \frac{1}{137}\frac{k_0^2}{\mu^2}\frac{d\Omega}{4\pi}(1+\cos^2\theta)(1-\cos\theta)\log\frac{k_m}{\omega}, \tag{15}$$

where θ is the scattering angle of the main quantum. (15) is the probability for Compton scattering with an additional very soft photon

† References § and || of previous page.

‡ The infra-red problem has been studied extensively in particular by: F. Bloch and A. Nordsieck, ibid. **52** (1937), 54; W. Pauli and M. Fierz, *Nuov. Cim.* **15** (1938), 167. In these older papers a fundamental difficulty remained: although the compensation at low frequencies was exhibited, the radiative corrections appeared to diverge at high frequencies of the virtual photon. This difficulty is now overcome by the consistent renormalization of the mass and charge. In the present context compare H. A. Bethe and J. R. Oppenheimer, *Phys. Rev.* **70** (1946), 451; R. Jost, ibid. **72** (1947), 815.

$\omega < k_r < k_m$. We see now that (15) is identical, but with opposite sign, to the infra-red term of (13) or (14), arising from virtual quanta $\omega < k' < k_m$. (Write $\log(\mu/\omega) = \log(\mu/k_m)+\log(k_m/\omega)$.) Thus the double Compton effect with emission of a low-energy photon $k_r < k_m$ is compensated by the correction $\sim e^6$ to the single Compton effect. We choose an arbitrary low limit k_m, below which we consider the emission of a soft photon $k_r < k_m$ as *indistinguishable* from the case where no such photon is emitted. We can then allow ω to tend to 0, and no divergence arises. The scattering formula is then the Thomson formula corrected (i) by (13) with ω replaced by the finite, but arbitrary, limit k_m and (ii) by a real double process, physically different from the single process, with emission of a second quantum $k_r > k_m$. This is given by formula (22) of § 23 with $k_r > k_m$. That the limit k_m is arbitrary lies in the nature of the situation. We, or the experimentalist, must decide *when* a single extremely soft photon is physically different from the vacuum. In this sense no infra-red problem exists.

That the double effect and the corrections to the single effect compensate each other for $k_r < k_m$ is not accidental but follows directly from the general formulae of perturbation theory. We leave this to the reader to show.

4. *Radiative corrections for the scattering of an electron in a Coulomb field.* The same situation arises for the scattering of an electron in an external field. There is a diverging correction to the radiationless scattering calculated from § 30, and there is Bremsstrahlung with emission of a low energy photon k_r. The two processes compensate each other, for $k_r < k_m$, where k_m is again an arbitrary limit very small compared with the energy of the incident electron. For the sake of completeness we also give the formula for the radiative corrections to the elastic scattering of an electron in a pure Coulomb field. This is given to a first approximation by the Rutherford formula § 24 eq. (33):

$$d\phi_4 = \frac{e^4 Z^2\, d\Omega}{2p^2(1-\cos\theta)^2}\left(1+\cos\theta+\frac{2\mu^2}{p^2}\right),$$

where p is the momentum of the incident electron.

The radiative correction of order e^6 is† in the non-relativistic case $p/\mu \ll 1$ $(E^2 = p^2+\mu^2)$

$$d\phi_6 = -d\phi_4 \frac{4}{3\pi 137}\frac{p^2}{E^2}(1-\cos\theta)\left(\frac{19}{30}+\log\frac{\mu}{2k_m}\right) \qquad (p/\mu \ll 1). \quad (16)$$

† J. Schwinger, *Phys. Rev.* **76** (1949), 790; H. Fukuda, Y. Miyamoto, and S. Tomonaga, *Progr. Theor. Phys.* **4** (1949), 47 and 121.

Here k_m is the minimum energy of a soft photon emitted as Bremsstrahlung, that is regarded as distinguishing the process from purely elastic scattering. It is again seen that $d\phi_6$ is very small compared with $d\phi_4$, the factors being $1/137$ and $p^2/E^2 \ll 1$.

5. *Damping corrections.* To study the influence of the second term of (1) it is sufficient to replace K by its first term K_2.

We first consider the non-relativistic case $k_0 = k \ll \mu$. The non-relativistic interaction of light with a free electron at rest has a direct matrix element for scattering eq. (4), viz.:

$$H^{(2)}_{kk_0} = \frac{e^2}{\mu}(\mathbf{e}_0\,\mathbf{e})\frac{2\pi c^2\hbar}{\nu}, \tag{17}$$

noting that $k_0 = k = \hbar\nu$. If we neglect all states where the primary light quantum has split up into two quanta, etc., the states B on the right-hand side of (1) are those where a scattered quantum $\mathbf{k}'$ is present with a direction different from $\mathbf{k}_0$ and $\mathbf{k}$. Thus (1) becomes

$$U_{kk_0} = H^{(2)}_{kk_0} - i\pi H^{(2)}_{kk'}\,\rho_{k'}\,U_{k'k_0}, \tag{18}$$

where the summation is over all angles of $\mathbf{k}'$, and the two directions of polarization $\mathbf{e}'$. To solve (18) we put

$$U_{kk_0} = X(\mathbf{e}_0\,\mathbf{e}), \qquad U_{k'k_0} = X(\mathbf{e}_0\,\mathbf{e}'). \tag{19}$$

Since the averaging over all directions of $\mathbf{k}'$ and the two directions of $\mathbf{e}'$ yields

$$\overline{(\mathbf{e}'\mathbf{e})(\mathbf{e}'\mathbf{e}_0)} = \tfrac{1}{3}(\mathbf{e}_0\,\mathbf{e}),$$

we find from (18)

$$X = \frac{2\pi c^2\hbar}{\nu}\frac{e^2}{\mu}\frac{1}{1+i\kappa}, \qquad \kappa = \frac{2}{3}\frac{e^2\nu}{\mu c}. \tag{20}$$

For the cross-section of scattering, summed over the two polarizations $\mathbf{e}$, we obtain

$$d\phi = \frac{1}{2}\left(\frac{e^2}{\mu}\right)^2\frac{1+\cos^2\theta}{1+\kappa^2}\,d\Omega. \qquad \text{N.R.} \quad (21)$$

This is identical with the classical formula § 5 eq. (4′). We see that the second term of (1) just accounts for the classical damping force $\sim \ddot{\mathbf{v}}$. $\hbar$ does not occur in (21).

The formula (21) does not really have any scope of application. For κ becomes comparable to unity only if $k/\mu \sim \hbar c/e^2 = 137$, i.e. at such high energies that a relativistic treatment becomes imperative, which is already the case at $k \sim \mu$. (21) suggests that a modification of the Klein–Nishina formula would arise from the damping at energies $\hbar\nu > 137\mu$. We shall see, however, that this is not the case, and that the influence of damping remains negligible at all energies. To show this we have to treat the problem relativistically.

The solution of the integral equation (1) for the case of relativistic scattering is extremely difficult. We shall not attempt an exact solution but confine ourselves to showing that the effect is negligible.† If the effect of damping is small we can expand

$$U_{kk_0} = H_{kk_0} + U'_{kk_0}$$

and assume that U' is small compared with H. We can then replace (18) to a first approximation by

$$U_{kk_0} = H_{kk_0} - i\pi H_{kk'} \rho_{k'} H_{k'k_0}. \tag{22}$$

We have to compare the order of magnitude of the second term of (22) with that of the first term. It is convenient to do this in a Lorentz system where the total momentum is zero: $\mathbf{p}_0 = -\mathbf{k}_0$, $\mathbf{p} = -\mathbf{k}$. The scattering process then consists only in a change of direction of $\mathbf{k}_0$, $\mathbf{p}_0$ into $\mathbf{k}$, $\mathbf{p}$, without a change of frequency: $k_0 = k$; $p_0 = p$. Since we expect the effect to take place only if $k \gg \mu$, we confine ourselves to the E.R. case and neglect the rest energy of the electron throughout. A simple calculation similar to that of § 22 gives the matrix element

$$H_{kk_0} = \frac{e^2\pi\hbar^2c^2}{k}\left\{\frac{u^*\alpha\alpha_0 u_0}{k} + \frac{u^*\alpha_0[(\boldsymbol{\alpha}\mathbf{k}_0)+(\boldsymbol{\alpha}\mathbf{k})]\alpha u_0}{E_0 k + (\mathbf{k}_0\mathbf{k})}\right\}, \tag{23}$$

where $E_0 = \sqrt{(\mu^2+k_0^2)}$ is the initial energy of the electron. We see that H_{kk_0} is of the order e^2/k^2.‡ Since $\rho_{k'} \sim k^2$ we see at once that the order of magnitude of the second term of (22) is $e^2(e^2/\hbar c)(1/k^2)$, writing one factor $e^2/\hbar c$ to account for the fact that it has the same dimensions as the first term. Thus the damping term is 137 times smaller than the first term. Moreover, if we form the total cross-section integrated over the directions of $\mathbf{k}$

$$\rho_k |U_{kk_0}|^2 \sim |H_{kk_0}|^2\rho_k + |H_{kk'}\rho_{k'} H_{k'k_0}|^2\rho_k + \dots \tag{24}$$

the cross-terms cancel. The second term of (24) is therefore of the order of magnitude $1/137^2$ of the first term.

Thus the damping corrections are very small at all energies. Actually, the second term of (24) is of order e^8, whereas the first radiative corrections are of order e^6 and therefore much larger. In the same way one can show that no appreciable damping effects exist for Brems-

† S. Power, *Proc. Roy. Ir. Ac.* **50** (1945), 139.

‡ (23) is nearly singular for $(\mathbf{k}_0\,\mathbf{k}) = -k_0 k$ as E_0 is almost equal to k_0. This gives rise to the $\log(k_0/\mu)$ in the integrated cross-section. It can be shown that the damping term does not give rise to higher powers of $\log(k_0/\mu)$ than the first term.

strahlung, pair production, etc., in contrast to what might be expected from classical theory.

The result is again of importance in that it justifies the use of the first-order perturbation theory for all radiation processes in Chapter V. In particular the application of the theory to processes of very high energy (cascade showers, for example) rests on the fact that the damping as well as the radiative corrections are small even for the highest energies known.

We use this opportunity to add some general remarks about the asymptotic behaviour of cross-sections at very high energies and the influence of damping in this case. Consider the collision of any two particles in the centre of mass system again giving rise to two particles and let the energy of one of them be ϵ. Suppose now that the cross-section, calculated without damping, behaves asymptotically like $\phi \sim |K|^2\rho_\epsilon \sim \epsilon^\alpha$. Since $\rho_\epsilon \sim \epsilon^2$, $|K| \sim \epsilon^{\frac{1}{2}(\alpha-2)}$. For simplicity we assume that K does not depend strongly on the angle of 'scattering' although this condition is usually not quite fulfilled for scattering processes in the extreme relativistic case, even in the centre of mass system. In this case U is given by $U \sim K/(1+i\pi K\rho_\epsilon)$. It follows that

$$\phi \sim |K|^2\rho_\epsilon/(1+\kappa^2), \qquad \kappa = \pi|K|\rho_\epsilon = A\epsilon^{\frac{1}{2}\alpha+1},$$

where A is a small numerical constant which, for electromagnetic processes, contains as many factors 1/137 as is the order of K. Thus

$$\phi \sim \frac{\epsilon^\alpha}{1+A^2\epsilon^{\alpha+2}}.$$

It depends now on whether $\alpha \leqslant -2$ or > -2. If $\alpha \leqslant -2$, as is the case for the Compton effect, the denominator is at most constant and, since A is small, close to unity for all ϵ. The damping is then negligible. On the other hand, if $\alpha > -2$, the asymptotic behaviour of the cross-section is $\phi \sim \epsilon^{-2}$, independent of α. The damping then plays a very big role. As far as processes involving photons and electrons are concerned no case is known where $\alpha > -2$, but the situation may arise in processes involving mesons. The consideration shows that no cross-section can increase indefinitely with energy. Similar arguments can be applied for processes involving more than two particles. Here one case exists (pp. 265–6) where $\phi \to \infty$ for $E \to \infty$.

34. Radiative corrections to bound states

1. *General considerations.* The previous sections of this chapter were concerned with collisions of free particles and we now consider the radiative effects for discrete levels. There are several types of

corrections to be expected. Consider an excited state O of an atom with one electron which can go over to the ground state with emission of a photon. For simplicity we assume that no other (step-wise) transitions take place. It was shown in § 16 that the probability distribution of the final states $E_n = E_0 + \hbar\nu$ (E_0 = energy of ground state) is

$$|b_{n|O}(\infty)|^2 = \frac{|U_{n|O}(\tilde{E}_n)|^2}{(\tilde{E}_O - \tilde{E}_0 - \hbar\nu + \Delta E)^2 + \frac{1}{4}\hbar^2(\mathscr{R}\Gamma(\tilde{E}_n))^2}. \tag{1}$$

Here $\tilde{E}_O$, $\tilde{E}_0$ are the energies *displaced by their self-energies* (see § 16.3)

$$\tilde{E}_O = E_O - H^{(s)}_{O|O}, \qquad \tilde{E}_0 = E_0 - H^{(s)}_{0|0},$$
$$-H^{(s)}_{O|O} = \frac{H_{O|m}H_{m|O}}{E_O - E_m} + H_{O|O}. \tag{1'}$$

($H_{O|O}$ is due to the Coulomb interaction only.) Actually, an additional displacement ΔE occurs in (1). This is (we omit the $\sim$ sign)

$$\Delta E = \tfrac{1}{2}\hbar\mathscr{I}\Gamma_{O|O}(E_n).$$

Whilst in the representation used (displaced levels) $\mathscr{I}\Gamma(E_O) = 0$, $\mathscr{I}\Gamma(E_n)$ for $E_n \neq E_O$ does not vanish exactly but depends in fact on the frequency ν. Since the line breadth is very small E_n is practically equal to E_O and $\mathscr{I}\Gamma(E_n)$ is very small. Inasmuch as $\mathscr{R}\Gamma$ can be regarded as constant it follows that the *maximum of the line lies at the energy difference of the displaced levels*, with the displacement calculated by (1').

In Chapter V (in particular in § 18, also in § 20) U and $\mathscr{R}\Gamma$ were treated in first approximation and the displacements of the levels neglected. Two types of radiative corrections now occur:

(i) The level displacements $-H^{(s)}_{O|O}$, etc., which in the first non-vanishing approximation are the self-energies of the bound states. We shall see (subsections 2, 3) that these, in contrast to the case of a free electron, do not merely reduce to a mass correction but are finite and observable, after mass and charge renormalization.

(ii) The radiative corrections to the transition probability given by $|U|^2$ and the line breadth $\mathscr{R}\Gamma$. Both quantities are connected by an *exact* relation: $\mathscr{R}\Gamma_{O|O}$ is the total transition probability from O. The radiative corrections to $\mathscr{R}\Gamma$ are treated in subsection 4, where the question of the additional displacement ΔE is also considered. The corrections (ii) are outside the present accuracy of measurements. Experimentally, the shifts are far more important.

It is of interest whether the line shift or the line breadth is the bigger quantity. Anticipating the results of subsections 2 and 3, we find that the situation for hydrogen-like atoms is the following. For P-levels which all have allowed transitions to the ground state the displacement is very much smaller than the level width. For the $2S$-state, which is metastable and has a negligible width, the displacement is much larger than the width. The same is true for the ground state, of course, which is sharp. For the higher S-levels, although allowed transitions to lower P-states occur, the transition probabilities as well as the level shift decrease rapidly with the main quantum number n and the level shift remains in general much bigger than the width.

For the three hydrogen-levels with $n = 2$, the situation is demonstrated in Table IX.

TABLE IX

Level shifts and level widths of the $2S$ and $2P$ levels of hydrogen (in frequency units ν, energy $= \hbar\nu$)†

Level	*Unperturbed energy relative to* $2S_{\frac{1}{2}}$	*Radiative shift (theoretical)*	*Level width* $\mathscr{R}\Gamma$
$2S_{\frac{1}{2}}$	0	$+1040\times 2\pi.10^6$	~ 7
$2P_{\frac{1}{2}}$	0	$-17\times 2\pi.10^6$	$6{\cdot}3\times 10^8$
$2P_{\frac{3}{2}}$	$1{\cdot}1\times 2\pi.10^{10}$	$+8\times 2\pi.10^6$	$6{\cdot}3\times 10^8$

2. *Calculation of the level shift.*‡ The explicit calculation of the level shift is naturally very lengthy. In the following we shall lay emphasis on all the important theoretical points but omit one part of the explicit calculation. The major contribution to the shift will come from a non-relativistic region (see below) for which the calculation is almost trivial, and we shall give this. Our main interest must lie in the question of how an unambiguous and finite result is derived. This is possible by a consistent application of the methods already developed in the preceding sections, namely, by eliminating effects due to the self-mass and self-charge.

Consider an electron in a bound state with potential energy V and wave function ψ_0. We call the self-energy in this state $W \equiv -H^{(s)}_{0|0}$.

† The unit 'megacycle' used in the experimental literature is the above unit divided by $2\pi.10^6$. For a better comparison this factor was left standing in the third column. $\mathscr{R}\Gamma$ is the transition probability per sec.

‡ The level shift of hydrogen was first calculated by H. A. Bethe, *Phys. Rev.* **72** (1947), 339, using a strictly non-relativistic method and 'cutting-off' at virtual states with energy $> mc^2$. Consistent relativistic calculations were made independently by: J. B. French and V. F. Weisskopf, ibid. **75** (1949), 1240; N. M. Kroll and W. E. Lamb, ibid. (1949), 388; R. P. Feynman, ibid. **74** (1948), 1430; H. Fukuda, Y. Miyamoto, and S. Tomonaga, *Progr. Theor. Phys.* (Japan), **4** (1948), 47 and 121. Below we mainly follow the paper by French and Weisskopf. A very interesting semi-classical derivation of the line shift is due to H. A. Kramers (see *Report of the Solvay Conference 1948*, Brussels 1950). This in fact preceded the experimental discovery. (Further literature quoted there.)

From the point of view of hole theory, all the negative energy states are filled, these states being the exact ones of negative energy in the potential V. As in § 29, the self-energy of the electron will in the first place be due to virtual transitions with emission of a transverse photon k into any other states n, whereby transitions to negative energy states are forbidden. In addition, the self-energy of the vacuum is modified in that no negative energy electron can go over virtually into the occupied state ψ_0. For a bound state no conservation of momentum is required. The electron (also a negative energy electron) can emit or absorb a photon without changing its state. Thus the following virtual sequence is also possible. The electron in ψ_0 emits a photon, remaining in ψ_0, and the photon is absorbed by a negative energy electron, remaining in its state or vice versa. We write these terms separately and call them, for reasons to be explained below, polarization terms. Then the part of the self-energy due to virtual photons, $W^{(tr)}$, say, is (in analogy to § 29 eq. (2)):

$$W^{(tr)} = W_Q^{(tr)} + W_P^{(tr)},$$

$$W_Q^{(tr)} = \sum_{k,E_n>0} \frac{H_{0|nk}H_{nk|0}}{E_0-E_n-k} - \sum_{k,E_n<0} \frac{H_{n|0k}H_{0k|n}}{E_n-E_0-k}, \qquad (2)$$

$$W_P^{(tr)} = -\sum_{k,E_n<0} \frac{1}{k}[H_{n|nk}H_{0k|0}+H_{0|0k}H_{nk|n}],$$

$$H_{nk|0} = -e\hbar c \sqrt{\left(\frac{2\pi}{k}\right)} \int d\tau \; (\psi_n^\dagger(\boldsymbol{\gamma}\mathbf{e})e^{-i(\mathbf{kr})/\hbar c}\psi_0), \text{ etc.} \qquad (2')$$

The two terms of $W_P^{(tr)}$ are in fact equal.

We use Coulomb gauge, as the handling of the auxiliary condition for bound states is difficult. Then the photons k occurring in (2) are transverse and in addition the static Coulomb energy contributes to W (term $H_{0|0}$ of (1′)). This is the quantum analogue of the classical self-energy $\lim_{r\to 0} e^2/r$.

In quantum theory it is given by

$$W^{(c)} = \tfrac{1}{2}e^2 \iint d\tau d\tau' \frac{(\psi^*(\mathbf{r})\psi(\mathbf{r}))(\psi^*(\mathbf{r}')\psi(\mathbf{r}'))}{|\mathbf{r}-\mathbf{r}'|}, \qquad (3)$$

where ψ is the electron field subjected to second quantization. We can expand ψ and ψ^*:

$$\psi = \sum_r a_r \psi_r, \qquad \psi^* = \sum a_r^* \psi_r^*,$$

but in contrast to § 12 the ψ_r are here the *exact* 1-*electron wave functions* *in the potential V.* a_r, a_r^* are the absorption and emission operators respectively and the sums extend over both positive and negative energy states. It is more convenient here to express this in terms of negative energy electrons rather than in terms of positrons. We have then to take the diagonal matrix element of the operator (3) referring to the state where one electron is in ψ_0 and all negative energy states are filled, and subtract (as for $W^{(tr)}$) the corresponding diagonal element of the vacuum. One easily finds:

$$W^{(c)} = W_Q^{(c)} + W_P^{(c)},$$

$$W_Q^{(c)} = \tfrac{1}{2}e^2\Big(\sum_{E_n>0} - \sum_{E_n<0}\Big) \iint d\tau d\tau' \frac{(\psi_0^*(\mathbf{r})\psi_n(\mathbf{r}))(\psi_n^*(\mathbf{r}')\psi_0(\mathbf{r}'))}{|\mathbf{r}-\mathbf{r}'|},$$

$$W_P^{(c)} = e^2 \sum_{E_n<0} \iint d\tau d\tau' \frac{(\psi_0^*(\mathbf{r})\psi_0(\mathbf{r}))(\psi_n^*(\mathbf{r}')\psi_n(\mathbf{r}'))}{|\mathbf{r}-\mathbf{r}'|}. \tag{4}$$

(4) can now be combined with (2) in a relativistic fashion. We observe that

$$\frac{(\psi_0^*(\mathbf{r})\psi_n(\mathbf{r}))(\psi_n^*(\mathbf{r}')\psi_0(\mathbf{r}'))}{|\mathbf{r}-\mathbf{r}'|} = \frac{1}{2\pi^2\hbar c}\int \frac{d^3k}{k^2}(\psi_0^*(\mathbf{r})e^{i(\mathbf{kr})/\hbar c}\psi_n(\mathbf{r}))(\psi_n^*(\mathbf{r}')e^{-i(\mathbf{kr}')/\hbar c}\,\psi_0(\mathbf{r}')) \tag{4'}$$

and similarly in $W_P^{(c)}$. When n is a negative energy state, the sign of $\mathbf{k}$ in the exponentials will be reversed. Furthermore, let $H = V+(\boldsymbol{\alpha}\mathbf{p})+\beta\mu$ be the energy operator, $H\psi_n = E_n\psi_n$, $\psi_n^* H = E_n\psi_n^*$. H does not commute with $\exp(\pm i(\mathbf{kr})/\hbar c)$, because of the term $(\boldsymbol{\alpha}\mathbf{p})$. It follows that

$$(E_n-E_m)(\psi_n^* e^{\pm i(\mathbf{kr})/\hbar c}\psi_m) = (\psi_n^*[H, e^{\pm i(\mathbf{kr})/\hbar c}]\psi_m) = \pm k(\psi_n^\dagger \gamma_3\, e^{\pm i(\mathbf{kr})/\hbar c}\psi_m)$$
$$(\gamma_3 = (\boldsymbol{\gamma}\mathbf{k})/k). \tag{5}$$

The right-hand side is essentially the matrix element $m \to n$ with absorption or emission of a *longitudinal* photon. Similarly $(\psi_n^\dagger \gamma_4\, e^{\pm i(\mathbf{kr})/\hbar c}\psi_m)$ is the matrix element for a scalar photon ($\psi^\dagger\gamma_4 = i\psi^*$).

We now use the identity:

$$\frac{1}{k} = \pm\Big(\frac{(E_0-E_n)^2}{k^2}-1\Big)\frac{1}{E_0-E_n\mp k} \mp \frac{E_0-E_n}{k^2}. \tag{6}$$

When $1/k$ in (4′) is replaced by (6), and (5) used, it is seen that (4), apart from the last term of (6), reduces to

$$W^{(c)} = \sum_{\lambda=3,4}\sum_k \Big\{ \sum_{E_n>0}' \frac{H_{0|nk\lambda}H_{nk\lambda|0}}{E_0-E_n-k} - \sum_{E_n<0} \frac{H_{n|0k\lambda}H_{0k\lambda|n}}{E_n-E_0-k} - $$
$$ - \sum_{E_n<0} \frac{H_{n|nk\lambda}H_{0k\lambda|0}+H_{0|0k\lambda}H_{nk\lambda|n}}{k}\Big\}, \tag{7}$$

where the matrix elements $H_{nk\lambda|0}$ are given by (2′), replacing $(\boldsymbol{\gamma}\mathbf{e})$ by γ_3 and γ_4 respectively.

The last term of (6) gives no contribution:

$$\int d^3k \sum_n (E_0-E_n)(\psi_0^* e^{i(\mathbf{kr})/\hbar c}\psi_n)(\psi_n^* e^{-i(\mathbf{kr}')/\hbar c}\psi_0)$$
$$\simeq \sum_n \int d^3k\ (\psi_0^*(\boldsymbol{\gamma}\mathbf{k})e^{i(\mathbf{kr})/\hbar c}\psi_n)(\psi_n^* e^{-i(\mathbf{kr}')/\hbar c}\psi_0) = \int d^3k\ (\psi_0^*(\boldsymbol{\gamma}\mathbf{k})\psi_0) = 0$$

on account of the completeness theorem, and because the angular integration of $(\boldsymbol{\gamma}\mathbf{k})$ vanishes. In (7) $H_{nk\lambda|0}$ is the matrix element for emission of a longitudinal ($\lambda = 3$) or scalar ($\lambda = 4$) photon, and

$$H_{0|nk\lambda} = H^*_{nk\lambda|0}$$

that for absorption. In the polarization term γ_3 does not actually give a contribution.

Combining now with (2) we see that the total self-energy $W^{(c)}+W^{(lr)}$ becomes

$$W_Q = \sum_{\mu=1}^{4}\sum_k \left\{\sum_{E_n>0}\frac{H_{0|nk\mu}H_{nk\mu|0}}{E_0-E_n-k} - \sum_{E_n<0}\frac{H_{n|0k\mu}H_{0k\mu|n}}{E_n-E_0-k}\right\},$$
$$W_P = -\sum_{\mu=1}^{4}\sum_k\sum_{E_n<0}\frac{1}{k}\{H_{n|nk\mu}H_{0k\mu|0}+H_{0|0k\mu}H_{nk\mu|n}\}. \qquad (8)$$

The result is the same as if we had used Lorentz gauge and ignored the Lorentz condition.† The relativistic combination works here in the same way as for free particles (compare §§ 10 and 24.1).

If (8) is worked out explicitly it is found to diverge logarithmically just as the self-energy of a free electron does. The remedy is obvious; we have to separate off the parts of (8) due to the self-mass and the self-charge.

We consider first the contribution from the polarization terms W_P. This is a purely relativistic effect and due to vacuum polarization. We can see this as follows. Let us imagine that the wave functions ψ_n of the negative energy states as well as ψ_0 are expanded into plane waves. Since ψ_n is a negative energy wave function in the potential V, the expansion into free particle waves will also contain plane waves with positive energy, or in other words, pairs are created. If V is a weak field (as is certainly the case in a hydrogen atom), the first terms in V are just the free pairs created by the external field. Since ψ_0 is practically non-relativistic, the expansion of ψ_0 will only give rise to plane waves with positive energy and small momentum. We write

$$\psi_0 = \sum_P c_P u_P e^{i(\mathbf{Pr})/\hbar c}, \qquad \psi_0^* = \sum_{P'} c^*_{P'} u^*_{P'} e^{-i(\mathbf{P'r})/\hbar c}. \qquad (9)$$

† This will not be true for $\Gamma(E)$ at any point $E \neq E_Q$. See subsection 4.

W_P is due to a virtual emission of a photon by the electron in ψ_0 and reabsorption by a negative energy electron, or vice versa. Looked upon after the resolution into plane waves this process means that the electron with momentum P $(E > 0)$ emits a photon, whereby it changes into P' $(E' > 0)$ and the photon is reabsorbed by the pairs created by the field V. This is precisely the process considered in § 32 and called there vacuum polarization. It can be visualized by the interaction diagram Fig. 21 (*b*), p. 283. It appears therefore that W_P is nothing but a superposition of matrix elements of the type calculated already in § 32,

$$W_P = \sum c^*_{P'} c_P \, G^{(e)}_{3(b)P'P}. \tag{10}$$

$G^{(e)}_{3(b)}$ was given in § 32 in terms of the Fourier frequencies K_μ of the field. For an electrostatic field $K_4 = 0$, $A^{(e)}_i = 0$, and $A^{(e)}_4 = iV(K)/e$. $V(K)$ is the Fourier coefficient of V. Then the formulae (1), (2), (7) of § 32 reduce to $G^{(e)}_{3(b)P'P} = L'(K^2)V(K)(u^*_{P'} u_P)$.

L' is calculated explicitly in § 32. It was found that $L'(0)$ *describes the self-charge and is to be omitted.* Thus we should replace L' by $L'(K^2) - L'(0)$. This quantity is given by eq. (18 c, d), § 32. Since V varies slowly in space it is sufficient to use the first term of the expansion in K^2, giving

$$W_P = \sum_{P,P'} c^*_{P'} c_P (u^*_{P'} u_P) \frac{1}{15\pi 137} \frac{K^2}{\mu^2} V(K). \tag{11}$$

Going over to coordinate representation $K^2 V(K) \rightarrow -\hbar^2 c^2 \nabla^2 V$, (11) can again be expressed by the original wave function

$$W_P = -\frac{1}{15\pi 137}\left(\frac{\hbar}{mc}\right)^2 \langle \nabla^2 V \rangle_0, \qquad \langle \nabla^2 V \rangle_0 \equiv \int (\psi^*_0 \nabla^2 V \psi_0)\, d\tau. \tag{12}$$

$\langle \nabla^2 V \rangle_0$ is the expectation value of $\nabla^2 V$. This will be given for hydrogen below. (12) is approximate in that (i) V is considered small, (ii) terms $\sim K^4$, or $\nabla^4 V$ are neglected. Both are quite justified, especially since (12) gives only a minor contribution to the whole effect.

Next we consider W_Q. This contribution also diverges when worked out explicitly, but the reason is a different one, namely, the self-mass. For a free electron we saw (§ 29) that the *entire* self-energy is due to a self-mass. This will not be so for a bound electron but the divergent part will disappear when the self-mass operator is subtracted. This operator, however, requires a generalization here. For a bound particle the Hamiltonian is $H = V + (\boldsymbol{\alpha}\mathbf{p}) + \beta\mu$. A mass correction must exhibit itself in the last term only and therefore the addition to the energy for a particle in ψ_0 must be

$$(\delta_m H)_{0|0} = \int (\psi^*_0 \beta \psi_0)\, d\tau \, \delta\mu. \tag{13}$$

If the theory is to be consistent, $\delta\mu$ must be the *same* invariant *quantity* as for a free electron. We thus require an operator, $\delta_m H$, such that the expectation value is of the form (13). For a free electron $\delta_m H$ had diagonal elements PP only (apart from the special non-diagonal elements of § 29.2), but if ψ_0 is expanded as in (9), it is seen that in general elements $(\delta_m H)_{P'|P}$ will also occur. The required generalization is quite easy to find, following § 29. For a free particle with momentum P, and energy E, the self-energy was given by § 29 eq. (4) which we can write

$$W = (u_P^\dagger(\delta_m H)_P u_P), \qquad (\delta_m H)_P = 2\pi\hbar^2c^2e^2\sum_{k,p'}\frac{1}{k}\frac{\gamma_\alpha u'u'^\dagger\gamma_\alpha}{E-E'-k\delta'}, \tag{14}$$

where $\delta' = \pm 1$, if the virtual state u' has $\pm$ energy. The sum is then over positive as well as negative energy states n'. We next consider the required operator $\delta_m H$ acting on ψ_0. We leave the first factor ψ_0^* out of consideration. We define now, with the help of (9),

$$\delta_m H\psi_0 = \sum_P c_P(\delta_m H)_P u_P e^{i(\mathbf{Pr})/\hbar c}, \tag{15}$$

where $(\delta_m H)_P$ is given by (14). This can also be expressed in the more concise form

$$\delta_m H = 2\pi\hbar^2c^2e^2\sum\frac{\gamma_\alpha u'u'^\dagger\gamma_\alpha}{k(H-E'-k\delta')} \qquad (H = (\boldsymbol{\alpha}\mathbf{p})+\beta\mu). \tag{15'}$$

Next we follow the derivation in § 29.1 step by step, carry out the summation over the intermediate states, and finally go over to the invariant variables defined there. In this derivation, although use is made of $(\gamma\cdot P)u = i\mu u$, the first factor $u_P^\dagger$ or ψ_0^* is never used. We can therefore quote § 29 eq. (13) immediately and find

$$\delta_m H\psi_0 = \frac{\mu}{2\pi 137}\sum_P c_P\beta u_P e^{i(\mathbf{Pr})/\hbar c}\int_1^\infty d\epsilon\,\frac{3\epsilon^2-1}{\epsilon^3} = \beta\psi_0\,\delta\mu. \tag{16}$$

If $\delta\mu$ is to be independent of P for a free particle, the integral over ϵ must be independent of P. This is evident from the limits of integration. Multiplying (16) by ψ_0^* we see that (15) does indeed give the desired result (13), and $\delta\mu$ is the same as for a free particle.‡

The observable part of the self-energy is now to be defined by§

$$\tilde{W}_Q = W_Q-(\delta_m H)_{0|0}. \tag{17}$$

‡ In searching for the correct operator $\delta_m H$, French and Weisskopf encountered an ambiguity which could be removed only by elaborate relativistic arguments (considering an external magnetic field also). From the above argument it is, however, clear that no addition to $\delta_m H$ can be made without either jeopardizing the invariance or the value of $\delta\mu$.

§ For the representation actually used when the mass correction is made, see subsection 4, p. 350, below.

For the explicit evaluation of (17) it is practical to divide the intermediate states into two classes, according to whether k is 'large or small'. Fortunately, the ionization energy I is separated from the quantity $\mu = mc^2$, where relativistic effects set in, by the small factor $1/2.137^2$. We can then define an intermediate energy $\alpha\mu$ where $\alpha \sim 1/137$, say, so that $I \ll \alpha\mu \ll \mu$. During the photon emission, momentum is conserved to within the momentum distribution in ψ_0 which is of order I. Therefore, if $k > \alpha\mu$, $|E_n| > \alpha\mu \gg I$. In this region of virtual energies the electron is practically free. On the other hand, for $k < \alpha\mu$, $|E_n| < \alpha\mu$, the electron can be treated entirely non-relativistically. We denote the contributions by $\tilde{W}_{\text{QR.}}$ and $\tilde{W}_{\text{QN.R.}}$ respectively.

In the R-region, the wave function of the intermediate electron, ψ_n, can be expanded according to the Born expansion, in powers of V and only terms up to V need be considered. (The V^2-terms actually vanish.) We shall omit this calculation, which is very lengthy. It gives a relatively small contribution. One finds

$$\tilde{W}_{\text{QR.}} = \frac{1}{3\pi 137}\left(\frac{\hbar}{mc}\right)^2\left\{\left[\log\frac{1}{2\alpha}+\frac{11}{24}\right]\langle\nabla^2 V\rangle_0 - \tfrac{3}{4}\langle SV\rangle_0\right\},$$
$$SV \equiv \frac{1}{\hbar c}(\text{grad}\, V[\boldsymbol{\sigma}\mathbf{p}]) - \tfrac{1}{2}\nabla^2 V. \tag{18}$$

The term $\langle SV\rangle_0$ arises from the relativistic interaction of the spin with the static field.

Finally we are left with the non-relativistic contribution $\tilde{W}_{\text{QN.R.}}$ This region gives the bulk of the effect and the calculation is very simple. Only positive energy intermediate states need be considered (first term of (8)) and also only transverse photons because the longitudinal and scalar parts cancel completely with the corresponding part of the mass operator $e^2/r|_{r\to 0}$. ψ_0 extends over a region a_0 (Bohr radius), and $(\mathbf{kr})/\hbar c \ll 1$ when $k \ll \alpha\mu$. We can replace the exponentials $\exp i(\mathbf{kr})/\hbar c$ which occur in the matrix elements by unity (an error is caused, however, at $k \sim \alpha\mu$, but this must be small since the result will join correctly with the R-region). Then (8) reduces to

$$W_{\text{QN.R.}} = \frac{2\pi\hbar^2 c^2 e^2}{\mu^2} \sum_{e,k,n} \frac{1}{k}\frac{|p_{e0|n}|^2}{E_0 - E_n - k} = \frac{2}{3\pi 137\mu^2}\sum_n \int\limits_0^{\alpha\mu} \frac{k\,dk|\mathbf{p}_{0|n}|^2}{E_0 - E_n - k}. \tag{19}$$

From this we have now to subtract the N.R. part of the mass operator. This is found from (14), (15). When $k \ll \mu$, $E \sim \mu$, E' is also $\sim \mu$.

u' refers to positive energies only and $(u^\dagger(\boldsymbol{\gamma}\mathbf{e})u')$ becomes $(u^* p_e u')$. The summation over u' is immediately carried out and with the help of (15) we obtain

$$(\delta_m H)_{0|0} = -\sum_{k,e} \frac{|p_e|^2_{0|0}}{\mu^2 k^2} 2\pi\hbar^2 c^2 e^2 = -\frac{2}{3\pi 137\mu^2}\int dk\ |\mathbf{p}|^2_{0|0}.$$

For $|\mathbf{p}|^2_{0|0}$ we can also write $\sum_n |\mathbf{p}_{0|n}|^2$. Thus, combining with (19), we obtain

$$\tilde{W}_{\text{QN.R.}} = \frac{2}{3\pi 137\mu^2}\sum_n \int_0^{\alpha\mu} dk \frac{|\mathbf{p}_{0|n}|^2(E_0-E_n)}{E_0-E_n-k}$$

$$= -\frac{2}{3\pi 137\mu^2}\sum_n |\mathbf{p}_{0|n}|^2(E_0-E_n)\log\frac{\alpha\mu}{|E_n-E_0|}, \tag{20}$$

neglecting E_n-E_0 compared with $\alpha\mu$. If it were not for the log the sum over n could be evaluated in closed form, for elementary wave mechanics yields

$$-\frac{1}{\mu^2}\sum_n |\mathbf{p}_{0|n}|^2(E_0-E_n) = \frac{1}{2}\left(\frac{\hbar}{mc}\right)^2 \langle\nabla^2 V\rangle_0. \tag{21}$$

As the logarithm varies slowly with E_n one may well replace E_n-E_0 by some average value. Then

$$\tilde{W}_{\text{QN.R.}} = \frac{1}{3\pi 137}\left(\frac{\hbar}{mc}\right)^2 \langle\nabla^2 V\rangle_0\left[\log\alpha+\log\frac{\mu}{\langle E_n-E_0\rangle}\right]. \tag{22}$$

Adding up the contributions (12), (18), and (22) we find for the change of energy

$$\tilde{W} = \frac{1}{3\pi 137}\left(\frac{\hbar}{mc}\right)^2\left\{\langle\nabla^2 V\rangle_0\left[\log\frac{\mu}{\langle E_n-E_0\rangle}-\log 2+\frac{31}{120}\right]-\tfrac{3}{4}\langle SV\rangle_0\right\}. \tag{23}$$

The limit α drops out. The average value $\langle E_n-E_0\rangle$ ($n = 0$ does not contribute) has been worked out numerically for hydrogen. Clearly it must be of order I. When ψ_0 is a state with main quantum number $n = 2$ one finds

$$\log\frac{\mu}{\langle E_n-E_0\rangle} = 7{\cdot}6876. \tag{24}$$

The term (24) is the predominant part of (23). For $Z \neq 1$ the numerical value of the log is slightly different.

3. *Results and experiments.* The expectation values occurring in (23) are easily worked out for hydrogen-like atoms. We have

$$\nabla^2 V = +4\pi e^2 Z\delta(\mathbf{r})$$

and

$$\left(\frac{\hbar}{mc}\right)^2 \langle \nabla^2 V\rangle_0 = 4\pi e^2 Z\left(\frac{\hbar}{mc}\right)^2 |\psi_0(0)|^2 = \begin{cases} \dfrac{8IZ^2}{137^2 n^3}, & \text{for } l = 0, \\ 0, & \text{for } l \neq 0, \end{cases} \tag{25}$$

$$\left(\frac{\hbar}{mc}\right)^2 \langle SV\rangle_0 = \frac{4IZ^2}{137^2 n^3(2l+1)} \times \begin{cases} -\dfrac{1}{l+1}, & \text{for } j = l+\frac{1}{2}, \\ +\dfrac{1}{l}, & \text{for } j = l-\frac{1}{2}. \end{cases} \tag{26}$$

$I = \frac{1}{2}Z^2e^2/a_0$ is the ionization energy. All the average values are therefore proportional to Z^4. The largest contribution to (23), namely, (25), exists only for S-states. For P-states $\tilde{W}$ is therefore very much smaller than for S-states. The results for the three states $n = 2$ of hydrogen are given in Table IX. These include a correction of $+6\times 2\pi.10^6$ for $2S_{\frac{1}{2}}$ due to the next higher order of the expansion in V and also to the finite nuclear mass†. This gives a slightly different value for deuterium. An important consequence of this theory is the fact that the degeneracy of the $2S_{\frac{1}{2}}$ and $2P_{\frac{1}{2}}$ states which, according to the Dirac equation, should coincide, is now removed. The two states are now separated by an amount $\nu = 1057\times 2\pi.10^6$ sec.$^{-1}$

The calculations have been extended to include the fourth order of the self-energy and a correction due to the anomalous value of the electron magnetic moment.‡ The precise theoretical separation, in units $2\pi.10^6$ sec^{-1} is

$$E_{2S_{\frac{1}{2}}} - E_{2P_{\frac{1}{2}}} = 1057{\cdot}2 \text{ (for H)},\ 1058{\cdot}5 \text{ (for D)}.$$

Attempts at measuring the fine structure of hydrogen by spectroscopical means had already led Pasternack§ in 1938 to the conclusion that the two states are not degenerate but separated by an amount of about $1000.2\pi.10^6$ sec.$^{-1}$ The spectroscopical resolution was, however, not good enough to measure the separation more accurately. This has become possible through the development of the short-wave radio technique by which wave-lengths of the order of a few cm. can

† M. Baranger, *Phys. Rev.* **84** (1951), 866. R. Karplus, A. Klein, and J. Schwinger, ibid. **86** (1952), 288.

‡ For a summary see E. E. Salpeter, ibid. **89** (1953), 92.

§ S. Pasternack, *Phys. Rev.* **54** (1938), 1113. Later spectroscopic measurements (H. Kuhn and G. W. Series, *Proc. Roy. Soc.* A, **202** (1950), 127) gave the correct shift with an error of only 5 per cent.

be used. In the important experiment of Lamb and Retherford† the resonance absorption of a beam of excited hydrogen atoms is measured directly. To increase the accuracy the atoms pass through a magnetic field H and the complete Zeeman pattern is measured. The limit $H \rightarrow 0$ gives the desired energy difference.

Precision measurements led to the result

$$E_{2S_{\frac{1}{2}}} - E_{2P_{\frac{1}{2}}} = 1057{\cdot}8 \pm 0{\cdot}1 \text{ (for H)},\ 1059{\cdot}0 \pm 0{\cdot}1 \text{ (for D)}.$$

This is in excellent agreement with the theoretical values. The very small discrepancy, if genuine, may well be due to higher orders.

Measurements have also been made in ionized He. The theoretical shift of the $2S$-level is $13{,}800 \times 2\pi . 10^6$ (roughly $Z^4 = 16$ times bigger than for H) whereas the experiments gave $(14{,}000 \pm 100) \times 2\pi . 10^6$ sec.$^{-1}$‡

The experimental confirmation of the theory of the self-energy of a bound electron is a most valuable result. Together with the anomalous magnetic moment of the electron it is the only verification so far of what we have called the radiative reaction, but the agreement between theory and experiment is so good that we can be confident that the radiative corrections, at least for moderate energies, are rendered correctly by the theory.

4. *Radiative corrections to the line breadth.*§ We now consider the corrections to the line shape and breadth due to the higher orders of radiation theory. For simplicity we consider an atom in the first excited level, so that only one transition can take place. We shall also consider here the effects due to the E-dependence of U and Γ as given by the exact theory of § 16. Although, at the time of writing, the accuracy of the measurements does not suffice to measure the natural line with great precision, it may be hoped that, with the development of the radio-frequency methods, a detailed and accurate study of the line breadth may become possible. Above all, the question is of considerable theoretical importance. If the claim is to be justified that quantum electrodynamics is a consistent theory, it should be possible to solve all problems to any desired accuracy. This also includes the time dependence of the emission act. Therefore we should be able to compute $U(E)$, $\Gamma(E)$ for any value of E, and to any order of approximation.

† W. E. Lamb and R. C. Retherford, *Phys. Rev.* **72** (1947), 241; **75** (1949), 1325, 1332; **79** (1950), 549; **81** (1951), 222. W. E. Lamb, *Reports of Progress in Physics*, **14** (1951), 19; *Phys. Rev.* **85** (1952), 259. S. Triebwasser, E. S. Dayhoff, and W. E. Lamb, ibid. **89** (1953), 98 and 106.

‡ W. E. Lamb and M. Skinner, *Phys. Rev.* **78** (1950), 539.

§ E. Arnous and K. Bleuler, *Helv. Phys. Acta*, **25** (1952), 581. E. Arnous, ibid. 631.

We have seen, however, that in the higher orders ambiguities arise which can only be removed by the additional demand of relativistic invariance. For bound state problems this cannot always be readily applied and it must be shown that in fact unambiguous results are obtained.

A problem that will have to be solved now is the elimination of the virtual field of a bound electron in an excited state. We have already mentioned (§ 16) that the exact definition of an excited state cannot be unambiguous but must, when higher orders are considered, depend on the way in which the atom was excited. We shall show that a crude canonical transformation which reflects these excitation conditions suffices to calculate the corrections to the line breadth.

In the following we omit all explicit calculations and confine ourselves to explaining the chief theoretical points.

For free particles the elimination of the virtual field was carried out by a canonical transformation (§ 15) such that the transformed Hamiltonian K only had matrix elements for which the energy was strictly conserved. The width of the 'energy shell' ϵ could be taken infinitely small. We can adapt this transformation to the present purpose by making use of the fact that the level width is small compared with the level distances. This makes it possible to define a *finite energy shell* of width 2ϵ such that ϵ is large compared with the level width but small compared with the level distances. For the ground state $\epsilon \to 0$. Processes taking place outside ϵ are virtual processes, and the smallness of ϵ ensures that practically all virtual processes are included. Processes taking place within ϵ are real transitions and they cover practically the whole of the line breadth. When the virtual states are thus eliminated an excited atomic state is *nearly stationary*, except for the relatively slow real transitions. The line is then, however, cut off at a distance ϵ on both sides of the maximum. This shows (§ 20) that ϵ must be interpreted as the *width of the incident spectrum* exciting the atom. The arbitrariness of the canonical transformation expressed by ϵ reflects therefore the variety of excitation conditions. We shall see that all corrections depending on ϵ are proportional to $1/\epsilon$ and vanish for $\epsilon \to \infty$. The line breadth will be independent of ϵ and only the numerator depends on the excitation conditions. Since this is already the case in the first approximation at large distances from the maximum it is not surprising that part of the corrections to the line shape also depend on them.

The transformed Hamiltonian is obtained in much the same way as in § 15. A difference is that for bound particles H also has elements on the energy shell. We use the representation where the displaced levels again appear but incorporate the mass renormalization from the start. The unperturbed Hamiltonian corrected by the change of mass is $H_0+\delta_m H$. The finite level displacement (subsection 2) is the diagonal part of $-(H^{(s)}+\delta_m H)_D$. $H^{(s)}$ and $\delta_m H$ are given by (1′) and (15′). (D means diagonal in all variables, $N.D.$ non-diagonal in at least one variable.) Thus the representation used is that where the eigenvalues of $H_0+\delta_m H-(H^{(s)}+\delta_m H)_D \equiv \tilde{H}_0$ appear. The interaction is then $H_{\text{int}}-\delta_m H+(H^{(s)}+\delta_m H)_D$. H_{int} includes the Coulomb interaction $H^{(c)}$, $H_{\text{int}} = H^{(tr)}+H^{(c)}$, where $H^{(tr)}$ is the interaction with the transverse field. $H_{n|m}$ is the matrix element of H_{int}.

The transformed Hamiltonian then becomes, up to the third order:

$$(\tilde{K}_{A|B})_{\text{d}} = (H-\delta_m H)_{A|B}+(H^{(s)}+\delta_m H)_{A|A}\,\delta_{AB}+$$

$$+\frac{(H-\delta_m H)_{A|n}((H-\delta_m H)_{n|B})_{\text{n.d.}}}{E_B-E_n}+$$

$$+\frac{((H-\delta_m H)_{A|n})_{\text{n.d.}}((H-\delta_m H)_{n|B})_{\text{d.}}}{E_A-E_n}-$$

$$-\tfrac{1}{2}(E_A-E_B)\frac{(H^{(tr)}_{A|n})_{\text{n.d.}}\,(H^{(tr)}_{n|B})_{\text{n.d.}}}{(E_A-E_n)(E_B-E_n)}+\text{terms } H^{(tr)}H^{(tr)}H^{(tr)}. \quad (27)$$

The third-order terms depending on $H^{(tr)}$ only are rather lengthy and are not written down. They are needed for $\mathscr{R}\Gamma_4$ but only when $E_A = E_B$. (exactly). We give below the expression for $\mathscr{R}\Gamma_4$ explicitly. The energies are all the *displaced* energies; d, n.d. mean on or outside the energy shell in the above sense respectively. A term of type $(H^{(tr)}_{\text{n.d.}}\,H^{(tr)}_{\text{d.}})_{\text{d.}}$ would vanish if the energy shell were infinitely small but does not vanish for finite ϵ. The same is true for the term $\sim (E_A-E_B)$.

This Hamiltonian $\tilde{K}$ has now to be substituted for H in the theory of § 16. We wish to calculate $\mathscr{R}\Gamma$ up to the fourth order and $\mathscr{I}\Gamma$ up to the second order. For this purpose we require an expansion of U.† From § 16 eq. (10) we obtain

$$U_{A|O}(E) = \tilde{K}_{A|O}+\tilde{K}_{A|B}\,\zeta(E-E_B)\tilde{K}_{B|O}+$$

$$+\tilde{K}_{A|B}\,\zeta(E-E_B)\tilde{K}_{B|C}\,\zeta(E-E_C)\tilde{K}_{C|O}+\ldots \quad (B,\, C \neq O,\, B \neq C). \quad (28)$$

† The expansion (28) is suitable only when U is not strongly dependent on E. This is the case for a single transition $O \to 0$. Otherwise U itself has a further resonance denominator (as in the example of § 20) and the expansion is unsuitable.

This U is now to be inserted in § 16 eq. (12) for Γ (again replacing H by $\tilde{K}$). We first calculate $\mathscr{I}\Gamma_2(E)$. It is easy to see that the contributions from $H^{(c)}$ cancel We further observe that

$$H-H_{\mathrm{d.}} = H_{\mathrm{n.d.}}, \qquad (H_{\mathrm{d.}}H)_D = (H_{\mathrm{d.}}H_{\mathrm{d.}})_D, \qquad (H_{\mathrm{n.d.}}H_{\mathrm{d.}})_D = 0, \text{ etc.}$$

We then find, using also (1′),

$$\tfrac{1}{2}\hbar\mathscr{I}\Gamma_{2O|O}(E) \equiv \Delta E = (H_{O|B})_{\mathrm{d.}}(H_{B|O})_{\mathrm{d.}}\left(\frac{1}{E-E_B}-\frac{1}{E_O-E_B}\right). \tag{29}$$

This vanishes for $E = E_O$, as a consequence of the representation used. (Compare § 16.3.) The matrix elements $H_{O|B}$ are now restricted to the energy shell. If O is the ground state, the width of the shell is zero and then (29) vanishes for any E. Through the above canonical transformation therefore the ground state has become stable and is no longer modified by the interaction with radiation (the condition § 16 eq. (32) is fulfilled).† For the excited levels ΔE is at any rate finite (because no infinite integrations occur any more) but depends on the width of the energy shell. The evaluation is very simple. The only contribution arises from the transition from O to the ground state 0 with emission of k, all other transitions give only a negligible contribution from extremely soft quanta. The result is, when we put $E = E_0+k$ (line shape after an infinitely long time)

$$\Delta E = -\frac{\hbar\gamma}{2\pi}\frac{k}{k_0}\log\left|\frac{k_0-k+\epsilon}{k_0-k-\epsilon}\right|, \qquad k_0 = E_O-E_0, \qquad \gamma \equiv \mathscr{R}\Gamma_{2O|O}(E_O). \tag{30}$$

γ is the line breadth in the second approximation as calculated in § 18. Only those frequencies $|k_0-k| \ll \epsilon$ should be considered. Then

$$\Delta E \simeq -\hbar\gamma(k_0-k)/\pi\epsilon. \tag{30′}$$

Since ϵ was assumed $\gg \hbar\gamma$, $\Delta E \ll \hbar\gamma$.

A further remark concerning ΔE: Only the interaction with the transverse field contributes to this quantity, whereas the part due to the Coulomb interaction is compensated by the (subtracted) line shift. This shows that for the computation of $\mathscr{I}\Gamma(E)$ the four types of photons do not occur in a symmetrical way as was the case for the self-energy. This is due to the Lorentz condition which cannot be ignored in this general case.

† This condition is also fulfilled when we go over to the case of free particles in which case $\epsilon \to 0$. Then $\mathscr{I}\Gamma(E) = 0$ for all E.

In the second order the real part of Γ becomes, for a transition $E_O \to E_0+k$,

$$\mathscr{R}\Gamma_{2O|O}(E) = \frac{2\pi}{\hbar}\sum_k |(H_{O|0k})_{\text{d.}}|^2\delta(E-E_0-k) = \frac{2\pi}{\hbar}|(H_{O|0k})_{\text{d.}}|^2\rho_k\Big|_{k=E-E_0}. \tag{31}$$

Since for the transition in question energy is conserved within the line breadth the d-restriction plays no role at all. $\mathscr{R}\Gamma_{2O|O}(E_O) = \gamma$ is the expression of §§ 17, 18. Since (31) is proportional to k we have

$$\mathscr{R}\Gamma_2(E) = \gamma\frac{E-E_O+k_0}{k_0} \to \gamma\frac{k}{k_0}, \tag{31'}$$

if $E = E_n = E_0+k$. For values of E differing from E_O by an order $\hbar\gamma$ the difference $\mathscr{R}\Gamma(E)-\gamma$ is of the order $\gamma\hbar\gamma/k_0 \ll \gamma$.

We consider the fourth order $\mathscr{R}\Gamma_{4O|O}$ at the point $E = E_O$ only because the calculations (which are very lengthy) are very much simplified in this case and $\mathscr{R}\Gamma_4(E)$ certainly differs very little from $\mathscr{R}\Gamma_4(E_O)$. This quantity is a typical radiative correction. We obtain

$$\begin{aligned}\frac{\hbar}{2\pi}\mathscr{R}\Gamma_{4O|O}(E_O) = \{&(H^{(tr)}PH^{(tr)}+H^{(c)}-\delta_m H)\delta(H^{(tr)}PH^{(tr)}+H^{(c)}-\delta_m H)_{N.D.}-\\ &-\pi^2H^{(tr)}\delta H^{(tr)}\delta(H^{(tr)}\delta H^{(tr)})_{N.D.}+[H^{(tr)}\delta\{H^{(tr)}P(H^{(tr)}\}_{N.D.}PH^{(tr)})_{N.D.}+\\ &+H^{(tr)}\delta H^{(tr)}P(H^{(c)}-\delta_m H)_{N.D.}+H^{(tr)}\delta(H^{(c)}-\delta_m H)_{N.D.}PH^{(tr)}+\\ &+\tfrac{1}{2}H^{(tr)}\delta\frac{\partial}{\partial E_O}(H^{(tr)}PH^{(tr)})_D H^{(tr)}-\tfrac{1}{2}H^{(tr)}\delta H^{(tr)}(H^{(tr)}P^2H^{(tr)}_{\text{n.d.}})_D+\\ &+\text{compl. conj.}]\}_{O|O}\end{aligned} \tag{32}$$

where $$\delta \equiv \delta(E_O-\tilde{H}_0), \qquad P \equiv \frac{\mathscr{P}}{E_O-\tilde{H}_0},$$

and that energy which corresponds to the position of the operator $((H\delta H)_{A|B} \equiv H_{A|n}\delta(E_O-E_n)H_{n|B}$, etc.), is to be inserted for $\tilde{H}_0$. Some terms which give no contribution owing to the n.d. and $N.D.$ restrictions are omitted. The width of the energy shell has dropped out everywhere except in one renormalization term. (32) includes the probability of the emission of two quanta $k_1+k_2 = E_O-E_0$.

$\mathscr{R}\Gamma_{4O|O}(E_O)$ can be worked out completely. The following theoretical points are of interest. (i) All terms can be combined in a 4-dimensional manner as if Lorentz gauge were used and the Lorentz condition ignored, with the exception of one contribution which depends on ϵ and is non-relativistic and unambiguous. For $E \neq E_O$ this relativistic combination does not work. Therefore the main part of $\mathscr{R}\Gamma_{4O|O}(E_O)$ (namely γ_4 of (33)) appears as proper Lorentz invariant and no question of ambiguity arises. (ii) $\mathscr{R}\Gamma_{4O|O}(E_O)$ is finite and unambiguous, owing to the proper incorporation of the mass correction, provided that regard is also taken of the charge renormalization, much in the same way as in the calculation of the line shift.† Altogether, the calculation follows very much the line of subsection 2.

The order of magnitude of $\mathscr{R}\Gamma_4$ for allowed transitions turns out to be

$$\mathscr{R}\Gamma_4 \sim \mathscr{R}\Gamma_2 \frac{1}{137}\frac{p^2}{\mu^2} - \frac{\hbar\gamma^2}{\pi\epsilon} \simeq \frac{\gamma}{137^3} - \frac{\hbar\gamma^2}{\pi\epsilon} \equiv \gamma_4 - \frac{\hbar\gamma^2}{\pi\epsilon}, \tag{33}$$

where p is the average momentum in the atom.

We now insert our results (30′), (31′), and (33) in the resonance denominator of (1). Since $\Delta E \sim k_0 - k$, this can be rewritten, up to the fourth order, in the form

$$\frac{1}{k_0 - k + \Delta E + \frac{1}{2}i\hbar(\gamma k/k_0 + \mathscr{R}\Gamma_4)} \simeq \frac{1 + \hbar\gamma/\pi\epsilon}{k_0 - k + \frac{1}{2}i\hbar(\gamma k/k_0 + \gamma_4)}. \tag{34}$$

The two *ε-dependent corrections cancel* in the denominator. This has almost the classical form. Instead, the numerator receives a correction depending on ϵ and of at least third order. This should be combined with the higher orders of U which we have not calculated. In first approximation the numerator is $|H_{0k|O}|^2 \rho_k \simeq \text{const.}\,k$.

The fourth order line breadth γ_4 is exceedingly small. The factor k/k_0 of γ and also the factor k of the numerator give rise to a small distortion of the line which includes an *additional shift* of the maximum of order γ^2/k_0. This is the analogue of the classical frequency shift of a damped oscillator (see footnote on p. 33). For optical transitions this is exceedingly small.‡

Thus it appears that the theory of §§ 18, 20 is an *excellent approximation* to the *exact line shape*. From the theoretical point of view it can be stated that the concepts of mass and charge renormalization also lead to *finite and unambiguous results for problems with finite line breadth*, including the time dependence of the emission act, etc. The ambiguity due to the finite ϵ can be avoided by treating the emission in connexion with the excitation, but this is a more complicated problem.§

† $\mathscr{R}\Gamma_4(E)$, $E \neq E_O$, can be computed by expansion according to $E - E_O$ and the difference is finite and unambiguous (as in the case of $\mathscr{I}\Gamma_2(E)$).

‡ For radio-frequency transitions with small k_0 (e.g. $2P_{\frac{3}{2}} \to 2S_{\frac{1}{2}}$ of hydrogen) the distortion is noticeable but only the numerator contributes to it. If k_0 is small, γ is even smaller (see § 18) and γ^2/k_0 remains small. If from O several transitions to levels m occur, (31′) is to be replaced by

$$\gamma k/k_0 \to \sum_m \gamma_m(E_1 + k - E_m)/(E_O - E_m), \tag{35}$$

where γ_m is the transition probability to m and E_1 the final level considered. Only transitions with large $E_O - E_m$ contribute and therefore (35) is practically equal to the total width γ of O and independent of k.

§ See E. Arnous and W. Heitler, *Proc. Roy. Soc.* A, **220** (1953), 290.

35. Further outlook

1. *The present situation.* Reviewing the results so far obtained, including in particular the results of this chapter also, we find that quantum electrodynamics is in a curious position. On the one hand, we can state that the present theory cannot possibly be final. We have found a number of divergent quantities, although all of them are unobservable in principle. These are: (i) the self-mass, (ii) the self-charge of the electron (both diverging logarithmically), and (iii) vacuum effects, e.g. the self-energy of the vacuum electrons, vacuum fluctuations, etc. Furthermore, we found in some cases that even observable effects are described by ambiguous mathematical expressions (cf. the integral I_2 in § 31, magnetic moment). The ambiguities can always be settled by applying a certain amount of 'wishful mathematics', namely, by using additional conditions for the evaluation of such ambiguous integrals. Unless such conditions are used, the results of the theory may contradict its very foundations, e.g. Lorentz and gauge invariance (cf. the photon self-energy, § 32). Clearly such a mathematical situation is unacceptable.

On the other hand, these difficulties do not prevent us from giving a theoretical answer to every legitimate question concerning *observable* effects. These answers are, wherever they can be tested, always in excellent agreement with the facts, and no serious discrepancy exceeding the limits of accuracy of the calculation has so far been discovered. This applies to the two cases of radiative corrections (magnetic moment and line shift of hydrogen §§ 31, 34) with very great accuracy. It applies to all kinds of radiation processes with great accuracy ($\sim 1\%$) up to energies of $\sim 600mc^2$ (see, for example, the absorption coefficient of γ-rays, § 36), and to much higher energies at least as far as order of magnitude is concerned (see cascade theory, § 38). Finally it applies to problems concerned with line breadth (including their time dependence, see § 34) that can, in principle, be computed to any approximation. Measurements of these phenomena are, however, few.†

† A problem that has not been treated in this book (because its main importance lies in the field of nuclear physics and hyperfine structure) is that of the bound states of two particles with retarded interaction. Whereas in the corresponding collision problem (§ 24) an expansion in e could be used, the Coulomb interaction must here be treated exactly, but for the retardation effect, due to the exchange of transverse photons, only expansions exist. However, for a purely electromagnetic interaction, retardation is small and the e^2-term will in general suffice. In contrast to § 24, the interaction is here also needed outside the energy shell, but this is easily obtained by a simple generalization of § 24.2. Also in this problem all higher terms will be finite with the help of the renormalization technique.

The situation described above does not only apply to the low orders of radiative corrections to which we have mainly limited ourselves in the preceeding sections. It has been proved,† at least for collisions between free particles, that in all approximations $\sim e^n$ no further divergent quantities occur than those mentioned (of course, with further divergent contributions of higher orders).

A further question arises when we consider the complete series of contributions to an observable quantity. Each term of this series is finite and the series progresses according to powers 1/137. The first and second radiative corrections have, indeed, been found to become progressively smaller (see, for instance, § 31, magnetic moment $\sim e^4$). It may thus be regarded as probable that the e^2-expansion for observable quantities is at least semi-convergent. Nevertheless, it is very improbable that the series converges, because the number of interaction diagrams contributing to each higher term increases enormously with the power of e^2.‡

There may be two different reasons for the mathematical inconsistencies of the theory. (i) It is quite possible that the expansion in powers of e^2 is not permissible. Owing to the great mathematical complications of the theory it has not yet been possible to decide whether quantum electrodynamics as it stands has exact non-divergent solutions or not. One may well visualize a situation where the self-energy of an electron turns out to depend on e in the form of a non-expandable function, and any attempt at an expansion would lead to a divergent result. (ii) It is equally possible that the basic ideas and formulation of the theory must be modified at some fundamental point. The second alternative suggests itself in view of the fact that some future theory must be expected to account for the ratio of the masses of the fundamental particles and the numerical value of $e^2/\hbar c$, problems which are quite outside the scope of the present theory. Any such modifications, whether it be in the mathematical solutions or of a more basic nature, must, of course, leave the useful results of the theory practically unaltered.

There is one point where an obvious, though very small, modification of all results has to be made. The electromagnetic field interacts with all fundamental particles and even if we only consider electron-photon processes, these particles will occur in intermediate states (pairs of

† F. J. Dyson, *Phys. Rev.* **75** (1949), 1736.

‡ For a simplified case (scalar field) it is proved that the series diverges for all values of the coupling constant; W. Thirring, *Helv. Phys. Acta*, **26** (1953), 33.

mesons, etc.). The following charged particles besides electrons are known to exist. (i) μ-mesons, spin $\frac{1}{2}$, mass $210m$, magnetic moment unknown. (ii) π-mesons, spin 0, mass $276m$. (iii) heavy mesons, several types, masses 900–1,500m, spin, magnetic moment (and higher moments, if spin $\geqslant 1$) unknown. (iv) Protons, 1,835m, spin $\frac{1}{2}$, magnetic moment 2·78 nuclear magnetons. This, however, is of mesonic origin. The list may be very incomplete.

Neutral particles with electromagnetic properties must also, in principle, be considered: neutron, spin $\frac{1}{2}$, magnetic moment $-1{\cdot}93$ nuclear magnetons; a neutral particle, mass 2,200, half-integral spin. The neutretto (neutral π-meson), mass $264m$, has spin 0 and therefore no electromagnetic properties. The same applies probably to the neutrino (spin $\frac{1}{2}$, mass 0, probably no magnetic moment).

Owing to their great mass, the influence of these charged and neutral particles on observable results concerning electrons and photons is extremely small. For example, pair creation of particles with mass M is $(m/M)^2$ times smaller than for electrons and is insignificant when this occurs as an intermediate state in a complicated electron-photon process. On the other hand, this argument does not apply to the divergent features of the theory and it may well be asked whether the existence of such fields can be of any help in removing the inconsistencies of the theory.

In what follows, this question, together with some formal considerations pointing in a similar direction, will be examined. It will be found that a mere superposition of fields with different masses does not suffice to remove all difficulties, although some indications for an improvement exist. A treatment of the electromagnetic properties of particles with spin 0, 1,... is outside the scope of this book and we discuss this here only in so far as such particles may influence the theory of electrons and photons.

No serious attempt has yet been made at removing the limitations set by the expansion according to e^2, although we have, in some places (for the discussion of damping phenomena) departed from the expansion. Further mathematical investigations in this direction are very much needed.

2. *Formal regularization.*† This is a purely formal device without direct physical significance, that will permit us to remove ambiguities

† W. Pauli and F. Villars, *Rev. Mod. Phys.* **21** (1949), 434. Regularization is a generalization of methods already used by: E. C. G. Stückelberg and D. Rivier, *Phys. Rev.* **74** (1948), 218 and 986; D. Rivier, *Helv. Phys. Acta*, **22** (1949), 265; R. P. Feynman, *Phys. Rev.* **74** (1948), 1430.

such as we have encountered in §§ 31 and 32. It rests on the following observations.

Ultimately all divergencies and ambiguities are due to the singular character of the Δ- and D-functions (§ 8) which occur in the commutation relations of the field quantities. We have seen in § 8 eq. (33) that the D and D_1 functions have singularities of the type $\delta(x_\mu^2)$ and $\mathscr{P}/x_\mu^2$ respectively on the light cone, plus lesser singularities. In the neighbourhood of the light cone these two functions are (§ 8 eqs. (33), (35)):

$$D \simeq \frac{1}{2\pi}\delta(x_\mu^2)\epsilon(x_0)+\begin{cases} 0 \quad (r > |x_0|), \\ -\dfrac{\eta^2}{8\pi}\epsilon(x_0) \quad (r < |x_0|), \end{cases} \tag{1 a}$$

$$D_1 \simeq \frac{1}{2\pi^2}\frac{\mathscr{P}}{x_\mu^2}+\frac{\eta^2}{4\pi^2}\log\frac{\eta\gamma}{2}\sqrt{|x_\mu^2|}, \tag{1 b}$$

η is essentially the rest mass of the particles of the field. We see that the strongest singularities are independent of η, the weaker singularities are $\sim \eta^2$. On the other hand, when η is very large, both D and D_1 vanish, everywhere except for the singularities, as is immediately seen from § 8 eqs. (30). The idea of regularization is the following. Suppose we replace D and D_1 by a superposition of such functions with various masses m_i (using m for η) $D \rightarrow \sum_i c_i D_{m_i}$, $D_1 = \sum c_i D_{1m_i}$. The singularities can be completely removed if

$$\sum_i c_i = 0, \qquad \sum c_i m_i^2 = 0. \tag{2}$$

These functions (or their Fourier transforms) occur in the computations wherever virtual fields occur. (Compare the developments in § 28.) For example, in the self-energy of the electron a virtual photon field occurs, or in the problem of vacuum polarization there is a virtual electron field. The above replacement means that we add formally further virtual fields with different masses and with 'weights' c_i. In order that these should have no observable physical effects these 'auxiliary masses' must all be allowed to go to infinity, whereas the original virtual field (photon or electron field) must have a coefficient 1, and the mass (0, or m) must remain unaltered. If we denote the field that actually occurs by an index 0, the auxiliary fields by indices $j = 1, 2,...$, and the auxiliary masses by M_j, (2) becomes:

$$1+\sum_j c_j = 0, \qquad m_0^2+\sum_j c_j M_j^2 = 0, \qquad M_j \rightarrow \infty. \tag{3}$$

$m_0 = 0$ for a virtual photon field, $m_0 = m$ for a virtual electron field.

At least two auxiliary fields are necessary to fulfil (3). The unphysical nature of this procedure is clear from the fact that the coefficients c_j must be partly negative.

The effect of this procedure on any particular result can be obtained very easily. Suppose a virtual photon field occurs, with virtual photons $\mathbf{k}$ (energy k), resulting, for example, in a matrix element $\int f(\mathbf{k}, k)\, d^3k$. We have then to add contributions arising from auxiliary 'photon' fields, which have rest energy M_j. Such fields are, of course, not really photon fields but are known as neutral vector meson fields and differ from the photon field essentially by the finite rest mass (and consequently the lack of gauge invariance, but this is here irrelevant).

Thus we have to replace

$$\int f(\mathbf{k}, k)\, d^3k \rightarrow \sum_i c_i \int f(\mathbf{k}, \sqrt{(k^2+m_i^2)})\, d^3k$$
$$= \int f(\mathbf{k}, k)\, d^3k + \sum_j c_j \int f(\mathbf{k}, \sqrt{(k^2+M_j^2)})\, d^3k. \quad (4)$$

Note that $\mathbf{k}$ is the momentum variable, which remains unaltered.

As an example we consider the integral I_2 of § 31, eq. (9_2) which we found to be ambiguous. The value that conforms with Lorentz invariance is $I_2 = 0$, but other values can be obtained according to evaluation. After integration over w and k_0 (in the complex plane) I_2 becomes

$$I_2 = \frac{\pi}{8i\mu^2} \int d^3k \left(\frac{1}{k} - \frac{k_n^2}{k^3} - \frac{1}{\sqrt{(k^2+\mu^2)}} + \frac{k_n^2}{\sqrt{(k^2+\mu^2)^3}}\right) \quad (5)$$

written now in ordinary k-space. k_n is a spatial component of $\mathbf{k}$. When I_2 is regularized in accordance with (4) we have to replace k by $\sqrt{(k^2+m_i^2)}$, whereas k_n remains unaltered. The same is true for the volume element in momentum space d^3k. Thus, if we denote the regularized value by I_{2R}, and integrate over the angles

$$I_{2R} = \frac{\pi^2}{2i\mu^2} \sum_i c_i \int k^2\, dk \left(\frac{1}{\omega_i} - \frac{1}{3}\frac{k^2}{\omega_i^3} - \frac{1}{\Omega_i} + \frac{1}{3}\frac{k^2}{\Omega_i^3}\right),$$
$$\omega_i = \sqrt{(k^2+m_i^2)}, \qquad \Omega_i = \sqrt{(k^2+\mu^2+m_i^2)}. \quad (6)$$

The integration is elementary and yields, if we let the upper limit k tend to ∞ before settling the values of m_i,

$$I_{2R} = \frac{\pi^2}{6i\mu^2} \sum_i c_i \left\{k\omega_i - m_i^2 \frac{k}{\omega_i} - k\Omega_i + (m_i^2+\mu^2)\frac{k}{\Omega_i}\right\}_{k\to\infty} = \frac{\pi^2}{12i} \sum_i c_i. \quad (7)$$

If we do not regularize, the value obtained is given by putting $c_0 = 1$ and all other $c_j = 0$, i.e. we obtain the finite value $\pi^2/12i$. This is just one of the ambiguous values that can be found for I_2. However, after regularization (7) vanishes, according to the first condition (2). The value zero is what conforms with the requirements of Lorentz invariance. The integration is carried out here in ordinary k-space, without using any additional arguments, and is now quite unambiguous.

In much the same way it can be shown that the non-gauge and Lorentz invariant part of the induction tensor $\bar{\bar{L}}_{\mu\nu}$ vanishes, using in this case both conditions (2) and regularizing with respect to the electron field (i.e. introducing 'auxiliary electron fields').† Also the self-stress of the electron vanishes unambiguously (appendix 7).

When regularization is applied to the diverging quantities $\delta\mu$ and δe, it is found that these would be finite if the auxiliary masses M_j were allowed to remain large but finite. In this case regularization is nothing but a kind of invariant 'cut-off'. In the limit $M_j \to \infty$, the expressions diverge again.

Regularization is a convenient tool for settling ambiguities without being forced to use the (often inconvenient) invariant variables. It has, however, hardly any physical significance.

3. *Superposition of real fields.* The formal success, limited though it is, of the regularization procedure, raises the question whether any substantial progress can be made in the theoretical situation by considering the superposition of actual fields with their known properties. As our knowledge of the particles that occur in nature is incomplete, we consider this question from a purely theoretical point of view.

We apply here the term meson for any charged particle with mass higher than m, and neutretto for a neutral particle with finite mass. Mesons and neutrettos may exist with spins $0, \frac{1}{2}, 1, \ldots$. Two distinctly different questions must be considered.

(i) The existence of meson fields will influence the electromagnetic properties of electrons in so far as they can occur as virtual fields. This is the case for all problems where, normally, virtual electron pairs occur, above all, in vacuum polarization.

(ii) In problems where the electromagnetic field is the virtual field (self-energy), virtual mesons will not produce any modification except

† In more complicated cases one must consider not only the regularization of a single D or D_1-function, but also products of two such functions. Owing to the purely formal character of this procedure we do not go into further details (see Pauli and Villars, loc. cit.)

in higher orders. We can expect a change only if we add to the electromagnetic field a virtual neutretto field to which the electron must be coupled directly. This would correspond to the formal regularization of the photon field, subsection 2. Although one type of neutretto with mass almost equal to that of the π-meson and spin 0 exists in nature, there is no evidence at all that this, or any other neutral field, is coupled to electrons directly. It is primarily coupled to nucleons. There is therefore little hope that real neutretto fields can help in removing the difficulties connected with the self-energy of electrons. We refrain therefore from a discussion of (ii).†

Turning to (i) we first consider the case where mesons with spin $\frac{1}{2}$ and 0, in addition to electrons, are drawn into consideration. Particles with these spins really exist. The quantum electrodynamics of these particles resembles that of electrons in so far as all divergent quantities are unobservable. In the case of spin $\frac{1}{2}$ this is only true when the magnetic moment is the Bohr magneton (corrected by radiative corrections), but no intrinsic moment exists.

Vacuum polarization due to a superposition of virtual meson fields has been investigated‡ including also the higher approximations.§ The following is the result. Let n^0, $n^{\frac{1}{2}}$ be the number of different meson fields (additional to the electron field) with spin 0 and $\frac{1}{2}$ respectively, and M_j^0, $M_j^{\frac{1}{2}}$ ($j = 1,...,n$) their masses, m = electron mass. Then the non-gauge and Lorentz invariant part of the induction tensor $\bar{\bar{L}}_{\mu\nu}$ vanishes (in the higher approximations also) if the following conditions are fulfilled:

$$n^0 = 2(1+n^{\frac{1}{2}}),$$
$$\sum_1^{n^0} (M_j^0)^2 = 2m^2+2\sum_1^{n^{\frac{1}{2}}} (M_j^{\frac{1}{2}})^2. \tag{8}$$

These rather resemble the regularization conditions (3). Whether these conditions are likely to be fulfilled by the actual particles existing in nature is impossible to state. They are certainly not fulfilled by the mesons known so far, but they may be fulfilled by supposing the existence of very heavy mesons (heavier than protons, say) which are beyond the means of detection up to the present.

† The coupling of electrons to neutral fields has been investigated by A. Pais, *Verh. Kon. Nederl. Ak. v. Wetenschappen*, **19** (1947), No. 1; D. Ito, Z. Koba, and S. Tomonaga, *Progr. Theor. Phys.* **3** (1948), 276 and 325 and subsequent papers.

‡ R. Jost and J. Rayski, *Helv. Phys. Acta*, **22** (1949), 457; D. Feldmann, *Phys. Rev.* **76** (1949), 1369; H. Umezawa and R. Kawabe, *Progr. Theor. Phys.* **4** (1949), 423, 443; H. Umezawa and S. Kamefuchi, ibid. **6** (1951), 543.

§ G. Källén, *Helv. Phys. Acta*, **22** (1949), 637.

On the other hand, no improvement is reached in the question of the self-charge. Virtual fields with spin 0 give rise to a *self-charge* (change of the interaction between electron and external field) with the *same sign* as fields with spin $\frac{1}{2}$. The self-charge cannot therefore be made finite by a superposition of such fields.

We may next ask whether mesons with higher spins, for example spin 1, could help. This is not possible for a different reason. In the case of spin 1 further quantities, and indeed observable quantities, diverge. For example, in vacuum polarization also the term $L_2 K^2$ (eq. (18 d), § 32), which is otherwise observable, diverges. In fact there is an infinite number of divergent quantities.†

In view of this situation it has been argued that charged particles with spin 1 cannot exist in nature. While this may be the case, the argument carries little weight. Whatever the fault of the present theory may be, whether it be the impermissible expansion, or a fundamental physical point, the (unknown) remedy is also bound to affect fields with spin 1, and it is impossible to foretell whether the electrodynamics of particles with spin 1 will then be convergent or whether difficulties will still persist.

Summarizing, we can say that the mere superposition of fields does not suffice to remove the difficulties of the present theory. There are some indications of an improvement, and possibly the final solution can only be found by including all existing fields, but an additional, mathematical or physical, new idea is definitely needed.

† It is interesting, in spite of this situation, to consider the self-charge due to virtual particles with spin 1. This consists of a logarithmically divergent part similar to that for electrons but with opposite sign, and an ambiguous quadratically divergent part which has the same sign as δe for electrons or mesons with spin 0. The quadratically divergent part can (but need not) be evaluated in such a way that it vanishes. So, with the help of fields with spin 1, it might be possible to make the self-charge finite or vanish, although some arbitrariness must then be applied. Cf. J. McConnell, *Phys. Rev.* 81 (1951), 275.

VII

PENETRATING POWER OF HIGH-ENERGY RADIATION

36. Absorption coefficient of γ-rays

1. *Theoretical values.* If a beam of X- or γ-rays passes through matter, its intensity decreases owing to absorption and scattering. Disregarding all kinds of selective absorption, i.e. processes by which only radiation of a single frequency is absorbed (excitation into discrete levels), we found in Chapter V that the following three processes give rise to *continuous absorption*:

(1) Photoelectric effect.

(2) Compton scattering by free electrons.

(3) Creation of pairs.

The intensity I of a *monochromatic* beam passing through matter decreases *exponentially*:

$$I = I_0 e^{-\tau x}, \tag{1}$$

where the *absorption coefficient* τ represents the average number of absorption and scattering processes which a single light quantum undergoes per cm. path. τ is composed additively of three parts referring to the three processes above:

$$\tau = \tau_{\text{phot}} + \tau_{\text{Compt}} + \tau_{\text{pair}}. \tag{2}$$

If ϕ represents the cross-section of one atom for one of these three processes, the corresponding value of τ is given by

$$\tau = N\phi, \tag{3}$$

where N is the number of atoms per cm.^{3}

The numerical values of the cross-sections for the three processes, as calculated from the theory, are given in this book as follows:

(1) For the photoelectric effect of the K-shell† in Tables II, III, and Fig. 9, pp. 208, 210.

(2) For the Compton scattering in Table IV and Fig. 11, pp. 221, 223. The values refer to a single electron and have to be multiplied by Z to give the cross-section of an atom.

(3) For the creation of pairs in Tables VI and VII and Fig. 17, pp. 262, 263.

† The higher shells give only a small contribution amounting to about 25 per cent. of the effect of the K-shell. In the following we have multiplied ϕ_K by a factor 5/4. (Cf. § 21.)

Since the quantities are expressed in units $\phi_0 Z^5/137^4$, $\phi_0 (= 8\pi r_0^2/3)$, $\bar{\phi} (= r_0^2 Z^2/137)$ respectively, the figures in the tables and graphs mentioned above have to be multiplied respectively by $N\phi_0 Z^5/137^4$, $NZ\phi_0$, $N\bar{\phi}$ to give the corresponding term in τ. The values of these quantities for several materials are given in appendix 8.

For the photo-electric contribution the exact values (Table III) will be used rather than those of the Born approximation, but for pair creation the exact values available are too scanty and we shall use the results of the Born approximation. This would cause a considerable error only for low energies and heavy elements, but here the contribution to the total absorption coefficient from pair creation is quite small compared with that from the Compton and photo-electric effects. At energies above $30mc^2$ we expect the theoretical absorption coefficient to be lower by about 10 per cent. for lead (less for light elements) than the values given below; the error is more or less independent of the energy (see § 26).

A certain contribution to pair creation also arises from the atomic electrons. For this a crude curve was drawn in Fig. 17. This contribution is negligible for energies up to, say, $10mc^2$ and rises roughly to $0{\cdot}8/Z$ of the nuclear contribution at higher energies. We take this fact into account by multiplying the pair contribution by $(Z+0{\cdot}8)/Z$ for all energies from $20mc^2$ upwards. The error caused by the uncertainty of this contribution is nowhere large (< 3 per cent.), even for light elements, because at low energies the other processes are far more important and at high energies the contribution cannot be very far from that assumed.

The three processes are not equally important in different energy regions. For small energies $\hbar\nu$ the photo-electric absorption gives the main contribution. For higher energies the Compton scattering becomes more important, and finally, for very high energies, the absorption is entirely due to pair production. The regions where these three effects give the largest contribution are roughly as follows:

	Photoelectric effect	*Compton effect*	*Pair formation*
Pb	$\hbar\nu/mc^2 < 1$	~ 1–10	> 10
Al	$\hbar\nu/mc^2 < 0{\cdot}1$	$\sim 0{\cdot}1$–30	> 30

As a function of the atomic number Z, the absorption coefficient behaves most simply in the region where the Compton scattering gives the only contribution. τ in this region is simply proportional to the total number of electrons NZ per cm.^{3} or, since Z is roughly proportional to the atomic weight, proportional to the density ρ. Thus the *total intensity absorbed is proportional to the total mass per cm.*^{2} which the γ-ray has passed. We can then define a *mass absorption coefficient* τ/ρ (gm.$^{-1}$ cm.^{2}) which is almost constant for all elements at a given energy. Such a mass absorption coefficient, however, does *not exist* in the region where pair production is effective, because the latter is proportional to Z^2.

Collecting our data, we find the following values for the total absorption coefficient.

TABLE X

Absorption coefficient τ per cm. for γ-rays (theoretical)†

$\hbar\nu/mc^2$	0·1	0·5	1	2	5	10	
air	2·16	1·44	1·11	0·803	0·501	0·347	$\times 10^{-4}$
H_2O	1·87	1·25	0·955	0·698	0·431	0·298	$\times 10^{-1}$
Al	0·96	0·297	0·224	0·164	0·104	0·0755	
Pb	—	$\sim$6	1·70	0·78	0·477	0·473	

$\hbar\nu/mc^2$	20	50	100	1,000	10,000	
air	0·262	0·210	0·206	0·23	0·24	$\times 10^{-4}$
H_2O	0·222	0·176	0·169	0·184	0·191	$\times 10^{-1}$
Al	0·0628	0·0596	0·0641	0·0804	0·084	
Pb	0·604	0·850	1·04	1·40	1·49	

The dependence of τ on the frequency and on Z is shown for several metals in Fig. 25. For small $\hbar\nu$, τ decreases rapidly owing to the rapid falling off of the photo-electric absorption. For higher frequencies τ decreases more slowly. This is due to the fact that here the main contribution arises from Compton scattering. For energies higher than $2mc^2$, *τ rises again*, because of *creation of pairs*. Finally, it tends to a constant asymptotic value which is greatest for heavy elements. Thus, there is a region of *greatest transparency* of all materials which lies at $\hbar\nu = 5$–$20mc^2$.

2. *Experiments. Critical remarks.* In Fig. 25 we have inserted the results of measurements for five energies: 0·835 MeV ($= 1{\cdot}635mc^2$),

† The figures for air refer to atmospheric pressure and 0° C. For air and H_2O the photo-electric absorption has been neglected; it will slightly affect the value for $0{\cdot}1mc^2$. For Pb, the value 6 for $\hbar\nu = 0{\cdot}5mc^2$ is only an estimate, since the photo-electric effect can hardly be calculated exactly for this energy. A value for $0{\cdot}1mc^2$ has not been inserted because this energy is smaller than the K-absorption edge.

The figures in Table X are calculated to an accuracy of 0·5–1 per cent. Owing to many inaccuracies in the basic theory and the material constants no higher accuracy is warranted although in the experimental literature more decimals are sometimes given.

2·76 MeV ($= 5\cdot40mc^2$), 17·6 MeV ($= 34\cdot4mc^2$), 88 MeV ($= 172mc^2$), and 280 MeV ($= 550mc^2$).†

We see that the agreement is perfect (within 3 per cent.) for light elements (Al, Cu) and also for heavy elements for energies up to $10mc^2$ or so. A certain discrepancy for higher energies is noticeable for Sn and quite conspicuous ($\sim$ 11 per cent.) for Pb.

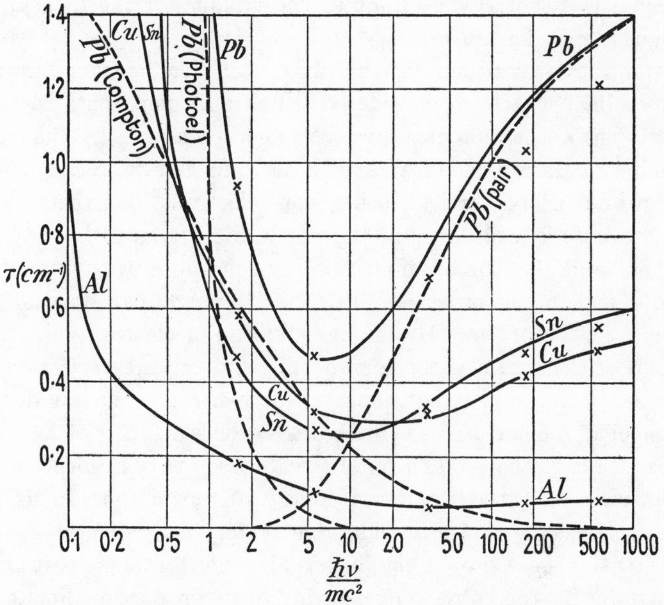

FIG. 25. Absorption coefficient τ (cm.$^{-1}$) for γ-rays in Al, Cu, Sn (density 7·00), Pb, as a function of the primary energy $\hbar\nu$ on a logarithmic scale. The dotted curves show the three components of τ for Pb. Measurements for $\hbar\nu = 1\cdot635, 5\cdot40, 34\cdot4, 172$, and $550mc^2$.

The discrepancy is due to the departure from the Born approximation of the cross-section for pair creation. (All other contributions to the absorption coefficient are small at the energies in question.) It is quantitatively explained by eq. (15′), § 26. For $\hbar\nu = 550mc^2$ the cross-section for pair creation in Pb is $10\cdot3\bar{\phi}$, according to the Born approximation. The correction in question is $-1\cdot04\bar{\phi}$. This means a decrease of 10 per cent., in very good agreement with the observations.

It can thus be concluded that the theory of pair creation is correct

† C. M. Davisson and R. D. Evans, *Phys. Rev.* **81** (1951), 404 (0·835 and 2·76 MeV); R. L. Walker, ibid. **76** (1949), 527 (17·6 MeV); J. L. Lawson, ibid. **75** (1949), 433 (88 MeV); J. W. de Wire, A. Ashkin, and L. A. Beach, ibid. **82** (1951), 447 (280 MeV).

up to an energy of at least $600mc^2$, and that with an accuracy of at least 1 per cent. A higher accuracy can hardly be expected for various reasons (see below) and also because the Thomas–Fermi model used in the calculations cannot claim perfect accuracy.

For energies higher than $550mc^2$ no direct measurements of τ are available, but from the cascade phenomena resulting from the penetration of high energy γ-rays through matter (§ 38) it can be concluded that there is essentially no limit to the validity of the theory, up to the highest energies known.

In the above treatment of the absorption coefficient a number of processes have not been considered which also contribute to a small extent to the absorption of γ-rays. We mention a few. (i) The creation of double pairs by a γ-ray and other multiple-processes; (ii) the creation of pairs of particles other than electrons, in particular μ, π, or heavy mesons, protons; (iii) the scattering of γ-rays by protons or neutrons; and (iv) meson production in collisions with a proton or neutron. (i) is of the order $\sim 1/137$th of the single pair creation, (ii) is smaller by a factor $(m'/m)^2$ than the creation of electron pairs if m' is the mass of the particles considered. (iii) and (iv) are very interesting processes. It is known that a proton or neutron is surrounded by a mesonic charge cloud. This contributes to the scattering at all energies and also leads to the production of a real meson with absorption of the photon when $\hbar\nu$ exceeds the rest energy of the meson. In hydrogen both effects may be quite considerable at very high energies and comparable with other γ-ray effects but in all other materials pair creation is always by far the largest effect owing to its proportionality to Z^2.

3. *Diffusion of γ-rays through matter.* The absorption coefficient was defined so that any photon which is either absorbed or scattered outside the direction of the primary, even by a small angle, is considered as removed from the primary beam. It is clear that the phenomena connected with the penetration of γ-rays through finite thicknesses of matter are much more complicated. At very high energies cascade phenomena occur which will be considered in § 38. When the energy is lower, the scattering is the most important process and a number of photons will be scattered by small angles with a small decrease of their energy. Thus a primary beam that is originally monochromatic and strictly linear passing through matter will be widened into a cone with small diverging angle and the radiation in this cone will become increasingly softer as the angle increases. Moreover, of course, the pair- and photo-electrons produce Bremsstrahlung and the positrons give

rise to annihilation radiation. Here we consider briefly the penetration phenomena connected with small angle scatterings and neglect all other processes.

Supposing a photon $\hbar\nu$ ($\gamma = \hbar\nu/mc^2$) is scattered through a small angle θ. After the scattering its energy γ_1 is given by (§ 22 eq. (4)) $1/\gamma_1 = 1/\gamma+\frac{1}{2}\theta^2$. When n such scattering acts have taken place $1/\gamma_n = 1/\gamma+\frac{1}{2}n\theta^2$. Since these scattering acts are statistically independent, the angle formed with the direction of the primary will be on the average $\sqrt{(n)}\theta = \theta'$. Hence it follows that when photons are observed at an angle θ' from the primary direction their average energy is given by $1/\gamma' = 1/\gamma+\frac{1}{2}\theta'^2$. In fact, if the thickness is so large that several scatterings have taken place, the distribution in angle will be Gaussian. The number of photons at a thickness x of material with energy γ' will thus be given roughly by:

$$F(x,\gamma',\theta') = f(x,\gamma')e^{-\frac{1}{2}\theta'^2\{\gamma\gamma'/(\gamma-\gamma')\}}e^{-\tau x}, \tag{4}$$

where τ is the total absorption coefficient. τ is assumed to be independent of energy here, within the range of energy $\gamma-\gamma'$ considered. The factor $f(x,\gamma')$, i.e. the energy distribution at depth x, can only be obtained from a more elaborate treatment of this diffusion problem. For details we refer the reader to the literature.†

If we fix our attention on the degradation of energy rather than on the spread in angle the diffusion of a photon through matter can be described as an energy loss. If NZ is the number of electrons per cm.^{3} the average energy loss per cm. path is, according to § 22 eq. (41),

$$-\left(\frac{\partial k_0}{\partial x}\right)_{\text{Compton}} = NZ \int\limits_{k_0\mu/(\mu+2k_0)}^{k_0} (k_0-k)\phi_k\, dk = NZ\phi_0\mu\frac{3}{8}\left[\log\frac{2k_0}{\mu}-\frac{5}{6}\right] \qquad \text{E.R.} \quad (5)$$

for $k_0 \gg \mu$. The formula is rather similar to that giving the energy loss of a fast electron by collisions (§ 37 eq. (8)) but (5) is numerically much smaller (for electrons the ratio μ/IZ occurs under the log which is very large). (5) also includes the contribution from large angle scattering.

37. Stopping power of matter for fast particles

A particle passing through matter can loose its energy chiefly in two different ways.

1. It transfers energy directly to an atom by exciting or ionizing it (inelastic collisions).

† L. L. Foldy, *Phys. Rev.* **81** (1951), 395; L. L. Foldy and R. K. Osborn, ibid. **81** (1951), 400.

2. The particle is deflected in the field of an atom and emits radiation (Bremsstrahlung). This process was studied in detail in § 25.

In the older theories of the stopping of fast particles, the first process only was taken into account. For electrons this is justifiable for energies up to about $5mc^2$, but for higher energies the Bremsstrahlung gives rise to an extraordinarily high energy loss which considerably outweighs the energy loss due to inelastic collisions.

1. *Average energy loss by inelastic collisions (gases).* The energy loss of particles passing through matter was first calculated by Bohr,† using the classical theory. In quantum mechanics the problem has been investigated by several authors‡ in a satisfactory way, using various atomic models.

The calculations are rather lengthy and in this book we must confine ourselves to a statement of the results. Energy is transferred to the atomic electrons by raising them to higher levels including the continuous spectrum. When the energy of the primary particle is large compared with the ionization energy of the electrons, the energy transferred to the atomic electrons is still in most cases only a few times the ionization energy. Larger energy losses occur but are comparatively rare. For the latter, the atomic electrons can be considered as free and one can use the theory of § 24 (eqs. (28) and (34)) which gives the cross-section for a certain energy transfer q in a collision with a free electron.

The basic formula for the energy lost per cm. path by a particle travelling with velocity $v = c\beta$ is

$$-\left(\frac{dE}{dx}\right)_{\text{coll}} = NZ\phi_0\mu\frac{3}{4}\frac{z^2}{\beta^2}\left[\log\frac{2\mu\beta^2W_m}{I^2Z^2(1-\beta^2)}-2\beta^2\right] \tag{1}$$

$$= NZ\phi_0\mu\frac{3}{4}\frac{z^2}{\beta^2}B,$$

where N = number of atoms per cm.^{3}; $\phi_0 = (8\pi/3)r_0^2$ unit cross-section; μ = rest energy of electron; z = charge of primary particle; Z = atomic number. I is some average ionization energy, which is not quite the same for all elements but certainly lies in the range 11–14 eV. Later we shall use an experimental determination for Al

$$I = 11{\cdot}5 \text{ eV}.$$

† N. Bohr, *Phil. Mag.* **25** (1913), 10; **30** (1915), 581.

‡ The most important papers are: Ch. Møller, *Ann. Phys.* **14** (1932), 531; H. Bethe, *Hdb. d. Phys.* XXIV (1), 521; E. J. Williams, *Proc. Roy. Soc.* A, **135** (1932), 108; **139** (1933), 163; **169** (1939), 531; *Phys. Rev.* **58** (1940), 292; F. Bloch, *Ann. Phys.* **16** (1933), 285; *Zs. Phys.* **81** (1933), 363; M. S. Livingston and H. A. Bethe, *Rev. Mod. Phys.* **9** (1937), 245; N. Bohr, *Kgl. Dansk. Vid. Selsk.* **18** (1948), No. 8. (See also references below.)

W_m is the *maximum energy* that can be transferred to a free electron by the primary. If this has mass M, energy E (including rest energy), the conservation laws show that

$$W_m = 2\mu \frac{E^2 - M^2c^4}{2E\mu + M^2c^4 + \mu^2}. \tag{2}$$

If the colliding particle is an electron ($M = m$) the faster of the two electrons after the scattering is to be defined as the primary and only half of (2) is to be taken, i.e.

$$W_m = \tfrac{1}{2}(E - \mu). \tag{2'}$$

When the primary is a heavy particle, $M \gg \mu$, (2) can be replaced by

$$W_m = \frac{2\mu\beta^2}{1-\beta^2}, \quad \text{if} \quad E \ll Mc^2\frac{M}{m}, \tag{2''}$$

$$W_m = E, \qquad \text{if} \quad E \gg Mc^2\frac{M}{m}. \tag{2'''}$$

For nearly all energies of interest (2″) holds. If W_m is inserted in (1), the energy loss also includes the *rare large transfers*.

Formula (1) holds if the colliding particle is heavy (not an electron) and if

$$\frac{z}{137\beta} \ll 1. \tag{3}$$

Otherwise the following corrections are to be made:

(i) (3) is really fulfilled for all velocities in which we are interested here. For very slow or highly charged particles, Bloch has given a correction (to be added to (1))

$$B' = 2\left(\Psi(0) - \mathscr{R}\Psi\left(\frac{iz}{137\beta}\right)\right). \tag{3'}$$

Here Ψ is the logarithmic derivative of the factorial function

$$\Psi(x) = \frac{d}{dx}\log\Gamma(x+1),$$

and $\mathscr{R}$ denotes the real part.† For large values of $z/137\beta$, (1) then goes over into the well-known classical formula of Bohr. For our purposes the correction is quite negligible.

(ii) If the primary particle is an electron, a modification ought to be made owing to the fact that for large energy transfers exchange and

† For large x the expansion

$$\Psi(x) = \log x + \frac{1}{2x} - \frac{1}{12x^2} + \ldots$$

can be used. $\Psi(0) = -0{\cdot}577\ldots$. For small x: $\Psi(x) = -0{\cdot}577\ldots + x\,1{\cdot}645\ldots - x^2 1{\cdot}20\ldots + \ldots$.

spin effects play some role (see § 24, difference of formulae (28) and (34)). According to Bethe, this correction results in an addition to (1)

$$B'_{\text{el}} = -[\beta^2+2\sqrt{(1-\beta^2)}]\log 2+1+\beta^2. \tag{4}$$

This correction is likewise small compared with B, eq. (1), for all values of β and it is doubtful whether it really goes beyond the limits of accuracy of the whole theory. (4) will, however, be included in the numerical results quoted below. For further refinements of formula (1) we refer the reader to the literature.

The following limiting cases are important:

(i) Heavy particles, non-relativistic energies, $E-Mc^2 \equiv T \ll Mc^2$,

$$-\left(\frac{dE}{dx}\right)_{\text{coll}} = NZ\phi_0\,\mu z^2\frac{3Mc^2}{4T}\log\frac{4Tm}{IZM}. \qquad \text{N.R.} \quad (5)$$

(ii) Electrons, non-relativistic energies, $E-\mu \equiv T \ll \mu$,

$$-\left(\frac{dE}{dx}\right)_{\text{coll}} = NZ\phi_0\,\mu\frac{3\mu}{4T}\left[\log\frac{T}{IZ\sqrt{2}}+\frac{1}{2}\right]. \qquad \text{N.R.} \quad (6)$$

In (5) and (6) it is understood that, although $\beta \ll 1$, $z/137\beta \ll 1$.

(iii) Heavy particles, extreme relativistic energies $E \gg Mc^2$, but $E \ll Mc^2(M/m)$:

$$-\left(\frac{dE}{dx}\right)_{\text{coll}} = NZ\phi_0\,\mu\tfrac{3}{2}z^2\left[\log\frac{2E^2m}{IZM^2c^2}-1\right]. \qquad \text{E.R.} \quad (7)$$

(iv) Electrons, extreme relativistic energies $E \gg \mu$:

$$-\left(\frac{dE}{dx}\right)_{\text{coll}} = NZ\phi_0\,\mu\tfrac{3}{4}\log\frac{E^3}{2I^2Z^2\mu}. \qquad \text{E.R.} \quad (8)$$

Essentially, the energy loss by collisions depends on the velocity of the primary particle and is practically independent of its mass. If expressed in terms of the energy of the primary the curves representing $-(dE/dx)_{\text{coll}}$ are nearly the same for all particles but are shifted to higher energies for higher masses. A small dependence on the mass is due to the addition (4) for electrons and for heavy particles at extremely high energies, to the dependence of W_m on M, when (2″) is not valid.

As a function of the material $-(dE/dx)_{\text{coll}}$ is nearly proportional to NZ, i.e. the number of electrons per cm.^{3} An additional dependence is due to the occurrence of Z under the log in (1).

The energy loss is large for small velocities and decreases $\sim 1/v^2$ until v is comparable with c. It reaches a minimum when the kinetic energy is of the order of the rest energy. The energy loss then increases again logarithmically but this increase is so slow that for many purposes it can be said that the energy loss is minimum when E is larger than $2Mc^2$ or so.

It should be noted that the above expressions represent the *average energy* loss, and this also includes the rare cases of large energy transfers up to W_m. In a not too thick layer of material these large energy transfers may be very improbable and the *most probable energy* loss (this is what is usually measured) is less than (1). It is obtained by replacing the upper limit W_m for the energy transfer by a smaller quantity W, which is determined such that in the thickness Δx, say, just one transfer (on the average) of magnitude W occurs. W depends then on Δx. The distribution of energy losses (straggling), when a fast particle passes through a finite thickness of matter, has been determined by Landau,† but since the deviation from the average loss is not large, we refrain from a detailed discussion.

The average energy loss is also, with fairly great accuracy, representative of the *average primary ionization* of the particle. The fractional number of cases in which a collision with an atom results in ionization (rather than excitation to a discrete level) is almost independent of the energy of the primary, and the same is true for the average energy transferred to the ionized electron. Thus the number of primary ion pairs formed per *cm*. path is very nearly proportional to the average loss by collisions.‡ The number of primary ion pairs can be inferred from the fact that about one ion pair is formed when the primary loses an amount of 32 eV. This figure is practically independent of the nature of the particle and its energy (at least if the latter is not excessively small). It should be noticed, however, that the secondary particles produce further ion pairs and the total ionization differs from the primary ionization.

2. *Polarization effect.* All the above expressions hold for isolated atoms, i.e. when the atoms are so far apart from each other that to a good enough approximation the primary interacts with only one atom at a time. This is the case for gases at very low pressure only. In dense materials a further important correction has to be made. Suppose a fast charged particle travels through a solid or liquid. The surrounding medium will then be polarized along the path of the particle. This polarization diminishes the effective field of the particle that may act on a particular atom. The effect is noticeable in particular for relatively

† L. Landau, *J. Phys. U.S.S.R.* 8 (1944), 201.

‡ For electrons it is, of course, the collision loss only (not the radiation loss) which is proportional to the primary ionization.

distant collisions. When the particle carries out a close collision with an atomic electron the surrounding atoms are far away from the particle and the polarization effect is small, but when the impact parameter is comparable with the interatomic distances the field is reduced considerably by the polarized medium around. Now such distant collisions only contribute appreciably to the energy loss when the particle is fast, in fact, as a more quantitative estimate shows, only in the relativistic region. We therefore expect that it is just the logarithmic increase of the energy loss which is partly compensated by the polarization effect in dense media. In gases the reduction of the energy loss sets in at still higher energies. The effect depends on the dielectric and dispersion properties of the medium traversed.

The polarization effect is closely connected with a peculiar phenomenon that occurs during the passage of very fast particles through polarizable media. Let the particle be so fast that v is larger than the velocity of light in the medium c/n ($n =$ diffraction index). It can then be shown from Maxwell's theory that the passage of the particle is accompanied by the emission of a fairly soft radiation (including visible light). This radiation has been observed by Čerenkov.† The question of the energy loss in dense media of very fast particles together with the Čerenkov radiation has been examined by several authors.‡ The result is not readily expressible in terms of a general formula and numerical results only will be quoted below. It is worth noting that whatever is left of the logarithmic increase of the energy loss after the minimum is entirely due to the emission of Čerenkov radiation.

In Figs. 26 and 27 the average energy loss in H_2O and Pb due to collisions is plotted as a function of energy $E-Mc^2$ for electrons, μ-mesons (mass $210m$), π-mesons (mass $276m$), and protons. The polarization effect is shown for electrons and μ-mesons. It is seen that it is just the logarithmic increase for high energies which is largely reduced in dense media. The curves for H_2O are also valid for air (Z is practically the same and only occurs under the log), but here, of course, the polarization effect sets in at higher energies: the ionization becomes constant at $1{\cdot}2.10^4mc^2$ for μ-mesons and at $600mc^2$ for electrons.

† P. A. Čerenkov, *C.R. Ac. Sc. U.S.S.R.* **2** (1934), 451; **20** (1938), 651; **21** (1938), 116 and 339; theoretical explanation by I. Frank and I. Tamm, ibid. **14** (1937), 109; I. Tamm, *Journ. Phys. U.S.S.R.* **1** (1939), 439. For a simple account see A. Sommerfeld, *Optik* (Wiesbaden 1950), § 47.

‡ O. Halpern and H. Hall, *Phys. Rev.* **57** (1940), 459; **73** (1948), 477; E. Fermi, *Phys. Rev.* **57** (1940), 485; C. Wick, *Ric. Scient.* **11** (1940), 273, and **12** (1941), 858; *Nuov. Cim.* **1** (1943), 302; A. Bohr, *Kgl. Dansk. Vid. Selsk.* **24** (1948), No. 19; M. Schönberg, *Nuov. Cim.* **8** (1951), 159.

The theory of the energy loss by collisions and of the ionization has been verified by numerous experiments and the agreement found to be as good as one can expect. We only mention that the logarithmic

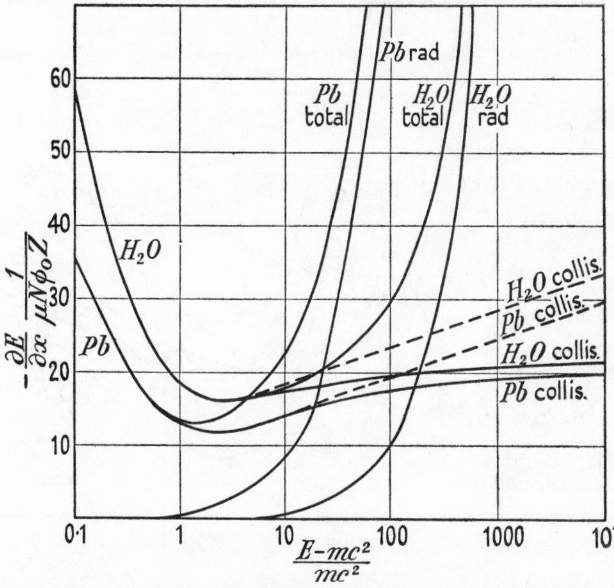

FIG. 26. Average energy loss in units $N\phi_0 Z\mu$ for electrons in H_2O and Pb as function of the kinetic energy $E-\mu$. Curves for loss by collisions, by radiation, and total loss. Dotted curves without polarization effect, full curves with polarization effect.

increase of the ionization after the minimum has also been verified for μ-mesons, in gases for energies up to $10{,}000mc^2$, where the setting in of the polarization effect has been observed.† The lack of a substantial increase of ionization in dense materials is also well established.‡

3. *Total energy loss.* The second way in which an electron loses energy in passing through matter is by emission of radiation. The average energy loss per cm. has been calculated in § 25. It is given by the formula

$$\left(-\frac{dE}{dx}\right)_{\text{rad}} = NE\phi_{\text{rad}}, \tag{9}$$

where ϕ_{rad} represents a cross-section which is plotted in Fig. 13 and Table V (pp. 252, 253).

† J. Becker, P. Chanson, E. Nageotte, P. Treille, B. T. Price, and P. Rothwell, *Proc. Phys. Soc.* **65** (1952), 437; S. K. Ghosh, G. M. D. B. Jones, and J. G. Wilson, ibid. (1952), 68.

‡ See, for example, E. Pickup and L. Voyvodic, *Phys. Rev.* **80** (1950), 89.

As may be seen from eq. (9) and from Fig. 13, the energy loss due to Bremsstrahlung behaves entirely differently from that due to inelastic collisions.

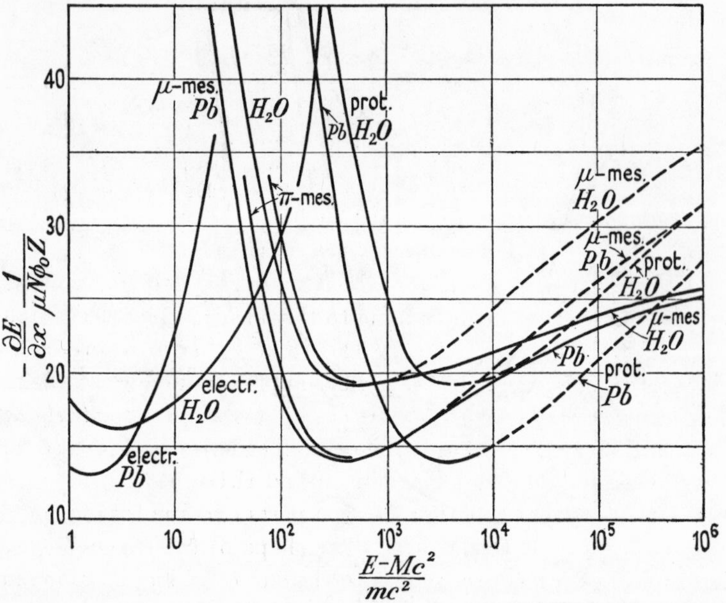

FIG. 27. Average energy loss by collisions in units $N\phi_0\, Z\mu$ for μ-mesons ($210m$), π-mesons ($276m$, partially shown), and protons in H_2O and Pb. Dotted curves without polarization effect, full curves (μ-mesons) with polarization effect. Also for comparison total energy loss for electrons. The scale is the kinetic energy of the particle $E-Mc^2$ in units of the electron rest energy ($mc^2 = 0{\cdot}51$ MeV).

(i) It is roughly proportional to E. This means that the energy lost per cm. path is constant up to energies of the order mc^2 and then *increases* proportionately to E itself. The increase for high energies is even more rapid because ϕ_{rad} also increases from a value of about $5\bar{\phi}$ for $E-\mu < \mu$ to a value $\sim 15\bar{\phi}$ at very high energies.

(ii) The energy loss is not proportional to Z, viz. to the mass per cm.^{2} which the electron has traversed, but to Z^2. It is therefore relatively more important for heavy elements than for light elements.

In both points the energy loss due to Bremsstrahlung (divided by the primary energy E) behaves very similarly to the absorption coefficient of γ-rays due to creation of pairs.

To obtain the numerical values of $(dE/dx)_{\text{rad}}$ we have to multiply the figures given in Table V and Fig. 13 by $EN\bar{\phi}$, where $N\bar{\phi}$ is given in appendix 8 for several materials.

The total energy loss per cm. is given by the sum

$$-\frac{dE}{dx} = \left(-\frac{dE}{dx}\right)_{\text{rad}} + \left(-\frac{dE}{dx}\right)_{\text{coll}}. \tag{10}$$

We find the following values for the total energy loss:

TABLE XI

Average energy loss per cm. path of a fast electron in units of mc^2

$(E-mc^2)/mc^2$	0·01	0·1	1	10	10^2	10^3	10^4
air†	0·0845	0·0145	0·0046	0·0049	0·0087	0·039	0·33
H_2O	78	13·1	4·08	4·00	6·56	29·6	262
Al	147	26·9	8·65	8·98	19·1	116	1,095
Pb	255	64·5	24·5	40·5	204	1,910	19,300

† Normal temperature and pressure.

For the radiative loss the contribution from the atomic electrons has been taken into account by adding to the nuclear effect a contribution $0{\cdot}8\phi_{\text{rad}}/Z$ (see § 25.5 and § 26.2). The value 0·8 is, of course, not very accurate. The departure from the Born approximation is not taken into account. *At very high energies the energy loss is almost entirely due to Bremsstrahlung.* The energy where this begins to be the case lies at about $20mc^2$ for Pb and $200mc^2$ for air and H_2O.

In Fig. 26 we have plotted the two parts and the total energy loss of an electron in H_2O and lead. The shape of the curves is, as it is seen from the graph, very similar to the curves in Fig. 25 representing the absorption coefficient of γ-rays.

For a heavy particle the energy loss is due to inelastic collisions only. The radiative part is smaller by a factor $(m/M)^2$ than for an electron of the same energy ($\bar{\phi}$ is proportional to $1/M^2$ where M is the mass of the colliding particle). Thus for sufficiently high energies a heavy particle is stopped more slowly than an electron. For protons this is the case when $E > 100mc^2$ in Pb, or $> 200mc^2$ in H_2O or air.

4. *Average range.* A fast particle traversing matter has a more or less well defined range only when it loses energy steadily and the fluctuations of energy loss (straggling) in a definite length of path are small. This is the case for the energy loss by collisions but not for that by radiation. Nevertheless, in order to survey the penetrating power of fast electrons, it will be convenient to define as average range the distance which the particle would travel if it always suffered just the *average energy loss* (10). This average range is then defined by

$$R(E_0) = \int\limits_{\mu}^{E_0} \frac{dE}{-dE/dx}. \tag{11}$$

As long as only collision loss matters, this is the actual range of the particle with initial energy E_0, apart from a very small straggling. The integration in (11) can only be carried out numerically, but a number of features can be read off from (11), (1), and (9) immediately:

(i) If only collision loss matters, i.e. for heavy particles, and for electrons with $E_0 < 2\mu$, say, dE/dx depends on the velocity only, and since the same is true for E/M, it follows that R/M, if expressed as a function of E_0/M (or the initial velocity v_0) is the same for all particles.†

(ii) In the non-relativistic region $\beta \ll 1$, $-dE/dx \sim 1/(E-Mc^2)$, apart from a slowly varying logarithm. R therefore increases roughly as $(E_0-Mc^2)^2$. For heavy particles $-dE/dx$ is varying slowly in the extreme relativistic region and therefore $R(E_0)$ increases there $\sim E_0$.

(iii) When the radiation loss for electrons has reached its full value, i.e. for $E_0 > 100\mu$, say, $-(dE/dx)_{\text{rad}}$ is proportional to E. For very fast electrons therefore $R(E_0) \sim \log E_0$. This means that a very fast electron cannot penetrate through thick layers of material, however high its energy.

(iv) If only collision loss matters, R is roughly proportional to $1/NZ$, i.e. inversely proportional to the density of electrons in the material. An additional dependence on the material arises, however, through the Z under the log in (1) and the latter is not very small when Z ranges, say, from 6 to 82. On the other hand, for very fast electrons $\phi_{\text{rad}} \sim Z^2$ and RNZ varies $\sim 1/Z$.

The average range of a fast electron is given in Table XII for various

TABLE XII

Average range (in cm.) of a fast electron

$(E-\mu)/\mu$	0·1	1	10	100	1,000
air†	3·9	$1{\cdot}55.10^{2}$	$2{\cdot}2.10^{3}$	$1{\cdot}5.10^{4}$	$6{\cdot}3.10^{4}$
H_2O	$4{\cdot}7.10^{-3}$	0·18	2·6	19	82
Al	$2{\cdot}5.10^{-3}$	$8{\cdot}5.10^{-2}$	1·15	7·8	28
Pb	$1{\cdot}0.10^{-3}$	$3{\cdot}1.10^{-2}$	0·33	1·25	2·4

† Atm. pressure, 0° C.

materials. Accurate data concerning the range of heavy particles are given in numerous books and need not be reproduced here.‡ For a

† A small dependence on M arises at very high energies through the dependence on M of W_m. For electrons there is also a different numerical factor under the log. Cf. (5) and (6).

‡ See, for example, D. J. X. Montgomery, *Cosmic Ray Physics*, Princeton 1949.

general survey we have plotted in Fig. 28 the $\log_{10}$ of the range, multiplied by the unit $N\phi_0 Z$, for various particles in air and Pb as a function of their energy. The initial slope corresponds to the $(E_0 - Mc^2)^2$ increase, which, for heavy particles gradually goes over into the $\sim E_0$ dependence. In contrast to this the range of electrons tends almost to a constant for very high energies.

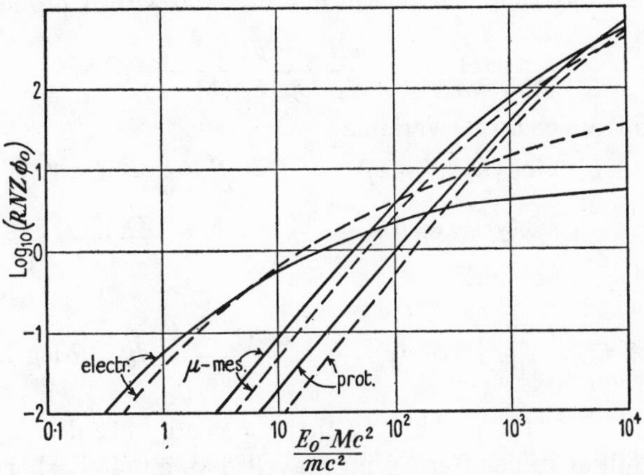

FIG. 28. $\log_{10}$ of the average range (multiplied by $NZ\phi_0$) for electrons, μ-mesons, protons in air (dotted), and Pb (full drawn) as function of the energy.

5. *Straggling.* In subsections 1–3 we have calculated the *average* energy loss. This average loss may differ considerably from the actual amount lost in a particular case when a fast particle travels through a thickness of matter. Only if the particle loses its energy in the form of a large number of small portions will the effect of straggling be small. This is the case for a heavy particle. An electron may, however, lose a large fraction of its energy in the form of a single hard γ-ray. According to § 25 (see Fig. 12) the radiative loss is due equally to the emission of large quanta, with energy comparable to that of the electron, and to small quanta. After having traversed a sheet of matter the electron will have emitted only few quanta of large energy and the fluctuations of the energy loss will be very large.

To obtain a rough idea of this straggling we neglect the collision loss and represent† the intensity distribution curves Fig. 12 (p. 250) by a

† H. Bethe and W. Heitler, *Proc. Roy. Soc.* A, **146** (1934), 83. The straggling near the critical energy (see below) has been studied by L. Eyges, *Phys. Rev.* **76** (1948), 1113; **77** (1950), 81.

rough but convenient formula:

$$\frac{k\phi_k}{E_0\bar{\phi}} = \frac{k}{E_0}\frac{a}{\log\{E_0/(E_0-k)\}}, \qquad a \sim 20\text{–}23. \tag{12}$$

(12) represents the curves of Fig. 12 fairly well for energies $> 50mc^2$. a is about 20 for Pb and 23 for H_2O.

Thus the probability that the electron has lost an energy k in travelling an infinitely short distance dl in a material with N atoms per cm.^{3} is given by

$$w(k)\,dk = \frac{a\bar{\phi}N}{E_0}\frac{dkdl}{\log\{E_0/(E_0-k)\}},$$

or if we introduce a new variable

$$y = \log\{E_0/(E_0-k)\}, \qquad E = E_0-k = E_0 e^{-y}, \tag{13}$$

$$w(y)\,dy = b\,dl\frac{e^{-y}\,dy}{y} \qquad (b = a\bar{\phi}N). \tag{14}$$

With (12), ϕ_{rad} becomes

$$N\phi_{\text{rad}} = \frac{N}{E_0^2}\int\limits_0^{E_0} k\phi_k\,dk = b\int\limits_0^{\infty}\frac{e^{-y}(1-e^{-y})}{y}\,dy = b\log 2. \tag{14'}$$

(14) represents the probability that the energy has decreased to e^{-y} times its initial value after having travelled an infinitely short distance dl. We wish to know the probability for a certain decrease in energy after traversing matter of a finite thickness l. For this probability we prove the following formula

$$w(y)\,dy = \frac{e^{-y}y^{bl-1}}{\Gamma(bl)}\,dy. \tag{15}$$

(15) becomes identical with (14) for small l ($\Gamma(bl) \sim 1/bl$). It is normalized to unity, $\int\limits_0^{\infty} w(y)\,dy = 1$. To prove that (15) is correct for finite l we let the electron travel first a distance l_1 then a distance l_2. The probability that the energy decreases in the first part of the path to e^{-y} times the initial value may be denoted by $w_1(y_1)\,dy_1$. If (15) is assumed to be correct for the two parts of the path the probability of a decrease to e^{-y} times the initial value becomes, according to the general rules of calculus of probabilities,

$$w(y)\,dy = dy\int\limits_0^{y} w_1(y_1)w_2(y-y_1)\,dy_1$$

$$= \frac{dy\,e^{-y}}{\Gamma(bl_1)\Gamma(bl_2)}\int\limits_0^{y} y_1^{bl_1-1}(y-y_1)^{bl_2-1}\,dy_1 = \frac{dy\,e^{-y}y^{bl_1+bl_2-1}}{\Gamma(bl_1+bl_2)}. \tag{16}$$

Therefore, if (15) is valid for l_1 and l_2, it is also valid for l_1+l_2, and since we have shown that it is valid for an infinitely short path we have proved that (15) is valid for any length of the path.

The probability distribution given by formula (15) is plotted in Fig. 29 for several values of bl. The constant b can be taken from (12), (14),

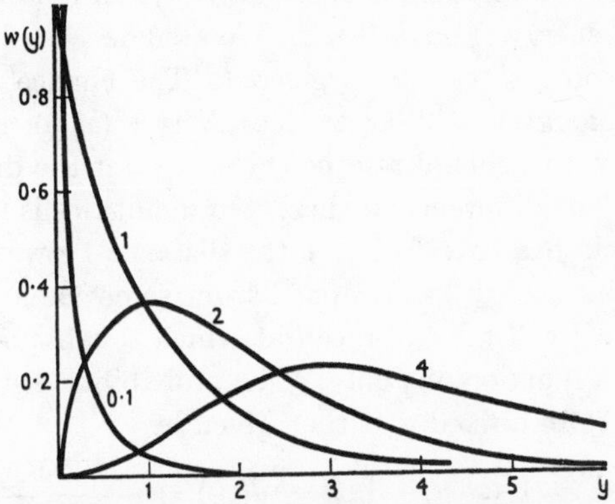

FIG. 29. Straggling. $w(y)$ represents the probability that the energy of an electron which has traversed a sheet l cm. thick has decreased by the factor e^{-y}. The numbers affixed to the curves are the thickness bl, where b is defined in equations (12) and (14). (For Pb $b = 2{\cdot}6$ cm.$^{-1}$)

and from appendix 8 where the values of $N\bar{\phi}$ are given for several materials. For a sheet of lead of 1 cm. thickness bl amounts to 2·6. The graph shows that the probable energy losses are distributed over a broad range. For a sheet of Pb of 8 mm. thickness ($bl = 2$) the probability that the energy decreases by a factor between $e^{-0{\cdot}5} = 0{\cdot}6$ and $e^{-2} = 0{\cdot}135$ is almost the same.

The probability that an electron still has an energy larger than e^{-y_0} times the initial energy after traversing the sheet is given by the integral

$$W(bl, y_0) = \int\limits_0^{y_0} w(y)\,dy = \int\limits^{y_0} \frac{e^{-y}y^{bl-1}}{\Gamma(bl)}\,dy = \frac{(bl-1, y_0)!}{\Gamma(bl)}, \tag{17}$$

where $(bl-1, y_0)!$ is the 'incomplete gamma function'.†

$W(bl, y)$ is plotted in Fig. 30 as a function of bl for various values of y. It is seen that the curves spread over a wide range of l, except if y is small, i.e. E_0/E close to one. A particle entering a material with initial energy E_0 will have a chance of 80 per cent. of having retained

† Numerical values of this function are given in Jahnke–Emde, *Tables of Functions* (2nd ed., 1933), p. 96, and Pearson, *Tables of the Incomplete Γ-function.*

an energy larger than $E_0 e^{-3} = E_0/20$ after a thickness $bl = 2$. After a thickness $bl = 4$ this chance is still 35 per cent. Keeping l fixed, the curves give the energy distribution of a beam of electrons entering with the same initial energy E_0, after the thickness l.

We can now also give some meaning to the *range*, and its fluctuations, of an electron. We first define a *partial range* as follows: consider an electron losing energy from E_0 to E. We assume E still to be so large that the collision loss can be neglected. The chance that after the distance l the energy is still larger than E is $W(bl,y)$, $y = \log(E_0/E)$. The chance that this should also be the case after the distance $l+dl$ is $W(bl+b\,dl,y)$. The difference of these two quantities is the probability that the particle has travelled just the distance between l and $l+dl$, within which its energy has dropped from some value $> E$ to some value $< E$.† We call the distance l at which this happens the *partial range* $l(y)$ (it is a function of y only). The probability distribution of the various values l, for a fixed y, is then given by

$$U(l)\,dl = W(bl,y) - W(bl+b\,dl,y) = -\frac{\partial W(bl,y)}{\partial l}\,dl. \tag{18}$$

$U(l)$ is correctly normalized to unity,

$$\int_0^\infty U(l)\,dl = -W(\infty,y)+W(0,y) = 1, \tag{18'}$$

because $W(\infty,y) = 0$ and $W(0,y) = 1$ (see Fig. 30).

$U(l)$ is also plotted in Fig. 30 for a few values of y. We see that the distribution of ranges is very broad indeed, and is broader the larger the drop in energy (i.e. y) is. For $E_0/E = 20$ ($y = 3$), for example, the probable ranges extend from $bl = 1$ to $bl = 5$, say. The most probable range is always $bl \sim y$.‡

† Note that the electron almost never reaches the energy E precisely, because energy is lost discontinuously in large portions.

‡ It should be noted that the average range, as defined in subsection 4, is not quite identical with the mean value of l. The *average partial range* is defined by (11), if the lower limit is replaced by E, and $-dE/dx$ is the *average* energy loss. With (12), the average partial range $R(y_0)$ would be given by

$$R(y_0) = \frac{1}{N\phi_{\mathrm{rad}}}\,y_0.$$

On the other hand the mean value of l is

$$\bar{l} = \int_0^\infty l\,U(l)\,dl = \frac{1}{b}\int_0^\infty W(bl,y)\,d(bl) = \frac{1}{b}\int_0^{y_0} dy\,e^{-y}\int_0^\infty \frac{y^{x-1}}{\Gamma(x)}\,dx.$$

For large x, the integral over x can be approximated by $\sum_0^\infty (y^n/n!) = e^y$ ($x = 0$ does not contribute since $\Gamma(0) = \infty$) and $\bar{l} = \frac{1}{b}y_0 = \frac{1}{b}\log\frac{E_0}{E}$, differing from $R(y_0)$ by a factor log 2.

The above considerations are limited owing to the neglect of the collision loss and therefore E must not be allowed to go to zero. To obtain the total range, and its probability distribution, the collision loss must be included. This is extremely complicated in the region where both types of loss are important, especially because the cross-section for Bremsstrahlung then decreases and the simple law (12) breaks down.

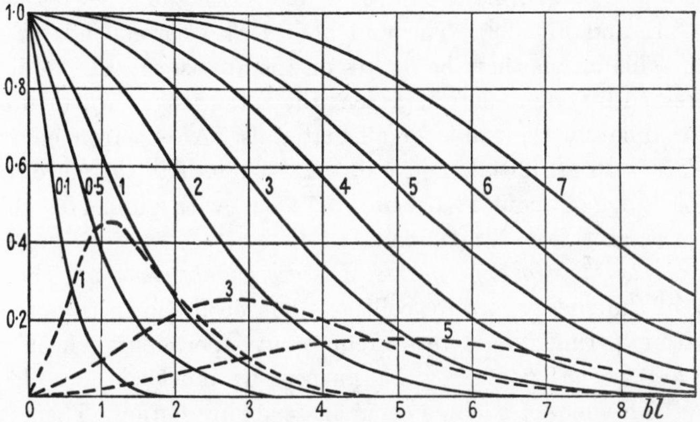

FIG. 30. Full curves: $W(bl, y)$ is the probability that an electron, having traversed a known thickness l of matter, has retained an energy $> E_0 e^{-y}$. Values of y attached. Dotted curves: probability distribution of partial ranges $U(l(y))$, i.e. probability that the electron has travelled a distance between l and $l+dl$ while its energy has decreased by the known factor $\exp(-y)$. Values of y attached.

We can, however, idealize the situation by assuming that the asymptotic radiation laws, viz. (12), hold when E is larger than a certain critical energy E_c say, and that below E_c the particle only loses energy by collisions. This crude assumption will only lead to reasonable results when $E_0 \gg E_c$. For E_c we may take some energy where radiation and collision losses become equal. This critical energy will play an important role in the cascade theory (§ 38), and we refer for numerical values to Table XIII, p. 391.† The actual total range, R, say, is then given by substituting in the argument of $l(y)$ the value $y_c = \log(E_0/E_c)$, $l_c \equiv l(y_c)$, and by adding the residual range R_c say, which the particle has if its initial energy is E_c.‡ The latter does not fluctuate (or very little)

† The definition of E_c will be slightly different in § 38.

‡ Actually this is not quite correct. l_c was defined as the distance at which the last jump of energy from $E > E_c$ to $E < E_c$ took place. So at l_c, some particles may have reached a considerably lower energy than E_c, and R_c should then be smaller. However, in the neighbourhood of E_c our assumptions are too crude to justify a more detailed statistical investigation. The above considerations concerning the total range R are not meant to be more than a crude approximation.

because energy is only lost by collisions. Thus the total range will be

$$R = l_c + R_c.$$

The probability distribution of actual total ranges $u(R)\,dR$, say, is then the same as (18), with l_c substituted for l. Thus

$$u(R)\,dR = U(R - R_c)\,dR.$$

For example: For Pb, $E_c \sim 14\mu$ (by Table XIII) and $R_c \sim 0{\cdot}4$ cm. (by Table XII and Fig. 28). The unit $bl = 1$ also corresponds to about 0·4 cm. This means that the figures on the abscissa in Fig. 30 have to be increased by one unit to represent the total range. R_c, in units bl, is approximately the same for all materials. An electron entering a lead plate with an initial energy of $e^3E_c \sim 280\mu$ has therefore a most probable range of about $4 \times 0{\cdot}4$ cm. $= 1{\cdot}6$ cm., with a broad distribution ranging roughly from 0·8–2·5 cm.

6. *Comparison with experiments. Discovery of the μ-meson.* The above theory of the energy loss by collisions has been shown to agree very well with experiments. Here we require an experimental check on the *energy loss by radiation.* For this purpose we must consider electrons of very high energies such as occur in cosmic radiation. Their energy loss can be measured in the following way. A plate of metal of a suitable thickness l (1 cm. Pb, for instance) is placed in a cloud chamber. A strong magnetic field is applied which deflects the electrons. By observing the tracks of electrons which have traversed the plate and measuring the curvature on both sides we obtain the energy before and after the passage through the plate, E_1 and E_2 respectively.

$(E_1 - E_2)/l$ must not, however, be identified with the theoretical average loss $-dE/dx$ given in Table XI. The straggling explained in subsection 5 must be taken into account. Instead of $E_1 - E_2$ we consider the quantity

$$Y = \frac{1}{l}y = \frac{1}{l}\log\frac{E_1}{E_2}.$$

The average value of Y follows from (15):

$$\overline{Y} = \frac{1}{l}\int_0^\infty y\,w(y)\,dy = \frac{1}{l}\frac{\Gamma(bl+1)}{\Gamma(bl)} = b. \tag{19}$$

$b = 2{\cdot}6$ for Pb. $\overline{Y}$ is therefore a constant, independent of the thickness and the primary energy of the electron. The deviation of Y from the average value $\overline{Y}$ is, however, expected to be very large in any individual case. Indeed, the mean deviation is

$$(\overline{Y^2} - \overline{Y}^2)/\overline{Y} = 1/l. \tag{19'}$$

In a 1 cm. Pb plate we therefore expect an average Y of 2·6 and an average deviation from it of 1.

Measurements of the energy loss of cosmic ray particles were first made by Anderson and Neddermeyer,† and they have been able to verify the theoretical predictions for energies up to a few hundred mc^2. More extensive measurements were made soon afterwards by Blackett and Wilson.‡ We represent their original results in Fig. 31.§

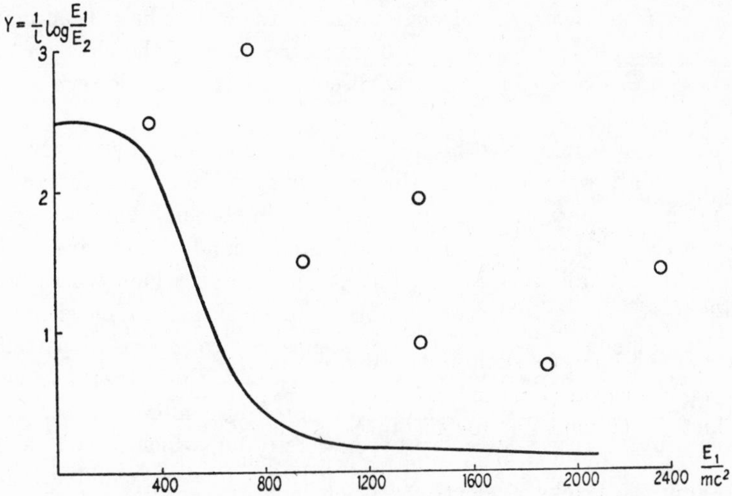

FIG. 31. The curve shows the measured average energy loss of cosmic ray particles (E_1 = initial energy, E_2 = energy after traversal of lead plate of thickness l cm.). Particles with $E_1 < 400mc^2$ are all electrons, those with $E_1 > 400mc^2$ nearly all μ-mesons. A few electrons in the high energy region (circles) show the correct energy loss $\overline{Y} = 2{\cdot}6$ with large straggling.

They were obtained with a 0·33 cm. Pb plate for energies up to $1{,}000mc^2$ and a 1 cm. Pb plate for higher energies. The curve represents the average value of Y obtained from a *large* number of cosmic ray particles. We see first that up to about $400mc^2$ the experiments confirm the theoretical predictions ($\overline{Y} = 2{\cdot}6$) very well. For higher energies, however, the curve drops down suddenly to a very small value. At first sight it would seem that this is in striking contradiction to the

† C. D. Anderson and S. H. Neddermeyer, *Phys. Rev.* **50** (1936), 263.

‡ P. M. S. Blackett, *Proc. Roy. Soc.* A, **165** (1938), 11; J. G. Wilson, ibid. **166** (1938), 482. The first to find the existence of two groups of particles with different penetrating power were S. H. Neddermeyer and C. D. Anderson, *Phys. Rev.* **51** (1937), 884. The measurements described above are the more accurate ones.

§ In representing the data corrections have been made to account for the loss by collisions.

theory. However, in addition to the bulk of particles with small energy loss a few have been found which show a large loss, of the right order of magnitude and with a large straggling. No doubt these particles must be electrons behaving exactly as the theory predicts. But if this is so, the particles with the small energy loss *cannot be electrons*, for it is inconceivable that a particle can behave so differently in the same conditions.

We know now that the particles with small energy loss are different in nature from electrons. Measurements of their mass have been made and it was found to be about 210 times the mass of the electron. In this way a *new type of particle* was discovered. It is now known as the μ-meson.

The meson would have been discovered more easily by energy-loss measurements were it not for the following situation: As we can see from Fig. 31, there are very few electrons in the high energy region, most of the particles there are mesons. And secondly, there are hardly any mesons at all with $E_1 < 400mc^2$. The energy region where both kinds of particles exist in comparable numbers is small. This situation made it difficult to establish the existence of two groups of particles with very different penetrating powers. In fact, the actual discovery of the meson has been very much promoted in a more indirect way, namely, by the cascade theory of showers (§ 38), which, based on the theory of radiative energy loss and pair production, could give an excellent explanation of a large group of cosmic ray phenomena.

7. *Annihilation probability of positrons.* Finally, we consider briefly the penetration of positrons through matter. The first question is whether our calculations of the energy loss have to be modified. This is obviously not the case in the region where Born's approximation can be applied. Here the transition probability is proportional to the square of the interaction of the electron with the other particles and the sign of the charge does not therefore make any difference. This remains true for the correction to Bremsstrahlung in the extreme relativistic region (whereas at low energies the correction is different for positrons, see § 25, eq. (19); ξ_0 and ξ both change sign for positrons).

The only difference is that a positive electron can be *annihilated* while passing through matter. In § 27 we have calculated the probability that a positive electron is annihilated in a collision with a free negative electron. If we denote the cross-section for this process† by $\phi(E')$, the probability that the positive electron is annihilated while

† We consider only the two-quanta annihilation.

travelling a distance dx is given by

$$w\,dx = NZ\phi(E')\,dx.$$

For E' we have to insert the energy which the positive electron has at the particular point of its path considered. Since ϕ is given as a function of E', it is convenient to introduce the probability of annihilation per energy interval dE'. This probability is given by†

$$\frac{NZ\phi(E')\,dE'}{-dE'/dx} = w(E')\,dE', \tag{20}$$

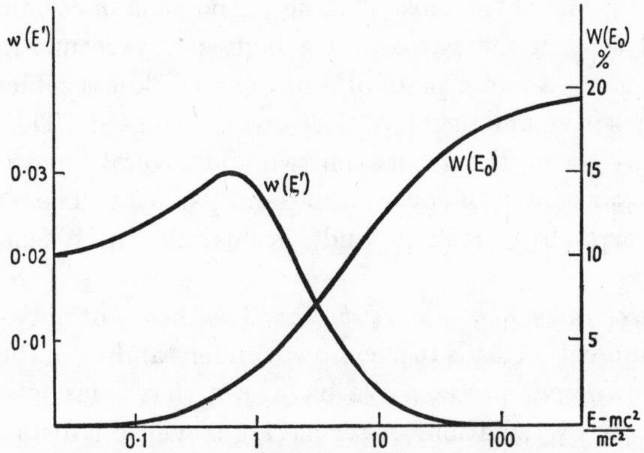

FIG. 32. $w(E')\,dE'$ represents the probability of annihilation in lead of a positive electron while having an energy between E' and $E'+dE'$. $W(E_0)$ is the total probability of annihilation while in motion of a positive electron with an initial energy E_0 in lead.

where $-dE'/dx$ represents the energy loss per cm. as calculated in subsections 1–3.

If we insert for $\phi(E')$ the function § 27 eq. (10) and for dE'/dx the values given in Fig. 26, we obtain for the '*differential annihilation probability*' per energy interval dE' in lead the function plotted in Fig. 32. We see that the probability of annihilation has a maximum for $E'-mc^2 \sim mc^2$. At higher energies and at smaller energies the positive electron loses its energy so quickly that it has no time to be annihilated. $w(E')$ also gives, in connexion with the results of § 27, the intensity distribution of the continuous annihilation radiation.

From (20) we obtain the 'total probability of annihilation while in motion' by integrating over the whole range:

$$W(E_0) = \int\limits_0^{E_0} w(E')\,dE'. \tag{21}$$

† Cf. H. Bethe, *Proc. Roy. Soc.* A, **150** (1935), 129.

$W(E_0)$ is also plotted in Fig. 32 for lead. The total annihilation probability increases slowly with the initial energy E_0 and reaches an asymptotic value for very high energies. In lead it amounts to about 20 per cent. Thus 20 per cent. of a beam of very fast positive electrons are annihilated while in motion; the rest, 80 per cent., come safely to the end of their range. They are then eventually annihilated at the rate calculated in § 27 eq. (11). One obtains similar values for air.

38. Cascade showers

Showers are one of the most striking phenomena in cosmic radiation. It is found that in the passage of a high-energy cosmic-ray electron through matter—a metal plate of a few cm. thickness suffices—a large number of positive and negative electrons are created. The number of particles may be anything between two and several hundred, and the largest showers consist of several thousand particles. The phenomenon was discovered by Blackett and Occhialini† in Wilson chamber experiments.

1. *The mechanism of a cascade shower.* The theory of radiative energy loss and pair creation leads to a complete understanding of these showers as a kind of *cascade process*. We have seen that a fast electron loses energy quickly by radiation. After having travelled a distance $1/N\phi_{\text{rad}}$ (or $1/b\log 2$, eq. (14′), § 37) the energy of the electron has, on the average, decreased to e^{-1} times its initial value E_0 (the distance where the energy has decreased to $\frac{1}{2}E_0$ is $1/b$). A large fraction of this energy is emitted in the form of light quanta with energy k comparable with E_0 itself. It is indeed quite probable that a light quantum of, say, $k = \frac{1}{2}E_0$ is emitted. Only a small fraction of the energy is lost in the form of soft quanta with, say, $k < 20\mu$, and that fraction is smaller the larger E_0. The distance $1/b\log 2$ is (using the exact results of § 25 now rather than the simplified eq. (12), § 37) for very high energies given by

$$b\log 2 = N\phi_{\text{rad}} = cN\bar{\phi}, \qquad c = \begin{cases} 15{\cdot}2 \text{ (Pb)}, \\ 18{\cdot}3 \text{ (H}_2\text{O)}. \end{cases} \tag{1}$$

Secondly, we have seen in § 36 that a high-energy photon has a very large absorption coefficient due to pair creation. In the limit of very high energies we have from § 26

$$\tau = c'N\bar{\phi} \text{ cm.}^{-1}, \qquad c' = \begin{cases} 11{\cdot}5 \text{ (Pb)}, \\ 13{\cdot}9 \text{ (H}_2\text{O)}. \end{cases} \tag{2}$$

† P. M. S. Blackett and G. P. S. Occhialini, *Proc. Roy. Soc.* A, **139** (1933), 699, and (with J. Chadwick) **144** (1934), 235. For later experiments and photographs see any book on cosmic radiation.

The mean free path of such a photon, i.e. the distance where the chance for a conversion into a pair is $1-e^{-1}$, is τ^{-1}, and this is of the same order as $1/b \log 2$. The ratio is practically the same,

$$c'/c = 0{\cdot}76, \tag{3}$$

for all materials. Fig. 16 on p. 261 shows that the energy of the light quantum in creating a pair is, on the average, shared in equal parts by the positive and negative electron. Both electrons therefore still have an energy of the order of magnitude E_0 (though, of course, it is smaller by a factor 4 or so) and are still capable of emitting high-energy quanta in traversing matter further. These in turn produce further pairs. And so the process repeats itself, more and more pairs being created. The whole 'cascade' multiplication takes place in a few times the length b^{-1}. Eventually, the energy of each individual particle and light quantum has become so small, by repeated division of the original energy, that no further emission of hard quanta and no more pair creation can take place. For this critical energy E_c we may take the value at which the energy loss by radiation becomes of the same order of magnitude as the energy loss by ionization. Looking at Fig. 26 we see that it is about $15mc^2$ for Pb and $200mc^2$ for H_2O. At approximately the same limit the rate of pair production by a light quantum also rapidly decreases (Fig. 17, p. 262).

The cascade theory of showers was developed independently on these lines by Carlson and Oppenheimer and by Bhabha and Heitler,† using different mathematical methods. Later improvements and amplifications with regard to the physical and mathematical approximations have been made by several authors.‡ Naturally, the working out of this mechanism is bound to be complicated and we cannot go into the details here. The most important features can, however, be seen from a very crude model.

The cascade process takes place in the same way in all materials if the thickness of the material is not measured in cm. but in the characteristic length $1/b \log 2$. We denote the thickness of matter measured

† J. F. Carlson and J. R. Oppenheimer, *Phys. Rev.* **51** (1936), 220; H. J. Bhabha and W. Heitler, *Proc. Roy. Soc.* A, **159** (1936), 432.

‡ The major publications are: N. Arley, *Proc. Roy. Soc.* A, **168** (1938), 519; *Theory of Stochastic Processes*, Copenhagen 1943; N. Arley and B. Eriksen, *Kgl. Dansk. Vid. Selsk.* **17** (1940), No. 11; L. Landau and G. Rumer, *Proc. Roy. Soc.* A, **166** (1938), 213; B. Rossi and K. Greisen, *Rev. Mod. Phys.* **13** (1941), 240; H. J. Bhabha and S. K. Chakrabarty, *Proc. Roy. Soc.* A, **181** (1943), 267; *Proc. Ind. Ac. Sc.* **15** (1942), 464; *Phys. Rev.* **74** (1948), 1352; S. K. Chakrabarty, *Proc. Nat. Inst. Sc. India*, **8** (1942), 331; L. Jánossy and H. Messel, *Proc. Phys. Soc.* **64** (1951), 1; I. Tamm and S. Belenky, *Phys. Rev.* **70** (1946), 660. See also references below.

in these units by t. The unit thickness l_0 is given below (Table XIII) for several materials. We now make the following crude assumptions. Whenever an electron passes through the thickness $t = \log 2$ (this is the distance where the energy has on the average dropped by half) it emits a photon of half its energy and retains the other half. In reality the energy $\frac{1}{2}E_0$ will be more or less divided up between several photons, and our model somewhat overrates the number of particles. Whenever a photon passes through $t = \log 2$ (ignoring the fact that the mean free path of a photon is slightly larger, eq. (3)) it is transformed into a pair, each electron receiving half the energy of the light quantum. The number of electrons+light quanta at a thickness t is therefore e^t, and each has energy $E_0 e^{-t}$. When this energy approaches the limit E_c, energy is lost in the form of soft quanta which no longer produce pairs and also by ionization. We take these facts into account by assuming that further multiplication ceases when $E_0 e^{-t}$ becomes κE_c, where κ is a number of order unity. The maximum number of particles+light quanta is therefore reached at a thickness

$$t_m \simeq \log \frac{E_0}{\kappa E_c}. \tag{4}$$

It is easily seen that, in this model, at any not too small t, the number of light quanta is about one-third of the total. Hence the number of particles in the maximum is

$$N_m \simeq \frac{2}{3\kappa} \frac{E_0}{E_c}. \tag{5}$$

After the maximum number of particles is reached the particles are gradually absorbed by ionization and emission of soft quanta, while only a few low-energy pairs are still created. Eventually the shower dies out.

The formulae (4) and (5) can, of course, make no claim to accuracy, and none of the details of the shower can be derived from such a crude model, but the most important features are well represented by (4) and (5). So we see:

(i) The maximum number of shower particles is proportional to the energy E_0 of the particle initiating the shower.

(ii) The thickness at which the maximum occurs is proportional to the logarithm of E_0 and equal to a few times the 'cascade unit' l_0.

(iii) The development of a shower is independent of the material, provided that the thickness is measured in cascade units and the initial energy in units of the critical energy E_c. As is seen from (1) and (2), the cascade unit is inversely proportional to NZ^2. The critical energy is roughly proportional to $1/Z$.

(iv) It is clear that the angular spread of such a shower is not large because emission of light quanta as well as pair production takes place at small angles if the energy is high.

These general conclusions will be corroborated substantially by the detailed theory.

2. *Results of cascade theory: average numbers.* The development of an electron-photon cascade is a very complicated problem of probability calculus that one cannot hope to solve exactly. Actually what one wishes to know is this. Given a primary electron with energy E_0 entering a sheet of matter, what is the probability of finding, say, N electrons with energy $> E$ at any given depth t, say, and what is the probability of finding a certain number of photons there? In all the work on cascade theory this problem is divided into two parts. I. First the question is answered what is the *average number* $\bar{N}(E_0, E, t)$ of electrons and photons $\bar{N}_p(E_0, k, t)$ with energy $> E$, or k, at depth t. Then, II, the question of the fluctuations is raised. It is clear that II is a much more difficult problem than I, and, so far, has been solved only partially (see subsection 3).

The problem I presents itself as a diffusion problem for $\bar{N}$, $\bar{N}_p$. Even this problem can only be solved if a number of physical simplifications are made. Complete solutions have been found in the following approximation (approximation A):

(i) For the cross-sections for Bremsstrahlung and pair creation the asymptotic formulae for very high energies and complete screening are used, i.e. the formulae § 25 eqs. (26) and (34); § 26 eqs. (13) and (15).

(ii) No other processes are taken into account, in particular the energy loss by collisions and the Compton effect are neglected.

(iii) It is assumed that all particles are emitted in the forward direction and no lateral spread of the shower is taken into account.†

One can then proceed to an improvement (approximation B), by taking into account the collision loss of the electrons. It is then assumed

(iv) that the energy loss by collisions is independent of the energy.

It will appear that when the problem A is solved the results in approximation B are obtained, at least to a reasonably good approximation by a redefinition of the variables occurring in A.

It is evident that all these assumptions are adapted to very high energies. The most serious approximation is (i). This is really valid

† The lateral spread has been calculated by J. Roberg and L. W. Nordheim, *Phys. Rev.* **75** (1949), 444. It is mainly due to multiple small angle scattering.

with reasonable accuracy only for $E > 200\mu$. For lower energies the rate of multiplication is overrated. However, even if ϕ_{rad} is taken as constant, the radiative energy loss is proportional to E and decreases with decreasing E. If collision loss is taken into account, this will in any case become larger than the radiative loss, at an energy $< E_c$, say (see below), and below E_c the electrons are stopped so quickly that an error in ϕ_{rad} can have no serious effect. We expect, therefore, to incur, in approximation B, some error in the region E_c which, however, cannot be too large for the total number† if only E_0 is large compared with E_c. As to (ii) and (iv) we remark that the neglect of the Compton effect is justified. In § 36 (eq. (5)) it was seen that a photon suffers a loss of energy by collisions, through repeated Compton processes, similar to that of an electron, but this is much smaller than that of electrons. Since the number of photons will be comparable with that of electrons, this loss of energy is irrelevant. On the other hand, the neglect of the increase of energy loss by collisions for low energies is serious only when $E < 2\mu$ and we shall not consider such low energies. (iii) is justified whenever $E \gg \mu$.

As has already been shown in subsection 1, the development of a cascade is independent of the material, if the thickness is measured in 'cascade units'. This follows from the fact that the lengths responsible for Bremsstrahlung and for the absorption of the photons are in a constant ratio that is practically independent of Z. The inclusion of the collision loss (in approximation B (iv)) will not alter this fact. Moreover, the differential cross-sections for Bremsstrahlung and pair creation only depend on the ratios E_0/E, E_+/k, etc. It follows, therefore, that in the problem A, $\bar{N}$ and $\bar{N}_p$ can only depend on the ratio of energies E_0/E, E_0/k, but not on the energies themselves. Thus in problem A we shall find functions $\bar{N}(E_0/E, t)$ and $\bar{N}_p(E_0/k, t)$.

In problem B a further parameter enters. Let the collision loss be

$$-\left(\frac{dE}{dx}\right)_{\text{coll}} = \beta.$$

Let l_0 be the cascade unit (in cm.) defined by

$$(N\phi_{\text{rad}}(\infty))^{-1} = l_0, \tag{6}$$

where $\phi_{\text{rad}}(\infty)$ is the value of ϕ_{rad} for $E \to \infty$. We then define as the

† In the work of Arley (loc. cit.) the following approximations are made. For $E > E_c$ collision loss is neglected, for $E < E_c$ Bremsstrahlung and pair creation is neglected. For a suitably defined E_c this gives results that are not very different from those derived by the above assumptions. His method is an extension of § 37.5.

critical energy E_c the energy which is lost by collisions in one cascade unit:

$$E_c = \beta l_0 \quad \text{or} \quad -\left(\frac{dE}{dx}\right)_{\text{coll}} = N\phi_{\text{rad}}(\infty)E_c. \tag{7}$$

This shows that E_c is also the energy where collision and radiation loss would be equal, if ϕ_{rad} were constant. Actually, ϕ_{rad} is smaller than $\phi_{\text{rad}}(\infty)$.† The difference is not great, however, because E_c will always be at least 15μ or so. For ϕ_{rad} the Born approximation is used.

The cascade units l_0 and E_c are given in Table XIII for several materials. In the computation of l_0 allowance is made for the contribution of the atomic electrons to Bremsstrahlung and pair creation by replacing Z^2 (in $\bar{\phi}$) by $Z(Z+0{\cdot}8)$ (see §§ 25, 26). E_c can hardly be defined very accurately, because the collision loss depends after all, to some extent, on energy. Since it matters most for relatively slow particles we take β from Fig. 26 for an energy in the neighbourhood of E_c itself, i.e. $\beta/NZ\phi_0\mu = 15$ for Pb, 20 for H_2O, taking the polarization effect into account for all dense materials.

TABLE XIII

The characteristic cascade units

	Air	H_2O	C	Al	Fe	Cu	Sn	Pb
Unit of length l_0 (cm.)	29,800	38	20	9·1	1·70	1·42	1·21	0·51
Critical energy (in mc^2)	155	170	180	90	45	42	24	14

The mathematical procedure for solving problems A and B is described in Jánossy's book‡ and will not be reproduced here. First we consider problem A (ionization loss neglected). Here the average number of particles is a function of t and of the ratio of their energy E_0/E only. The problem A is solved exactly, apart from the final evaluation of certain integrals for which the saddle-point method is used. The latter involves errors of at most 10–15 per cent.

In Figs. 33, 34, 35, full curves, the average number of electrons (including positive, negative, and the primary particles) with *energy greater*

† In the data given by some authors E_c is defined as the energy where

$$(dE/dx)_{\text{coll}} = (dE/dx)_{\text{rad}}.$$

This gives slightly different values for E_c. Note that some authors give figures for $l_0 \log 2$ as 'cascade unit'. The figures of Table XIII should still be corrected for the departure from the Born approximation (§ 25 eq. (35), § 26 eq. (15′)). This leads to an increase of l_0 and E_c by 10 per cent. for Pb, and less for light elements.

‡ L. Jánossy, *Cosmic Rays*, Oxford 1948. Extensive numerical data have been published by Leonie Jánossy and H. Messel, *Proc. Roy. Ir. Ac.* **54** (1951), 217. The curves in Figs. 33–35 are constructed from their tables.

than E is shown as function of the thickness t, for various ratios E_0/E. In Fig. 33 the number of photons with energy $k > E_c$ is also plotted. It is seen that this is almost the same as that of the electrons with the same energy.† In each case the maximum is reached after a few cascade units, and the position of the maximum moves slowly to higher thicknesses as E_0/E increases. The maximum number of particles increases very rapidly with E_0/E. This is all quite as we have been led to expect from our simple considerations in subsection 1. Actually the maxima of the full curves of Figs. 33–35 are quite well represented by

$$t_m = \log_e \frac{E_0}{3E}, \qquad \bar{N}_m = \frac{1}{\kappa'}\frac{E_0}{E}, \qquad \kappa' = 12\text{–}16. \tag{8}$$

Next we consider approximation B. Roughly, what one expects is this. An electron that is created with energy E' will have energy $E = E' - E_c t$ after a thickness t if it loses no energy by radiation. On the average it travels a distance $t \sim 1$ before it loses energy substantially by radiation. If we look for the electrons with energy E found at t, the majority of them will have been created at $t-g$ and then had energy $E' = E + gE_c$, say, where g is of order of magnitude unity. Now the actual distribution function must be more or less the same as if collision loss were neglected, but the energy at creation $E + gE_c$ substituted for E. Actually this is what turns out to be the case. However, g is not quite a constant, but depends slightly on t and E as well. To a reasonably good approximation (the solution is no longer exact), one finds

$$\bar{N}_B(E_0, E, t) = \bar{N}_A(E_0, E + gE_c, t), \qquad g = g\left(\frac{E_0}{E}, t\right), \tag{9}$$

where $\bar{N}_A$, $\bar{N}_B$ are the average numbers of electrons in problems A and B respectively. Thus $\bar{N}_B$ now depends on the two ratios E_0/E and E/E_c. g is shown in Fig. 36 and it is seen that it does not really depend very strongly on either E_0/E or t.

In Fig. 33 the typical development of a shower initiated by an electron is shown for a fixed primary energy $E_0 = 100E_c$. It is seen that collision loss reduces the number of the relatively fast electrons $E > E_c$ roughly by half. The number of electrons with $E > 0{\cdot}1E_c$ (collision loss taken into account) is only slightly larger than $\bar{N}_A$ $(E > E_c)$. So if we classify the electrons roughly as fast and slow according to whether $E \gtrless E_c$, we obtain more or less the same number of fast and slow

† There is, however, a large number of low-energy photons. See J. A. Richards and L. W. Nordheim, *Phys. Rev.* **74** (1948), 1106.

electrons, except at small thicknesses where, naturally, few slow electrons are found.

In Figs. 34, 35 three types of curves are shown: (i) $\bar{N}_A(E_0/E)$ (full drawn), (ii) $\bar{N}_B(E_0/E,\, E = E_c)$ (broken curves), and (iii) $\bar{N}_B(E_0/E,\, E = 0{\cdot}1E_c)$ (dotted curves). The values of $\epsilon = \log_{10}(E_0/E)$ are attached. Note that $\bar{N}_B(\epsilon, E = E_c)$ and $\bar{N}_B(\epsilon+1, E = 0{\cdot}1E_c)$ refer to the same E_0. Again, the curves $\bar{N}_A(\epsilon)$ almost coincide with $\bar{N}_B(\epsilon+1,\, E = 0{\cdot}1E_c)$. $\bar{N}_B(\epsilon,\, E = E_c)$ is about half of $\bar{N}_A(\epsilon)$, except at relatively small thicknesses.

The energy spectrum of the electrons can be inferred from Figs. 33–35 as follows. If we are interested in really fast electrons with $E \gg E_c$, (9) shows that the collision loss can be neglected (as g is hardly ever greater than 1). The curves $\bar{N}_A$ (full curves) then give the integral spectrum. Near the maximum, for example, we have $\bar{N}_m(E) \sim 1/E$ and therefore the differential spectrum

$$\bar{n}(E)\,dE \sim \frac{dE}{E^2}. \tag{10}$$

This law still holds roughly down to $E \sim E_c$. In the neighbourhood of and below E_c the spectrum flattens out. In the interval $0{\cdot}1E_c$—E_c there are only about as many particles as in the interval E_c—E_0. However, the statements of the theory in the low-energy region must be taken to be more qualitative than quantitative, owing to the crude approximation made there. Owing to this flattening out of the spectrum, the number of electrons increases only slowly if E is allowed to decrease further. E should not, however, be allowed to decrease below 2μ, as the collision loss then increases rapidly.

From the figures shown, and with the help of g, we can easily construct $\bar{N}_B$ for all values of E_0, E, t, by interpolation.

Showers can, of course, also be initiated by photons. The average number of electrons in a photon initiated shower is also plotted in Fig. 34 for two cases $\epsilon = 2$ and 4. What we expect is this: it takes about one cascade unit before the first pair is created. Therefore within the first unit we find practically no electrons. Afterwards we must expect that a shower develops, much in the same way as if it were initiated by an electron. It cannot make much difference in the later stages whether a shower is produced by a single electron with energy E_0 or by two electrons sharing the same energy between themselves. Thus we find that the curves for photon initiated showers are much the same as for electron initiation, only they are shifted by about one unit to larger thicknesses.

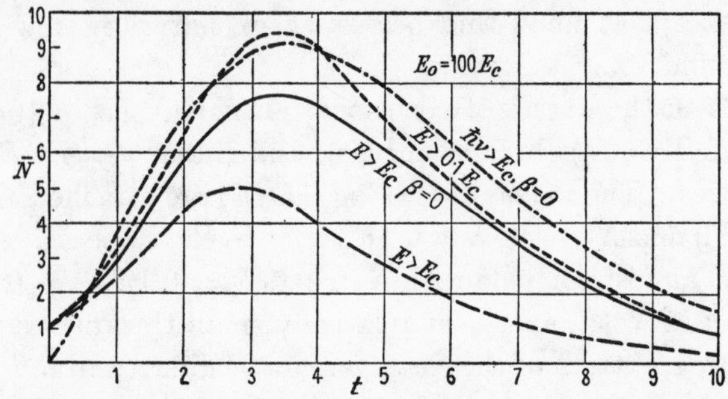

FIG. 33. Typical cascade shower initiated by electron with energy $E_0 = 100E_c$. Average number of electrons (+, − and primary) as function of thickness t in cascade units. Full curve: number of particles with energy $E > E_c$, collision loss neglected. Curve — — —: the same with inclusion of collision loss. Curve - - - -: number of particles with $E > 0{\cdot}1E_c$ (with collision loss). Curve -·-·-: number of photons with $k > E_c$, collision loss neglected.

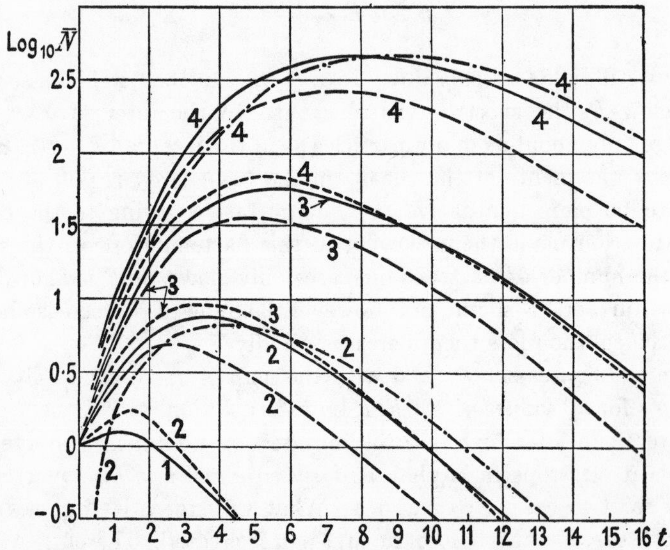

FIG. 34. Cascade shower initiated by electron with energy E_0. $\log_{10}$ of the average number of particles, with energy greater than E, as function of t. The numbers attached to the curves are $\epsilon = \log_{10} E_0/E$. Full curves calculated without collision loss are valid when $E \gg E_c$. Curves ——— with collision loss and $E = E_c$. Curves - - - - with collision loss and $E = 0{\cdot}1E_c$. Curves -·-·- represent photon initiated showers (primary energy k_0), $\epsilon = \log_{10} k_0/E$, numbers of electrons with energy $> E$, collision loss neglected.

3. *Fluctuations*. We now turn to the question of how far the actual number of electrons deviates from the calculated average number $\bar{N}$. If the emission of Bremsstrahlung and pair creation were all statistically

independent acts, the Poisson distribution would be valid. The probability of finding N particles, when the average is $\bar{N}$ would be

$$W(N) = \frac{e^{-\bar{N}}\bar{N}^N}{N!}. \tag{11}$$

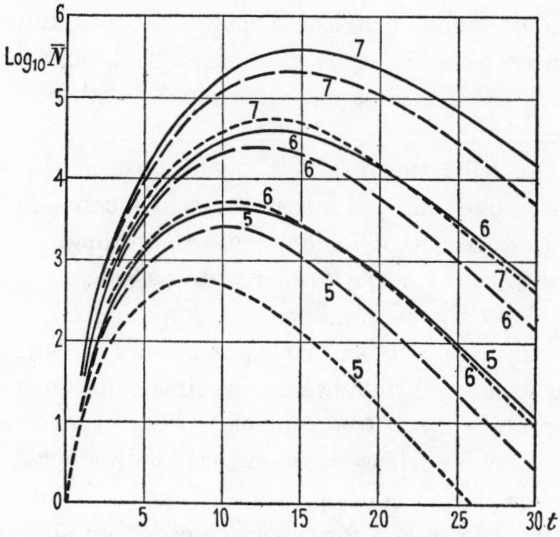

Fig. 35. Same as Fig. 34. Higher values of ε.

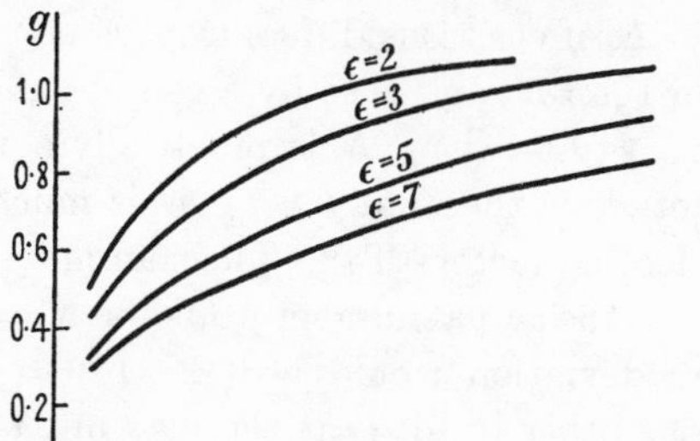

Fig. 36. The auxiliary function g used in calculation of collision loss. $\epsilon = \log_{10} E_0/E$. (See eq. (9).)

It is a property of this simple distribution law that the mean square deviation is given by

$$\overline{\delta N^2} \equiv \overline{(N-\bar{N})^2} = \overline{N^2} - \bar{N}^2 = \bar{N}. \tag{12}$$

However, in a shower, the various emission acts are not statistically independent and the Poisson law does not hold. It will be convenient to compare the actual fluctuations with the Poisson fluctuation (11), (12).

A complete solution of the fluctuation problem that would give $W(N)$ has not yet been found. Substantial progress has been made, however, by Bhabha and Ramakrishnan and by Jánossy and Messel.† From their theory it is, in principle, possible to evaluate all the *moments* of the distribution, $\overline{N}$, $\overline{\delta N^2}$, $\overline{\delta N^3}$,.... As is well known, $W(N)$ would be determined completely, if *all* these moments were known. Actually, it has only been possible to evaluate the most important one, $\overline{\delta N^2}$. In all this work the simpler approximation A (neglect of collision loss) has been used.

Before we describe the results a qualitative consideration will be helpful. There is one source of fluctuation which can easily be surveyed. In § 37.5 we have seen that for an electron, although it travels on the *average* a distance $t = 1$, while its energy drops to $E_0 e^{-1}$, the fluctuations of this *range* about the average are very large indeed, and it is almost as likely that this distance is in fact $\frac{1}{2}$ or 2 (see Fig. 30). Now consider the initiating process of a shower. The first collision may take place with a large probability at $t = 2$ instead of $t = 1$, for example. If this should be the case, and if we assume that the further development of the shower takes place according to the average schedule, the transition curve would be the same as for the average $\overline{N}$, but shifted to the right by one cascade unit. If then we fix our attention on a fixed distance t, the actual number N of electrons differs from $\overline{N}$ as follows: for $t < t_m$, N is much smaller than $\overline{N}$, and is so by a considerable factor owing to the steep slope of $\overline{N}$; at $t \sim t_m$, N is practically equal to $\overline{N}$, because the curve does not slope there; at $t > t_m$, N is much larger than $\overline{N}_m$, again by a considerable factor. Take, for example, $\epsilon = 3$, $t_m = 6$. At $t = 12$, $\overline{N} = 14$, but the actual number would be $N = \overline{N}(t = 13) = 19$. The mean square deviation would be $(19-14)^2 = 25$, almost twice Poissonian. On the other hand, near the maximum the fluctuation is much *smaller than Poissonian*. The fluctuations are large in the opposite direction, when the range of the first collision is $t = \frac{1}{2}$ instead of $t = 1$.

Fluctuations in range also take place at the later stages of the process. However, these have little influence. If, for example, a photon somewhere near t_m travels $t = \frac{1}{2}$ instead of 1 and creates a pair, this would only influence the subsequent generations derived from this

† H. J. Bhabha and A. Ramakrishnan, *Proc. Ind. Ac. Sc.* **32** (1950), 141; A. Ramakrishnan, *Proc. Camb. Phil. Soc.* **46** (1950), 595; L. Jánossy, *Proc. Phys. Soc.* **63** (1950), 241; L. Jánossy and H. Messel, ibid. **53** (1950), 1101; H. Messel, ibid. **64** (1951), 807. See also W. T. Scott, *Phys. Rev.* **82** (1951), 893. For earlier work on the fluctuation problem see in particular: W. T. Scott and G. E. Uhlenbeck, ibid. **62** (1942), 497, and the excellent book by Arley, *Theory of Stochastic Processes*, loc. cit.

particular photon and that is only a small fraction of the shower. Moreover, there are many photons there, whose ranges *fluctuate independently* in both directions.

We expect, therefore, that the largest source of fluctuations is a fluctuation in range of the first few collisions initiating the shower, with the effect just described.

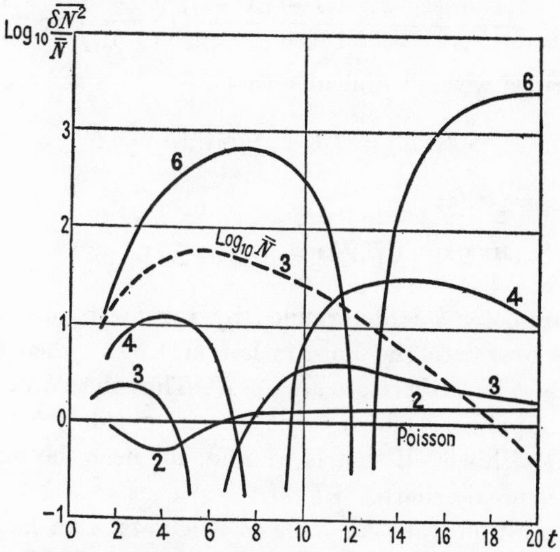

FIG. 37. Fluctuations of cascade showers (electron initiated, collision loss neglected). $\log_{10}\overline{\delta N^2}/\overline{N}$ as function of t for various $\epsilon = \log_{10} E_0/E$. $\overline{\delta N^2} = \overline{N^2} - \overline{N}^2$. For comparison also average number $\overline{N}$ for $\epsilon = 3$.

In Fig. 37 the $\log_{10}$ of the mean square deviation $\overline{\delta N^2}$ (divided by $\overline{N}$) is plotted against t for various values of $\epsilon = \log_{10} E_0/E$. The Poisson fluctuation would be $\log_{10}\overline{\delta N^2}/\overline{N} = 0$. There are always two regions where $\overline{\delta N^2}$ is much larger than $\overline{N}$, and that is before and after the shower maximum where the $\overline{N}$-curve has its greatest slope. Near the maximum $\overline{\delta N^2}$ drops to a very small value, indistinguishable from 0 (the calculations become very inaccurate there). The fluctuations here are much smaller than Poissonian. All this is qualitatively in accord with the above considerations.

There is a region where $\overline{\delta N^2}/\overline{N}$ becomes even larger than $\overline{N}$, and that is at large distances behind the maximum, where $\overline{N}$ is one or two or smaller.

It may now be asked what the actual probability distribution $W(N)$ is. Clearly this cannot be derived from $\overline{\delta N^2}$ alone. However, as a working hypothesis, a plausible suggestion may be made.† Pólya has generalized the Poisson distribution in such a way that it can be adapted to *any* mean square deviation $\overline{\delta N^2}$, as well as $\overline{N}$, provided that $\overline{\delta N^2} > \overline{N}$. Both these average values occur in the Pólya distribution as parameters, whereas in the Poisson law only $\overline{N}$ occurs and $\overline{\delta N^2}$ is fixed by $\overline{N}$. The distribution is given by

$$W(N) = \frac{1(1+b)(1+2b)...(1+b(N-1))}{N!}\left(\frac{\overline{N}}{1+b\overline{N}}\right)^N \frac{1}{(1+b\overline{N})^{1/b}}. \quad (13)$$

It can be verified without difficulty that

$$\sum_0^\infty W(N) = 1, \qquad \sum_0^\infty NW(N) = \overline{N}$$

and, furthermore, that

$$\sum_0^\infty N^2W(N) - \overline{N}^2 \equiv \overline{\delta N^2} = \overline{N}(1+b\overline{N}). \quad (14)$$

Thus the parameter b is determined by the mean square deviation. (13), (14) go over into the Poisson law (11), (12) when $b \to 0$. It is necessary that $b \geqslant 0$, therefore $\overline{\delta N^2} \geqslant \overline{N}$. Thus the Pólya distribution can only be applied some distance on either side from the shower maximum, but this is all that is wanted; for near the maximum the distribution is nearly sharp.

As an illustration, take the case of a shower at a large thickness behind the maximum. For example, for $\epsilon = 3$, $t = 20$, according to Fig. 37, $\overline{N} = 0{\cdot}385$, $\overline{\delta N^2}/\overline{N} = 1{\cdot}85$, hence $b = 2{\cdot}2$. Poisson and Pólya distributions give for $W(N)$:

N	0	1	2	3	4
Poisson	0·68	0·26	0·049	0·0063	0·0006
Pólya	0·755	0·157	0·052	0·0194	0·008

We see that while there is no great difference between the two distributions for $N = 0, 1, 2$, the Pólya distribution gives much higher probabilities for the rare cases of a fairly large number of particles. Whilst here the average value is only $\sim 0{\cdot}4$ there is still a chance of almost 1 per cent. of observing four particles.

The true distribution can deviate from the suggested Pólya distribution only by the higher moments $\overline{\delta N^3}$, etc. These could show up only in certain irregularities—which are unlikely—or in the values of $W(N)$ for very improbable cases, where $W(N)$ is in any case very small.

† The Pólya distribution was first suggested by Arley (loc. cit.), who also gives a detailed mathematical treatment of it.

Thus one may hope that the Pólya distribution is fairly good except for very improbable cases. (For the really probable cases $N = 0, 1, 2$ even the Poisson distribution is not too wrong.)

4. *Experimental evidence. Conclusions.* Qualitatively, the cascade theory has been verified completely. Photographs of showers containing any number up to several hundred particles are well known and cases have been found where the total number of particles runs into thousands. In the big air showers the total number of particles cannot be far from 10^5. Before, however, these can be called cascade showers it has to be proved that we are in fact dealing with a step-by-step process and not (as was in fact assumed at one stage) with a multiple process or a series of multiple processes where several particles are created in one quantum act. That we are in fact dealing with a cascade has been shown by placing a series of thin Pb-plates in a cloud chamber. Between the plates the particles are visible, and it can then be seen how they multiply and how the shower is gradually being built up, and eventually dies out again. Even in this experiment, if one wishes to be critical, one can hardly exclude the possibility that the shower may be of a mixed nature, a series of multiple processes each with small multiplicities, superposed on a cascade process. Perhaps the most direct evidence for the pure cascade nature of these showers† has come from experiments in photographic plates where every part of the tracks is visible. In such photographs it is seen that the *whole* phenomenon consists of 2-branch fork-tracks, i.e. *single* pairs of particles originating along the shower axis.‡

Quantitative tests are more difficult to obtain. There are countless experiments which verify the predictions of the theory in a semi-quantitative way.§ The following points, for example, have been verified. The shower develops to a maximum at a thickness of a few cascade units, and this thickness increases slowly with the size of the shower (i.e. the primary energy); the cascade unit is approximately that given in Table XIII, and depends on Z in that way. The majority of the particles in a fully developed shower have energies $\sim E_c$, and E_c is also approximately what is given in Table XIII. For $E > E_c$, the energy distribution is roughly (10). Further multiplication takes place in the transition of an already developed cascade from a medium with

† The occurrence of some multiple processes, at the rate of $1/\pi 137$ of the single process, must of course be expected. (See § 23.)

‡ Photographs have been obtained by J. E. Hooper, D. T. King, and A. H. Morrish, *Can. Journ. Phys.* **29** (1951), 545.

§ We must refer the reader for details to books on cosmic radiation.

low Z to a medium with high Z (because E_c is larger for small Z), and this transition effect is also what the theory predicts. By adding up the energies of the shower particles (and allowing for the energy contained in photons and low-grade energy losses) the order of magnitude of the primary energy E_0 can at least be estimated and the relation between E_0 and the number of shower particles be verified approximately.

At the time of writing no really quantitative tests at very high energies have been published. From the evidence, however, the following conclusions are well justified.

There must be at least approximate agreement between theory and experiment up to energies which give rise to showers of several hundred or thousand particles, that is roughly 10^5–10^6 MeV. This estimate is conservative. If the cross-sections for pair creation were wrong by an order of magnitude at 10^6 MeV, it would then be very hard to understand how a shower of a thousand particles could ever be created in a small thickness of lead. Naturally, a discrepancy by a factor 2 or 3 could not yet be detected.

We therefore come to the conclusion that the *theory of radiative energy loss and pair creation*, and presumably also of all other radiation processes, i.e. the *quantum twin theories of radiation and the positron, are essentially correct up to the highest energies known, and there is no visible limit* (as yet) *to the validity of these theories.*

APPENDIX

1. The angular momentum of light (§ 7)

IN Maxwell's theory the Poynting vector **S** (divided by c^2) is interpreted as the density of momentum of the field. We can then also define an angular momentum relative to a given point O or to a given axis,

$$\mathbf{M} = \frac{1}{c^2}\int [\mathbf{rS}]\,d\tau, \tag{1}$$

where $\mathbf{r}$ is the distance from O.

A plane wave travelling in the z-direction and with infinite extension in the xy-directions can have no angular momentum about the z-axis, because **S** is in the z-direction and $[\mathbf{rS}]_z = 0$. However, this is no longer the case for a wave with finite extension in the xy-plane. Consider a cylindrical wave with its axis in the z-direction and travelling in this direction. At the wall of the cylinder $r = R$, say, we let the amplitude drop to zero. It can then be shown that the wall of such a wave packet gives a finite contribution to M_z.

If in the interior $r < R$ the wave is a circularly polarized plane wave with frequency ν (or very nearly so, because the building up of the wave packet requires many frequencies) M_z turns out to be proportional to the total energy U,

$$M_z = \pm U/\nu, \tag{2}$$

where the sign depends on the direction of polarization. So in spite of the fact that M_z is a wall effect, its value is proportional to the volume.†

(2) is a purely classical relation but can also be interpreted as a quantum relation. If n is the number of photons, $U = n\hbar\nu$, we have $M_z = \pm n\hbar$. This means that each circularly polarized photon contributes an angular momentum $\pm\hbar$. In a sense we can say that the photon has spin 1 with two components ± 1. The third component, $M_z = 0$, however, which one might expect for spin 1, does not exist. This is a consequence of the transversality relation $\operatorname{div}\mathbf{A} = 0$, which in turn is connected with the gauge invariance. If the photon had a finite rest mass, three independent polarizations would exist, including a longitudinal polarization for which $M_z = 0$. For linear polarization, of course, the angular momentum vanishes classically. In quantum theory M_z is not sharp, and has the expectation value zero.

We consider more explicitly the case of spherical waves with singularity at the origin O. We may consider such waves to be emitted by an electric or magnetic multipole situated in O. Such spherical waves can generally be classified according to their angular dependence which is that of spherical harmonics $Y_l^m(\theta,\phi)$. Let **A** be the vector potential satisfying

$$\operatorname{div}\mathbf{A} = 0, \qquad \Box\mathbf{A} = 0, \qquad \mathbf{E} = -\frac{1}{c}\dot{\mathbf{A}}, \qquad \mathbf{H} = \operatorname{curl}\mathbf{A}. \tag{3}$$

There will be two types of spherical waves, corresponding to the two polarizations of a plane wave, which may be called electric and magnetic multipole

† Compare C. G. Darwin, *Proc. Roy. Soc.* A, **136** (1932), 36; J. Géhéniau, *Actualités Scientifiques*, No. 778, Paris 1939; J. Humblet, *Physica*, **10** (1943), 585.

waves respectively. Each type can be further classified by the numbers l, m, $m = -l, \ldots, +l$, of the spherical harmonics.†

We shall not derive the explicit expressions for **E**, **H** but merely quote the result of this straightforward but tedious calculation.

The angular dependence of each component of **E**, **H** is given by a combination of spherical harmonics Y_l^m, normalized to

$$\int Y_l^{m*} Y_{l'}^{m'}\, d\Omega = \delta_{ll'}\,\delta_{mm'}. \tag{4}$$

The radial dependence is given by Bessel functions with half-integer index

$$f_l(\kappa r) = C_l \frac{J_{l+\frac{1}{2}}(\kappa r)}{\sqrt{(\kappa r)}}. \tag{5}$$

We may conveniently assume f to be normalized to a sphere of radius R

$$\int_0^R f_l^2 r^2\, dr = 4\pi c^2. \tag{6}$$

Below, the factor of $f_{l\pm 1}$ is also C_l, not $C_{l\pm 1}$. **E** and **H** are given by:

Electric multipole:

$$E_z = \frac{i\nu}{c} e^{-i\nu t} a_l^m [q_{l-1}^m Y_{l-1}^m f_{l-1} + p_{l+1}^m Y_{l+1}^m f_{l+1}] + \text{compl. conj.,} \tag{7a}$$

$$E_x \pm iE_y = \mp \frac{i\nu}{c} e^{-i\nu t} a_l^m [Q_{l-1}^{\pm m} Y_{l-1}^{m\pm 1} f_{l-1} - P_{l+1}^{\pm m} Y_{l+1}^{m\pm 1} f_{l+1}] \mp$$
$$\mp \frac{i\nu}{c} e^{+i\nu t} a_l^{m*} [Q_{l-1}^{\mp m} Y_{l-1}^{m\mp 1 *} f_{l-1} - P_{l+1}^{\mp m} Y_{l+1}^{m\mp 1 *} f_{l+1}]; \tag{7b}$$

$$H_z = -\frac{i\nu}{c} e^{-i\nu t} a_l^m \frac{m}{\sqrt{[l(l+1)]}} Y_l^m f_l + \text{compl. conj.,} \tag{7c}$$

$$H_x \pm iH_y = +\frac{i\nu}{c} e^{-i\nu t} a_l^m \sqrt{\left(\frac{(l\mp m)(l\pm m+1)}{l(l+1)}\right)} Y_l^{m\pm 1} f_l -$$
$$-\frac{i\nu}{c} e^{+i\nu t} a_l^{m*} \sqrt{\left(\frac{(l\pm m)(l\mp m+1)}{l(l+1)}\right)} Y_l^{m\mp 1 *} f_l. \tag{7d}$$

Magnetic multipole: interchange $\mathbf{E} \to \mathbf{H}$, $\mathbf{H} \to -\mathbf{E}$.

In (7), a_l^m is an independent amplitude and q, Q,... are abbreviations:

$$q_{l-1}^m = \sqrt{\left(\frac{(l+1)(l^2-m^2)}{l(2l+1)(2l-1)}\right)}, \qquad p_{l+1}^m = \sqrt{\left(\frac{l[(l+1)^2-m^2]}{(l+1)(2l+1)(2l+3)}\right)}, \tag{8a}$$

$$Q_{l-1}^{\pm m} = \sqrt{\left(\frac{(l+1)(l\mp m)(l\mp m-1)}{l(2l+1)(2l-1)}\right)}, \qquad P_{l+1}^{\pm m} = \sqrt{\left(\frac{l(l\pm m+1)(l\pm m+2)}{(l+1)(2l+1)(2l+3)}\right)}. \tag{8b}$$

We shall also require the radial components. For an electric multipole these are:

$$H_r = 0, \qquad E_r = ie^{-i\nu t} a_l^m Y_l^m \sqrt{[l(l+1)]}\, \frac{f_l}{r} + \text{compl. conj.} \tag{9}$$

By an investigation of the singularity at $r = 0$ it can, indeed, be verified that the sources of these waves are electric and magnetic $2l$-poles.

In deriving the above expressions frequent use is made of the well-known relations between spherical harmonics and between Bessel functions. For the

† See A. W. Conway, *Proc. Roy. Ir. Ac.* **41** (1932), 8; W. Heitler, *Proc. Camb. Phil. Soc.* **37** (1936), 112; W. Franz, *Z. f. Phys* **127** (1950), 363.

convenience of the reader we summarize them† in the form in which they occur here:

$$\frac{z}{r}Y_l^m = \sqrt{\left(\frac{l+1}{l}\right)}p_{l+1}^m Y_{l+1}^m + \sqrt{\left(\frac{l}{l+1}\right)}q_{l-1}^m Y_{l-1}^m, \tag{10a}$$

$$\frac{1}{r}(x \pm iy)Y_l^m = \pm\sqrt{\left(\frac{l+1}{l}\right)}P_{l+1}^{\pm m} Y_{l+1}^{m\pm 1} \mp \sqrt{\left(\frac{l}{l+1}\right)}Q_{l-1}^{\pm m} Y_{l-1}^{m\pm 1}, \tag{10b}$$

$$\frac{\partial}{\partial z}Y_l^m f_l = \sqrt{\left(\frac{l+1}{l}\right)}p_{l+1}^m Y_{l+1}^m \alpha_l + \sqrt{\left(\frac{l}{l+1}\right)}q_{l-1}^m Y_{l-1}^m \beta_l, \tag{11a}$$

$$\left(\frac{\partial}{\partial x} \pm i\frac{\partial}{\partial y}\right)Y_l^m f_l = \pm\sqrt{\left(\frac{l+1}{l}\right)}P_{l+1}^{\pm m} Y_{l+1}^{m\pm 1}\alpha_l \mp \sqrt{\left(\frac{l}{l+1}\right)}Q_{l-1}^{\pm m} Y_{l-1}^{m\pm 1}\beta_l, \tag{11b}$$

$$\alpha_l \equiv \frac{\partial f_l}{\partial r} - l\frac{f_l}{r} = -\kappa f_{l+1}, \qquad \beta_l = \frac{\partial f_l}{\partial r} + (l+1)\frac{f_l}{r} = +\kappa f_{l-1}. \tag{12}$$

We shall also require integrations over r. The following relation holds:

$$\int \{lf_{l+1}^2 + (l+1)f_{l-1}^2\}r^2\,dr = (2l+1)\int f_l^2 r^2\,dr + \frac{r^2 f_l}{\kappa}[(l+1)f_{l-1} - lf_{l+1}]. \tag{13}$$

In order to obtain an orthogonal enumerable set of waves it will be convenient to impose some boundary condition at $r = R$, say. We shall require that the tangential components of E vanish at $r = R$ (reflecting mirror). This expresses itself in the relation

$$(l+1)f_{l-1}(\kappa R) = lf_{l+1}(\kappa R). \tag{14}$$

If, in (13), the integration is extended between 0 and R the last term on the right vanishes, and hence

$$\int_0^R [lf_{l+1}^2 + (l+1)f_{l-1}^2]r^2\,dr = (2l+1)\int_0^R f_l^2 r^2\,dr = (2l+1)4\pi c^2. \tag{15}$$

We now calculate (i) the total energy

$$U = \frac{1}{8\pi}\int (E^2 + H^2)\,d\tau \tag{16}$$

and (ii) the z-component of $\mathbf{M}$:

$$M_z = \frac{1}{4\pi c}\int d\tau\, r(E_r H_z - E_z H_r). \tag{16'}$$

It is clear that the result is the same for magnetic and electric multipole radiation. On account of the orthogonality of the Y_l^m all cross products belonging to waves with different l or m vanish.‡

For a single partial wave l, m, we obtain:

$$U = \frac{2\nu^2}{4\pi c^2}|a_l^m|^2\{[(p_{l+1}^m)^2 + \tfrac{1}{2}(P_{l+1}^m)^2 + \tfrac{1}{2}(P_{l+1}^{-m})^2]\int_0^R f_{l+1}^2 r^2\,dr +$$

$$+[(q_{l-1}^m)^2 + \tfrac{1}{2}(Q_{l-1}^m)^2 + (\tfrac{1}{2}Q_{l-1}^{-m})^2]\int_0^R f_{l-1}^2 r^2\,dr\} = 2\nu^2|a_l^m|^2$$

† The sign of Y_l^m is such that $Y_l^{-m} = (-1)^m Y_l^{m*}$.

‡ Although Y_l^m is not orthogonal to Y_l^{-m} in the sense that $\int Y_l^m Y_l^{-m}\,d\Omega \neq 0$, no cross products occur on account of (14).

on account of (8) and (15). Similarly,

$$\mathbf{M}_z = \frac{2\nu}{4\pi c^2} m|a_l^m|^2 \int\limits_0^R f_l^2 r^2\, dr = 2\nu m|a_l^m|^2 = m\frac{U}{\nu}. \tag{17}$$

Also, for a single wave l, m, $\qquad M_x = M_y = 0.$

We see again that U and M_z are in a fixed relationship. If U is composed of contributions from photons $n\hbar\nu$ (see below), $M_z = nm\hbar$, which shows that each photon has an angular momentum $m\hbar$, as would be the case for an electron (without spin). However (17) is again a purely classical relation.

It is interesting to note that the angular momentum is not contained in the pure wave zone, where the field strengths are perpendicular to $\mathbf{r}$ and behave like

$$f_l \sim \frac{1}{\kappa r}\cos[\kappa r - \tfrac{1}{2}\pi(l+1)].$$

In this zone, indeed, M_z vanishes, as is seen from (16′) and (9): M_z is proportional to E_r and $E_r \sim 1/r^2$. The contributions to M_z arise from a subtle interference effect† where $E \sim 1/r^2$ and $H \sim 1/r$ (for magnetic multipole waves $E \sim 1/r$, $H \sim 1/r^2$). On the other hand, U derives its main contribution from the wave zone. Nevertheless the *exact* proportionality (17) is valid.

The spherical waves can be quantized in much the same way as plane waves (§ 7). We subject the amplitudes a_l^m to the commutation relations (for each frequency ν),

$$a_l^{m*} a_{l'}^{m'} - a_{l'}^{m'} a_l^{m*} = \frac{\hbar}{2\nu}\delta_{ll'}\delta_{mm'}, \tag{18}$$

from which it follows that $\qquad a_l^{m*} a_l^m = n_l^m \dfrac{\hbar}{2\nu};$

n_l^m is the number of light quanta, with quantum numbers l, m. Here a photon is not, as in § 7, represented by a plane wave, but by a spherical wave. Then $U = n_l^m \hbar\nu$, $M_z = n_l^m m\hbar$. It is not difficult to verify also that the components of $\mathbf{M}$ then satisfy the commutation relations

$$[M_x M_y] = i\hbar M_z, \qquad [M_x U] = 0,$$
$$M_x^2 + M_y^2 + M_z^2 = \hbar^2 l(l+1)$$

as should generally be the case.

We see that the angular momentum M_z can assume all integral values m. A clear-cut distinction between spin and orbital momentum cannot, however, be made.‡

The angular momentum of a circularly polarized wave has been measured§ by letting it pass through a doubly refracting plate which alters the polarization. The plate was allowed to rotate about the axis of the light wave. It was found that the plate, indeed, begins to rotate as soon as the light passes through, and the angular momentum acquired by the plate was in agreement with (17) and proportional to the number of photons that have passed.

† M. Abraham, *Phys. Zs.* **15** (1914), 914.

‡ In the case of cylindrical waves solutions also exist with $M_z = \pm|m|\hbar$ ($|m| > 1$). These are not plane waves but depend on r. The result (2) refers to a wave packet for which the field is independent (or practically so) of r for $r < R$.

§ R. A. Beth, *Phys. Rev.* **50** (1936), 115.

2. Commutation relations of the vector potentials in Coulomb gauge (§§ 9–13)

To derive the commutation relations for the space components $A_i(\mathbf{r}, t)$ of the vector potential, when Coulomb gauge is used, we start from the expansion § 7 eq. (3), using interaction representation

$$\mathbf{A} = \sqrt{(4\pi c^2)} \sum_\lambda \mathbf{e}_\lambda \{q_\lambda e^{i(\boldsymbol{\kappa}_\lambda \mathbf{r}) - i\nu_\lambda t} + q_\lambda^* e^{-i(\boldsymbol{\kappa}_\lambda \mathbf{r}) + i\nu_\lambda t}\}, \tag{1}$$

where $\mathbf{e}_\lambda$ is perpendicular to $\boldsymbol{\kappa}_\lambda$. This automatically satisfies the condition $\operatorname{div} \mathbf{A} = 0$ as an operator identity. q_λ, q_λ^* satisfy

$$[q_\lambda q_\lambda^*] = \hbar/2\nu_\lambda \tag{2}$$

and all other commutators vanish. Going over to the integral in $\boldsymbol{\kappa}$-space ($\nu = \kappa c$), we obtain

$$[A_i(\mathbf{r}_2, t_2) A_k(\mathbf{r}_1, t_1)] = -\frac{4\pi i\hbar c}{(2\pi)^3} \int \frac{d^3\kappa}{\kappa} \left(\delta_{ik} - \frac{\kappa_i \kappa_k}{\kappa^2}\right) e^{+i(\boldsymbol{\kappa}, \mathbf{r}_2 - \mathbf{r}_1)} \sin \nu(t_2 - t_1) \tag{3}$$

observing that $$\sum_{\text{pol}} e_{i\lambda} e_{k\lambda} = \delta_{ik} - \kappa_{\lambda i} \kappa_{\lambda k}/\kappa_\lambda^2.$$

The first integral $\sim \delta_{ik}$, gives the relativistic Δ-function § 8 eq. (26a). This term is $-4\pi i\hbar c \Delta(\mathbf{r}, t)$, $\mathbf{r} \equiv \mathbf{r}_2 - \mathbf{r}_1$, $t \equiv t_2 - t_1$. The second integral can be written

$$\int \frac{d^3\kappa}{\kappa^3} \kappa_i \kappa_k e^{+i(\boldsymbol{\kappa}\mathbf{r})} \sin \nu t = \frac{\partial^2}{\partial x_{i2} \partial x_{k1}} \int \frac{d^3\kappa}{\kappa^3} e^{+i(\boldsymbol{\kappa}\mathbf{r})} \sin \nu t$$

$$= 4\pi \frac{\partial^2}{\partial x_{i2} \partial x_{k1}} \frac{1}{|\mathbf{r}|} \int_0^\infty \frac{d\kappa}{\kappa^2} \sin \kappa |\mathbf{r}| \sin \kappa c t. \tag{4}$$

The integral in (4) has the value $\frac{1}{2}\pi ct$ when $c|t| < |\mathbf{r}|$; $\frac{1}{2}\pi |\mathbf{r}| \epsilon(t)$ when $c|t| > |\mathbf{r}|$ ($\epsilon(x) = \pm 1$, for $x \gtrless 0$). This can be written in the condensed form

$$-\tfrac{1}{4}\pi[(r - ct)\epsilon(r - ct) - (r + ct)\epsilon(r + ct)].$$

Thus, defining the function

$$H(\mathbf{r}, t) = \frac{1}{8\pi r}[(r - ct)\epsilon(r - ct) - (r + ct)\epsilon(r + ct)], \tag{5}$$

we obtain

$$[A_i(\mathbf{r}_2, t_2) A_k(\mathbf{r}_1, t_1)] = -4\pi i\hbar c\left\{\delta_{ik} \Delta(\mathbf{r}, t) + \frac{\partial^2}{\partial x_{i2} \partial x_{k1}} H(\mathbf{r}, t)\right\}. \tag{6}$$

The derivative of $\epsilon(x)$ is $\epsilon'(x) = 2\delta(x)$. It follows that the function H has the properties ($\epsilon(r) = 1$)

$$\frac{1}{c^2} \frac{\partial^2 H}{\partial t^2} = \nabla^2 H = \Delta(\mathbf{r}, t), \qquad H(r, 0) = 0, \qquad \frac{\partial H}{\partial t}(r, 0) = -\frac{c}{4\pi r}. \tag{7}$$

(6) differs essentially from the commutation relations for Lorentz gauge by the last term involving H. This term vanishes inside the light cone $c|t| > r$ because then $H(\mathbf{r}, t) = -(1/4\pi)\epsilon(t)$ and this gives no contribution to the space derivatives in (6). It is different from zero outside the light cone. Clearly this addition only has formal significance. It shows that $\mathbf{A}$ cannot be a measurable quantity.

Nevertheless, the second term of (6) is essential in order to obtain the correct commutation relations for the field strengths. We form, for example,

$$[E_i(\mathbf{r}_2,t_2)E_k(\mathbf{r}_1,t_1)] = -\frac{4\pi i\hbar c}{c^2}\frac{\partial^2}{\partial t_1\,\partial t_2}\left\{\delta_{ik}\,\Delta(\mathbf{r},t)+\frac{\partial^2}{\partial x_{i2}\,\partial x_{k1}}H(\mathbf{r},t)\right\}$$

$$= -4\pi i\hbar c\left\{\frac{1}{c^2}\frac{\partial^2}{\partial t_1\,\partial t_2}\delta_{ik}-\frac{\partial^2}{\partial x_{i2}\,\partial x_{k1}}\right\}\Delta(\mathbf{r},t) \qquad (8)$$

on account of (7). The derivatives of H, as they occur in (8), reduce to a contribution on the light cone only, as should be the case. (8) is identical with the commutation relations § 9 eqs. (8a), (8b). Similarly the commutation relations for the magnetic field strengths, etc., can be rederived.

3. The Lorentz condition in the presence of charges (§ 10, 13)

It was shown in § 10 that the Lorentz condition $\partial A_\alpha/\partial x_\alpha = 0$ cannot be fulfilled as an operator identity when Lorentz gauge is used. We saw that in the absence of charges this relation takes the form of a condition to be imposed on the state vector (valid at each space-time point)

$$\frac{\partial A_\alpha^-}{\partial x_\alpha}\Psi = 0. \qquad (1)$$

The scalar field is quantized by indefinite metric. This condition singles out certain permissible state vectors Ψ_L, called the Lorentz set. (1) then ensures that the expectation value of $\partial A_\alpha/\partial x_\alpha$ vanishes

$$\left\langle\frac{\partial A_\alpha}{\partial x_\alpha}\right\rangle \equiv \left(\Psi_L^*\eta\frac{\partial A_\alpha}{\partial x_\alpha}\Psi_L\right) = 0. \qquad (2)$$

For a pure field Ψ_L is independent of time, the time dependence being transferred to the operators by going over to interaction representation. (1) can be regarded as an initial condition at $t = t_0$, and then (1) and (2) are valid for all times.

When charges are present Ψ depends on t, and in interaction representation fulfils

$$i\hbar\frac{\partial\Psi}{\partial t} = H(t)\Psi(t), \qquad H(t) = -\frac{1}{c}\int i_\mu A_\mu\,d\tau \qquad (3)$$

($H(t) \equiv H'_{\text{int}}$, in interaction representation).

It is easy to see that, if (1) were required to be valid for all times, this would contradict the wave equation (3). We therefore require a generalization of the initial conditions that is valid when charges and currents are present or are taken into account as operators in the interaction $H(t)$. We shall give this generalization not in a form analogous to (1) but only in the form of initial conditions. This is all that is needed.

What we must have, in order that Maxwell's equations should hold as relations between expectation values, is that (2) should continue to hold for *all times*

$$\left\langle\frac{\partial A_\alpha}{\partial x_\alpha}(t)\right\rangle = \left(\Psi^*(t)\eta\frac{\partial A_\alpha}{\partial x_\alpha}(t)\Psi(t)\right) = 0. \qquad (4)$$

To express this in the form of an initial condition at $t = t_0$ we form the time derivative of (4). This must vanish also because (4) must hold for all t. For $\dot\Psi$ and $\dot\Psi^*$ the wave equation (3) is to be used. $H(t)$ is not hermitian, because the term $i_4 A_4 = ic\rho A_4$ is antihermitian (A_4 is hermitian). However, this term anticommutes with η also and therefore

$$-i\hbar \frac{\partial \Psi^*}{\partial t} \eta = \Psi^*(t) H^\dagger(t) \eta = \Psi^* \eta H(t). \tag{5}$$

We then obtain

$$\frac{\partial}{\partial t}\left\langle \frac{\partial A_\alpha}{\partial x_\alpha} \right\rangle = \left\langle \frac{\partial}{\partial t} \frac{\partial A_\alpha}{\partial x_\alpha} \right\rangle + \frac{1}{i\hbar} \Psi^*(t)\eta \left[\frac{\partial A_\alpha}{\partial x_\alpha}(t), H(t)\right] \Psi(t). \tag{6}$$

If we insert the integral (3) for $H(t)$ (and denote the integration variable by $\mathbf{r}'$) the commutator can be worked out. According to § 13 eq. (16)

$$[A_\alpha(\mathbf{r}, t) A_\mu(\mathbf{r}', t')] = -4\pi i\hbar c \delta_{\alpha\mu} \Delta(\mathbf{r}-\mathbf{r}', t-t')$$

with $\quad \Delta(\mathbf{r}-\mathbf{r}', 0) = 0, \qquad \left.\frac{\partial \Delta(\mathbf{r}-\mathbf{r}', t-t')}{\partial t}\right|_{t=t'} = c\delta(\mathbf{r}-\mathbf{r}').$

Thus

$$\left[\frac{\partial A_\alpha(\mathbf{r}, t)}{\partial x_\alpha}, H(t')\right]_{t'=t} = 4\pi i\hbar \int d\tau' i_\mu(\mathbf{r}', t') \frac{\partial}{\partial x_\mu} \Delta(\mathbf{r}-\mathbf{r}', t-t')\Big|_{t'=t}$$

$$= 4\pi i\hbar c \int d\tau' \rho(\mathbf{r}', t) \delta(\mathbf{r}-\mathbf{r}') = 4\pi i\hbar c \rho(\mathbf{r}, t) \tag{7}$$

($i_4 = ic\rho$). The space derivatives in (7) vanish for $t' = t$. Thus (6) becomes

$$\frac{\partial}{\partial t}\left\{\Psi^*(t)\eta \frac{\partial A_\alpha}{\partial x_\alpha}(t)\Psi(t)\right\} = \Psi^*(t)\eta \left[\frac{\partial}{\partial t}\frac{\partial A_\alpha}{\partial x_\alpha}(t) + 4\pi c\rho(t)\right] \Psi(t) = 0. \tag{8}$$

We can now show that it is sufficient to require that (8) *and* (4) should hold at an initial time t_0 only (but, of course, for all $\mathbf{r}$):

$$\left\langle \frac{\partial A_\alpha}{\partial x_\alpha} \right\rangle(t_0) \equiv \Psi^*(t_0)\eta \frac{\partial A_\alpha}{\partial x_\alpha}(t_0)\Psi(t_0) = 0, \tag{9 a}$$

$$\frac{\partial}{\partial t}\left\langle \frac{\partial A_\alpha}{\partial x_\alpha} \right\rangle\Big|_{t=t_0} = \Psi^*(t_0)\eta \left\{\frac{\partial}{\partial t}\frac{\partial A_\alpha}{\partial x_\alpha} + 4\pi c\rho\right\}_{t_0} \Psi(t_0) = 0 \tag{9 b}$$

and then (4) (and (8)) will always be valid. For this purpose it is only necessary to prove that $\langle \partial A_\alpha / \partial x_\alpha \rangle$ satisfies the second order equation $\Box \langle \partial A_\alpha / \partial x_\alpha \rangle = 0$, as a consequence of the wave equation (3) and of

$$\Box A_\alpha = 0. \tag{10}$$

(10) holds in interaction representation as an operator equation; the interaction with the currents (classically $-(4\pi/c) i_\alpha$ on the right-hand side) is expressed in this representation by the wave equation (3) for the state vector. As Ψ is independent of $\mathbf{r}$ we obtain from (8) in the same manner as above, making use of (10):

$$\Box \left\langle \frac{\partial A_\alpha}{\partial x_\alpha} \right\rangle = -\frac{1}{c^2}\left\langle 4\pi c \frac{\partial \rho}{\partial t} + \frac{1}{i\hbar}\left[\frac{\partial}{\partial t}\frac{\partial A_\alpha}{\partial x_\alpha}, H(t)\right] + \frac{4\pi c}{i\hbar}[\rho, H(t)] \right\rangle. \tag{11}$$

In (11) the same time t is always to be inserted in all factors. To obtain the commutators we refer to the specialization of the anticommutation relations § 12 eq. (20) for $t = t'$:

$$\{\psi^\dagger_{\rho'}(\mathbf{r}', t) \psi_\rho(\mathbf{r}, t)\} = i\gamma_{4\rho\rho'} \delta(\mathbf{r}-\mathbf{r}').$$

This follows by using the properties of the D-function § 8 eqs. (32). Only the time derivative $\gamma_4 \partial/\partial x_4$ in § 12 eq. (20) contributes. Hence the commutator of two current components i_μ, i_ν taken at the same time is ($i_\mu = ec\psi^\dagger \gamma_\mu \psi$)

$$[i_\mu(\mathbf{r}, t), i_\nu(\mathbf{r}', t)] = ie^2c^2 \psi^\dagger (\gamma_\mu \gamma_4 \gamma_\nu - \gamma_\nu \gamma_4 \gamma_\mu) \psi \delta(\mathbf{r}-\mathbf{r}') \tag{12}$$

as a simple calculation shows. If one of the factors in (12) is $i_4 = ic\rho$, the commutator vanishes. Hence the last term of (11) vanishes, because ρ commutes with A_α also. The second term of (11) is

$$\left[\frac{\partial}{\partial t}\frac{\partial A_\alpha}{\partial x_\alpha}(\mathbf{r},t), H(t')\right]_{t'=t} = 4\pi i\hbar \int d\tau' i_\mu(\mathbf{r}',t') \frac{\partial}{\partial t}\frac{\partial}{\partial x_\mu}\Delta(\mathbf{r}-\mathbf{r}',t-t')\bigg|_{t'=t}$$

$$= 4\pi i\hbar c \int d\tau'\, i_k(\mathbf{r}',t)\frac{\partial}{\partial x_k}\delta(\mathbf{r}-\mathbf{r}') = +4\pi i\hbar c\frac{\partial i_k(\mathbf{r},t)}{\partial x_k} = -4\pi i\hbar c\dot{\rho} \quad (13)$$

on account of the conservation of charge, and because

$$\frac{\partial^2}{\partial t^2}\Delta(\mathbf{r},t)\bigg|_{t=0} = 0.$$

(Δ is an odd function of t.) Thus the second term of (11) cancels with the first and

$$\Box\left\langle\frac{\partial A_\alpha}{\partial x_\alpha}\right\rangle = 0. \quad (14)$$

It follows then that the equations (4) and (8) hold for any time t when they are satisfied at the initial time t_0. Thus two initial conditions are required. (9) was quoted in § 13 and is the quantum analogue of the classical initial conditions § 6 eq. (41'). Note in particular that the second equation (9 b) or (8) can be stated as

$$\langle -\text{div}\,\mathbf{E}+4\pi\rho\rangle = 0 \quad (15)$$

either at $t = t_0$ or for all times t. For $\rho = 0$ the initial conditions reduce to

$$\left\langle\frac{\partial A_\alpha}{\partial x_\alpha}\right\rangle = 0 \quad \text{and} \quad \left\langle\frac{\partial}{\partial t}\frac{\partial A_\alpha}{\partial x_\alpha}\right\rangle = 0, \quad \text{at } t = t_0.$$

Both are satisfied by the *one condition* (1)

$$\left(\frac{\partial A_\alpha^-}{\partial x_\alpha}\Psi\right)_{t=t_0} = 0.$$

For a free field $(\partial/\partial t)(\partial A_\alpha^-/\partial x_\alpha)$ only differs from $\partial A_\alpha^-/\partial x_\alpha$ by the trivial factor $-i\nu$ for each partial wave, and (1) holds separately for each ν. Thus from (1) we also have automatically

$$\frac{\partial}{\partial t}\frac{\partial A_\alpha^-}{\partial x_\alpha}\Psi = 0 \quad \text{and hence} \quad \Psi^*\eta\frac{\partial}{\partial t}\frac{\partial A_\alpha}{\partial x_\alpha}\Psi = 0.$$

There are also various other forms in which the auxiliary condition can be expressed. In particular it is possible to split up the condition in a form analogous to (1), viz. $L^-\Psi = 0$, where L^- differs from $\partial A_\alpha^-/\partial x_\alpha$ by an additional term containing ρ. We shall not require any of these forms and refer the reader to the literature quoted in § 10.

We now also show explicitly that Maxwell's equations are satisfied as relations between expectation values. Always taking the same time t everywhere, we have

$$\left\langle\frac{\partial A_\alpha}{\partial x_\alpha}\right\rangle = 0. \quad (15\text{a})$$

Furthermore for $\langle A_\alpha\rangle$ itself:

$$\frac{\partial}{\partial x_i}\langle A_\alpha\rangle = \left\langle\frac{\partial}{\partial x_i}A_\alpha\right\rangle, \qquad \nabla^2\langle A_\alpha\rangle = \langle\nabla^2 A_\alpha\rangle, \quad (15\text{b})$$

$$\frac{\partial}{\partial t}\langle A_\alpha\rangle = \left\langle\frac{\partial}{\partial t}A_\alpha\right\rangle + \frac{1}{i\hbar}\langle[A_\alpha, H(t)]\rangle = \left\langle\frac{\partial}{\partial t}A_\alpha\right\rangle. \quad (15\text{c})$$

(15c) follows because $A_\alpha(t)$ commutes with $A_\mu(t)$ and therefore with $H(t)$, on account of $\Delta(t=0)=0$.

Hence:

$$\frac{1}{c^2}\frac{\partial^2}{\partial t^2}\langle A_\alpha\rangle = \frac{1}{c^2}\frac{\partial}{\partial t}\left\langle\frac{\partial A_\alpha}{\partial t}\right\rangle = \frac{1}{c^2}\left\langle\frac{\partial^2 A_\alpha}{\partial t^2}\right\rangle+\frac{1}{i\hbar c^2}\left\langle\left[\frac{\partial A_\alpha}{\partial t}, H(t)\right]\right\rangle$$
$$= \frac{1}{c^2}\left\langle\frac{\partial^2 A_\alpha}{\partial t^2}\right\rangle+\frac{4\pi}{c}\langle i_\alpha\rangle.$$

Since $\Box A_\alpha = 0$, identically, we obtain

$$\left(\nabla^2-\frac{1}{c^2}\frac{\partial^2}{\partial t^2}\right)\langle A_\alpha\rangle \equiv \Box\langle A_\alpha\rangle = -\frac{4\pi}{c}\langle i_\alpha\rangle. \tag{16}$$

(16) and (15a) are equivalent to Maxwell's equations for the expectation values (see § 1). For this purpose note also that, for the field strengths which are *first* derivatives of A,

$$\langle f_{\alpha\beta}\rangle = \left\langle\frac{\partial A_\beta}{\partial x_\alpha}-\frac{\partial A_\alpha}{\partial x_\beta}\right\rangle = \frac{\partial}{\partial x_\alpha}\langle A_\beta\rangle-\frac{\partial}{\partial x_\beta}\langle A_\alpha\rangle.$$

We shall not attempt to derive the state vectors of the Lorentz set that satisfy (9) explicitly, as we did in § 10 for a pure field. This would be rather complicated, and is not necessary for the applications in this book. For collisions between free particles we shall throughout consider the interaction between charges and field as a perturbation, and therefore regard field and charges as free and non-interacting at the initial time $t_0=-\infty$. Then, as explained in § 13, the initial conditions reduce to those for a free field with $\rho=0$, and the Lorentz condition is satisfied if initially no longitudinal and scalar photons are present. For problems involving bound states, where the full conditions (9) would have to be used, it is simpler to use Coulomb gauge.

4. Iterated damping equation, transition to free particles (§§ 16, 34)

In § 16 a general theory for problems involving a finite line breadth was developed. The essential quantities determining the probability amplitudes for various states are U, eq. (10), and Γ, eq. (12). U has no diagonal elements and Γ is diagonal. Both equations are combined in eq. (10′) or

$$U = H+H\zeta U+\tfrac{1}{2}i\hbar\Gamma. \tag{1}$$

E is a variable energy on which U and Γ depend, but H is independent of E. It was explained in § 16 that a transformed interaction Hamiltonian should be substituted for the interaction H. This should be such that the virtual states are included in the definition of the atomic states but, for bound states, this transformation is not unambiguous. In the following we operate with the 'naked' states and use a representation where the unperturbed Hamiltonian $\tilde{H}_0 = H_0-H^{(s)}$, corrected by the level displacements, is diagonal (§ 16.3). $-H^{(s)}$ is the self-energy of the bound states, assumed here to be diagonal simultaneously with H_0. Then in (1) $\zeta \equiv \zeta(E-\tilde{H}_0)$, and the interaction is $H = H_{\text{int}}+H^{(s)}$.

We show that the theory goes over into that for free particles, § 15, when a corresponding limiting procedure is carried out. This will be the case for the low orders of approximation (up to the third inclusive) but in the higher orders a discrepancy will be found which is just due to the use of naked states. The discrepancy disappears under the same condition as was obtained in § 16, eq. (32), for the canonical transformation that eliminates the virtual processes. The condition is fulfilled for the transformation of § 15 or that of § 34.4.

We first change (1) into another form closely resembling eq. (33), § 15, that is valid for free particle collisions. For this purpose we split up the ζ-function in (1) into its principal value and δ-function parts. We write for short (omitting the $\sim$ on $\tilde{H}_0$)

$$P \equiv \frac{\mathscr{P}}{E-H_0}, \qquad \delta \equiv \delta(E-H_0). \tag{2}$$

Neither P nor δ of course, commutes with H:

$$(PH)_{n|m} = \mathscr{P}/(E-E_n)\cdot H_{n|m}, \qquad (HP)_{n|m} = H_{n|m}\,\mathscr{P}/(E-E_m)$$

(E_n = eigenvalue of $\tilde{H}_0$). Then

$$U = H+HPU+\tfrac{1}{2}i\hbar\Gamma-i\pi H\,\delta U. \tag{3}$$

We leave the last term unaltered but replace U in the second term HPU by what follows from the equation for U:

$$U = H+HPH+HPHPU+\tfrac{1}{2}i\hbar(1+HP)\Gamma-i\pi(H+HPH)\,\delta U. \tag{4}$$

We continue the same process of iteration in the term $HPHPU$, and so on in all similar terms. Ultimately we find†

$$U(E) = K(E)-i\pi K(E)\,\delta U(E)+\tfrac{1}{2}i\hbar(1+K(E)P)\Gamma(E), \tag{5}$$

$$K(E) = H+HPH+HPHPH+.... \tag{5$'$}$$

(5) and (5′) are exact and equivalent to (1). $K(E)$ given by (5′) is a function of E. For example, the fourth-order term of K written out explicitly is, if $H_{A|n} \equiv H_{\text{int}A|n}$,

$$K_{4A|B}(E) = \sum_{n,m,l} \frac{H_{A|n}H_{n|m}H_{m|l}H_{l|B}}{(E-E_n)(E-E_m)(E-E_l)} \tag{6}$$

apart from terms containing $H^{(s)}$. The denominators are all principal values.

Now consider the case of free particles. It is evident physically that the transition probability, which is $\mathscr{R}\Gamma$, goes to zero when the volume L^3 in which the process is considered to take place, goes to infinity (see § 14, p. 141). On the other hand, $\mathscr{I}\Gamma$ is a self-energy and is independent of L^3.

When $\mathscr{R}\Gamma \to 0$, a transition probability per sec. $w_{A|O}$ exists for the transition $O \to A$ and this concept then has exact validity. It was shown in § 16.2 that $w_{A|O}$ is given by

$$w_{A|O} = \frac{2\pi}{\hbar}\,|U_{A|O}(E_A)|^2\delta(E_A-E_O). \tag{7}$$

Thus U is only needed for $E = E_A = E_O$. We then obtain the case of free particles from (5) by (i) letting $\mathscr{R}\Gamma \to 0$, (ii) putting $E_A = E_O$ and letting $E \to E_O$ or E_A.

We take the AO matrix element of (5), and put $E_A = E_O$ *exactly*. Then we let E tend to E_O. This is straightforward as long as only the lower orders of K, up to the third inclusive (i.e. $HPHPH$), are considered. The term $\sim\Gamma$ has $\Gamma_{O|O}(E_O)$ as factor, and we have shown in § 16 that $\mathscr{I}\Gamma_{O|O}(E_O) = 0$ when the level displacements are included in $\tilde{H}_0$. Also $\mathscr{R}\Gamma \to 0$. Thus the third term of (5) vanishes. We obtain immediately

$$U_{A|O} = K_{A|O}-i\pi K_{A|B}\,\delta(E_O-E_B)U_{B|O}. \tag{8}$$

$K_{A|B}$ is identical with the expressions in § 15 and (8) identical with eq. (33), § 15.

In the fourth and higher orders of K it may happen that the second denomi-

† Compare W. Pauli, *Meson Theory of Nuclear Forces*, New York 1946, pp. 41 ff.

nator of (6) vanishes when $E = E_O$. The transition $E \to E_O$ must then be carried out more carefully. The situation arises when the state m in (6) is identical with either A or O.† Similar singularities occur in those fourth-order terms dependent on $H^{(s)}$ which are

$$H^{(s)}PH^{(s)}+H^{(s)}PH_{\text{int}}PH_{\text{int}}+H_{\text{int}}PH^{(s)}PH_{\text{int}}+H_{\text{int}}PH_{\text{int}}PH^{(s)}. \tag{9}$$

To obtain the right limiting value we multiply (5) by $\delta(E-E_O)$ and integrate over E. For the regular parts of $K(E)$ we just obtain $K(E_O)$. For this singular factor the same representation must be used as for ζ, P, δ in (5), i.e. when $\zeta(x) = 1/(x+i\sigma)$, we must put $\delta(x) = (\sigma/\pi)/(x^2+\sigma^2)$ with the same σ. Then P and δ combine to give

$$\frac{\mathscr{P}}{E-E_O}\delta(E-E_O) = -\tfrac{1}{2}\delta'(E-E_O) \tag{10}$$

and the integration over E can be performed. We obtain for the fourth-order contribution to $K_{4A|O}$ after a short calculation, taking into account that

$$-H^{(s)}_{A|A} = \frac{H_{A|n}H_{n|A}}{E_A-E_n} = K_{2A|A}, \qquad K_{2A|O} = \frac{H_{A|n}H_{n|O}}{E_O-E_n},$$

$$K_{4A|O} = \int K_{4A|O}(E)\delta(E-E_O) = -\tfrac{1}{2}\frac{H_{A|n}H_{n|A}}{(E_A-E_n)^2}K_{2A|O}-\tfrac{1}{2}K_{2A|O}\frac{H_{O|m}H_{m|O}}{(E_O-E_m)^2} \tag{11}$$

+contributions from non-singular terms.

The terms written out in (11) are the renormalization terms. Altogether, (11) is identical with the expression $\tilde{K}_{4A|O}$, § 15 eq. (21_4) (the self-energies are subtracted by $H^{(s)}$). In § 15 the limiting process (10) has already been anticipated. We see now that it arises automatically when we start from the theory with finite line breadth and go over to $\mathscr{R}\Gamma \to 0$. In the same way, it is seen that the second part of (5) yields

$$\int K_{4A|B}(E)\delta(E-E_B)U_{B|O}(E)\delta(E-E_O) = K_{4A|B}(E_B)\delta(E_O-E_B)U_{B|O}(E_O), \tag{12}$$

where $K_{4A|B}(E_B)$ is the expression (11) with O replaced by B. The terms that would arise from the differentiation of $U_{B|O}(E)\delta(E-E_O)$ with respect to E are proportional to $K_{2A|B}(K_{2B|B}+H^{(s)}_{B|B}) = 0$, etc., and vanish.

Thus the first two terms on the right of (5) reduce to (8) with K given precisely by the expressions of § 15. If it were not for the last term of (5) we would obtain the correct theory of free particle collisions. Up to the third order in K this last term vanishes because $\Gamma_{O|O}(E_O) = 0$. In the fourth order, however, a contribution occurs:

$$\frac{i\hbar}{2}\int dE\,\delta(E-E_O)(H_{\text{int}}PH_{\text{int}}+H^{(s)})P\Gamma_{O|O}(E) = -\frac{\hbar}{4}K_{2A|O}\mathscr{I}\frac{\partial\Gamma_{O|O}(E)}{\partial E}\bigg|_{E=E_O}. \tag{13}$$

$\mathscr{I}\Gamma$ alone contributes. Thus in order to obtain the theory of free particle collisions correctly in the limit $\mathscr{R}\Gamma = 0$ we must require that

$$\mathscr{I}\partial\Gamma_{O|O}(E)/\partial E|_{E=E_O} = 0. \tag{14}$$

† When $m \neq A$, O but $E_m = E_O$ it can be seen that the contribution to K_4 vanishes when $L^3 \to \infty$. The reason is that for $m \neq A$, O only isolated intermediate states l, n occur, whereas for $m = O$, the states contain, for example, a photon with variable momentum $\mathbf{k}$ and the integration over $\mathbf{k}$ involves another factor L^3. This is also the reason why $\mathscr{I}\Gamma_{O|O}$ is finite whereas $\mathscr{R}\Gamma_{O|O} \to 0$ for $L^3 \to \infty$.

This condition is not satisfied as long as we operate with the naked states and calculate transitions between them.

(14) is identical with the condition eq. (32), § 16, which we found to be required if the states are to include the virtual admixtures. The condition is to be understood in the sense that a transformation of H should be carried out first to include the virtual field and then when $\mathscr{I}\Gamma$ is expressed by the transformed H in the same way as it now is expressed by H, (14) must be satisfied. Then the theory goes over into that for free particles in the limit $\mathscr{R}\Gamma \to 0$.

It is shown in § 34.4 that such a transformation can be carried out. It is unambiguous for stable states only, e.g. for an atomic ground state or for free particles. For these (14) is then fulfilled.

5. The principle of detailed balance (§ 15)

If collisions between two free particles are treated in first approximation the transition probability from the state O to A is, apart from the density function ρ_A, given by $|K_{A|O}|^2$ where K is a hermitian matrix. It follows that the probability for the reverse process $A \to O$ is the same

$$|K_{O|A}|^2 = |K_{A|O}|^2. \tag{1}$$

Here A and O may be specified in all details, including polarizations, spin directions, etc. (1) is called the 'principle of detailed balance' and is usually quoted in statistics as a sufficient condition for the increase of entropy and the attainment of statistical equilibrium (see below).

As an exact statement, however, the principle cannot be valid. The transition probability is not given by $|K|^2$ but by $|U|^2$, and owing to the damping terms in eq. (33), § 15, U is not hermitian. In fact $-2\pi iU$ is the non-diagonal part of the unitary $\mathscr{S}$-matrix. Apart from the statistical implications, it is of considerable interest to know in which way a process $O \to A$ is connected with its reverse process $A \to O$. In fact a certain general relationship differing in some points from (1) can be derived from quantum mechanics. The principles to be used are (i) the invariance against a reversal of time, and (ii) the invariance against space reflections.† To illustrate the point we consider the non-relativistic wave equation for a particle with spin $\frac{1}{2}$ in a given field with potentials ϕ, $\mathbf{A}$:

$$i\hbar\dot{\psi} = H\psi \equiv \left\{\frac{1}{2\mu}(\mathbf{p}-e\mathbf{A})^2+e\phi-\frac{e\hbar}{2mc}(\boldsymbol{\sigma}\mathbf{H})\right\}\psi, \qquad \mathbf{p} = \frac{\hbar c}{i}\,\text{grad}. \tag{2}$$

We now inquire into the invariance properties of (2) for a time reversal $t \to -t$. If we ignore the spin term in (2) it is seen at once that a new wave function

$$\psi'(t') = \psi^*(t), \qquad t' = -t \tag{3}$$

can be defined which satisfies as function of t' the same wave equation as is satisfied by $\psi(t)$. For this purpose it is only necessary to observe that by the substitution $t \to -t$ the sign of the currents (but not of the charges) producing ϕ, $\mathbf{A}$ is also reversed, resulting in the substitution

$$\mathbf{A}' = -\mathbf{A}, \qquad \phi' = +\phi, \qquad e' = e. \tag{4}$$

(Note that the term $(\mathbf{pA})$ is imaginary.) Thus

$$i\hbar\frac{\partial\psi'}{\partial t'} = H'\psi', \tag{5}$$

† F. Coester, *Phys. Rev.* **84** (1951), 1259; S. Watanabe (to be published).

where H' arises from H by substituting $\mathbf{A}'$ for $\mathbf{A}$, etc. However, the spin term in (2) destroys this simple relationship because $\boldsymbol{\sigma}$ has real as well as imaginary elements and $\mathbf{H}$ changes its sign by (4). To restore a wave equation of the type (5) we generalize (3) by putting

$$\psi'(t') = \tau\psi^*(t), \tag{6}$$

where τ is an operator with the property

$$\tau\boldsymbol{\sigma} = -\boldsymbol{\sigma}^*\tau. \tag{7}$$

$\boldsymbol{\sigma}^*$ is the matrix vector complex conjugate to $\boldsymbol{\sigma}$ (not the adjoint). Reversing the time in (2) and multiplying by τ from the left it is seen that $\psi'(t')$ again satisfies the wave equation (5). If § 11 eq. (3') is used as representation for $\boldsymbol{\sigma}$, τ is represented by

$$\tau = \begin{pmatrix} 0 & -1 \\ 1 & 0 \end{pmatrix}. \tag{8}$$

It is seen that τ acting on ψ *changes the spin direction.* If ψ_+, ψ_- are the two solutions $\begin{pmatrix} 1 \\ 0 \end{pmatrix}$ and $\begin{pmatrix} 0 \\ 1 \end{pmatrix}$ with $\sigma_z = \pm\frac{1}{2}$,

$$\tau\psi_+ = +\psi_-, \qquad \tau\psi_- = -\psi_+. \tag{9}$$

Consider now the collision of two particles 1 and 2 and assume for simplicity that particle 1 has spin $\frac{1}{2}$, but particle 2 is spinless. A state A or O is then specified by the momenta $\mathbf{p}_1$, $\mathbf{p}_2$ and by the spin direction of particle 1, say, $+$, $-$. There will be probabilities for collision processes of the type (writing w for $|U|^2$) $w_{p_1'p_2'+|p_1p_2+}$, $w_{p_1'p_2'-|p_1p_2+}$, etc. If we now reverse the time direction, a solution is again obtained by (6). Reversal of t means that (i) initial and final states are interchanged, (ii) the sign of all momenta are reversed (replace ψ by ψ^*), and the occurrence of τ means that the spin directions are to be reversed. The signs in (9) are, of course, irrelevant when the probabilities are formed. Thus we can state generally that

$$w_{p_1'p_2'\pm|p_1p_2\pm} = w_{-p_1-p_2\mp|-p_1'-p_2'\mp}. \tag{10}$$

As a further step we now apply a reflection of all space coordinates, $x \to -x$, $y \to -y$, $z \to -z$. During this transformation all momenta change sign, but the spin directions remain the same. The wave equation remains invariant during space reflections. It follows that $w_{p_1'p_2'|p_1p_2} = w_{-p_1'-p_2'|-p_1-p_2}$ for each specified spin direction. Hence combining with (10) we also obtain

$$w_{p_1'p_2'\pm|p_1p_2\pm} = w_{p_1p_2\mp|p_1'p_2'\mp}. \tag{11}$$

If we disregard the spin directions, this is indeed the principle of detailed balance. However, (11) differs from the detailed balance in that the spin directions are changed in the reverse process. *Detailed* balance would require

$$w_{p_1'p_2'\pm|p_1p_2\pm} = w_{p_1p_2\pm|p_1'p_2'\pm}.\dagger$$

If we sum over the spin directions of both the initial and the final states a certain balance is restored:

$$\sum_{\text{spin}} w_{p_1'p_2'\pm|p_1p_2\pm} \equiv w_{p_1'p_2'|p_1p_2} = w_{p_1p_2|p_1'p_2'}. \tag{12}$$

This balance is, however, only 'semi-detailed'. The fact that no detailed balance holds when the spin is taken into consideration has a classical analogue. Boltzmann, in his original work on statistics, found that no detailed balance holds for collisions between molecules with a non-spherical shape.

† An example for such a violation of the detailed balance has been found in meson theory by J. Hamilton and H. W. Peng, *Proc. Roy. Ir. Ac.* **49** (1944), 197.

The above considerations can be extended to particles with spin 1 (photons!), etc. In general one cannot expect the principle of detailed balance to hold when angular momenta, polarizations, etc., are involved, except to a first approximation.

The lack of detailed balance raises the question of statistical equilibrium and the validity of the H-theorem. Detailed balance is a sufficient but by no means necessary condition for these theorems to hold. In order that the entropy should increase until statistical equilibrium is attained, it is quite sufficient† if for all possible transitions $A \rightleftarrows B$

$$\sum_{B\neq A} \rho_B |U_{A|B}|^2 = \sum_{B\neq A} \rho_B |U_{B|A}|^2. \tag{13}$$

This is much less than detailed balance. The lack of detailed balance is made up by more complicated cycles of processes. In quantum electrodynamics (13) is always satisfied, at least for free particles. This follows from the fact that U is the non-diagonal part of the $\mathscr{S}$-matrix which is unitary. Hence, if we use $\mathscr{S}$ in the form § 15, eq. (35″),

$$\sum_B |\mathscr{S}'_{A|B}|^2 = \sum_B |\mathscr{S}'_{B|A}|^2$$

and since the diagonal elements $A = B$ are equal on both sides (13) follows immediately. No doubt the H-theorem also holds when atomic bound states are involved, but the question does not seem to have been examined from the point of view of exact quantum electrodynamics.

6. The Weizsäcker–Williams method (§§ 25, 26)

The computation of radiative processes occurring in high-energy collisions often turns out to be prohibitively complicated. This is the case, for example, for Bremsstrahlung emitted in electron-electron collisions, pair production by charged particles, etc. In such cases the following simplified method‡ is often of great help and yields results quite easily, at some cost of accuracy.

Consider a fast charged particle moving with velocity very nearly equal to c. The field of this particle is almost identical with that of a set of light waves with various frequencies. This is the more accurately true, as we shall see, the smaller $1-v^2/c^2$ is. The electromagnetic action of this particle on another charged particle, say at rest, is then equivalent to that of these 'virtual' light waves. For example, one of these photons, $\mathbf{k}$, may be scattered by the particle at rest into a secondary photon $\mathbf{k}'$, and $\mathbf{k}$ is thus removed from the spectrum of virtual photons of the fast particle. The particle at rest receives the recoil $\mathbf{k}-\mathbf{k}'$. The process appears as Bremsstrahlung $\mathbf{k}'$ emitted in the collision, with energy k lost by the fast particle. The cross-section is that for scattering, multiplied by the number of virtual quanta k.

In order that this method should be applicable it is necessary that the motion of the fast particle can be treated classically and remains practically a straight line during the process. Thus if M is the mass of the fast particle and E its energy, we have the conditions

$$E/Mc^2 \equiv \gamma \gg 1, \qquad k \ll E. \tag{1}$$

† This condition is due to E. C. G. Stückelberg, *Helv. Phys. Acta*, **25** (1952), 577.

‡ E. J. Williams, *Kgl. Dansk. Vid. Selsk.* **13** (1935), No. 4; C. F. v. Weizsäcker, *Z. Phys.* **88** (1934), 612.

The second condition may seem severe, for it is just large energy losses $k \sim E$ which are the most important, for example in Bremsstrahlung. However, there is another case which can be treated equally well. Suppose the particle at rest (mass M') only receives a small recoil energy: $(\mathbf{k}-\mathbf{k}')^2/2M'c^2 \ll M'c^2$. We then consider the process in the 'opposite Lorentz system' where the particle actually at rest is moving fast and the particle moving fast is transformed to rest. The fact that the particle M' receives a small recoil means that it departs little from a straight line motion in the opposite Lorentz system. We then consider its equivalent photon field and proceed as above. In the end we transform back to the original Lorentz system. In this case, it is the particle actually at rest whose virtual field is considered to be scattered by the particle actually moving fast. In the opposite Lorentz system the virtual photons, k^* say, must fulfil (1), but after transforming back to the original frame the scattered photon k' will have an energy comparable with E. This will be explained in the example treated below. The intermediate cases where both particles suffer considerable changes of energy cannot be treated accurately in this way, but one may well approximate the actual process by adding the two contributions.

The field of a moving point charge was calculated in § 3 eqs. (10). Let the particle move along the z-axis with constant velocity, $\dot{\mathbf{v}} = 0$, such that it passes through $z = 0$ at $t = 0$. Consider the field at a point $z = 0$ and at a distance $b = \sqrt{(x^2+y^2)}$ from the line of motion. The retarded distance from the particle at time t is then given by

$$\mathbf{r} = \mathbf{b}+(t-r/c)\mathbf{v}, \qquad r^2 = b^2+v^2(t-r/c)^2 \tag{2}$$

from which r is determined explicitly. It follows that

$$s \equiv r+(\mathbf{r}\mathbf{v})/c = [v^2t^2+b^2(1-\beta^2)]^{\frac{1}{2}}.$$

Inserting this in the equations (10) § 3, we obtain

$$\mathbf{E} = -eZ\frac{(1-\beta^2)(\mathbf{b}+\mathbf{v}t)}{[v^2t^2+b^2(1-\beta^2)]^{\frac{3}{2}}}, \qquad \mathbf{H} = \frac{eZ}{c}\frac{(1-\beta^2)[\mathbf{b}\mathbf{v}]}{[v^2t^2+b^2(1-\beta^2)]^{\frac{3}{2}}}. \tag{3}$$

This is still exact for any $\mathbf{v}$. We now decompose $\mathbf{E}$ and $\mathbf{H}$ into Fourier components. We write

$$f(t) = \int\limits_0^\infty f(\nu)\begin{Bmatrix}\cos\nu t\\ \sin\nu t\end{Bmatrix}\,d\nu. \qquad f(\nu) = \frac{1}{\pi}\int\limits_{-\infty}^{+\infty} f(t)\begin{Bmatrix}\cos\nu t\\ \sin\nu t\end{Bmatrix}\,dt \tag{4}$$

according to whether $f(t)$ is an even or odd function of t. $\mathbf{H}$ is an even function of t, and we denote the even and odd parts of $\mathbf{E}$ by $\mathbf{E}^{\text{even}}$ and $\mathbf{E}^{\text{odd}}$ respectively. The integration leads to Hankel functions:

$$\left.\begin{aligned} \mathbf{E}^{\text{even}} &= \frac{eZ\nu}{v^2\gamma}\frac{\mathbf{b}}{b}H_1^{(1)}(iz) \\ \mathbf{E}^{\text{odd}} &= -\frac{eZ\nu\mathbf{v}}{v^3\gamma^2}\,iH_0^{(1)}(iz), \qquad z \equiv \frac{b\nu}{v\gamma} \\ \mathbf{H} &= -\frac{eZ\nu}{cv^2\gamma b}\,[\mathbf{b}\mathbf{v}]H_1^{(1)}(iz), \qquad \gamma = \frac{1}{\sqrt{(1-\beta^2)}} \end{aligned}\right\}. \tag{5}$$

Now when $v \to c$, $\gamma \gg 1$. $\mathbf{E}^{\text{odd}}$ is negligible compared with $\mathbf{E}^{\text{even}}$, and $\mathbf{E}^{\text{even}}$ and $\mathbf{H}$ are in the relationship corresponding to a light wave of frequency ν travelling in the direction of z. To obtain the number of equivalent photons $p(\nu)\,d\nu$ with frequency ν, passing at the distance b per unit area, we form the Poynting vector:

$$\int_{-\infty}^{+\infty} S_z\, dt = \int_0^\infty p(\nu)\hbar\nu\, d\nu = \tfrac{1}{4}c\int_0^\infty d\nu\, [\mathbf{E}^{\text{even}}(\nu), \mathbf{H}(\nu)]_z.$$

Comparison with (5) gives

$$p(\nu)\, d\nu = \frac{e^2 Z^2 \nu\, d\nu}{4c^3\gamma^2\hbar}\{H_1^{(1)}(iz)\}^2, \qquad z = \frac{b\nu}{c\gamma}. \tag{6}$$

We are not interested in any particular impact parameter b. We therefore integrate (6) over the area perpendicular to z. This integration must not, however, be extended to $b \to 0$. The limitation is imposed by quantum theory. Both particles are in fact represented by wave packets. In order to speak of an impact parameter b the sideways extension of these wave packets Δx must be smaller than b. On the other hand, we must not allow the particles to have a great sideways velocity v_x. During the time of collision, which is of order b/c, the distance b must not change much, so $bv_x/c \ll b$, or $v_x \ll c$. By the uncertainty relation, therefore, the extension of each wave packet must be $\Delta x > \hbar/mc$, where m is the mass of the particle considered. At the same time $b > \Delta x$. It follows that the impact parameter b must be larger than

$$b_{\text{min}} = \alpha\frac{\hbar}{mc}, \tag{7}$$

where α is a constant of order unity. For m we should take the mass of the lighter of the two particles (electron mass if one of the particles is an electron).

Integrating (6) over the area $b > b_{\text{min}}$, we obtain for the total number of equivalent photons ($k = \hbar\nu$)

$$q(k)\, dk = 2\pi \int_{b_{\text{min}}}^{\infty} p(\nu)\, d\nu\, b\, db = \tfrac{1}{2}\pi \frac{Z^2e^2}{\hbar c}\frac{dk}{k}P(z_m), \tag{8}$$

$$P(z_m) = -\tfrac{1}{2}z_m^2(H_1^{(1)2}+H_0^{(1)2})-iz_m H_0^{(1)}H_1^{(1)}, \qquad z_m = \frac{b_{\text{min}}k}{\hbar c\gamma}.$$

The argument of the Hankel functions is iz_m.

When $k/\gamma mc^2 = \dfrac{Mk}{mE} \ll 1$ $(M \gg m)$, $z_m \ll 1$ and $P(z_m)$ reduces to a simple form. In this case

$$q(k)\, dk = \frac{2}{\pi}\frac{Z^2e^2}{\hbar c}\frac{dk}{k}\left[\log\frac{E}{\alpha k} - 0{\cdot}38\right], \qquad \frac{Mk}{mE} \ll 1. \tag{9}$$

In fact $P(z_m)$ decreases rapidly when $z_m \sim 1$ so that the condition (1) is automatically satisfied by (8). (9) can always be used for the applications.

As an application we consider the Bremsstrahlung emitted in a collision between two electrons. There will be two contributions I and II according to whether the change of momentum of the fast or slow particle is negligible. For the contribution II we have to consider the process in the opposite Lorentz system. Let $\phi(k,k')\,dk'$ be the cross-section for Compton scattering, § 22 eq. (41), by an electron at rest of a photon k into a secondary photon k'. Then the first contribution to the cross-section for Bremsstrahlung with emission of k' is given by

$$\phi_{\text{I}}\, dk' = dk' \int_{k'} q(k)\phi(k,k')\, dk. \tag{10}$$

The upper limit is $\mu k'/(\mu - 2k')$ when $k' < \frac{1}{2}\mu$, and E when $k' > \frac{1}{2}\mu$. On account of (1), we should, however, impose a smaller limit, say E_m where $E_m < E$, but we may well assume $E_m \gg \mu$. We confine ourselves to the case $k' \gg \mu$, which is the most interesting. One easily finds

$$\phi_{\mathrm{I}}\, dk' = \frac{8r_0^2\mu}{3.137}\frac{dk'}{k'^2}\left\{\log\frac{E}{\alpha k'} - 1{\cdot}2 - O\left(\frac{k'}{E_m}\right)\right\}. \tag{11}$$

It is seen that small k' are the most probable, and ϕ_{I} decreases rapidly with k'. We therefore commit a small error when we neglect the terms of order k'/E_m and extend the integration to infinity instead of to E_m.

Comparing (11) with the results of § 25 we see that ϕ_{I} is smaller by an order μ/k'. However, contribution II will give a larger order of magnitudes as we shall now show.

We consider next the process in the opposite Lorentz system, where the fast particle is transformed to rest. We denote all quantities referring to this Lorentz system by an asterisk. The cross-section for emission of k'^* due to an equivalent photon k^* is again given by (10) with k, k' replaced by k^*, k'^* respectively, and this is valid for $k^* \ll E$. We then transform back to the original Lorentz system. Since k^* has the direction of the fast particle we have

$$k^* = \frac{2Ek}{\mu}, \qquad k'^* = \frac{2Ek}{\mu}\frac{E-k'}{E-k}, \tag{12}$$

and the condition $k^* \ll E$ becomes

$$k \ll \mu. \tag{13}$$

Hence k is also small compared with E, and (12) simply becomes

$$k'^* = \frac{k^*}{E}(E-k'). \tag{12'}$$

We keep k' fixed, and express the Klein–Nishina formula by k' instead of k'^*, but use k^* as the integration variable. For fixed $k' < E$ the limits of integration are

$$\left.\begin{aligned} k'^* \leqslant k^* &\leqslant \frac{\mu k'^*}{\mu - 2k'^*} \\ k'^* \leqslant k^* &< E_m \end{aligned}\right\} \quad \text{when} \quad k'^* \lessgtr \tfrac{1}{2}\mu, \quad \text{or} \quad \left\{\begin{aligned} \frac{\mu k'}{2(E-k')} < k^* &< \frac{\mu E}{2(E-k')} \\ k^* &> \frac{\mu E}{2(E-k')} \end{aligned}\right.$$

viz.,

$$\frac{\mu k'}{2(E-k')} < k^* < E_m. \tag{14}$$

Expressed in terms of k^*, k', the Klein–Nishina formula is

$$\phi(k', k^*)\, dk' = \pi r_0^2 \mu \frac{dk'}{Ek^*}\left[\frac{E}{E'} + \frac{E'}{E} - \frac{2\mu k'}{E'k^*} + \frac{\mu^2 k'^2}{k^{*2}E'^2}\right], \qquad E' \equiv E - k'. \tag{15}$$

This has to be multiplied by $q(k^*)\, dk^*$ and integrated between the limits (14). There is again only a small error involved if we replace E_m by ∞. We readily obtain

$$\phi_{\mathrm{II}}\, dk' = \frac{4r_0^2}{137}\frac{dk'}{k'}\frac{E'}{E}\left\{\left[\log\frac{2EE'}{\alpha\mu k'} - 1{\cdot}4\right]\left[\frac{E}{E'} + \frac{E'}{E} - \frac{2}{3}\right] - \frac{1}{9}\right\}. \tag{16}$$

This is not restricted to $k' \ll E$. The neglected terms which depend on E_m are of order $dk'\mu/E_m E$. We see that for $k' \gg \mu$ the contribution II is larger than I by a factor k'/μ. ϕ_{I} is thus negligible.

If one of the two particles (e.g. the particle at rest) is a heavy particle with mass $M \gg m$, the present considerations are valid *a fortiori*. Contribution II gives the major effect. In the opposite Lorentz frame the fast particle moves with energy γMc^2 and the condition for k^* is $k^* \ll \gamma Mc^2$; this means that the upper limit E_m may be chosen much higher than if the particle is an electron, and the replacement $E_m \to \infty$ is well justified. Indeed (16) (when multiplied by Z^2) is practically identical with the formula § 25 eq. (21) for Bremsstrahlung of a fast electron in the field of a nucleus. The additional term $-1/9$ is quite insignificant. To obtain otherwise the correct result we have only to put $\log 1/\alpha = 0{\cdot}9$, or $\alpha = 0{\cdot}4$.

We see now that practically the same formula holds for Bremsstrahlung in electron–electron collisions. However, some numerical difference is to be expected. The replacement of E_m by ∞ may not be very serious as long as $E_m \gg \mu$, but the numerical value of α may be different. The impact parameter was limited by the sideways extension of the wave packets of both particles. If one of the particles is heavy, the extension of its wave packet is negligible and only the electron gives rise to the limitation (7). If, however, both particles are electrons, one must expect α to be larger. Perhaps we do not go very far wrong if we say that α is then larger by a factor 2. This means that the numerical values for the Bremsstrahlung of electrons are slightly smaller than in electron–nucleus collisions. For example, the formula for radiative energy loss (§ 25 eq. (33)) would, for electron–electron collisions, be

$$\phi_{\text{rad}} \simeq \frac{4r_0^2}{137}\left(\log\frac{2E}{\mu}-\frac{1}{3}-\log 2\right), \tag{17}$$

and presumably a similar correction is to be made in the case of complete screening. For $E = 100\mu$ the correction would amount to about 15 per cent. No great weight can, however, be put on this numerical value. A correction of this order of magnitude has been taken into account in §§ 25, 26, 36–38, for Bremsstrahlung and pair creation in the field of electrons and the experiments, as well as the direct calculations, quoted there seem to confirm the order of magnitude.

The present method can be applied to numerous other processes which are difficult to treat exactly. The cross-sections for pair creation by particles referred to in § 26 have been obtained in this way.

7. The energy-momentum tensor and the self-stress (§§ 4, 29)

In § 4 and again in § 29 the question has been raised whether an electron accompanied by its electromagnetic field can behave relativistically like a particle described by an energy-momentum 4-vector. The electromagnetic field itself has an energy-momentum tensor $T_{\mu\nu}^{\text{rad}}$ and it has been shown in § 2 that $\int T_{4x}\,d\tau$ and $\int T_{44}\,d\tau$ (total momentum and total energy) are only components of a 4-vector when $\partial T_{\mu\nu}/\partial x_\nu = 0$. For $T_{\mu\nu}^{\text{rad}}$ this is classically in general not the case when a point charge is present. On the other hand, it has been shown in § 29 that the correct transformation properties of the electromagnetic self-energy W can be achieved by prescribing an invariant way in the evaluation of the diverging integral representing W. Then W reduces to a contribution to the mass of the electron. It may be expected that the question of the transformation properties of the complete energy tensor of the electron can be solved similarly.

The properties of $T_{\mu\nu}^{\text{rad}}$ required are such that, if $T_{\mu\nu}^{\text{rad}}$ is combined with the energy and momentum of the electron, both transform together like a 4-vector. Classically, the electron itself has an energy-momentum 4-vector. In quantum theory, however, electrons are represented by a ψ-field, subjected to second quantization. This field, like any other field, will have an energy-momentum *tensor*, $T^{\mathrm{el}}_{\mu\nu}$, say.‡ The energy of the electron is the space integral of an *energy density* $\mathscr{H}$ (given by § 13 eq. (25)), and the latter must be the 44-component of $T^{\mathrm{el}}_{\mu\nu}$.

Although there is a general method of deriving $T_{\mu\nu}$ from the Lagrangian L for any field, it is very easy to find $T^{\mathrm{el}}_{\mu\nu}$ for the electron field directly. (L is actually given in § 13.) Like its 44-component, $T^{\mathrm{el}}_{\mu\nu}$ will be bilinear in $\psi^\dagger$ and ψ. It must, for a pure electron not interacting with the electromagnetic field, satisfy $\partial T^{\mathrm{el}}_{\mu\nu}/\partial x_\nu = 0$, in order that the total energy and momentum should form a 4-vector. It must also be a symmetrical 4-tensor.§ Thus we put

$$-T^{\mathrm{el}}_{\mu\nu} = \tfrac{1}{4}\psi^\dagger(\gamma_\mu p_\nu+\gamma_\nu p_\mu)\psi-\tfrac{1}{4}\psi^\dagger(\gamma_\mu p_{\nu\mathrm{op}}+\gamma_\nu p_{\mu\mathrm{op}})\psi, \tag{1}$$

$$p_\mu \equiv \frac{\hbar c}{i}\frac{\partial}{\partial x_\mu}, \qquad \psi^\dagger p_{\mu\mathrm{op}} \equiv \frac{\hbar c}{i}\frac{\partial \psi^\dagger}{\partial x_\mu},$$

treating ψ and $\psi^\dagger$ symmetrically. $p_{\nu\mathrm{op}}$ is understood to act backwards on $\psi^\dagger$ and not on ψ. The 44-component is

$$T^{\mathrm{el}}_{44} = \frac{\hbar i}{2}(\psi^*\dot\psi-\dot\psi^*\psi) = \tfrac{1}{2}\psi^*((\boldsymbol{\alpha}\mathbf{p})+\beta\mu)\psi+\tfrac{1}{2}\psi^*(-(\boldsymbol{\alpha}\mathbf{p}_{\mathrm{op}})+\beta\mu)\psi \tag{2}$$

by virtue of the Dirac equation. This is a symmetrical form of the Hamiltonian density, § 13. Upon integration over the whole of space

$$\int \psi^*\mathbf{p}_{\mathrm{op}}\psi\, d\tau = -\int \psi^*\mathbf{p}\psi\, d\tau,$$

and (2) reduces to the usual form. Similarly, the $4x$-component becomes

$$-\frac{1}{i}\int T_{4x}\, d\tau = \frac{1}{4i}\int d\tau\{\psi^\dagger(\gamma_4 p_x+\gamma_x p_4)\psi-\psi^\dagger(\gamma_4 p_{x\mathrm{op}}+p_{4\mathrm{op}}\gamma_x)\psi\} = \int d\tau\, \psi^* p_x \psi, \tag{3}$$

again with the help of the Dirac equation for $p_4\psi$ and $\psi^\dagger p_{4\mathrm{op}}$. (3) is the total x-momentum of the electron. The 4-divergence of (1) vanishes

$$-\frac{4\hbar c}{i}\frac{\partial T^{\mathrm{el}}_{\mu\nu}}{\partial x_\nu} = \psi^\dagger(\gamma_\mu p_\nu^2+\gamma_\nu p_\mu p_\nu)\psi+\psi^\dagger(\gamma_\mu p_\nu p_{\nu\mathrm{op}}+\gamma_\nu p_\mu p_{\nu\mathrm{op}})\psi-$$
$$-\psi^\dagger(\gamma_\mu p_{\nu\mathrm{op}} p_\nu+\gamma_\nu p_{\mu\mathrm{op}} p_\nu)\psi-\psi^\dagger(\gamma_\mu p_{\nu\mathrm{op}}^2+\gamma_\nu p_{\mu\mathrm{op}} p_{\nu\mathrm{op}})\psi = 0,$$

because $\gamma_\nu p_\nu\psi = i\mu\psi$, $\psi^\dagger\gamma_\nu p_{\nu\mathrm{op}} = -i\mu\psi^\dagger$, $p_\nu^2\psi = -\mu^2\psi$, $\psi^\dagger p_{\nu\mathrm{op}}^2 = -\mu^2\psi^\dagger$. This means that, for a free electron, not interacting with radiation, $\int T^{\mathrm{el}}_{4x}\, d\tau$ and $\int T^{\mathrm{el}}_{44}\, d\tau$ form a 4-vector, the energy-momentum vector.

When the electron interacts with radiation, (1) is evidently to be generalized as follows:

$$-T^{\mathrm{el}}_{\mu\nu} = \tfrac{1}{4}\psi^\dagger[\gamma_\mu(p_\nu-eA_\nu)+\gamma_\nu(p_\mu-eA_\mu)]\psi-\tfrac{1}{4}\psi^\dagger[\gamma_\mu(p_{\nu\mathrm{op}}+eA_\nu)+\gamma_\nu(p_{\mu\mathrm{op}}+eA_\mu)]\psi. \tag{4}$$

The total energy tensor for the electron radiation system will thus be

$$T_{\mu\nu} = T^{\mathrm{rad}}_{\mu\nu}+T^{\mathrm{el}}_{\mu\nu}, \qquad T^{\mathrm{rad}}_{\mu\nu} = \frac{1}{4\pi}[-f_{\mu\lambda}f_{\nu\lambda}+\tfrac{1}{4}\delta_{\mu\nu}f^2_{\lambda\rho}], \tag{5}$$

where $f_{\mu\nu}$ is the electromagnetic field strength.

‡ This tensor has nothing to do with the 'internal stress' hinted at in § 4.

§ The requirement of symmetry is necessary for the existence of an angular momentum; this will not be shown here. The analogy to the symmetrical tensor $T^{\mathrm{rad}}_{\mu\nu}$ of the electromagnetic field may suffice to justify this requirement.

When the fields are quantized $T_{\mu\nu}$ becomes an operator and the time derivatives $\dot{\psi}$ and $\dot{A}_\mu$, wherever they occur, are to be interpreted in the appropriate way, namely, as the *total* time derivatives defined by § 13 eq. (14b).

We now consider the energy tensor for an electron at rest, when only the electron with its accompanying virtual field but no light quanta are present. We call the expectation values of the components of $T_{\mu\nu}$ in this case $\langle T^0_{\mu\nu}\rangle$. It is to be understood that these expectation values are to be formed with the *correct* state vector of quantum electrodynamics including the admixtures from virtual states, up to the first radiative corrections, say. Since the T^0_{i4} are momentum densities, their expectation values surely vanish. For reasons of spatial isotropy we also have

$$\langle T^0_{11}\rangle = \langle T^0_{22}\rangle = \langle T^0_{33}\rangle, \qquad \langle T^0_{i4}\rangle = \langle T^0_{4i}\rangle = 0. \tag{6}$$

We now ask what is required in order that the energy and momentum of the electron, including its field, should have particle properties. Instead of using the general condition $\partial T_{\mu\nu}/\partial x_\nu = 0$ we consider the tensor in a Lorentz frame moving along the x-axis with velocity $c\beta$. We form the space integrals of the $\langle T_{\mu\nu}\rangle$, taking into account (6) and the fact that the volume element transforms like $d\tau = d\tau^0\sqrt{(1-\beta^2)}$.

Then from the transformation properties of 4-tensors we find

$$\begin{aligned} \int \langle T_{44}\rangle\, d\tau &= \frac{1}{\sqrt{(1-\beta^2)}}\left\{\int \langle T^0_{44}\rangle\, d\tau^0 - \beta^2 \int \langle T^0_{xx}\rangle\, d\tau^0\right\}, \\ i\int \langle T_{x4}\rangle\, d\tau &= -\frac{\beta}{\sqrt{(1-\beta^2)}}\left\{\int \langle T^0_{44}\rangle\, d\tau^0 - \langle T^0_{xx}\rangle\, d\tau^0\right\}. \end{aligned} \tag{7}$$

We see that $\int \langle T_{44}\rangle\, d\tau$ and $i\int \langle T_{x4}\rangle\, d\tau$ transform like energy and momentum of a particle if and only if $\int \langle T^0_{xx}\rangle\, d\tau = 0$.‡ We call this integral the ***self-stress.*** The problem now reduces to the question of whether or not the self-stress for a free particle at rest vanishes when the accompanying field of the electron is included. We shall see that this can be decided directly from the explicit expressions for the self-energy.§ For this purpose we form the spur of $T^0_{\mu\nu}$ $\left(\equiv \sum_\mu T^0_{\mu\mu}\right)$. It follows from (5) that $\mathrm{Sp}\, T^{\mathrm{rad}} = 0$ and from (4)

$$\mathrm{Sp}\langle T^{\mathrm{el}}\rangle = -\tfrac{1}{2}\langle \psi^\dagger(\gamma_\mu p_\mu - \gamma_\mu p_{\mu\mathrm{op}} - 2e\gamma_\mu A_\mu)\psi\rangle = -i\mu\langle\psi^\dagger\psi\rangle = \mu\langle\psi^*\beta\psi\rangle.$$

Thus, using (6), we obtain

$$\mathrm{Sp}\langle T^0\rangle = 3\langle T^0_{xx}\rangle + \langle T^0_{44}\rangle = \mu\langle\psi^*\beta\psi\rangle. \tag{8}$$

On the other hand, $\int \langle T^0_{44}\rangle\, d\tau^0$ is nothing but the total energy of the electron at rest including its self-energy:

$$\int \langle T^0_{44}\rangle\, d\tau^0 = \mu + W^0. \tag{9}$$

‡ For a moving electron $-\int \langle T_{xx}\rangle\, d\tau = \dfrac{\beta^2}{\sqrt{(1-\beta^2)}}\int \langle T^0_{44}\rangle\, d\tau_0$ is different from zero, even if the self-stress at rest vanishes. This is even so for an electron without radiation field, as can easily be verified from (1). The stress is then merely a function of the momentum and does not interfere with the particle properties discussed above. $T_{xy} = T_{yy} = \ldots = 0$ if the motion is along the x-axis.

§ For the following argument see A. Pais and S. T. Epstein, *Rev. Mod. Phys.* **21** (1949), 445 (these authors arrived at the wrong result $\int \langle T^0_{xx}\rangle\, d\tau^0 \neq 0$, using an unsuitable expression for the self-energy, see below); F. Rohrlich, *Phys. Rev.* **77** (1950), 357; F. Villars, ibid. **79** (1950), 122; K. Sawada, *Prog. Theor. Phys.* **5** (1950), 117.

W^0 is calculated in § 29. The expectation value $\langle\psi^*\beta\psi\rangle$ refers to an electron at rest but is to be formed with the *exact* state vector of the system and differs from the value which it would have for an electron without interaction with the radiation field. We need not work out this expectation value explicitly. We observe that the operator $\mu(\psi^*\beta\psi)$ occurs in the total Hamiltonian and is the only term which contains the mass μ explicitly. It can be singled out by differentiation. Since the expectation value of the Hamiltonian is given by (9), we conclude that

$$\mu\langle\psi^*\beta\psi\rangle = \mu\frac{\partial}{\partial\mu}(\mu+W^0). \tag{10}$$

Hence it follows from (8)–(10) that

$$3\int \langle T^0_{xx}\rangle\, d\tau^0 = \mu\frac{\partial}{\partial\mu}(\mu+W^0)-(\mu+W^0) = \mu\frac{\partial W^0}{\partial\mu}-W^0. \tag{11}$$

All now depends on how W^0 depends on μ explicitly. For a particle at rest $(u^*\beta u) = 1$; thus, according to § 29 eq. (13),

$$W^0 = \frac{\mu}{2\pi.137}\int\limits_1^{\epsilon_{max}} \frac{3\epsilon^2-1}{\epsilon^3}\, d\epsilon, \tag{12}$$

where we have replaced the upper limit by ϵ_{max} to make the integral finite. ϵ is an invariant variable, connected with virtual photon and electron variables in a complicated way. If ϵ_{max} is independent of μ,

$$\mu\frac{\partial W^0}{\partial\mu} = W^0 \quad\text{and}\quad \int \langle T^0_{xx}\rangle\, d\tau^0 = 0. \tag{13}$$

Then the self-stress indeed vanishes. However, the statement 'ϵ_{max} independent of μ' is by no means automatic. We can equally well choose different variables (e.g. the momentum of the virtual photon **k**) where the integral (12) would depend on μ and then the self-stress would not vanish. In fact certain modes of evaluation of W^0 lead to a finite (but, as we see, ambiguous) result for the self-stress. The ambiguity is not very different from that which occurred in the magnetic moment (§ 31) or the vacuum polarization (§ 32). Evidently, for physical reasons, it must be required that ϵ_{max} is defined as independent of the mass μ of the particle. Only when this is so does the self-stress vanish and then the electron, including its field, has the transformation properties of a particle as is actually observed. It may be remarked finally that the regularization procedure described in § 35 also leads unambiguously to the result

$$\mu\frac{\partial W^0}{\partial\mu} = W^0.$$

8. Universal and material constants

UNIVERSAL LENGTHS

$a_0 = \hbar^2/me^2 = 0{\cdot}529\times10^{-8}$ cm. (radius of hydrogen atom).
$\lambda_0 = \hbar/mc = 3{\cdot}862\times10^{-11}$ cm. (Compton wave-length).
$r_0 = e^2/mc^2 = 2{\cdot}818\times10^{-13}$ cm. (classical electronic radius).

$$a_0 = 137\lambda_0 = 137^2 r_0.$$
$$\text{'137'} = \hbar c/e^2 = 137{\cdot}036.$$

ENERGIES AND WAVE-LENGTHS

$$e^2/a_0 = 2I_0 = 27{\cdot}09 \text{ e. volts } (I_0 \text{ ionization energy of hydrogen atom}).$$

$$I = Z^2 I_0 = \text{ionization energy of } K\text{-electron} = Z^2\mu/2\times 137^2.$$

$$\mu = mc^2 = 2I_0 \times 137^2 = 0{\cdot}511\times 10^6 \text{ e. volts.}$$

$$10^6 \text{ e. volts} = 1{\cdot}957\,\mu.$$

If $2\pi\lambda = 1$ X.U. $= 10^{-11}$ cm. then $\hbar\nu = 12{\cdot}40\times 10^6$ e. volts $= 24{\cdot}3\,\mu$.

If $\hbar\nu = \mu$ then $2\pi\lambda = 2\pi\lambda_0 = 2{\cdot}427\times 10^{-10}$ cm.

If $\hbar\nu = 10^6$ e. volts then $2\pi\lambda = 1{\cdot}240\times 10^{-10}$ cm.

UNIVERSAL CROSS-SECTION

$$\phi_0 = 8\pi r_0^2/3 \text{ (Thomson formula)} = 6{\cdot}653\times 10^{-25} \text{ cm.}^2$$

$$\bar{\phi} = Z^2 r_0^2/137 = Z^2\times 5{\cdot}795\times 10^{-28} \text{ cm.}^2$$

SOME MATERIAL CONSTANTS

	C†	Al	Fe	Cu	Sn‡	Pb	air§	H_2O‖	
N	11·28	6·02	8·46	8·46	3·56	3·30	0·0027	3·34	$\times 10^{22}$
$NZ\phi_0$	0·450	0·521	1·46	1·63	1·18	1·80	$2{\cdot}57\times 10^{-4}$	0·222	
$NZ^2\dfrac{r_0^2}{137}$	0·235	0·590	3·31	4·12	5·16	12·86	$1{\cdot}62\times 10^{-4}$	0·128	$\times 10^{-2}$
$N\phi_0\dfrac{Z^5}{137^4}$	0·165	4·22	190	327	2100	23100	—	—	$\times 10^{-5}$

N number of atoms (for air and H_2O molecules) per cm.^{3}, Z nuclear charge.

† Graphite, density 2·25. ‡ Sn, density 7·00.

§ Atmospheric pressure, 0° C. temperature.

‖ For air and water NZ means the total number of electrons per cm.^{3}, NZ^2 the sum of Z^2 of all nuclei per cm.^{3}

NOTATIONS

N.R. means non-relativistic case, energies $\ll mc^2$.

E.R. means extreme-relativistic case, energies $\gg mc^2$.

'Classical' always means $\hbar \to 0$.

All momenta p, etc., have dimensions of energy ($c\times$ usual momentum).

Space-time indices: Greek indices run from 1 to 4, Latin indices from 1 to 3. $x_4 = ix_0 = ict$, etc.

Space vectors are denoted by heavy letters, $\mathbf{k}$ and $k^2 \equiv |\mathbf{k}|^2$.

In Chapter VI, however, simplified notations are used for 4-products, etc. (see p. 276), and $k^2 \equiv |\mathbf{k}|^2 - k_0^2$.

LIST OF REFERENCES

Publications by several authors are quoted under the first name only. Technical terms connected with a name, like Lorentz force, are found in the subject index.

(i) THEORETICAL

(ii) EXPERIMENTAL

SUBJECT INDEX

A CATALOG OF SELECTED DOVER BOOKS IN ALL FIELDS OF INTEREST

A CATALOG OF SELECTED DOVER BOOKS IN ALL FIELDS OF INTEREST

25 KITES THAT FLY, Leslie Hunt. Full, easy-to-follow instructions for kites made from inexpensive materials. Many novelties. 70 illustrations. 110pp. 5⅜ × 8½. 22550-X Pa. $2.25

PIANO TUNING, J. Cree Fischer. Clearest, best book for beginner, amateur. Simple repairs, raising dropped notes, tuning by easy method of flattened fifths. No previous skills needed. 4 illustrations. 201pp. 5⅜ × 8½. 23267-0 Pa. $3.50

EARLY AMERICAN IRON-ON TRANSFER PATTERNS, edited by Rita Weiss. 75 designs, borders, alphabets, from traditional American sources. 48pp. 8¼ × 11. 23162-3 Pa. $1.95

CROCHETING EDGINGS, edited by Rita Weiss. Over 100 of the best designs for these lovely trims for a host of household items. Complete instructions, illustrations. 48pp. 8¼ × 11. 24031-2 Pa. $2.25

FINGER PLAYS FOR NURSERY AND KINDERGARTEN, Emilie Poulsson. 18 finger plays with music (voice and piano); entertaining, instructive. Counting, nature lore, etc. Victorian classic. 53 illustrations. 80pp. 6½ × 9¼. 22588-7 Pa. $1.95

BOSTON THEN AND NOW, Peter Vanderwarker. Here in 59 side-by-side views are photographic documentations of the city's past and present. 119 photographs. Full captions. 122pp. 8¼ × 11. 24312-5 Pa. $6.95

CROCHETING BEDSPREADS, edited by Rita Weiss. 22 patterns, originally published in three instruction books 1939-41. 39 photos, 8 charts. Instructions. 48pp. 8¼ × 11. 23610-2 Pa. $2.00

HAWTHORNE ON PAINTING, Charles W. Hawthorne. Collected from notes taken by students at famous Cape Cod School; hundreds of direct, personal *apercus*, ideas, suggestions. 91pp. 5⅜ × 8½. 20653-X Pa. $2.50

THERMODYNAMICS, Enrico Fermi. A classic of modern science. Clear, organized treatment of systems, first and second laws, entropy, thermodynamic potentials, etc. Calculus required. 160pp. 5⅜ × 8½. 60361-X Pa. $4.00

TEN BOOKS ON ARCHITECTURE, Vitruvius. The most important book ever written on architecture. Early Roman aesthetics, technology, classical orders, site selection, all other aspects. Morgan translation. 331pp. 5⅜ × 8½. 20645-9 Pa. $5.50

THE CORNELL BREAD BOOK, Clive M. McCay and Jeanette B. McCay. Famed high-protein recipe incorporated into breads, rolls, buns, coffee cakes, pizza, pie crusts, more. Nearly 50 illustrations. 48pp. 8¼ × 11. 23995-0 Pa. $2.00

THE CRAFTSMAN'S HANDBOOK, Cennino Cennini. 15th-century handbook, school of Giotto, explains applying gold, silver leaf; gesso; fresco painting, grinding pigments, etc. 142pp. 6⅛ × 9¼. 20054-X Pa. $3.50

FRANK LLOYD WRIGHT'S FALLINGWATER, Donald Hoffmann. Full story of Wright's masterwork at Bear Run, Pa. 100 photographs of site, construction, and details of completed structure. 112pp. 9¼ × 10. 23671-4 Pa. $6.50

OVAL STAINED GLASS PATTERN BOOK, C. Eaton. 60 new designs framed in shape of an oval. Greater complexity, challenge with sinuous cats, birds, mandalas framed in antique shape. 64pp. 8¼ × 11. 24519-5 Pa. $3.50

REASON IN ART, George Santayana. Renowned philosopher's provocative, seminal treatment of basis of art in instinct and experience. Volume Four of *The Life of Reason*. 230pp. 5⅜ × 8. 24358-3 Pa. $4.50

LANGUAGE, TRUTH AND LOGIC, Alfred J. Ayer. Famous, clear introduction to Vienna, Cambridge schools of Logical Positivism. Role of philosophy, elimination of metaphysics, nature of analysis, etc. 160pp. 5⅜ × 8½. (USCO) 20010-8 Pa. $2.75

BASIC ELECTRONICS, U.S. Bureau of Naval Personnel. Electron tubes, circuits, antennas, AM, FM, and CW transmission and receiving, etc. 560 illustrations. 567pp. 6½ × 9¼. 21076-6 Pa. $8.95

THE ART DECO STYLE, edited by Theodore Menten. Furniture, jewelry, metalwork, ceramics, fabrics, lighting fixtures, interior decors, exteriors, graphics from pure French sources. Over 400 photographs. 183pp. 8⅜ × 11¼. 22824-X Pa. $6.95

THE FOUR BOOKS OF ARCHITECTURE, Andrea Palladio. 16th-century classic covers classical architectural remains, Renaissance revivals, classical orders, etc. 1738 Ware English edition. 216 plates. 110pp. of text. 9½ × 12¾. 21308-0 Pa. $11.50

THE WIT AND HUMOR OF OSCAR WILDE, edited by Alvin Redman. More than 1000 ripostes, paradoxes, wisecracks: Work is the curse of the drinking classes, I can resist everything except temptations, etc. 258pp. 5⅜ × 8½. (USCO) 20602-5 Pa. $3.50

THE DEVIL'S DICTIONARY, Ambrose Bierce. Barbed, bitter, brilliant witticisms in the form of a dictionary. Best, most ferocious satire America has produced. 145pp. 5⅜ × 8½. 20487-1 Pa. $2.50

ERTÉ'S FASHION DESIGNS, Erté. 210 black-and-white inventions from *Harper's Bazar*, 1918-32, plus 8pp. full-color covers. Captions. 88pp. 9 × 12. 24203-X Pa. $6.50

ERTÉ GRAPHICS, Erté. Collection of striking color graphics: *Seasons, Alphabet, Numerals, Aces* and *Precious Stones*. 50 plates, including 4 on covers. 48pp. 9⅜ × 12¼. 23580-7 Pa. $6.95

PAPER FOLDING FOR BEGINNERS, William D. Murray and Francis J. Rigney. Clearest book for making origami sail boats, roosters, frogs that move legs, etc. 40 projects. More than 275 illustrations. 94pp. 5⅜ × 8½. 20713-7 Pa. $2.25

ORIGAMI FOR THE ENTHUSIAST, John Montroll. Fish, ostrich, peacock, squirrel, rhinoceros, Pegasus, 19 other intricate subjects. Instructions. Diagrams. 128pp. 9 × 12. 23799-0 Pa. $4.95

CROCHETING NOVELTY POT HOLDERS, edited by Linda Macho. 64 useful, whimsical pot holders feature kitchen themes, animals, flowers, other novelties. Surprisingly easy to crochet. Complete instructions. 48pp. 8¼ × 11. 24296-X Pa. $1.95

CROCHETING DOILIES, edited by Rita Weiss. Irish Crochet, Jewel, Star Wheel, Vanity Fair and more. Also luncheon and console sets, runners and centerpieces. 51 illustrations. 48pp. 8¼ × 11. 23424-X Pa. $2.00

CATALOG OF DOVER BOOKS

SOURCE BOOK OF MEDICAL HISTORY, edited by Logan Clendening, M.D. Original accounts ranging from Ancient Egypt and Greece to discovery of X-rays: Galen, Pasteur, Lavoisier, Harvey, Parkinson, others. 685pp. 5⅜ × 8½.
20621-1 Pa. $10.95

THE ROSE AND THE KEY, J.S. Lefanu. Superb mystery novel from Irish master. Dark doings among an ancient and aristocratic English family. Well-drawn characters; capital suspense. Introduction by N. Donaldson. 448pp. 5⅜ × 8½.
24377-X Pa. $6.95

SOUTH WIND, Norman Douglas. Witty, elegant novel of ideas set on languorous Meditterranean island of Nepenthe. Elegant prose, glittering epigrams, mordant satire. 1917 masterpiece. 416pp. 5⅜ × 8½. (Available in U.S. only)
24361-3 Pa. $5.95

RUSSELL'S CIVIL WAR PHOTOGRAPHS, Capt. A.J. Russell. 116 rare Civil War Photos: Bull Run, Virginia campaigns, bridges, railroads, Richmond, Lincoln's funeral car. Many never seen before. Captions. 128pp. 9⅜ × 12¼.
24283-8 Pa. $6.95

PHOTOGRAPHS BY MAN RAY: 105 Works, 1920-1934. Nudes, still lifes, landscapes, women's faces, celebrity portraits (Dali, Matisse, Picasso, others), rayographs. Reprinted from rare gravure edition. 128pp. 9⅜ × 12¼. (Available in U.S. only) 23842-3 Pa. $6.95

STAR NAMES: THEIR LORE AND MEANING, Richard H. Allen. Star names, the zodiac, constellations: folklore and literature associated with heavens. The basic book of its field, fascinating reading. 563pp. 5⅜ × 8½. 21079-0 Pa. $7.95

BURNHAM'S CELESTIAL HANDBOOK, Robert Burnham, Jr. Thorough guide to the stars beyond our solar system. Exhaustive treatment. Alphabetical by constellation: Andromeda to Cetus in Vol. 1; Chamaeleon to Orion in Vol. 2; and Pavo to Vulpecula in Vol. 3. Hundreds of illustrations. Index in Vol. 3. 2000pp. 6⅛ × 9¼. 23567-X, 23568-8, 23673-0 Pa. Three-vol. set $36.85

THE ART NOUVEAU STYLE BOOK OF ALPHONSE MUCHA, Alphonse Mucha. All 72 plates from *Documents Decoratifs* in original color. Stunning, essential work of Art Nouveau. 80pp. 9⅜ × 12¼. 24044-4 Pa. $7.95

DESIGNS BY ERTE; FASHION DRAWINGS AND ILLUSTRATIONS FROM "HARPER'S BAZAR," Erte. 310 fabulous line drawings and 14 *Harper's Bazar* covers, 8 in full color. Erte's exotic temptresses with tassels, fur muffs, long trains, coifs, more. 129pp. 9⅜ × 12¼. 23397-9 Pa. $6.95

HISTORY OF STRENGTH OF MATERIALS, Stephen P. Timoshenko. Excellent historical survey of the strength of materials with many references to the theories of elasticity and structure. 245 figures. 452pp. 5⅜ × 8½. 61187-6 Pa. $8.95

Prices subject to change without notice.

Available at your book dealer or write for free catalog to Dept. GI, Dover Publications, Inc., 31 East 2nd St. Mineola, N.Y. 11501. Dover publishes more than 175 books each year on science, elementary and advanced mathematics, biology, music, art, literary history, social sciences and other areas.